This book is about the way the coast evolves. It outlines the latest concepts in terms of deposition and erosion of shorelines and their dynamics over the last Ice Age. There are reviews of how deltas, reefs, estuaries, cliffs, polar and other coasts change with time, and how conceptual models can be extended to include shorelines modified by human influence.

Coastal Evolution is aimed at undergraduates studying coastal geomorphology, geologists who are mapping coastal sedimentary sequences, and environmental scientists, engineers, planners and coastal managers who need to understand the processes of change that occur on shorelines.

D0642872

COASTAL EVOLUTION
Late Quaternary shoreline morphodynamics

COASTAL EVOLUTION

Late Quaternary shoreline morphodynamics

Edited by

R. W. G. CARTER
University of Ulster at Coleraine

C. D. WOODROFFE
University of Wollongong, New South Wales

a contribution to IGCP Project 274: Coastal Evolution in the Quaternary

PUBLISHED BY THE PRESS SYNDICATE OF THE UNIVERSITY OF CAMBRIDGE
The Pitt Building, Trumpington Street, Cambridge CB2 1RP, United Kingdom

CAMBRIDGE UNIVERSITY PRESS
The Edinburgh Building, Cambridge CB2 2RU, United Kingdom
40 West 20th Street, New York, NY 10011-4211, USA
10 Stamford Road, Oakleigh, Melbourne 3166, Australia

First published 1994
First paperback edition 1997

Printed in the United Kingdom at the University Press, Cambridge

Typeset in Times 10/13pt

A catalogue record for this book is available from the British Library

Library of Congress Cataloguing in Publication data

Coastal evolution: Late Quaternary shoreline morphodynamics/
    edition by R. W. G. Carter, C. D. Woodroffe.
    p. cm.
    "A contribution to IGCP Project 274: Coastal Evolution in the Quaternary."
    Includes bibliographical references.
    ISBN 0 521 41976 X
    1. Paleogeography–Quaternary.   2. Coast changes.   I. Carter,
    Bill. (R. W. G.)   II. Woodroffe, C. D.   III. IGCP Project 274.
    QE501.4.P3C63   1994
    551.4'57-dc20   94-7763   CIP

ISBN 0 521 59890 7 paperback
ISBN 0 521 41976 X hardback

Contents

List of contributors   *page* ix

Dedication   J.D. Orford   xi

Foreword   O. van de Plassche   xvii

1   Coastal evolution: an introduction   *R.W.G. Carter and C.D. Woodroffe*   1

2   Morphodynamics of coastal evolution   *P.J. Cowell and B.G. Thom*   33

3   Deltaic coasts   *J.R. Suter*   87

4   Wave-dominated coasts   *P.S. Roy, P.J. Cowell, M.A. Ferland and B.G. Thom*   121

5   Macrotidal estuaries   *J. Chappell and C.D. Woodroffe*   187

6   Lagoons and microtidal coasts   *J.A.G. Cooper*   219

7   Coral atolls   *R.F. McLean and C.D. Woodroffe*   267

8   Continental shelf reef systems   *D. Hopley*   303

9   Arctic coastal plain shorelines   *P.R. Hill, P.W. Barnes, A. Héquette and M-H. Ruz*   341

10   Paraglacial coasts   *D.L. Forbes and J.P.M. Syvitski*   373

11   Coastal cliffs and platforms   *G.B. Griggs and A.S. Trenhaile*   425

12   Tectonic shorelines   *P.A. Pirazzoli*   451

13   Developed coasts   *K.F. Nordstrom*   477

     Index   511

Contributors

Peter W. Barnes
United States Geological Survey, Branch of Pacific Marine Geology, 345 Middlefield Road, MS999, Menlo Park, California 94025, USA

John Chappell
Division of Archaeology and Natural History, The Australian National University, Canberra, ACT 0200, Australia

J. Andrew G. Cooper
Environmental Studies, University of Ulster at Coleraine, Coleraine, Co. Londonderry, BT52 1SA, Northern Ireland

Peter J. Cowell
Coastal Studies Unit, Department of Geography, University of Sydney, Sydney, NSW 2006, Australia

Marie A. Ferland
Department of Geography, University of Sydney, Sydney, NSW 2006, Australia

Don L. Forbes
Geological Survey of Canada, Atlantic Geoscience Centre, Bedford Institute of Oceanography, PO Box 1006, Dartmouth, Nova Scotia, B2Y 4A2, Canada

Gary B. Griggs
Institute of Marine Sciences, University of California, Santa Cruz, California, 95064, USA

Arnaud Héquette
Centre d'études nordiques, Université Laval, Sainte-Foy, Quebec, G1K 7P4, Canada

Philip R. Hill
COR Département d'océanographie, Université du Québec à Rimouski, 300, allée des Ursuliners, Rimouski (Quebec), G5L 2A1, Canada

Contributors

David Hopley
Sir George Fisher Centre for Tropical Marine Studies, James Cook University of North Queensland, Townsville, Queensland 4811, Australia

Roger F. McLean
Department of Geography and Oceanography, Australian Defence Force Academy, Canberra, ACT 2601, Australia

Karl F. Nordstrom
Institute of Marine and Coastal Sciences, The State University of New Jersey, Rutgers, P O Box 231, New Brunswick, New Jersey 08903-0231, USA

Julian D. Orford
School of Geography, Archaeology and Palaeoecology, Queen's University, Belfast, BT7 1NN, Northern Ireland

Paolo A. Pirazzoli
Laboratoire de Géographie Physique, Centre National de la Recherche Scientifique, 1 Place Aristide Briand, 92195, Meudon-Bellevue, France

Peter S. Roy
Department of Mineral Resources, Department of Geography, University of Sydney, Sydney, NSW 2006, Australia

Marie-Hélène Ruz
Centre d'études nordiques, Université Laval, Sainte-Foy, Québec, G1K 7P4, Canada

John R. Suter
Exxon Production Research Company, P O Box 2189, Houston, TX 77252-2189, USA

James P. M. Syvitski
Geological Survey of Canada, Atlantic Geoscience Centre, Bedford Institute of Oceanography, P O Box 1006, Dartmouth, Nova Scotia B2Y 4A2, Canada

Bruce G. Thom
Vice-Chancellor, University of New England, Armidale, NSW 2351, Australia

Alan S. Trenhaile
Department of Geography, University of Windsor, Ontario, N9B 3P4, Canada

Orson van de Plassche
Faculty of Earth Sciences, Free University, P O Box 7161, 1007 MC Amsterdam, The Netherlands

Colin D. Woodroffe
Department of Geography, University of Wollongong, Northfields Ave, Wollongong, NSW 2522, Australia

Richard William (BILL) Gale CARTER
1946–1993

The sudden death of Bill Carter in July 1993 from cancer has robbed coastal geomorphology of one of its leading luminaries and taken from a wide circle of friends and colleagues a person of great stature, intellectual ability and unfailing support. His early death at 47 has deprived the discipline of someone who was just approaching the height of his powers and who had so much more potential to give.

Bill was born in the old coastal town of Bristol, south-west England, and remained a west-country man to the end. His childhood home at Portishead lay only a stone's throw from the Severn estuary and during his early years the shoreline was an ever-changing back-cloth to his deep interest in the natural world. Those who stayed with Bill at his mother's home, quickly sensed his delight in showing those sites along the estuary where he had first played and later studied as an undergraduate. Those early years also bred into him a deep and steadfast belief in family life which became a hallmark characteristic of Bill the person.

He studied Geography for his first degree at Aberystwyth, where he was influenced by Clarence Kidson and Brian McCann. He graduated in 1968, winning the University of Wales Prize for Geography *en route*. Due to personal reasons he was unable to take up a studentship offered to him to work under the late Prof. Joe Jennings at ANU Canberra, and instead had to scramble to find a research studentship in the UK towards the end of the yearly grant-round. After writing to several institutions he was offered a chance to 'do something geomorphological' by Frank Oldfield, at a multidisciplinary School of Biological and Environmental Science in a new university that had just opened in a small country town near the north coast of Northern Ireland. Bill went to Coleraine and, except for a brief excursion to London for a year, remained there as postgraduate student (1968–71), lecturer (1972), senior lecturer (1981), reader (1987), professor (1991) and head of department (1989–92). In many ways the strength and vivacity of the Environmental Studies department at Ulster University is a reflection of the input of Bill, who became a mainstay of departmental life and direction.

These early years in Ireland saw the foundations laid of his coastal life with his perceptive and transformational study of Magilligan Foreland for his doctoral thesis (1972). This formed the cornerstone of his later research, giving direction to mesoscale approaches to morphosedimentary environments, the appreciation of shoreline development through morphodynamics, and the stimulus for his eolian work: research that is also a foundation for this present volume on coastal evolution. Those early years in Ireland also saw the establishment of the other force in his life, that of his family, when he met and married Clare Binney, also a postgraduate student at Coleraine.

It is no overstatement to say that Bill transformed coastal studies in Ireland. Before him, the coasts were regarded as stratigraphical sections to aid interpretation of Quaternary sea-level movement as a key to glaciation chronology. His interests in sea-level change and his realisation that the coastal stratigraphy and beach state had a process story to tell that was explicitly important for coastal studies were witnessed by the string of publications in the 1970s and 1980s which identified these themes. Magilligan, the north coast of Ulster and even the freshwater source of Lough Neagh offered a variety of unstudied shoreline features for Bill to work on. In particular, the work that Bill undertook on the structure and dynamics of swash bars and nearshore bars at a time preceding the emergence of the morphodynamics models of Short and Wright is an instructive example of how he recognised that the dynamics of coastal systems could be coupled with morphology in other than a descriptive fashion. His work on process did not deny his long-standing interest in Quaternary-scale coastal change. He was a strong contributor to the UK element of the IGCP 200 programme and his establishment of a sea-level curve for Northern Ireland set a standard for an approach to coastal problems. Although not a palynologist or palaeoecologist by training, he had the ability to synthesise the essentials of their work and had the foresight to be able to recognise the reality of coastal environments, rather than 'hopeful' thinking, when translating the results of such work into the context of actual sea-level indicators. In this respect his ability to recognise the three-dimensionality and continuity of coastal process and responses gave him a head start over other more traditional workers on the Irish coastline.

The work on the northern coast of Ireland was followed by a successful phase in the early 1980s when the mixed sand and gravel barriers of south-east Ireland were studied in association with me. The realisation that the plethora of publications on US coastal barriers during the 1970s excluded a wide range of coarse clastic barriers, only served to entice him into studying the wider world of barrier features occurring around the mid- and upper-latitudes as reworked remnants of the Pleistocene inheritance. Bill's main international co-operation dates from this period when he and I were lured to Canada by Don Forbes and Bob Taylor (Bedford Institute of Oceanography) to involve ourselves in the marvellous Holocene shorelines of the Eastern Shore, Nova Scotia. Bill readily identified with both this coastal environment and with the welcoming hearths of many Nova Scotian homes. He returned every year, sometimes with his growing family (son Ben and daughter Helen), for nearly a decade, always discovering something new each time. A painting of Chezzetacook Inlet (at the heart of the Eastern Shore study) still takes pride of place over his family hearth. Much of the success in unravelling the bewildering context of Holocene shoreline development of this barrier-dominated coastline came from the incisive and constantly fertile mind of Bill. Indeed, much of the background to this volume comes from his attempts to organise the diversity of barrier and beach forms that he saw in Nova Scotia. The stability and dominance of certain forms, the way in which those forms attempted to minimise their variation through feedback structures and the possibilities of indeterminancy in a complex morphosedimentary environment, all arose through this period of close contact with the coarse clastic shorelines of Nova Scotia. This Ulster link with Nova Scotia was further cemented, to the delight of Bill, by the appointment of John Shaw, one of Bill's postgraduates, to the Bedford Institute.

It is difficult to state which coastal morphology intrigued Bill the most. He was unusual in some senses in that he had expertise in several areas of coastal studies, in that his name will be linked with gravel-dominated barriers, with sea-level studies, with sand beach morphodynamics and with coastal dunes. It was in the last field that he had other close international ties. In particular, Karl Nordstrom and Norbert Psuty and their associates through Rutgers University proved to be a fertile source of association for Bill in his dune studies.

Although keenly interested in dunes as a morphology interesting in its own right, Bill regarded them as an integral element of a dynamic coastal system and was just about to undertake a major research project on the interaction of beach, dune and sea-level forcing when his last illness occurred.

During 1981, Bill was unfortunately sidelined by the need for major surgery. While convalescing he started two long-term commitments; one was to discover that he enjoyed cooking 'real' bread, the other was to discover the enjoyment of writing a book. *Coastal Environments* (1988) will stand as a lasting testament to his pertinence, incisiveness and comprehension that became the hallmark of his work. This book departed radically from what had gone before in that it included an explicit identification of the problem and prospects presented by human involvement with the coast. This holistic theme was a product of the environmental tradition that Bill discovered at Coleraine and which he willingly embraced. He felt that coasts had to be managed if they were to survive as functioning elements of the natural world. It pained him that his own country, with such great past maritime allegiances, seemed to make all of the fundamental mistakes in using and abusing the coast, and despite being told still would not recognise this fact. He spent considerable time involving himself in unpaid coastal advisory work for the National Trust (the largest private charity landowner of UK coastline) and would always oblige coastal landowners and managers, regardless of status and size, with his views on management. His was a voice not always welcomed by coastal engineers in that he felt that coastal engineers were too partisan over engineering structures for the good of the coast, but his integrity and purpose were respected by them, and he kept being invited back to talk at their meetings.

On an individual basis, Bill was a professional to his finger-tips when it came to matters coastal. He often gave the appearance of being aloof and possibly cool towards others he did not know. This was because he had a shyness which he fought all the time. I can well remember my first meeting with him as an unexpected visitor to the north coast of Ireland, after ten minutes I wondered how on earth I was going to survive the rest of the day with him! After one day with him, I realised that I had found a colleague bursting with ideas and knowledge. After fifteen years of working closely with him, I know I had a friend for life. He was well known for his verbal explosions but it all related to one aspect; he admired all who tried honestly, but would not suffer fools and charlatans. If you admitted you did not know then he was the soul of help. I know for certain that he agonised over all the material that came his way to be read. Bill was a glutton for work, his output was prodigious, somewhat in excess of 120 papers and four books in two decades. He never stopped still for a moment, always jotting something down or planning the next paper. He was an avid traveller, always looking for new coasts to see. He kept the tourist card industry and postal services in business around the world! His early disappointment at missing out on Australia was somewhat mollified by the last trip he made when he visited there in November 1992 in order to work on the editing of manuscripts for this book. He took the opportunity to visit most of the centres of coastal excellence that have appeared over the last two decades and was greatly impressed with the variety of coasts he saw and the friendliness of all whom he met. One of his comments to me after this trip was his sadness that the UK had never recognised the importance of the coastal environment as had Australia.

Bill's illness was diagnosed early in 1993. His fortitude and courage would have been a revelation to those who did not know him before his illness. To those who did know him he was just the same Bill, going about life's adversities with his usual commonsense and down-to-earth approach. I doubt if he ever realised the admiration and awe that he engendered in his friends over the last few months of his life. He had to undergo a series of chemotherapy cycles which clearly took a toll of him physically, although mentally he was in total command, endeavouring to do more work than ever. He wrote several papers, two major end-of-grant reports, completed his share of editing papers for this IGCP 274 volume and answered numerous enquiries about his health from his very wide circle of correspondents. He often likened himself to the spider at the centre of a web, he only had to twitch and we came running! He talked often about new ideas about existing work and new projects. However, it was not to be, as eventually the cancer took its toll and Bill died on July 17th 1993. This book, which engenders so many of the ideas and themes upon which Bill worked so hard, is dedicated to the memory of him as an outstanding worker, a selfless colleague and a firm friend to all in the coastal environment.

Belfast, 1993   *Julian Orford*

Foreword

The International Geological Correlation Programme (IGCP) is concerned with networking and research projects of continental to global scale and scope. In the case of IGCP Project 274, 'Quaternary coastal evolution: case studies, models and regional patterns' (short title: 'Coastal evolution in the Quaternary'), much of the initial network and research focus came forth from two previous IGCP Projects: No. 61 (1974 to 1982) and, in particular, No. 200 (1983 to 1987). These two earlier Projects dealt primarily with the global variation in relative sea-level (RSL) changes during the last 15 000 and 200 000 years respectively. At the final meeting of Project 200 (held in Halifax and Ottawa Canada in 1987) an overwhelming majority favoured continuation of sea-level and coastal research under the stimulating IGCP umbrella of UNESCO and IUGS. Project proposal No. 274 was submitted in October 1987 and endorsed as Project No. 274 in March 1988 to run until 1992 (later extended to 1993). The inaugural meeting was held on 22 September 1988 during an International Symposium on 'Theoretical and applied aspects of coastal and shelf evolution, past and future', held in Amsterdam from 19 to 24 September 1988.

The topic 'coastal evolution' was generally considered the natural choice for a follow-up project given the focus and results of the two foregoing IGCP sea-level projects. While much of the RSL database was being put to good use in validating regional and global models of earth crustal movements and ice-melt histories, available RSL change records remained to be analysed with respect to the effects, in space and time, on different coastal environments and the importance of this factor relative to other controlling variables such as energy regime, sediment supply and characteristics, basement topography, freshwater input, and internal controls and feedback mechanisms operating within any evolving depositional system. Within two years of the initiation of the Project over 400 coastal researchers from more than 50 countries were involved in palaeo-, actuo- and/or predictive studies of a wide range of coastal environments and, equally important, the continental shelf which provides a framework for coastal evolution. As the Project ends, its membership list numbers well over 600 participants from 70 countries.

Essential for the growth and maintenance of this network were the efforts by the National Representatives and Correspondents, the organisers of the many national and international symposia, conferences, workshops, fieldtrips and business meetings (Fig. 1), and the editors of a significant number of acknowledged contributions to the Project (see list below). Several of the international meetings were co-sponsored by Sub-Commissions of the INQUA Commission on Quaternary Shorelines. The generous offer by 24 radiocarbon laboratories from 17 countries to carry out limited numbers of free datings for Project participants from developing nations provided a much-appreciated extra stimulus. An IGCP Project on Quaternary coastal evolution was justified for several reasons:

One, it would promote (a) further compilation and critical evaluation of existing RSL records and collection of new, much-needed, high-resolution sea-level data, and (b) analysis of coastal records for proxy data of past climate conditions and oceanic influences and of seismotectonic events. (A measure of the importance of coastal records research is found in the fact that UNESCO and the IUGS have endorsed, within their new joint international programme entitled 'Earth Processes in Global Change', a Pilot Project called 'Climates of the Past' (CLIP). The primary objective of CLIP is to determine and understand the natural variability of climate in the tropical belt, through

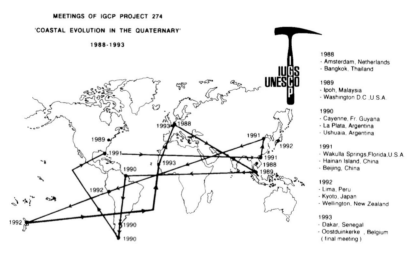

MEETINGS OF IGCP PROJECT 274

'COASTAL EVOLUTION IN THE QUATERNARY'

1988-1993

IUGS
UNESCO

1988
- Amsterdam, Netherlands
- Bangkok, Thailand

1989
- Ipoh, Malaysia
- Washington D.C ,U.S.A.

1990
- Cayenne, Fr. Guyana
- La Plata, Argentina
- Ushuaia, Argentina

1991
- Wakulla Springs,Florida,U.S.A
- Hainan Island, China
- Beijing, China

1992
- Lima, Peru
- Kyoto, Japan
- Wellington, New Zealand

1993
- Dakar, Senegal
- Oostduinkerke , Belgium
(final meeting)

Figure 1. Meetings of IGCP Project 274 'Coastal Evolution in the Quaternary', 1988–1993.

observation of high-resolution dating of geological records *in coastal areas,* noting that such records are one key to interaction of changing oceanic and atmospheric patterns.)

Two, in many parts of the world the coastal environment is under increasing pressure; the questions whether and, if so, how coastal environments should be managed must be based on the best possible understanding of shelf and coastal dynamics and interaction, as well as on adequate knowledge of the state or mode and sensitivity of a system, the controlling factors, the scale on which these operate (or operated) and the direction into which they are likely to change in the near future.

Three, the variety in coastal environments, the range in stages of research reached in different areas and the need for reliable prediction called for generalisation, coherent classification, and education and for further development of conceptual and numerical models of coastal/shelf processes and evolution.

On the basis of these considerations it was agreed that the primary objectives of Project 274 would be: (i) to document and explain local to global variations in coastal and continental-shelf evolution, incorporating knowledge of coastal and shelf processes and environments with geodynamic, climatic, oceanographic and other data to produce local and regional models, ranging from descriptive to numerical, leading to better understanding of interactive forces responsible for past, present, and future changes of coastlines; (ii) to promote specified thematic studies, which are necessary to solve problems of coastal change affecting human occupation of the coastal zone (e.g., assessment of the impacts of past and future sea-level change on coastal environments); (iii) to develop a globally coherent framework for integrated analysis and prediction of coastal change on different spatio-temporal scales, concentrating on the last 125 000 years; and (iv) to promote education on matters concerning coastal evolution.

The study of the evolution of a given coastal or shelf area requires independent and sufficiently accurate documentation of at least the RSL and tidal-range histories for that area and of the changes in the position and characteristics of the shoreline and coastal (sub)environments through space at time scales of millennia, centuries, and, if possible, decades with special attention to the influence of high-energy events. In view of the fact that large parts of the continental shelves and coastlines of the world have not been mapped in any detail and given the time-consuming and labour-intensive nature of coastal and shelf mapping and sample collection, it is no surprise that regional and global syntheses of coastal evolution remain a long-term goal. An important general achievement of IGCP Project 274 is, however, that through the many international meetings and the numerous individual and collective publications, coastal researchers the world over have obtained a supra-local scope of their field of research. At the same time it is now generally appreciated that while a coherent regional to global analysis of the spatiotemporal hierarchy of controlling factors is both possible and important, this cannot be achieved without carefully conducted local studies that can be compared and analysed for similarities and differences in influence of RSL change and other parameters.

One of the first major contributions to Project 274 (and to several other international activities) was the *World Atlas of Holocene Sea-level Changes* (Pirazzoli, 1991). This atlas of some 800 field and 100 model-predicted RSL curves is at once a milestone and a millstone in that it brings out not only the global effort of sea-level researchers over the past 30 years, but also the weight of the question as regards the independency and accuracy of each curve when it comes to explaining general or specific aspects of shelf or coastal evolution during the past 10 000 years. Indeed, a well-documented history of coastal evolution generates its own demands on (existing) sea-level, palaeo-tide, and other parametric data. Conversely, as this atlas invites a global first-order analysis of the relation between variation in RSL change and in coastal evolution during the Holocene, it challenges the coastal researcher to define accurately first-order features of shelf and coastal depositional histories.

Relative sea-level change never is, of course, the only operating variable. For a given rate and sign of RSL change the effect is strongly dependent on other system parameters, such as substrate topography (accommodation space), rate of sediment supply, biogenic productivity and fluvio-tidal balance. Thus, adjacent estuaries, which experienced the same RSL influence, nevertheless can have markedly different records of environmental change. On the other hand, patterns of Quaternary evolution of coastal and lowland riverine plains can be broadly similar across a region as large as Southeast Asia to northern Australia.

The natural variability and complex nature of coastal systems requires careful and systematic documentation and analysis of depositional records and calls for evolutionary morphosedimentary models and general concepts that may be applied or serve as useful working hypotheses in other, less-well-studied areas. Fine examples can be found in many of the Project 274 contributions listed below, or among the hundreds of individual papers produced by Project participants over the past six years. The present book on *Coastal Evolution,* born out of that very idea of model and concept transfer, will be a source of inspiration to many.

Project Leader   *Orson van de Plassche*
Amsterdam, 1993

Selected contributions to IGCP Project 274

Gayes, P.T., Lewis, R.S. & Bokuniewicz, H.S., eds. (1991). *Quaternary geology of Long Island Sound and adjacent coastal areas. Journal of Coastal Research, Special Issue* No 11, 227 pp. (12 contributions).

Pirazzoli, P.A. (1991). *World Atlas of Holocene Sea-level Changes.* Amsterdam: Elsevier Oceanography Series, 58. 300 pp.

Yunshan, Qin & Songling, Zhao, eds. (1991). *Quaternary coastline changes in China.* Beijing: China Ocean Press. 192 pp. (15 contributions).

Donoghue, J.F., Davis, R.A., Fletcher, C.H. & Suter, J.R., eds. (1992). *Quaternary coastal evolution. Sedimentary Geology,* Special Issue, Vol. 80, 3/4, 137–332 (13 contributions).

Fletcher, Ch. H. III & Wehmiller, J.F., eds. (1992). *Quaternary coasts of the United States: marine and lacustrine systems. Society of Economic Paleontologists and Mineralogists,* Special Publication No. 48. 450 pp. (39 contributions).

Ota, Y., Nelson, A.R. & Berryman, K.R., eds. (1992). *Impacts of tectonics on Quaternary coastal evolution. Quaternary International,* Vol. 15/16. 184 pp. (14 contributions) .

Prost, M.T., ed. (1992). *Evolution des littoraux de Guyane et de la zone Caraibe meridionale pendent le Quaternaire. Collections Colloques et Seminaires,* ORSTOM Editions. 578 pp. (25 contributions).

Shennan, I., Orford, J.D. & Plater, A.J., eds. (1992). IGCP Project 274, Quaternary Coastal Evolution: case studies, models and regional patterns. Final report of the UK Working Group. *Proceedings of the Geologists' Association,* London, **103**, 161–272. (7 contributions).

Woodroffe, C.D., ed. (1993). Late-Quaternary evolution of coastal and lowland riverine plains of Southeast Asia and northern Australia. *Sedimentary Geology,* Special Issue, Vol. 83, 163–358. (10 contributions).

1

Coastal evolution: an introduction

R.W.G. CARTER AND C.D. WOODROFFE

'if the environment is the theatre, then evolution is the play'
G. Evelyn Hutchinson

Studies of coastal evolution examine and explore the reasons why the position and nature of the shoreline alter from time to time. Although this type of approach has been practised for generations – by geomorphologists, geologists and engineers – events over the last two decades have brought a new immediacy to the subject. Generally there has been realisation that many of the world's coastlines are under 'threat' (see, for example the recent US Geological Survey Publication *Coasts in crisis* (Williams, Dodd & Gohn, 1990) or the Intergovernmental Panel on Climate Change (IPCC) report *Global climate change and the rising challenge of the sea* (1992)) and that environmental change is the consequence of human occupation of shorelines, to which adjustment (of some kind) is inevitable. Specifically, the spectre of rising sea levels has induced a strong political response as well as raising inevitable questions among scientists as to the state of our knowledge and understanding (processes that are not always convergent). It would not be hard to conclude that our knowledge is woefully thin. Despite many excellent studies from a wide range of environments, we are still well short of understanding how a coastline will respond or react to secular variations in forcing functions such as sea-level rise, storm intensity and magnitude variations or shifts in the sea state pattern. The commonest conclusion is to predict flooding or coastal erosion, yet such processes are clearly only part of much broader responses, which need to be viewed over a range of scales, in both space and time. It is important to remember that erosion simply releases sediment for deposition elsewhere, and that sediment budgets need to be considered in order to explain coastal evolution.

Current practices and paradigms

Most coastal scientists concerned with the processes, products and patterns of shoreline evolution are aware of the need to investigate the fundamental issues which control sediment transport. This can be done directly, by the association of fluid flows and fluid flow structures to the movement of individual particles or mass aggregates, or indirectly through the interpretation of sediment textures, structures and bedforms. A few studies combine both approaches, but regardless of which approach is used there are many drawbacks, particularly in developing robust, generic models of coastal behaviour that might be useful in predicting future evolution – certainly of the type requested by planners and shoreline managers. Part of the problem relates to the current database; with very few exceptions data are sparse, particularly on long-term change. Commonly, one is forced to use information of dubious or unreliable quality, in which error is often acknowledged but seldom analysed. All too often a coastal-change study will use a few historical maps, perhaps one or two aerial photographs and some ground surveys, developing a discursive discussion from this inherently unstable mix of information. If coastal change is obvious and explainable then this approach can be relatively powerful and indicate some underlying symptomatic process. However, in nature such examples are rare, although, paradoxically, they may be relatively abundant in the coastal literature, simply because both authors and editors would view them more favourably. No-one wishes to publish a paper in which the information (however carefully collected) fails to support a clear conclusion.

A second problem perhaps relates to a common scientific inability to break with convention – this type of reluctance to shift to new 'paradigms' has been discussed by Kuhn (1962). For many years, coastal evolution studies have tended to be stereotyped, assuming that there is some kind of observable and straightforward explanation for most changes (the example of the Bruun Rule is discussed below). The universal presence of simple linkages is manifestly not the case (just consider how many studies fail to find such links). This problem is not confined to coastal studies; recently, Schumm (1991) has examined a wide range of geomorphic examples and extracted many common failings of this type, which lead inevitably to misinterpretation. As a matter of some urgency, researchers concerned with coastal evolution should consider the alternative models, even if there are few supporting data. The ideas of non-linear response, stochastic development, deterministic chaos, catastrophism and criticality all deserve investigation.

One, almost Pavlovian, response to the dilemmas listed above, is to collect more and more data. There is some rationale in this as the bigger database may well provide better insights, although more data do not equal better data. Indeed, consistency of information, especially between different coastal workers and their projects, is rarely questioned, yet it is evident that a modicum of effort in standardisation might pay dividends in terms of better interpretation.

Another recent trend (epitomised by the International Geological Correlation Programme Project 274 which provided the impetus for this book) has been to both widen the context of coastal studies and to move towards integration between the two dominant spheres of research activity, namely process studies of wave–morphology interaction and Holocene-scale investigations of environmental change (especially charting the course of former sea levels). This consideration of the co-adjustments of process and form is termed morphodynamics (see Chapter 2).

The widening of context arises from the view of the shoreline as part of a much larger system, within which the parts cannot be easily divorced. Obviously, there are critical links between the oceans and the shore, but the exact nature of these was practically unexplored until studies, initiated by Swift, Hayes, Pilkey, Wright and others along the Gulf and eastern US seaboards in the 1970s and 1980s, began to reveal the very intimate connections between the beach, estuary, nearshore, shoreface and shelf. These have a profound effect on sediment exchange (see Wright, 1987, for review). Often, this type of sediment budget has developed on a long-term (Holocene or Pleistocene – ~10^3 to 10^6 years) time scale, although there is no guarantee that recent accelerations in the rates of global change will not reactivate and re-establish such mass movements. A further important broadening of scope comes from the view that terrestrial environments play a profound role in coastal evolution. In some ways, this view has been implicit in studies since the mid-nineteenth century, but nonetheless the increasing amount of empirical evidence, especially where off-land sediment yields have been decimated by such processes as shoreline 'protection' and river regulation, has brought a new recognition to the subject. Again, time scale is important, as there is evidence that such processes as sediment amount, transport and delivery to the coast may vary by several orders of magnitude both through natural, climatically controlled factors as well as through those mediated by human activity. For example, Mediterranean coastal evolution shows distinct episodes, which may be related not only to regional climatic change, but also to anthropogenic influences such as agricultural expansion and dam building (Vita-Finzi, 1964; Sestini, Jeftic & Milliman, 1989).

Over the last 15 000 years, first-order mediation of coastal evolution has been related to global sea-level history, which in turn is coupled to global climatic change and, to a lesser extent at this time scale, to tectonic activity (Carter, 1992). There are strong latitudinal differences in climate response as well as hemispheric ones, with the result that predictability of events is low. Nonetheless, it is possible to speculate that, for example, fluctuations in the glacial mass balance of Antarctica would have a worldwide impact. Anderson

& Thomas (1991) suggest such ice-balance changes may well lead to rapid, episodic sea-level variations around the world, which can be detected in coastal sediments and morphology from a range of environments, *assuming one is looking for such evidence.*

Coastal evolution also has an importance within the biological sciences. At one level coasts need to be viewed as functional ecosystems playing an important role in the recycling of nutrients and minerals (Brown & McLachlan, 1990). It is not too far fetched to describe some coastal environments, such as estuaries, as 'reactors', involved in energy production, which is then exported to sustain adjacent, energy-deficient areas. For this function alone it is necessary to maintain coastal dynamics. There is, however, a broader function, related to long-term biological diversity and genetic evolution. Woodroffe & Grindrod (1991) indicate that sea-level change and the consequent extent of intertidal habitat has been a major control on the distribution of mangrove forests, and perhaps other intertidal communities, which in turn gives rise to biodiversity. As such we have a vested interest in coastal evolution as one means of providing global stability! Deceleration of sea-level rise around 6000 years ago may have triggered Predynastic agriculture in Egypt (Stanley & Warne, 1993a), setting the scene for the dawn of western civilisation.

There is an opposite tendency to proselytise ideas which at one stage proved interesting, but which through time become largely unsustainable. Such concepts often have a large number of adherents, who are reluctant to abandon what, in truth, is a simple, easily applicable procedure. There are plenty of examples of such occurrences in the Earth Sciences – readers are referred to Ginsburg's (1973) book *Evolving Concepts in Sedimentology* for a series of case studies and Dury (1978) for a more cynical view focusing on geomorphology. Two examples in coastal evolution studies are the formula devised by Bruun (1962), which claims to predict coastline retreat under sea-level rise, and the 'equilibrium profile concept', which has a far longer history, but which was calculated by Dean (1977, 1991; Dean, Healy & Dommerholt, 1993) to be modelled by the expression $y = \rho x^{2/3}$. (It is worth noting that both Bruun and Dean are engineers, with an interest in quantifying clients' problems – an absolute value will always beat statistical caution.)

The original 1962 paper of Bruun's is almost certainly a citation 'classic' as it is widely and persistently quoted. Unfortunately, in the intervening 30 years the basic idea has never been fully vindicated (certainly not in the field), for the rather obvious reason that coastlines do not behave in a strict two-dimensional sense. Although Bruun himself (Bruun, 1988) has tried to widen the rule's applicability, it remains in most situations an inadequate oversimplification.

The equilibrium profile debate is a far more complex one. There is an intuitive feeling that if a wave-formed slope (like a hillslope) is given enough 'time' it will come into some kind of equilibrium with the forces applied to it. Most authors cite Cornaglia (1891) as being the first to introduce the concept. However, as with the Bruun problem, a wave-formed slope is subject to multidirectional forces of considerable spatial and temporal variability, so that the simple 'time to equilibrium' models of, for example, Schumm & Lichty (1965) and Carson & Kirkby (1972) cannot apply (see Fig. 1.1*a*). Alternatively, one anticipates a noisy model, in which there is constant movement across and along the coast.

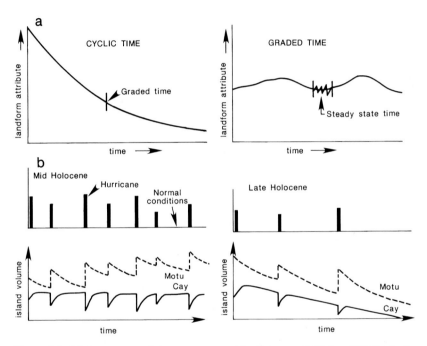

Figure 1.1. (*a*) the concept of time as expressed by Schumm & Lichty (1965). Cyclic time is represented by the evolution of landform attributes over geological time; graded time is a shorter timescale, over which landforms may be considered to be in a steady state. (*b*) An example of dynamic equilibrium of reef islands on a Pacific atoll (after Bayliss-Smith, 1988). Coarse-grained islands, termed motu, react to the passage of a storm (hurricane) through an increase in size as rubble ramparts are formed; these break down and are redistributed during the periods between storms. Cays, sandy islands, are eroded by storms, but gradually recover after the passage of the storm. In mid-Holocene times sand cays may have been in a steady state, and motu may have been undergoing gradual net enlargment. During the late Holocene, as a result of less-frequent storms, lower wave energy and reef productivity, less island vegetation and rising sea level, islands may have less capacity to recover and may gradually become smaller.

The recent work of Wright and others mentioned above (Wright, 1987, 1993; Young & Pilkey, 1992; Wright *et al.*, 1991) is very important in this context, as it begins to provide extensive data and new insights into the problem of long-term nearshore response of sandy beach and barrier systems. Pilkey *et al.* (1993) provide a recent review and analysis of the equilibrium slope debate, and interestingly show the almost complete inapplicability of the Dean formula to field data.

In many cases there is likely to be a dynamic equilibrium. One such example is shown in Fig. 1.1*b*. On Pacific atolls there are two types of islands; motu are islands composed of coral shingle and boulders on the higher-energy reef flats, and cays are sandy islands in less-exposed situations. These respond differently to the impact of tropical storms, with a net input of coarse material as rubble ramparts on motu, but storm cut on cays. Motu subsequently alter with redistribution of material between storms, while cays rebuild between storms. Bayliss-Smith (1988) has suggested that in mid-Holocene times these reef islands may have existed in a steady state, with storm effects averaging out over time, but in the late Holocene, under different conditions of storm occurrence and with reduced vegetation cover, the islands may be exhibiting a dynamic equilibrium, but gradually reducing in size (Fig. 1.1*b*; see also Chapter 7).

History of coastal evolution

For as long as shorelines have been occupied it is likely humans have taken note of coastal changes. The oral traditions (dreamtimes) of Australian aboriginals, perhaps extending back tens of millennia, identify clearly the early Holocene transgression and the consequences of the rising sea in destroying food sources and disrupting communications (Flood, 1993). Biblical evidence also emphasises the impact of coastal change, particularly associated with extreme events (Bentor, 1989). Geomorphological reconstructions indicate how the site at Troy, and other cities referred to in the Iliad, have altered (Kraft, Kayan & Erol, 1980). In this, and subsequent Old World periods, shoreline changes were especially noteworthy where they interfered with commercial or defensive activities around river mouths, harbours or ports (e.g. Inman, 1978; Masters & Flemming, 1983).

The development of geographical studies of coastlines in the mid-nineteenth century quickly acknowledged the need to understand how coastal changes occurred. The apotheosis of this early movement came with the works of Gilbert on Lake Bonneville, and Davis on Cape Cod, and later the studies of Johnson (1919), of which his book *The New England – Acadian Shoreline*, published in 1925, represents perhaps the finest example of the genre. In this

Johnson attempts to argue from morphological evidence how a complex shoreline has evolved. In the years immediately after Johnson's monograph was published, J. Alfred Steers began his career in Cambridge. Steers' impact on British – and to some extent Australian – coastal research was significant. Steers (along with Vaughan Lewis) began a detailed investigation of coastal changes around the British Isles, culminating in the publication of *The Coastline of England and Wales* in 1946 (later extensively revised in 1964). This work frequently refers to coastal evolution, as for instance in Steers' interpretation of the evolution of Dungeness in southern England. Steers was also responsible, along with biologists such as Frank Oliver, for the initiation of long-term studies on Scolt Head Island off the Norfolk coast, and the maintenance of records which have proved invaluable to present-day studies.

It is somewhat invidious to name only a few individuals, as many people have been influential in the development of coastal evolution studies. However, honourable mention should be made of Axel Schou who published his monograph *Det Marine Forland* immediately after the Second World War. This remarkable contribution – sadly in Danish – provides a concise and detailed exposition of the evolution of the Danish coast. Although Schou lacked much of the chronological support that modern coastal evolutionists take for granted, he still managed to develop coherent arguments, which stand the test of time. Slightly later comes the contribution of André Guilcher in France. Again, working under wartime restrictions, Guilcher was to bring a broader perspective to coastal evolution studies through his oceanographic (and to a lesser extent geological) interests. Guilcher's text *Coastal and Submarine Morphology* (published in French in 1954 and in English in 1958) marked an important diversification, which has continued to find echoes in many contemporary studies. Finally, it would be remiss not to mention the Russian work of V. P. Zenkovich. Although occasional Russian studies had appeared in the English literature, it was the 1967 translation and publication of Zenkovich's book which provided the first real insight into the parallel (and in places, divergent) developments in coastal evolution studies between the western and communist scientists. Many of Zenkovich's examples clearly found their original impetus in the pre-Revolution studies of Gilbert, Davis and Johnson, but nonetheless a very distinctive praxis had developed. For example, work on sediment tracing to ascertain sediment transport rates was progressing in the 1930s, well before similar studies were attempted in the USA or UK. Zenkovich's own ideas on shoreline evolution – often based on spectacular examples from inland sea coasts – are still worthy of consideration, although many were completed by the 1950s.

During the 1950s and early 1960s, coastal research was still influenced by a number of very active individuals, such as R.J. Russell who began the Coastal Studies Institute at Louisiana State University (Russell, 1967), C.A.M. King in Britain (King, 1972), and C.A. Cotton in New Zealand (Cotton, 1974). After the 1960s, it becomes harder to link influential individuals with specific developments. There are two reasons for this. One, coastal studies in general were expanding and, two, there was a shift of emphasis towards process research – and to a more limited extent towards an emphasis on management of shorelines. However, a major shift took place in the mid- to late-1970s with the emergence of the sea-level 'issue'. Sea-level research had been gaining momentum throughout the 1960s under the guidance of Rhodes Fairbridge, Walter Newman and Art Bloom, and workers were beginning to use the Carbon-14 timescale (basically the last 40 000 years) as a foundation for assessing the time/depth course of sea-level changes for a wide variety of environments from submerging coral atolls to Arctic raised shorelines. Bloom, in particular, was a key figure in setting up the global networks of sea-level researchers and was the coordinator of the first IGCP Project 61 on sea-level changes. The scares of the late 1980s and early 1990s, which suggested that sudden massive rises of sea level might occur as a consequence of global warming (with estimates of several metres) rapidly provided an unprecedented public and political awareness of coastal studies. Researchers who had worked for many years on coastal problems were suddenly 'in demand', both by the media and funding organisations. Needless to say they were joined by many others who recognised the opportunity to present themselves as coastal 'experts'. In some respects the normal scientific caution was ignored, as many reports – often illustrated with lurid and misleading maps – indicated areas likely to be flooded or submerged by rising water. Yet, the positive side has been a strong resurgence of coastal evolution studies, often from an interdisciplinary perspective. The completion of IGCP Project 200 under the direction of Paolo Pirazzoli and its successor IGCP Project 274 under Orson van de Plassche synergised coastal evolution research on a worldwide basis. Now, new research programmes such as the Land–Ocean Interactions in the Coastal Zone (LOICZ) core project of the International Geosphere Biosphere Programme (IGBP) (Holligan & de Boois, 1993) are providing clear signals as to future research directions.

Coastal behaviour

The coast 'behaves' in a wide variety of ways, depending on time scale, length scale, geological structure, tectonic setting, sediment type and availability, sea-level position, wave and current processes, and the adjacent terrestrial and oceanic environments (Fig. 1.2*a*). The coast must be considered as a system;

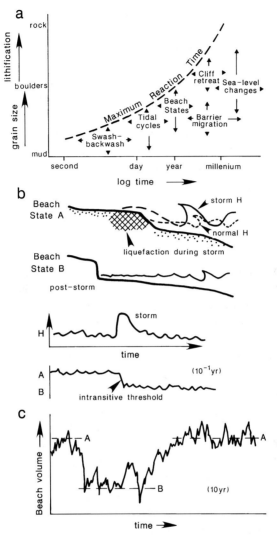

Figure 1.2. (*a*) Morphodynamic responses on different shorelines and coastal settings occur at different rates. Reaction time also depends upon the sediment size, and lithification. Thus, beaches respond over time scales of months, while barriers respond over time scales of decades or longer (after Carter, 1988). (*b*) Beach states A and B represent opposite ends of a continuum of beach states. A is accreted; however, a large storm may cause beach liquefaction and result in the eroded state B. This appears as an intransitive threshold over the time scale of months, with fluctuations occurring in response to regular wave processes (H, wave height). This is a metastable equilibrium in which there are apparently two equilibrium states. (After Chappell, 1983.) (*c*) Over the decadal time scale, the beach may recover from state B and the beach may again become accreted. This is the case at Moruya, southeast Australia (after Thom & Hall, 1991), where storms in the 1970s caused extensive beach cut, but in the 1980s the beach recovered to its pre-storm volume (see also Fig. 2.11).

its morphodynamics involve complex mutual coadjustment of processes and forms. For example, on many coasts, shoaling and nearshore transformation of waves reduce the energy incident on the beach by 95 to 99%. The frictional turbulence in the wave zone is largely 'lost', although at some levels a proportion leads to sediment entrainment and movement. In places, the coast is able to become organised, inasmuch as relatively resistant structures are built up which assist or alter the modes of energy, mass or information transfer. Perhaps the best known of these organisations is that inherent in beach morphodynamics (Wright & Thom, 1977; Wright *et al.*, 1979; Wright & Short, 1984; Short, 1991). From the mid 1970s, various models of beach stage were developed; basically, this involved exploring the morphological sequences exhibited by sandy beaches in terms of transition (the movement from one characteristic form to another). The appropriateness of this method can be judged by both the volume of research and the fact that 'stage' could be linked, quite successfully, to a wide variety of physical and, more recently, biological phenomena (McLachlan & Erasmus, 1983). The approach also thus allows a categorisation of sedimentary coasts at relatively short time scales – storm– fairweather oscillations, tidal cycles and seasonal, climatically induced variations (Fig. 1.2).

The morphodynamic approach relies on predictability along certain environmental gradients, such as tidal range, wave exposure and sediment type; morphodynamic behaviour varies in a deterministic manner. It is feasible to draw out invariants – for example, Huntley & Short (1992) identify the influence of sediment fall velocity on rip current spacing, which appears, at the site they examined, to be independent of wave period. Fig. 1.2*b* shows two end-members of a continuum of beach states. State A is accreted; however, under a severe storm, beach sediment may undergo liquefaction, and the beach may be cut back to an eroded beach state. The storm exceeds an intransitive threshold (Chappell, 1983); under the usual range of non-storm wave conditions the beach fluctuates around one state or the other. This represents an example of metastable equilibrium. Viewed over a longer term, however, there may be broad periods during which the beach oscillates around the erosional state with gradual replenishment of the beach to the accreted state (Fig. 1.2*c*). Reaction to the storm occurs rapidly, but relaxation and beach recovery take considerably longer, as is shown by long-term beach monitoring at Moruya in southeast Australia (Thom & Hall, 1991; see also Chapter 2).

As a coast evolves towards an organised morphodynamic system, then it begins to control its own environment. It may well move from a less-stable to a more-stable one, capable of absorbing a far greater range of energy, sediment or informational inputs (Fig. 1.2). Excellent examples of this type of system are provided by gravel (coarse clastic beaches) which evolve towards stable morphosedimentary forms (see Fig. 1.4), in which process, forms and sediments are coupled in both the longshore and the cross-shore directions to provide very efficient absorbers across a wide range of energy conditions (Carter & Orford, 1993).

The essential morphodynamic behavioural models of beaches and surf zones offer great scope for understanding the short-term (days, weeks, months) behaviour of the coast. Set transitions may be compared from place to place and even along extensive stretches of coast, in the manner described by Short (1992) for the Netherlands. However, the morphodynamic model does not necessarily hold for longer periods or larger stretches of coastal development. One alternative is to view the coast as a series of discrete cells, which are open to energy input, but relatively closed to inter-sediment transfers. At the smallest scale (perhaps 10^1 to 10^3 m shoreline) cells merge with morphodynamic units, but at a meso-length scale (10^3 to 10^5 m) the cell is significantly larger. Over time incident wave and current forces within a cell may come into balance such that erosion and deposition are effectively cancelled by equipotential gradients of littoral power. However, totally balanced cells are rare as there is almost always some loss or gain of sediment to the system. Negative budget cells (such as estuaries, where there is a net sink of sediment in the estuary; see Fig. 1.3*a*, and Chapter 5) are characterised by erosion (and shoreline transgression), although some may indicate 'arrested' development as the cell becomes armoured. A positive sediment budget (such as a delta, where riverine sediment reaches the coast; see Fig. 1.3*b*, and Chapter 3) will see gradual accumulation of material within the cell, which may in time renew inter-cell leakage and re-establish longshore transport. It is common to find cells that have (or had) positive sediment budgets evolving extensive beach ridge plains, based on persistent supplies from the shelf or from fluvial sources (Fig. 1.3*b*). Over Holocene time scales it is clear that many cells have shifted from positive to negative budgets, so that ones that were formerly accumulating are now eroding. Interestingly, some cells, especially those near circulatory fluvial/marine systems (at river and estuary mouths), may evolve as mass sediment attractors. On the east coast of the USA and elsewhere this results in huge ebb and flood tidal-delta systems, satiated with long phases of sediment input. Such systems are clearly too over-fit in the sense that they bear little relationship (in terms of size) to current inlet hydraulics. In southeast Australia, fluvial input is stored both in estuaries and along the shore in the form of massive inshore ramp structures (Roy, 1984). Cell development has allowed the gradual leakage of material from estuaries onto the shelf where it has accumulated, in what might be justifiably described as fossil forms (see Chapter 4). On many downwind ocean coasts, leakage and storage is often landward into coastal dunes. The gradual transfer of sediment

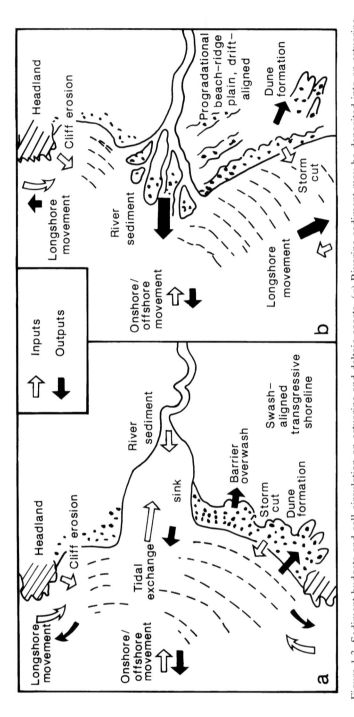

Figure 1.3. Sediment budgets and cell circulation on estuarine and deltaic coasts. (*a*) Riverine sediment may be deposited into an estuarine embayment, which may also receive sediment from the seaward direction through tidal processes. Note the swash-aligned sandy barrier which retreats landwards (transgresses) through washover and aeolian processes. (*b*) On a deltaic coast, riverine sediment contributes to coastal sediment budgets, and is moved up the drift-aligned coast, giving rise to a beach-ridge plain. Dashed lines represent rocky coast; stipples represent sandy coast.

from the marine to the aeolian system may be relatively slow, with net movement only a fraction of gross transport, but over a geological time span it can be hugely effective, especially if fuelled by a positive-turned-negative sediment budget, allowing the release of large quantities of sediment. On the coast of northwest Ireland, the present shelf and shoreface are almost devoid of large sand bodies as all the sediment has moved inland to form dunes and sand sheets (Carter & Wilson, 1993).

There are many different types of cell, including those associated with fixed and free boundary conditions, zeta-bays (log-spiral bays shaped by wave refraction patterns) formed under conditions of oblique, but persistent swell, spit cells maintained by counteracting waves and currents, and so on (Carter, 1988). The dominant attribute of cell structures is that they function as longer-term energy absorbers based on net sediment transport potential.

The cell structure may be extended to embrace the onshore and offshore boundaries, particularly onto the shoreface and inner shelf, and where subaerial sediment storage (in beach ridges, dunes and wetlands) occurs. Long-term exchange of material across and along the coast within defined cells provides one firm basis for analysis of secular coastal change. At a scale of 10^1 to 10^2 km cell definition breaks down, as the process field becomes too variable. At this length scale, the size of individual weather systems (particularly tropical cyclones or hurricanes) may be equal to or smaller than coastal compartments. A pertinent example of this is provided by the western Gulf of Mexico, where the Louisiana–Texas arc is over 1000 km, while individual hurricane landfalls may only occupy a tenth of this zone. As Keen & Slingerland (1993) have demonstrated, the impact at this scale is to create a far more complex pattern of sediment removal and redistribution which cannot be easily reconciled with any cellular structure. In this context sediment is being slowly transferred from the riverine terrestrial to the shelf via coastal processes. Evolution of the shore depends on both the balance between input and output of sediment, and a more complex probabilistic relationship linking storms to sediment transport.

In recent years, some attention has focused on the ideas of Large-Scale Coastal Behaviour (LSCB) (Terwindt & Battjes, 1991; Stive, Roelvink & de Vriend, 1991; de Vriend, 1992a, 1992b; see also Chapter 2), although the term is relative, as some author's viewpoints tend to reflect their previous scale of coastal research. The LSCB concept implies that coastal behaviour may be, in some way, non-linear. The last decade has seen a proliferation of possible non-linear models, ranging from deterministic to stochastic to chaotic, and it is not difficult to envisage situations that might fit certain coastal contexts (e.g. Phillips, 1992). There is reasonable evidence, for example, that the structure of turbulence is deterministically chaotic, and as a number of coastal phenomena develop under turbulent flow, this should (or could) impart a similar spatial pattern to the landforms. Coastal dunes are an obvious example; transverse dunefields, such as the Alexandria Dunefield east of Port Elizabeth in South Africa (Illenberger & Rust, 1988), are chaotic in terms of form. Certainly, given free expression, there is a myriad of superimposed sand forms at a variety of length scales, which may result in a chaotic landscape.

A bigger problem arises in trying to argue for complex non-linear behaviour over time, although in terms of understanding coastal evolution this is probably a key issue. Even if a system is driven by a relatively persistent forcing function – say sea-level rise – there is no guarantee the response will be straightforward. The classic earth science example of this phenomenon is where inter– and intra-plate stresses are resolved into a complex pattern of earthquakes, which have now been shown to be chaotically predictable (Turcotte, 1991). It may be simplistic to assume that sea-level rise will result in predictable responses. A more likely 'scenario' is that a simple forcing function will lead to a variety of responses, depending to a large extent on the ability of the coastline to absorb the stress. Thus, initially a cliff coast will resist (or at least show few signs of alteration), while a diverse, sedimentary coast may well show a range of rapid adjustments. In time, however, thresholds may be exceeded, so that widespread cliff failures occur; to the casual observer (and the scientist), these may be simply unexplainable. Adjustment thus proceeds in a stepwise fashion. The behaviour may also jump from one mode to another; Forbes *et al.* (1991) described the behaviour of a gravel barrier in eastern Nova Scotia, which prior to 1954 had built-up *in situ*, yet after 1954 (the date marks the passage of a major, but not exceptional storm) began to migrate landward at rates of over $10\,\text{m}\,\text{a}^{-1}$. Carter & Orford (1993) have speculated that this type of jump behaviour is not unusual and in fact approximates to a fold catastrophe model (Zeeman, 1977) in which one slow variable (sea-level rise) is associated with one or two fast variables (wave field variation, sediment supply). Existence of such behaviour has implications for adjoining environments, which may have to undergo controlled phases of change, associated with the dominant morphological structure. Antecedent morphology is almost certainly an important variable in this respect (see Orford, Carter & McCloskey, 1993).

Single events with a geological time scale return period

There remains the possibility, at the extreme end of the frequency/magnitude curve, that single, independent or isolated events may have an unduly important significance to coastal evolution. In the past few years evidence has been mounting from the North Sea (Long, Smith & Dawson, 1989), southwest Britain (Foster *et al.*, 1993), the Mediterranean (Andrade, 1992), the Pacific Northwest of North America (Atwater, 1987; Atwater & Moore, 1992), Japan (Minoura & Nakaya, 1991) and Australia (Young & Bryant, 1992; Bryant, Young & Price, 1992) that exceptional tsunamis may be capable of massive reorganisation of the coast. (It needs to be emphasised that tsunamis are relatively common on decadal scale (Watanabe, 1985), but in these cases the events can probably be assimilated into the high-frequency end of the spectrum rather in the manner of hurricanes.) This discussion centres on what might be described as 'mega-tsunamis', occurring so rarely that they are beyond the time scale of perhaps even secular fluctuations of Holocene sea levels. In many cases they centre on major epipeiric Earth movements, explosive volcanic eruptions (Santorini, Krakatoa) or perhaps meteorite impacts (i.e. Bilham & Barrientos, 1991). The environmental context of such perturbations is obviously difficult to assess objectively – although some authors see them as 'driving forces' for global change (Rampino & Self, 1992). For example, Young & Bryant (1992) consider that they have been able to identify an event, around 100 000 years ago, which devastated the coast of southeast Australia. They believe this may have been caused by a massive submarine landslip in the Hawaiian Islands 105 000 years BP, creating a group of tsunami waves which radiated circumferentially across the southwest Pacific. Young & Bryant are able to point to a wide range of what at first appear to be anomalous shore features, including rock channels across headlands, scoured and fluted rock platforms at a variety of levels, poorly-sorted high-level boulder deposits (some boulders weigh over 100 tonnes), and hummocky 'dump' deposits in sheltered areas. Taken individually such features would elicit as many explanations as there are geomorphologists in a field party, but put together they begin to reveal a pattern repeated over several hundred kilometres of shoreline. More controversially, Young & Bryant argue that this single spectacular event may have completely removed the outer coastal barriers and initiated a new phase of barrier, dune and beach development along the coast. Should this be the case, then the many long-held ideas about coastal evolution in this area will need to be revised (see Young *et al.*, 1993).

A more general point about the single, almost geologically isolated event is that the coast has to be in a receptive or sensitive condition in order to retain the record. The widespread evidence for the Storegga slide tsunami in the North Sea (Dawson, Long & Smith, 1988; Long *et al.*, 1989) around 7200 years BP coincides with the peak Holocene transgression in Scotland, so that the deposit – basically a sand layer – stood an excellent chance of preservation.

Moreover, the local timing is clearly important, as the arrival of a tsunami at high tide is likely to leave more evidence than if one arrives at low water. Similarly, a salt marsh or coastal lake only records storm surges or hurricane events in localised sites and when conditions are suitable for deposition of discrete sand or shell layers (i.e. Ehlers *et al.*, 1993; Liu & Fearn, 1993).

Models of coastal evolution

There have been various attempts to model the way that coasts have developed, ranging from highly subjective denudation chronologies proposed by W.M. Davis and those geomorphologists who viewed coasts as developing as part of a geographical cycle, to highly quantitative computer models (Lakhan & Trenhaile, 1989). A major objective of IGCP 274 was to develop models of coastal evolution. In most cases these are conceptual morphodynamic models based on detailed radiometric-dating reconstructions of what has occurred in the past, though as our understanding of processes involved increases it may be possible to develop simulation models, such as the shoreface translation model applied to the coast of New South Wales (see Chapters 2 and 4). Below we describe two conceptual models of coastal evolution from our own studies, from disparate field areas, and illustrate how several common themes may be identified between them.

Gravel barrier evolution in Nova Scotia

Nova Scotia provides a wide range of coastal environments (Stea *et al.*, 1992). Among the most distinctive is the drumlin coast of the Eastern Shore, where late-Wisconsinan deposits are being eroded and transported under a rapidly rising sea level (3.8 mm/year since 1917) (Orford *et al.*, 1992). As the sea level rises so drumlins are eroded in turn; the average 'life' of each eroding drumlin is probably about 2000 years (Carter *et al.*, 1990). The drumlins comprise a heterogeneous mix of coarse and fine sediments and as they are attacked by waves, released material is dispersed according to transportability. This leads to the development of a transport pathway network (which is interactive), moving different types of sediment onto the shelf, along the coast or inland through the estuaries or coastal dunes. Because each drumlin is in a different stage of evolution there is a competitiveness between them to control the transport pathway network, and indeed, such control is sequestered only reluctantly as the drumlin sediment is eventually depleted and the sea rises finally to drown it. Distinct morphosedimentary architectures and morpho-dynamic responses can be recognised for both sandy barriers (Nichol & Boyd,

1993) and gravel barriers (Carter & Orford, 1993) along this coast (see also Chapters 4 and 10).

Mobile gravel released from the drumlin is moved rapidly by waves along the drumlin flanks within what is a very abrupt wave power gradient; wave power may decrease by 90% in only a few hundred metres (Carter *et al.*, 1990). At first these flanking beaches are immature in the sense that the sediments show few signs of sorting or profile development (Fig. 1.4). However, in time

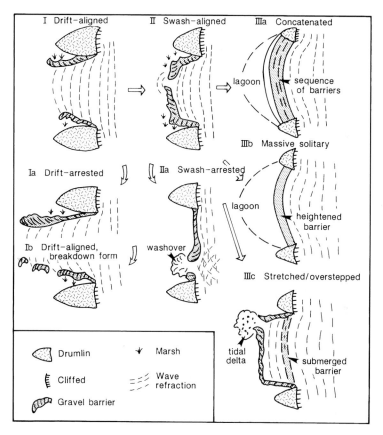

Figure 1.4. Evolutionary stages of gravel barriers in Nova Scotia (based on Orford *et al.*, 1991, and Carter & Orford, 1993). Three phases, linked with drumlin erosion, which provides a source of sediment, can be recognised. In the initial phase (I) barriers are drift-aligned; some may become drift-arrested (Ia) and can break down (Ib). Further sediment supply leads to a phase that is swash-aligned (II). In some cases barriers can become swash arrested (IIa). In the third phase there are either concatenated (IIIa), or a metastable equilibrium between massive solitary barriers (IIIb), where the gravel barrier becomes heightened, or a stretched and overstepped barrier (with a submerged barrier left behind) (IIIc).

an organisation begins to evolve. During the initial stages continued erosion of the source leads to the extension of the flanking barrier, which in time will lose contact with the drumlin and tend towards an equilibrium with the incident waves. This leads to a switch from drift-aligned barriers to swash-aligned barriers (Orford, Carter & Jennings, 1991). The gravel barrier then enters a new phase of evolution, that of a transgressive feature. In time, the barrier becomes stretched and a breach may occur. This breach may well provide a threshold for dramatic change, causing the realignment and breakdown of the swash-orientated barriers, and allowing the penetration of wave energy into previously sheltered back barrier zones. The swash-aligned barriers also breakdown under sea-level forcing. In time, there may be a catastrophic switch from a heightened, massive solitary barrier to an overstepped barrier (Carter & Orford, 1993) where elements of the upper and lower barrier can be disengaged (Forbes *et al.*, 1991).

Over the relatively short time scale of the last two millennia, the gravel barriers of Nova Scotia show a surprising range of behaviour, linked to both morphodynamic and morphosedimentary responses to changing sea level, sediment availability and local wave field variability (see Fig. 1.4). Many of these responses can be found in the records of adjacent environments, such as salt marshes (Carter *et al.*, 1992; Jennings, Carter & Orford, 1993.) Overall, the changes are decidedly non-linear, but within periods of decades it is possible to detect linearity between some variables (Orford *et al.*, 1991).

Tidal river and floodplain evolution in northern Australia

A number of northward-flowing rivers drain the sandstone uplands of the Arnhem Plateau and low rolling laterite hills of the Top End of the Northern Territory. These rivers are tidal as they enter macrotidal van Diemen Gulf, with tidal influence extending up to 100 km upstream as they meander across broad Holocene alluvial/estuarine plains. The tropical climate of this area is distinctly seasonal, with a dry season from June to October followed by a monsoonal, wet season from November to May.

Recent morphostratigraphic work on the macrotidal rivers and their floodplains has built upon initial observations on the Ord River in Western Australia (Wright, Coleman & Thom, 1973; Thom, Wright & Coleman, 1975). The most detailed morphodynamic reconstruction has been undertaken on the South Alligator River, but a similar pattern of development is indicated for other neighbouring river systems (Woodroffe *et al.*, 1989, 1993; Chappell, 1993, see also Chapter 5).

The basic sequence of coastal evolutionary events can be divided into three phases (Fig. 1.5). Phase 1 occurred when the prior valley was flooded by a

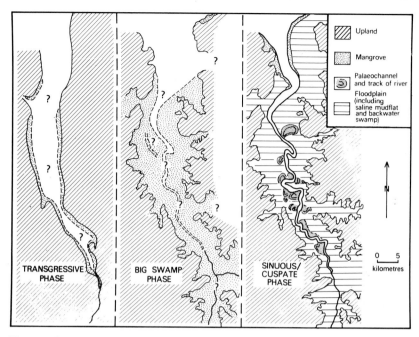

Figure 1.5. Estuarine plains flanking the macrotidal estuaries of the Top End of the Northern Territory, Australia can be shown to have undergone three phases of Holocene evolution. The first phase is the transgressive phase, 8000–6800 years BP, during which the prior valley was inundated by rapidly rising postglacial sea-level rise and organic muds were deposited beneath landward-migrating mangrove forests. The second phase, the big swamp phase, 6800–5300 years BP (on the South Alligator River), was a period of widespread mangrove forests formed on vertically accreting intertidal mudflats. This was replaced by a phase during which the river adopted a sinuous or cuspate meandering form, reworking part of the floodplain, which accreted vertically under freshwater wetland vegetation (from Woodroffe *et al.*, 1993).

rapidly rising postglacial sea level which reached around 14 m below the present level about 8000 years BP. This transgressive phase is represented by highly organic muds or peat and landward extension of mangrove forests, which grade upwards into vertically accreted estuarine muds deposited beneath mangroves.

Deceleration of sea level rise around 6000 years BP, plus a rapid and widespread growth of mangroves (both within the newly-forming estuaries and geographically along the coast of northern Australia), resulted in what has been termed the 'big swamp' phase (Woodroffe, Thom & Chappell, 1985). This 'big swamp' phase lasted from 6800 to 5300 years BP in the South Alligator estuary, and occurred contemporaneously in many other embayments in northern Australia (based on numerous ^{14}C dates – see Woodroffe, Mulrennan & Chappell, 1993).

Termination of the big swamp phase was rapid and it was replaced by vertical aggradation of alluvial plains beneath freshwater wetlands. Rapid progradation of the coastal plains associated with these estuarine systems occurred 5000–3000 years BP, followed by much slower progradation and the episodic development of chenier ridges. The channel morphology of the meandering river, and the extent to which it has reworked the floodplain (see Fig. 1.5), varies between rivers in response to fluvio-tidal parameters, and is examined in Chapter 5.

Similarities between these models

These models of very different landform suites from disparate parts of the world have several features in common. First, both are 'systems' which need to be assessed in terms of time and space, and which respond to marine and terrestrial processes. The systems become organised, absorbing and adjusting to environmental change, especially in becoming buffered against fluctuations in ambient energy, for example storms. For example, the gravels are able to develop specific clastic architecture, which is more resistant to energy perturbation than loose aggregates and the shoreline changes from the reflective to the dissipative end of the spectrum with time, while the expansion of shoreline vegetation (especially mangroves) on the tropical shoreline performs a similar function. As the organisational level increases (an entropy defying process, reliant on extracting and using incident energy), so the system becomes more robust, often developing quite sophisticated feedbacks between one sub-system and another. This leads to a situation of homeostasis, which is hard to break down, although such breakdown may ultimately be achieved by, for example, allogenic influences such as failures of sediment supply or extreme climatic conditions. Under such conditions the systems become stressed, the gravel barriers thin and eventually breach, the finer silt/clay deposition of the tropical estuaries fails and perhaps ceases, and, in time, the system becomes sensitive and suitable for sudden change, such as the rapid salinisation seen on several of the estuarine floodplains (see Chapter 5). The switch in sedimentation patterns in the estuaries of the northern Australian rivers marks transitions from one system state to another. The trigger for change in the Nova Scotia gravel systems appears to be storms, but only if antecedent conditions are suitable. Adjacent beaches and barriers have been observed to respond differently to what, in many cases, are isotropic sea-state conditions.

Three factors are important in both examples. First, the role of relative sea-level change; sea level rose rapidly flooding prior valleys in northern Australia, but stabilised around 6000 years ago, enabling extensive mangrove forest development. In contrast to this largely eustatically determined relative sea-level history, the Nova Scotian coast has experienced glacioisostatic adjustment with a more individualistic pattern of relative sea-level change that includes ongoing rapid sea-level rise to which gravel barriers continue to respond. The second factor is sediment supply. In the case of the Nova Scotian coast, sediment is supplied from the erosion of drumlins; input rates are highly variable in space and time, and exert important morphodynamic controls on barrier evolution. In northern Australia, muds have infilled the estuarine embayments largely from a seaward source which appears to have become depleted in recent millennia. Floodplains have since been reworked to different extents, depending upon catchment size and differing riverine sediment inputs. The third factor is geological inheritance. In Nova Scotia this is the drumlin landscape which is being reworked. In northern Australia, it is the prior valley topography which influences estuarine morphodynamics. These three factors, relative sea level, sediment supply and geological inheritance can also be seen to be important controls on coastal evolution in a range of settings in the chapters that follow.

The two examples above also serve to illustrate the approaches which can be used to construct conceptual models of coastal evolution and the scope and applicability of such models. Both models were developed on the basis of detailed field observations of morphosedimentary architecture, morpho-stratigraphy and morphodynamic response. They were developed on the basis of comparison of adjacent sites, and each has been found to have elements which may be applied outside the area of derivation; thus, the gravel barrier studies in Nova Scotia show many elements of similarity to gravel-dominated coasts in Ireland (Carter *et al.*, 1989), whereas the pattern of Holocene estuarine infill and coastal progradation in northern Australia is similar to that experienced through many parts of Southeast Asia (Woodroffe, 1993).

Human impact and coastal evolution

To ignore the role of humans and their impact on coastal evolution would be fallacious. Like it or not, almost all coastal systems have been influenced by human intervention and the last two centuries of evolution have been dominated by attempts to control the shoreline. Coastal resources are seen as valuable (for tourism, minerals, agriculture, aquaculture, etc.), although it is only since the 1970s that any serious resource conservation has been attempted, and often then on a limited scale. National initiatives are now emerging, for example in the USA, Australia, France, Spain, Denmark and the UK (Ruddle *et al.*, 1988), and further efforts are being made to bring developing countries (which perhaps have the most coastal resources still to lose) into various local and regional plans (Gomez, 1988; Paw & Thia-Eng, 1991). However, it is proving very difficult to break free of existing governmental and administrative structures, and in many places local interests still have over-riding influence.

There are two ways humans have affected coastlines. The first is purely exploitative, and the second is by trying to control processes and thus change. One can see such actions both as advertent and inadvertent. Very often the consequences of any particular action are unforeseen and may unfold over a long period of time. Legally, such actions may well fall outside any normal statute of limitation (say 25 years) and, in any case, it may be impossible to 'prove' cause and effect.

The history of coastal evolution is linked inextricably with the evolution of coastal catchments and drainage basins, and their ability to deliver sediment to the shoreline. Work by Millimam, Meade, Walling and others has shown that the rate of sediment yield from catchments has varied systematically over the last few centuries (Milliman & Syvitski, 1992), parallelling deforestation and agricultural expansion (and in places, contraction). The type and volume of sediment depends on the lithology, tectonics, climate and vegetation of the drainage basin. The Yellow River in southern China, which flows through thick loessic deposits, produces over $1000 \times 10^6 \, \text{t a}^{-1}$ of mainly fine silt, which both supplies the rapidly expanding and subsiding delta and produces dense turbidity currents taking sediment to deeper water (Wright *et al.*, 1990). In contrast, the Brahmaputra drains a tectonically active mountain region providing a much coarser sediment yield, although much of the material most suitable for coastal evolution tends to be sequestered by the subsiding, alluvial delta (Barua, 1991). In general, smaller, high-relief catchments deliver coarser sediment to the coast, partly due to the nature of the weathering process and partly because of the relatively high competence of flows in the lower reaches of the channel, enabling larger clasts to be moved.

The wholesale perturbation of coastal sediment budgets worldwide at local, regional and global scales is a matter for international concern. The contributions of both large (Milliman & Meade, 1983) and small rivers (Milliman & Syvitski, 1992) seem equally important, although data, particularly for the latter, are sparse. A conservative estimate would suggest a total annual sediment input to the coast around 10^{16-17} t, of which maybe 10 to 15% would be suitable for 'coastal' accumulation. In addition there are great worldwide disparities; for example, continental Europe yields only small amounts of sediment relative to southeast Asia.

The net result of changing catchment conditions has been a major influx of sediment to the coast over the last few hundred years. The timing of this event has varied from place to place, but the manifestation of the impact has been remarkably similar, with the progradation of river mouth deltas, the augmentation of littoral drift, the infilling of estuaries, the construction of beach ridge plains and the formation of dunes. In many places, such as the Nile or Ganges deltas, the expansion has been in the order of kilometres, creating large tracts of new land which has been voraciously settled, so that it now supports some of the densest rural settlement in the world (Stanley & Warne, 1993b) . These increases in coastal sedimentation brought problems, not least the need to abandon harbours as waterways became unnavigable. This led to the development of dredging techniques and later to harbour protection works, such as jetties.

In many places, the influx of coastal sediment masked the slow but much longer-term tendency towards coastal recession. In the twentieth century, far more attention has been paid to catchment management, as it was becoming obvious that the spread of soil erosion and loss of soil fertility could not continue unabated. This has been augmented through the extensive construction of dams. Although dams have been constructed throughout recorded history, their large-scale impact on the environment has only become apparent in the last few decades (Jeftic, Milliman & Sestini, 1992; Stanley & Warne, 1993b). Dams are being constructed for a variety of reasons, including power generation, regulation of water supply and, all too often, politics. The chief impact on the coast has been to reduce dramatically (often by several orders of magnitude) the amount of sediment. This has led, inevitably, to an increase in coastal recession. For example, in the Nile Delta, where the sediment supply has almost completely ceased following the completion of the Aswan High Dam in 1964 (Stanley & Warne, 1993b), this amounts to tens if not hundreds of metres a year (Milliman, Broadus & Gable, 1989). Coping with such erosion is not easy and the general policy in Egypt is to abandon the shoreline in favour of constructing solid defences inland.

The artificial removal of sediments from the coast is a worldwide phenomenon, practised on both large and small scales. Common reasons for removing shoreline materials include precious-mineral mining (diamonds, titanium, gold), building aggregates and agricultural soil improvers. In some examples, such practices have direct impacts on the shoreline; Msangi, Griffiths & Banyikwa (1988) recount how coastal sediments along the Tanzanian coast were used for concrete in beachfront hotel construction, resulting almost immediately in a severe erosion problem. Elsewhere the consequences are not so obvious or fast.

The potential impact of engineering works on coastal evolution is enormous, especially if coupled to sea-level change. Universal response to these alterations and fluctuations in the sediment budget has been to defend the coast using engineering. The scale and ingenuity of coastal engineering can be extraordinary. In the USA long stretches of the east coast are wholly artificial and have been through several phases of engineering over the last 120 years. The south and east coast of Britain is heavily 'defended' with an extensive mixture of old and modern protective works, many now in need of replacement, while the Belgium, Dutch, German and Danish coast is almost entirely groyned and has long stretches of sea-walls (see, for example, Ehlers, 1990). Recently, Miossec (1991) has produced a major summary (in French) of comparative approaches to coastal defence in Europe, which indicates how strongly regional and national policies have influenced the type of artificial shoreline that has been developed. Chapter 13 explores the concept of developed coasts, drawing examples particularly from New Jersey.

Overview

The chapters which follow develop the issues of Late Quaternary coastal evolution introduced above, and examine specific sites where detailed morphodynamic studies have been undertaken. Chapter 2 examines the concept of coastal morphodynamics; it reviews the approaches adopted by Wright & Thom (1977) in a benchmark paper, and extends these in the light of recent developments in the field.

Subsequent chapters examine specific coastal types. Each chapter demonstrates general principles by a detailed study of particular coasts on which it is possible to reconstruct something of the evolution of landforms, and then examines the extent to which such models can be extrapolated to other parts of the world. It is clear that even the detailed case studies require further research to refine the models; nevertheless, the extension of models of coastal evolution developed in one location to another geographical setting, recognising that that entails altered boundary conditions, has been very much a central focus of IGCP Project 274.

It has been recognised for some time that the morphology of clastic coasts responds to the relative dominance of river, wave and tidal factors (Wright & Coleman, 1973). This view has been extended by Boyd, Dalrymple & Zaitlin (1992) and Dalrymple, Zaitlin & Boyd (1992), and is shown schematically in Fig. 1.6. Deltas are examined in Chapter 3, with emphasis on the Mississippi deltaic plain. Chapter 4 describes wave-dominated coasts, with particular attention to the development of sandy barriers in southeast Australia, a location in which coastal evolution studies have been assisted by the application of a

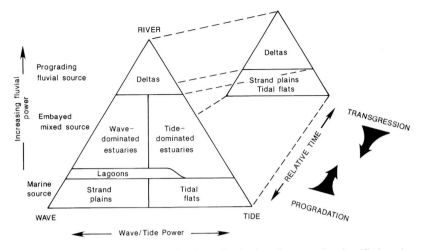

Figure 1.6. A ternary diagram showing how clastic shorelines may be classified on the basis of the extent to which they are dominated by river, wave or tidal factors (after Boyd *et al.*, 1992, and Dalrymple *et al.*, 1992).

simulation model of shoreline behaviour. Chapter 5 examines the morpho-dynamics of macrotidal estuaries, concentrating on tidal rivers and floodplains in northern Australia, where adjacent systems may vary in terms of important boundary conditions. Chapter 6 considers microtidal and lagoonal coasts, comparing such settings along the African coastline with similar settings elsewhere.

Carbonate coasts, such as reefs, where calcareous sediment is produced, transported, deposited, lithified and eroded, show some significant differences from clastic coasts. Chapter 7 examines coral atolls, comparing mid-ocean atolls from the Indian and Pacific Oceans. Chapter 8, on the other hand, describes the evolution of continental shelf reef systems, with particular emphasis on the Great Barrier Reef, where the influence of the continent in terms of water quality can be considerable.

High-latitude coasts differ from low- and mid-latitude coasts, both because of climate and hence processes, but also because glacioisostatic responses are still ongoing. Chapter 9 examines Arctic coastal plains, those polar shorelines on which permafrost is an important influence on sediments, and sea ice is seasonally extensive. Chapter 10 examines paraglacial coasts, using the example of the Nova Scotian shoreline to illustrate the diversity of landforms which can result from reworking of a complex, glacially influenced landscape.

Rocky coasts tend to respond more slowly than soft coasts to perturbations. Chapter 11 describes cliffs and shore platforms, demonstrating that a variety of controls operate to influence rate of cliff retreat. Chapter 12 considers tectonic shorelines, those coasts on which vertical movement is, or has been, significant. Tectonic coasts cover a wide range of different types of coast, and vertical movment may be expressed morphologically in different ways.

Finally, as emphasised above, consideration of coastal evolution without the human factor overlooks an increasingly important variable influencing processes and directions of change. Chapter 13 discusses developed coasts, using the highly urbanised and built up coast of the eastern United States as an example.

These chapters represent only a selection of the studies undertaken within the scope of IGCP Project 274. While they demonstrate something of our understanding of the morphodynamics of Late Quaternary shorelines, there are also many new questions raised as to how coasts adjust their form and evolve through time.

Acknowledgements

Much of this chapter was written by Bill Carter in the weeks before his death in July 1993. His inspiration lies not only behind this chapter, but in the shaping of this book, in the initiatives that the IGCP Project 274 has taken, and in new directions that coastal evolution studies will continue to take. Julian Orford has kindly read and commented on a draft of this chapter. David Martin and Richard Miller drew the diagrams.

References

Anderson, J.B. & Thomas, M.A. (1991). Marine ice-sheet decoupling as a mechanism for rapid, episodic sea level change; the record of such events and their influence on sedimentation. *Sedimentary Geology,* **70**, 87–104.

Andrade, C. (1992). Tsunami generated forms in the Algarve Barrier Islands. *Science of Tsunami Hazards,* **10**, 21–4.

Atwater, B.F. (1987). Evidence for great Holocene earthquakes along the outer coast of Washington State. *Science,* **236**, 942–4.

Atwater, B.F. & Moore, A.L. (1992). A tsunami about 1000 years ago in Puget Sound, Washington. *Science,* **258**, 1614–17.

Barua, D.K. (1991). The coastline of Bangladesh: an overview of processes and forms. *Coastal Zone '91,* American Society of Civil Engineers, 2284–301.

Bayliss-Smith, T.P. (1988). The role of hurricanes in the development of reef islands, Ontong Java Atoll, Solomon Islands. *Geographical Journal,* **154**, 377–91.

Bentor, Y. K. (1989). Geological events in the bible. *Terra Nova,* **1**, 326–38.

Bilham, R. & Barrientos, S. (1991). Sea-level and earthquakes. *Nature,* **350**, 386.

Boyd, R., Dalrymple, R. & Zaitlin, B.A. (1992). Classification of clastic coastal depositional environments. *Sedimentary Geology,* **80**, 139–50.

Brown, A.C. & McLachlan, A. (1990). *Ecology of sandy shores.* Amsterdam: Elsevier.

Bruun, P. (1962). Sea level rise as a cause of shore erosion. *American Society of Civil Engineers, Proceedings and Journal, Waterways and Harbors Division,* **88,** 117–30.

Bruun, P. (1988). The Bruun rule of erosion: a discussion on large-scale two and three dimensional usage. *Journal of Coastal Research,* **4,** 626–48.

Bryant, E.A., Young, R. and Price, D.M. (1992). Evidence of tsunami sedimentation on the southeastern coast of Australia. *Journal of Geology,* **100,** 753–65.

Carson, M.A. & Kirkby, M.J. (1972). *Hillslope form and process.* Cambridge University Press.

Carter, R.W.G. (1988). *Coastal environments: an introduction to the physical, ecological and cultural systems of coastlines.* London: Academic Press.

Carter, R.W.G. (1992). Sea level changes, past, present and future. In *Applications of Quaternary research,* ed. J.M. Gray, *Quaternary proceedings,* vol. 2, pp. 111–32. Cambridge: Quaternary Research Association.

Carter, R.W.G., Forbes, D.L., Jennings, S.C., Orford, J.D., Shaw, J. & Taylor, R.B. (1989). Barrier and lagoon coast evolution under differing relative sea-level regimes; examples from Ireland and Nova Scotia. *Marine Geology,* **88,** 221–42.

Carter, R.W.G., Orford, J.D., Forbes, D.L. & Taylor, R.B. (1990). Morphosedimentary development of drumlin flank barriers with rapidly rising sea-level, Story Head, Nova Scotia. *Sedimentary Geology,* **69,** 117–38.

Carter, R.W.G., Orford, J.D., Jennings, S.C., Shaw, J. & Smith, J.P. (1992). Recent evolution of paraglacial estuary conditions under conditions of rapid sea-level rise: Chezzetcook Inlet, Nova Scotia. *Proceedings of the Geologists Association, London,* **103,** 167–86.

Carter, R.W.G. & Orford, J.D. (1993). The morphodynaimcs of coarse clastic beaches and barriers: a short– and long-term perspective. In *Coastal morphodynamics,* ed. A. Short. *Journal of Coastal Research,* Special Issue, **15,** 158–79.

Carter, R.W.G. & Wilson, P. (1993). The geomorphological, ecological and pedological development of coastal foredunes at Magilligan Point, Northern Ireland. In *Coastal dunes: form and process,* ed. K.F. Nordstrom, N.P.Psuty & R.W.G. Carter, pp. 129–57. Chichester: John Wiley & Sons.

Cornaglia, P. (1891). *Sul regime della spiagge e sulla regulazione dei porti.* Turin: GB Paravia. 569pp.

Chappell, J. (1983). Thresholds and lags in geomorphological changes. *Australian Geographer,* **15,** 357–66.

Chappell, J. (1993). Contrasting Holocene sedimentary geologies of lower Daly River, northern Australia, and lower Sepik-Ramu, Papua New Guinea. *Sedimentary Geology,* **83,** 339–58.

Cotton, C.A. (1974). *Bold coasts: annotated reprints of selected papers on coastal geomorphology, 1916–1969.* Wellington: A.H. & A.W. Reed.

Dalrymple, R.W., Zaitlin, B.A. & Boyd, R. (1992). Estuarine facies models: conceptual basis and stratigraphic implications. *Journal of Sedimentary Petrology,* **62,** 1030–43.

Dawson, A.G., Long, D. & Smith, D.E. (1988). The Storegga slides: evidence from eastern Scotland for a possible tsunami. *Marine Geology,* **82,** 271–6.

Dean, R.G. (1977). *Equilibrium beach profiles: U.S. Atlantic and Gulf Coasts.* Department of Civil Engineering, Ocean Engineering Report 12. Newark, DE: University of Delaware.

Dean, R.G. (1991). Equilibrium beach profiles: characteristics and applications. *Journal of Coastal Research,* **7,** 53–84.

Dean, R.G., Healy, T.R. & Dommerholt, A.P. (1993). A 'blind-folded' test of equilibrium beach profile concepts with New Zealand data. *Marine Geology*, **109**, 253–66.

de Vriend, H.J. (1992a). Mathematical modelling and large-scale coastal behaviour. Part 1. Physical processes. *Journal of Hydraulic Research, Special Issue, Maritime Hydraulics*, 727–40.

de Vriend, H.J. (1992b). Mathematical modelling and large-scale coastal behaviour: Part 2. Predictive models. *Journal of Hydraulic Research, Special Issue, Maritime Hydraulics*, 741–53.

Dury, G.H. (1978). The future of geomorphology. In *Geomorphology: present problems and future prospects,* ed. C. Embleton, D. Brunsden & D.K.C. Jones, pp. 263–74. London: Oxford University Press.

Ehlers, J. (1990). *The morphodynamics of the Wadden Sea..* Rotterdam: Balkema. 397 pp.

Ehlers, J., Nagorny, K., Schmidt, P., Stieve, B. & Zietlow, K. (1993). Storm surge deposits in North Sea salt marshes dated by [134]Cs and ^{137}Cs determination. *Journal of Coastal Research*, **9**, 698–701.

Flood, J. (1993). *Archaeology of the dreamtime.* Sydney: Harper Collins.

Forbes, D.L., Taylor, R.B., Orford, J.D., Carter R.W.G. & Shaw, J. (1991). Gravel barrier migration and overstepping. *Marine Geology*, **97**, 305–13.

Foster, I.D.L., Dawson, A., Dawson, S., Lees, J.A. & Mansfield, L. (1993). Tsunami sedimentation sequences in the Scilly Isles, southwest England. *Science of Tsunami Hazards*, **11**, 35–46.

Ginsburg, R.N. (1973). *Evolving concepts in sedimentology, Studies in Geology,* vol. 21, Johns Hopkins University. 191 pp.

Gomez, E.D. (1988). Overview of environmental problems in the East Asian Seas region. *Ambio*, **17**, 166–9.

Guilcher, A. (1958). *Coastal and submarine morphology.* New York: John Wiley and Sons.

Holligan, P.M. & de Boois, H. (1993). *Land–ocean interactions in the coastal zone: science plan.* Stockholm: The International Geosphere–Biosphere Programme.

Huntley, D.A. & Short, A.D. (1992). On the spacing between observed rip currents. *Coastal Engineering*, **17**, 211–25.

Illenberger, W.K. & Rust, I.C. (1988). A sand budget for the Alexandria coastal dunefield, South Africa. *Sedimentology*, **35**, 513–21.

Inman, D.L. (1978). The impact of coastal structures on shorelines. In *Coastal zone '78,* pp. 2265–71. New York: American Society of Civil Engineers.

IPCC Response Strategies Working Group (1992). *Global climate change and the rising challenge of the sea.* Canberra: IPCC.

Jeftic, J.D., Milliman, J.D. & Sestini, G. (1992). *Climatic change and the Mediterranean: environmental and societal impacts of climatic change and sea-level rise in the Mediterranean Region.* Sevenoaks: Edward Arnold. 673 pp.

Jennings, S.C., Carter, R.W.G. & Orford, J.D. (1993). Late Holocene salt marsh development under a regime of rapid sea-level rise: Chezzetcook Inlet, Nova Scotia. Implications for the interpretation of palaeo-marsh sequences. *Canadian Journal of Earth Sciences*, **30**, 1374–84.

Johnson, D.W. (1919). *Shore processes and shoreline development.* New York: Prentice Hall.

Johnson, D.W. (1925). *The New England–Acadian shoreline.* New York: Wiley. 608 pp.

Keen, T.R. & Slingerland, R.L. (1993). Four storm-event beds and the tropical cyclones that produced them: a numerical hindcast. *Journal of Sedimentary Petrology*, **63**, 218–32.

King, C.A.M. (1972). *Beaches and coasts, 2nd edn.* London: Edward Arnold. 570 pp.

Kraft, J.C., Kayan, I. & Erol, O. (1980). Geomorphic reconstructions in the environs of Ancient Troy. *Science*, **209**, 776–82.

Kuhn, T.S. (1962). *The structure of scientific revolutions.* Chicago: University of Chicago Press.

Lakhan, V.C. & Trenhaile, A.S., eds. (1989). *Applications in coastal modeling.* Amsterdam: Elsevier. 387 pp.

Liu, K-B. & Fearn, M.L. (1993). Lake sediment record of late Holocene hurricane activities from coastal Alabama. *Geology*, **21**, 793–6.

Long, D., Smith, D.E. & Dawson, A.G. (1989). A Holocene tsunami deposit in eastern Scotland. *Journal of Quaternary Science*, **4**, 61–6.

McLachlan, A. & Erasmus, T. (1983). *Sandy beaches as ecosystems.* The Hague: Junk. 757 pp.

Masters, P.M. & Flemming, N.C. (1983). *Quaternary coastlines and marine archaeology: towards the prehistory of land bridges and continental shelves.* La Jolla: Scripps Institute of Oceanography.

Milliman, J.D., Broadus, J.M. & Gable, F. (1989). Environmental and economic implications of rising sea level and subsiding deltas: the Nile and Bengal examples. *Ambio*, **18**, 340–5.

Milliman, J.D. & Meade, R.H. (1983). Worldwide delivery of river sediments to the oceans. *Journal of Geology*, **91**, 1–21.

Milliman, J.D. & Syvitski, J.P.M. (1992). Geomorphic/tectonic control of sediment discharge to the ocean: the importance of small mountainous rivers. *Journal of Geology*, **100**, 525–44.

Minoura, K. & Nakaya, S. (1991). Traces of tsunami preserved in inter-tidal lacustrine and marsh deposits: some examples from northeast Japan. *Journal of Geology*, **99**, 265–87.

Miossec, A. (1991). *Defense des côtes et protection du littoral.* Université de Nantes.

Msangi, J.P., Griffiths, C.J. & Banyikwa, W.E. (1988). Man's response to change in the coastal zone of Tanzania. In *The coastal zones: man's response to change*, ed. K. Ruddle, W.B. Morgan & J.R. Pfafflin, pp. 37–60. Osaka, Japan: National Museum of Ethnology.

Nichol, S.L. & Boyd, R. (1993). Morphostratigraphy and facies architecture of sandy barriers along the eastern shore of Nova Scotia. *Marine Geology*, **114**, 59–80.

Orford, J.D., Carter, R.W.G. & Jennings, S.C. (1991). Coarse clastic barrier environments: evolution and implications for Quaternary sea-level interpretation. *Quaternary International*, **9**, 87–104.

Orford, J.D., Carter, R.W.G. & McCloskey, J. (1993). A method of establishing mesoscale (decadal to sub-decadal) domains in coastal barrier retreat rate from tide-gauge analysis. In *Large scale coastal behaviour '93*, ed. J.H. List, pp. 155–8. *US Geological Survey Open File Report* 93-381.

Orford, J.D., Hinton, A.C., Carter, R.W.G. & Jennings, S.C. (1992). A tidal link between sea-level rise and coastal response of a gravel-dominated barrier in Nova Scotia. In *Sea level changes: determination and effects. Geophysical Monograph*, 69, 71–9.

Paw, J.N. & Thia-Eng, C. (1991). Climate change and sea level rise: implications on coastal area utilisation and management in south-east Asia. *Ocean & Shoreline Management*, **15**, 205–32.

Phillips, J.D. (1992). Nonlinear dynamical systems in geomorphology: revolution or evolution? *Geomorphology*, **5**, 219–29.

Pilkey, O.H., Young, R.S., Riggs, S.R., Smith, A.W., Wu, H. & Pilkey, W.D. (1993). The concept of shoreface profile equilibrium: a critical review. *Journal of Coastal Research*, **9**, 255–78.

Rampino, M.R. & Self, S. (1992). Volcanic winter and accelerated glaciation following the Toba super-eruption. *Nature,* **359**, 50–2.

Roy, P.S. (1984). New South Wales estuaries: their origin and evolution. In *Coastal geomorphology in Australia,* ed. B.G. Thom, pp. 99–121. Sydney: Academic Press.

Ruddle, K., Morgan, W.B. & Pfafflin, J.R. (1988). *The coastal zones: man's response to change.* Osaka, Japan: National Museum of Ethnology.

Russell, R.J. (1967). *River plains and sea coasts.* Berkeley: University of California Press. 173 pp.

Schou, A. (1945). *Det marine forland: geografiske studier over Danske Fladkystlandskabers Dannelse og Formudvikling samt traek af disse Omraaders Kulturgeografi, med saerlig Hensyntagen til Sjaelland. Folia Geog. Danica.* vol. 4, 236 pp.

Schumm, S.A. (1991). *To interpret the earth: ten ways to be wrong.* Cambridge University Press.

Schumm, S.A. & Lichty, R.W. (1965). Time, space, and causality in geomorphology. *American Journal of Science,* **263**, 110–9.

Sestini, G., Jeftic, L. & Milliman, J.D. (1989). *Implications of expected climatic changes in the Mediterranean region: an overview.* UNEP Regional Seas Reports and Studies, Kenya, UNEP. 46 pp.

Short, A.D. (1991). Macro-meso tidal beach morphodynamics: an overview. *Journal of Coastal Research,* **7**, 417–36.

Short, A.D. (1992). Beach systems of the central Netherlands coast: processes and morphology and structural impacts in a storm-driven multi-bar system. *Marine Geology,* **107**, 103–37.

Stanley, D.J. & Warne, A.G. (1993a). Sea level and initiation of Predynastic culture in the Nile delta. *Nature,* **363**, 435–8.

Stanley, D.J. & Warne, A.G. (1993b). Nile Delta: recent geological evolution and human impact. *Science,* **260**, 628–34.

Stea, R.R., Mott, R.J., Belknap, D.F. & Radtke, U. (1992). The pre-late Wisconsinan chronology of Nova Scotia, Canada. *Geological Society of America, Special Paper,* **270**, 185–206.

Steers, J.A. (1946). *The coastline of England and Wales,* Cambridge University Press.

Stive, M.J.F., Roelvink, D.J.A. & de Vriend, H.J. (1991). Large-scale coastal evolution concept. In *Proceedings 22nd International Conference on Coastal Engineering,* pp. 1962–74. New York: American Society of Civil Engineers.

Terwindt, J.H.J. & Battjes, J.A. (1991). Research on large-scale coastal behaviour. In *Proceedings 22nd International Conference on Coastal Engineering,* pp. 1975–83. New York: American Society of Civil Engineers.

Thom, B.G. & Hall, W. (1991). Behaviour of beach profiles during accretion and erosion dominated periods. *Earth Surface Processes and Landforms,* **16**, 113–27.

Thom, B.G., Wright, L.D. & Coleman, J.M. (1975). Mangrove ecology and deltaic-estuarine geomorphology, Cambridge Gulf–Ord River, Western Australia. *Journal of Ecology,* **63**, 203–22.

Turcotte, D.L. (1991). Earthquake prediction. *Annual Review of Earth and Planetary Sciences,* **19**, 263–81.

Vita-Finzi, C. (1964). Synchronous stream deposition throughout the Mediterranean area in historical times. *Nature,* **202**, 1324.

Watanabe, T. (1985). *Comprehensive bibliography on tsunami of Japan.* University Press of the University of Tokyo. 260 pp. (In Japanese.)

Williams, S.F., Dodd, K. & Gohn, K.K. (1990). *Coasts in crisis.* U.S. Geological Survey Circular. 32 pp.

Woodroffe, C.D. (1993). Late Quaternary evolution of coastal and lowland riverine plains of Southeast Asia and northern Australia: an overview. *Sedimentary Geology*, **83**, 163–75.

Woodroffe, C.D., Chappell, J., Thom, B.G. & Wallensky, E. (1989). Depositional model of a macrotidal estuary and floodplain, South Alligator River, Northern Australia. *Sedimentology*, **36**, 737–56.

Woodroffe, C.D. & Grindrod, J. (1991). Mangrove biogeography: the role of Quaternary environmental and sea level change. *Journal of Biogeography*, **18**, 479–92.

Woodroffe, C.D., Mulrennan, M.E. & Chappell, J. (1993). Estuarine infill and coastal progradation, southern van Diemen Gulf, northern Australia. *Sedimentary Geology*, **83**, 257–75.

Woodroffe, C.D., Thom, B.G. & Chappell, J. (1985). Development of widespread mangrove swamps in mid-Holocene times in northern Australia. *Nature*, **317**, 711–13.

Wright, L.D. (1987). Shelf-surfzone coupling: diabathic shoreface transport. *Coastal Sediments '87*, 25–40.

Wright, L.D. (1993). Micromorphodynamics of the inner continental shelf: a middle Atlantic Bight case study. *Journal of Coastal Research*, SI **15**, 93–124.

Wright, L.D., Boon, J.D., Kim, S.C. & List, J.H. (1991). Modes of cross-shore sediment transport on the shoreface of the Middle Atlantic Bight. *Marine Geology*, **96**, 19–51.

Wright, L.D., Chappell, J., Thom, B.G., Bradshaw, M.P. & Cowell, P. (1979). Morphodynamics of reflective and dissipative beach and inshore systems: southeastern Australia. *Marine Geology*, **32**, 105–40.

Wright, L.D. & Coleman, J.M. (1973). Variations in morphology of major river deltas as functions of ocean wave and river discharge regimes. *American Association of Petroleum Geologists Bulletin*, **57**, 370–98.

Wright, L.D., Coleman, J.M. & Thom, B. G. (1973). Processes of channel development of a high-tide range environment: Cambridge Gulf–Ord River Delta, Western Australia, *Journal of Geology*, **81**, 15–41.

Wright, L.D. & Short, A.D. (1984). Morphodynamic variability of surf zones and beaches: a synthesis. *Marine Geology*, **56**, 93–118.

Wright, L.D. & Thom, B.G. (1977). Coastal depositional landforms: a morphodynamic approach. *Progress in Physical Geography*, **1**, 412–59.

Wright, L.D., Wiseman, W.J., Yang, Z-S., Bornhold, B.D., Keller, G.H., Prior, D.B. & Suhayda, J.N. (1990). Processes of marine dispersal and deposition of suspended silts off the mouth of the Huanghe (Yellow River). *Continental Shelf Research*, **10**, 1–40.

Young, R. & Pilkey, O. H. (1992). A critical review of the engineering concept of beach and shoreface profile of equilibrium. *Geological Society of America, Abstracts with Programs*, **24**, 74–5.

Young, R.W. & Bryant, E.A. (1992). Catastrophic wave erosion on the southeastern coast of Australia: impact of the Lanai tsunami ca 105 ka ? *Geology*, **20**, 199–202.

Young, R.W., Bryant, E.A., Price, D.M., Wirth, L.M. & Pease, M. (1993). Theoretical constraints and chronological evidence of Holocene coastal development in central and southern New South Wales, Australia. *Geomorphology*, **7**, 317–29.

Zeeman, E.C. (1977). *Catastrophe theory*. New York: Addison-Wesley.

Zenkovich, V.P. (1967). *Processes of coastal development*. Edinburgh: Oliver & Boyd.

2

Morphodynamics of coastal evolution

P.J. COWELL AND B.G. THOM

Introduction

Coastal evolution is the product of morphodynamic processes that occur in response to changes in external conditions (Wright & Thom, 1977). Coastal morphodynamics is defined as the 'mutual adjustment of topography and fluid dynamics involving sediment transport' (Wright & Thom, 1977) or, alternatively, the 'dynamic behaviour of alluvial boundaries' of fluid motions (de Vriend, 1991b). Sediment transport provides the time-dependent coupling mechanism by which this adjustment occurs (Fig. 2.1). Fluid dynamics drive sediment transport resulting in morphological change over time. Progressive modification of topography in turn alters boundary conditions for the fluid dynamics, which evolve to produce further changes in sediment-transport patterns and their depositional products. Sediment properties and abundance affect the process through their influence upon sediment transport and sediment budgets respectively.

The essential properties of coastal morphodynamic processes are attributable to the feedback loop between topography and the fluid dynamics that drive sediment transport producing morphological change (Fig. 2.1). The feedback can be either negative or positive. Negative feedback confers properties of *self regulation* in response to minor perturbations (Wright & Thom, 1977). Positive feedback signifies growth of an instability and confers properties of *self organisation*, which results in new modes of operation (Waldrop, 1992; Phillips, 1992). Feedback reversal marks thresholds in morphodynamic behaviour.

A fuller appreciation now exists of the complexity inherent in these morphodynamic processes (de Vriend, 1991b) following recent developments in non-linear dynamics (Gleick, 1988; Waldrop, 1992; Phillips, 1992). The complexity derives from the morphodynamic feedback that is responsible for *state-determining* behaviour or, to use the new language of *chaos theory*,

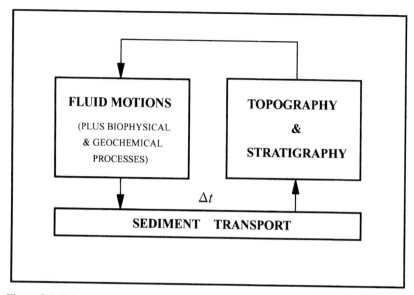

Figure 2.1. Primary components involved in coastal morphodynamics. The feedback loop between form and process is responsible for fundamental complexity in coastal evolution. Δt signifies the time dependence inherent in morphodynamic evolution of coasts.

'sensitive dependence upon initial conditions'. This, together with the strongly *stochastic* boundary conditions, gives rise to *Markovian inheritance* of antecedent morphology by succeeding morphodynamic states. Such inheritance is subject to additional memory effects associated with response-time characteristics inherent in morphological change. These characteristics are attributable to the finite rate of volume displacement involved in any sediment transport that produces morphological change.

Evolution of coastal landforms is inherently non linear and time dependent as a consequence of the combined effects of morphodynamic feedback and *Markovian* behaviour. Details of coastal evolution are therefore unpredictable, unrepeatable and irreversible, when subject to stochastic boundary conditions over the long term (decades to millennia). By definition, coastal evolution is also *nonstationary* in the statistical sense, making extrapolation based upon geostatistical relationships unreliable, especially given the complicating effects of non linearity stemming from the interconnectedness of processes (Swift & Thorne, 1991; Schumm, 1991). This complexity underlies the problem of *large-scale coastal behaviour* (LSCB) which signifies that coastal evolution is the outcome of multiplicative probabilities and therefore largely a question of historical accident (Terwindt & Battjes, 1991; Stive, Roelvink & de Vriend, 1991; de Vriend, 1991a, 1992a,b). Application of classical, reductionist methods of prediction and explanation based upon separate analysis of component processes is of limited use under these circumstances (Schumm, 1991). Nevertheless, new techniques are emerging for prediction and explanation of LSCB based upon *inverse modelling* principles that allow both a measure of quantitative rigour and the incorporation of uncodified qualitative expertise (see page 64). These models provide a general understanding and explanation of coastal evolution. Future advances in the mathematics of non-linear dynamics may also yield ways of undertaking more conventional *forward modelling* predictions for LSCB (de Vriend, 1991b).

The 'chicken-and-egg' nature of the mutual interaction between coastal topography and processes, as articulated formally by Wright & Thom (1977), involves changes to coastal landforms over a broad range of time and space scales. These dimensions tend to vary together, due to scale-related variation in response times. Scales at which morphodynamic processes operate are grouped into four classes in Fig. 2.2 based upon considerations similar to those presented by Stive *et al.* (1991) and de Vriend (1991b). '*Instantaneous*' time

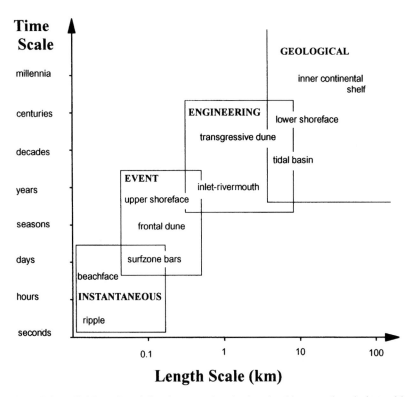

Figure 2.2. Definition of spatial and temporal scales involved in coastal evolution, with typical classes of sedimentary features used as illustrations throughout the text.

scales involve the evolution of morphology during a single cycle of primary forcing agents, such as waves or tides. The coupling between fluid flows and morphology relates directly to physical processes causing sediment transport. Morphodynamic adjustment can be followed in detail as a real-time process. Coastal evolution at *event* scales is a response to processes operating across time spans ranging from that of a single event, such as a storm, through to seasonal variations in environmental conditions. Morphological changes comprise time-averaged effects of *instantaneous* processes during a single fluctuation in boundary conditions. Depositional products correspond to *event beds*, which form the fundamental sedimentary units in lithofacies and depositional sequences (Swift, Phillips & Thorne, 1991a). *Engineering* (and *historical*) time scales involve composite evolution over many fluctuations in boundary conditions, each of which entails many cycles in the fundamental processes responsible for sediment transport. Evolutionary outcomes depend upon both the nature and sequencing of fluctuations in boundary conditions. The dependence on individual fluctuations becomes less important toward *geological scales* where, over millennia, evolution occurs more in response to mean trends in environmental conditions. At *geological scales* longer than millennia, fluctuations in boundary conditions again become significant, primarily driven by the effects of Milankovich cycles (Hays, Imbrie & Shackleton, 1976). The successively larger length scales in Fig. 2.2 represent the spatial counterpart of the time-related variability. The smallest scales range from grain dimensions to bedform lengths (which even so, may be in the order of 10^2 metres when associated with low-frequency water motions such as tides). At these scales, spatial uniformity or simple, systematic variation (e.g., over the length of a single bedform) allows localised consideration of processes. In contrast, the successively larger length scales in Fig. 2.2 are associated with an increasing degree of spatial variation (and possibly *non homogeneity*). Larger morphologies may incorporate a more extensive hierarchy of subordinate forms, signifying a greater level of spatial complexity in related geomorphic process (Schumm, 1991).

Morphodynamic studies during the past two decades have been concerned predominantly with *instantaneous* and *event* scales. However, the focus of this chapter is on coastal evolution over *engineering* and *geological* time spans. Such a focus is not only in keeping with the themes throughout this book; it also reflects the emphasis given at present to the study of *large-scale coastal behaviour*. The purpose of this chapter is to: (a) explore the essential properties of coastal morphodynamics, especially in relation to coastal evolution at *engineering* through to *geological* time scales, and (b) suggest an approach to morphodynamic modelling at these time scales. The chapter is divided into four sections dealing with (i) the organisation of morphodynamic systems, (ii) the properties of morphodynamic processes; (iii) the cumulative nature of coastal evolution attributable to morphodynamic behaviour; and (iv) the nature of LSCB and how it can be dealt with in modelling coastal evolution.

Morphodynamic systems: definitions

The purpose of this section is to outline the conceptual basis for understanding the complexity in coastal morphodynamics. The organisational properties and vernacular of general systems theory are employed to define the relationships between constituent processes. These relationships involve formal representation of the structural and functional logic of interconnections between processes (von Bertalanffy, 1968; Bennett & Chorley, 1978; Phillips, 1992). This approach allows for a manageable reduction of the problem into self-contained elements that are simple enough to analyse in isolation, whilst maintaining the context for each of the elements as well as their significance to the overall problem. Moreover, it provides a necessary methodology when interactions between component processes are not weak enough to be neglected and when relations describing behaviour of the components are not linear (von Bertalanffy, 1968). An overview of the general application of *systems dynamics* to coastal modelling is given by Lakhan & Trenhaile (1989).

Boundary conditions

The most fundamental step in defining a morphodynamic system is to identify the spatial and process boundaries of the system.

Spatial boundaries

Spatial boundaries are defined in Fig. 2.3. They most appropriately conform, for the purposes of studying coastal evolution, to the conventions set out by Inman & Brush (1973). These boundaries correspond to the limits to which coastal processes have extended during the Quaternary due to the large sea-level fluctuations, ranging over more than 100 metres vertically, associated with expansion and contraction of Quaternary ice sheets (Hays *et al.*, 1976; Chappell, *et al.*, 1982; Shackleton, 1987).

The landward limit encompasses the coastal depositional and marine-erosion surfaces formed during the interglacial highstands of the present and Quaternary sea levels. The seaward limit is defined by the edge of the continental shelf, which typically occurs in water depths from 100 to 200 metres (Inman & Brush, 1973). The reason that the coastal zone is so much

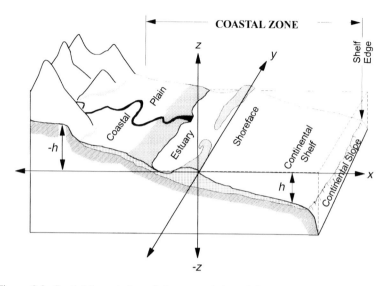

Figure 2.3. Spatial boundaries of the coast (adapted from Inman & Brush, 1973). h represents water depth; $-h$ represents elevation of land.

wider than the present-day littoral zone is twofold. First, the coastal water mass is usually distinct from the ocean water mass in terms of its physical and dynamic properties (Johnson & Rockcliff, 1986; GESAMP, 1991); ocean flows typically have vertical length scales an order of magnitude greater than coastal waters so that, for ocean flows, the continental slope effectively is the coast (e.g., Bane, 1983; Johnson & Rockliff, 1986; Brink, 1987; Blanton *et al.*, 1987), although surface manifestations of ocean currents can intrude into the coastal ocean (Cresswell, *et al.*, 1983; Blanton *et al.*, 1987; Huyer *et al.*, 1988). Second, the lowest sea levels probably placed littoral processes close to the continental-shelf edge in many parts of the world on several occasions during the Quaternary (Chappell *et al.*, 1982; Shackleton, 1987). Moreover, shorelines have swept across mid- and inner-continental shelf zones many times throughout most of this period. The evolutionary sequences for large-scale coastal deposits, such as barrier and deltaic plains, have histories covering such time spans (10^3 years). The early development of these features at lower sea levels has, in part, determined present-day morphology through the *Markovian inheritance* that is characteristic of morphodynamic processes.

Process boundaries

Process boundaries for the coastal system correspond to the environmental conditions in Fig. 2.4. Exogenous inputs from the environment are responsible for the geographic variation among coasts. The environment comprises

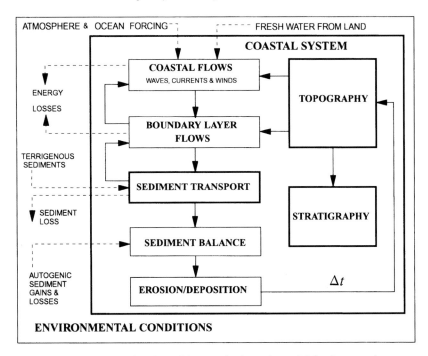

Figure 2.4. Structure and function of the morphodynamic model for the coastal system. The dashed arrows represent input–output between the coastal system and the environment. Sediment transport is highlighted because it provides the coupling mechanism between fluid dynamics and morphological change in the direction of forcing (feedforward); it thereby exerts frequency-response modulation.

climatic and geological controls (Davies, 1980). The geological controls govern the physiographic setting as well as the abundance and properties of sediments. The regional climates of the ocean and atmosphere determine the energy regime. The energy regime is responsible for introducing uncertainty into morphodynamic processes, and therefore coastal evolution, because it is stochastic.

Tectonic setting and denudational history of continental margins throughout the Cainozoic are responsible for the primary geological controls (Inman & Nordstrom, 1971) which determine continental-shelf dimensions and terrestrial sediment supply (Davies, 1980). Over Quaternary time scales, Milankovich cycles (Hays *et al.*, 1976) drove interaction between climate and geology causing glacio- and hydroisostatic adjustments that are responsible for the variations in relative sea level; a boundary condition of fundamental importance in coastal evolution from *geological* down to *engineering* time spans (Roy & Thom, 1987; Thom & Roy, 1988). Process boundary conditions dominating at time scales typically less than one year include oceanic and atmospheric inputs, although the importance of inter-annual and inter-decadal variations, such as the El Niño Southern Oscillation (ENSO) (Deser & Wallace, 1987), are becoming increasingly evident (Bryant, 1983; Komar & Enfield, 1987; Braatz & Aubrey, 1987; Thom & Hall, 1991). Ocean forcing results from: (i) direct and indirect intrusion of ocean-boundary currents and eddies (Curray, 1960; Johnson & Rockliff, 1986; Blanton, *et al.*, 1987; Flemming 1988; Lee *et al.*, 1989); (ii) astronomical and meteorological tides in internal and barotropic modes (Baines, 1986; Brink, 1987; Pugh, 1987; Blanton *et al.*, 1987; Griffin & Middleton, 1992); (iii) swell waves (Inman & Brush, 1973; Davies, 1980; Wright, 1976); and (iv) other sources of variation in the ocean water level (Komar & Enfield, 1987; Braatz & Aubrey, 1987). Winds transport sediment in coastal dunes (Hsu, 1987; Nordstrom, Psuty & Carter, 1990). They are also responsible in the coastal ocean for generation of wind waves (Bretshneider, 1966; Davies, 1980; Bishop & Donelan, 1989), and contribute to Eckman currents and internal flows (Csanady, 1977, 1982; Swift *et al.*, 1985; Atkinson *et al.*, 1989; Lee *et al.*, 1989; Wright *et al.*, 1991; Griffin & Middleton, 1992), as well as coastally trapped waves (Hamon, 1966; Huthnance, Mysak & Wang, 1986; Griffin & Middleton, 1992).

External inputs of fresh water as run-off from the land (Fig. 2.4) are estimated globally at $1.1 \times 10^8\,m^3\,s^{-1}$, carrying an estimated sediment load of $530\,t\,s^{-1}$ (Inman & Brush, 1973). These sediment inputs are supplemented by autogenic (mainly biogenic) production which yield peats and other organic accumulations in protected areas (Davies, 1980; Frey & Basan, 1985; Nichols & Biggs, 1985), with carbonate sediment supply predominating in arid and tropical environments (Davies, 1986).

Sediment losses from the coastal system in Fig. 2.4 mainly comprise fine material which is transported over the shelf break, out onto the abyssal plains fronting continental margins, especially during glacial low sea levels (Thorne & Swift, 1991a). Energy output (losses) mostly involves heat, being the ultimate product of dissipation in fluid and sedimentary processes, although some reradiation of energy back into the open ocean can be expected, particularly for lower-frequency water motions such as the tides. The global rate of energy dissipation in the coastal ocean is estimated to total $5 \times 10^9\,kW$, or $0.17\,W\,m^{-2}$ of continental-shelf surface area (Inman & Brush, 1973).

Organisation of morphodynamic systems

The process linkages, shown by the arrows in Fig. 2.4, chart the input–output relationships between each of the sub systems in the morphodynamic model. These relationships show that, to know about morphological change, it is necessary to know the sediment budgets during the time period of interest throughout the region. To know about sediment budgets, a model is required for sediment dynamics, and another for the near-bed boundary-layer flows which entrain and carry the sediments. In turn, the flows driving the boundary layer must be understood. Practical theory for each of these components is required before morphodynamic modelling can be undertaken. Identification of which flows are important, and how they might be modelled best, has occupied most of the effort in coastal-process studies over the past two decades (eg, Wright, 1987, 1993; Wright *et al.*, 1991) and is by no means complete (de Vriend, 1991b). The development of principles for integrated, time-dependent modelling of morphodynamics has occurred only recently. These principles, and their numerical implementation employing constituent models for each of the above processes, are reviewed by de Vriend (1991a) and GESAMP (1991).

Sediment-transport processes

These are highlighted in Fig. 2.4 because of their importance in providing the primary coupling mechanism between topography and fluid dynamics. However, it is the spatial budget, resulting from sediment-transport patterns, that causes erosion and deposition. This is described exactly by the continuity equation for sediment transport,

$$\frac{\partial h}{\partial t} = \varepsilon \left(\frac{\partial C}{\partial t} + \frac{\partial q_x}{\partial x} + \frac{\partial q_y}{\partial y} \right) - \frac{\partial V}{\partial t} \qquad (2.1)$$

in which the h is the bed elevation ($h = -z$) at each point (x, y) on the topographic surface in the coordinate system defined in Fig. 2.3, and ε accounts for sediment density and porosity. Each of the terms inside the brackets is depth integrated; C is the total concentration of sediment suspended above the bed, q_x and q_y are the sediment-mass fluxes normal and parallel to the shore respectively, and V (per unit surface area) represents autogenic sediment production and losses, as well as engineering interventions such as beach nourishment and sand mining. (A term like V also is required to include the effects of subsidence and isostatic or tectonic movements.) Generally, q_x and q_y are the important terms, except under special conditions such as during the formation of tempestites (Niedoroda, Swift & Hopkins, 1985) when $\partial C/\partial t$ is critical, or in low-energy and biologically rich environments where $\partial V/\partial t$ dominates (Frey & Basan, 1985; Nichols & Biggs, 1985; Davies, 1986).

Topography and stratigraphy

A primary objective of coastal morphodynamics is to predict coastal topography at *event* time scales, and its evolution over *engineering* and *geological* time scales. Topography is therefore highlighted in Fig. 2.4, along with stratigraphy. Stratigraphy is the integrated effect of topographic change through time in depositional environments. Stratigraphic sequences therefore become a partial record of coastal evolution (Kraft & Chrzastowski, 1985; Nummedal & Swift, 1987; Haq, Hardenbol & Vail, 1987; Boyd, Suter & Penland, 1989; Swift *et al.*, 1991a; Thorne & Swift, 1991a, Vail *et al.*, 1991). Topography itself represents a time integration of morphological changes, as inferred by the time loop, Δt in Fig. 2.4. Therefore, coastal evolution is expressed as

$$h = h_0 + \int \frac{\partial h}{\partial t}\, dt \qquad (2.2)$$

starting with an initial topography h_0. Topography controls coastal evolution in two ways. First, it governs boundary conditions for the fluid dynamics. Second, it determines the accommodation space available for deposition in relation to the rate at which sediments are supplied to the coastal cell from fluvial and biogenic sources, or from along the coast (Thorne & Swift, 1991a).

Morphodynamic systems – general concepts

The primary process–response linkages in Fig. 2.4 follow the forcing sequence from the environmental inputs, through the fluid and sediment dynamics, to drive morphological evolution. However, the opportunity for morphodynamic adjustment stems from the all-important feedback loop between topography and fluid dynamics (Wright & Thom, 1977; Hardisty, 1987).

Properties of morphodynamic processes

Topographic feedback provides the essence of morphodynamic behaviour and it arises through the effects of sediment transport (Fig. 2.1). Feedback is responsible for making evolution of coastal morphology partially self determining, since the operation of the system, and its state at any instant, is affected by the antecedent topography; ie, there is '*state dependence*' (Bennett & Chorley, 1978). The state of the system is defined by the values of all internal (state) variables at an instant. It is the inheritance by morphodynamic processes of their own past that is responsible for the so-called '*sensitivity to initial conditions*' (Phillips, 1992).

The essential properties of morphodynamic processes comprise: (i) *self regulation* (equilibrium tendencies); (ii) *self forcing* (which leads to *thresholds*, *self organisation* and regime changes); (iii) *Markovian inheritance* (which introduces uncertainty – Bennett & Chorley, 1978, p. 322); (iv) *hysteresis* (which causes a filtered response by morphology to changes in boundary conditions); (v) *non linearity*; (vi) *nonstationarity*; and (vii) *nonhomogeneity*. *Feedback* is responsible for each of these, except for *hysteresis* and *nonhomogeneity*. However, it is *self regulation* and *self organisation*, attributable to state-determined feedback, that bear the hallmark of morphodynamic processes.

Because of these properties, complexity is inherent in coastal evolution. The *non linearity* involves interdependence of morphodynamic states through space and time (i.e., they are '*state-determined*'). The spatial interdependence is complicated further by *nonhomogeneity*, which generally exists in morphodynamics since the coast is a transition zone between the land and deep ocean. Because of time dependence, evolutionary sequences feature cumulative rather than uncorrelated, random changes, which can result in strong *nonstationarity*. This compounds the filtering effects of *hysteresis* which occur due to the finite time required for the redistribution of sediment volumes in response to changing environmental inputs.

Feedback mechanisms

Morphodynamic feedback fundamentally occurs through the control that the solid boundary (i.e., the topographic surface) exerts over flows, across all scales from continental-shelf circulations (e.g., Csanady, 1982; Cannon & Lagerloef, 1983; Pettigrew & Murray, 1986) to shoaling wind waves (e.g., Svendsen & Buhr Hansen, 1978; Cowell, 1982; van Rijn, 1990). This control mechanism involves (i) the continuity principle for fluids; and (ii) the bottom-boundary condition of no flow at the bed. (The latter is not to be confused with the spatial and system boundary conditions.)

The fluid continuity principle is a statement that flows over an irregular boundary must experience spatial changes in their velocities, pressures, and/or densities in order to conserve their fluid mass. The continuity principle for fluids is expressed as

$$\frac{\partial \rho}{\partial t} + \rho \left(\frac{\partial u}{\partial x} + \frac{\partial v}{\partial y} + \frac{\partial w}{\partial z} \right) = 0 \qquad (2.3)$$

where ρ is the fluid density, and u, v, w are the velocity components in the x, y, z directions respectively (see Fig. 2.3).

The bottom-boundary condition (at $z = -h$) stipulates that the velocity vector $\bar{v} = 0$. This condition arises because there can be no effective flow through the solid boundary for velocity components (w) perpendicular to it, whilst friction causes tangential components (u, v) to vanish at the boundary. The boundary-normal condition ($w = 0$) prevails to preserve fluid continuity in the absence of cavitation over a bed that is impervious (or only weakly infiltrated). The tangential condition ($u = v = 0$) must exist to avoid shear at the wall becoming infinite in viscous fluids. The wall friction also provides an opposing force acting against pressure and inertia in flows. It affects the overall hydrodynamic behaviour through the momentum balance. This is expressed by the Navier–Stokes equation (in vector form), as

$$\rho \frac{\partial \bar{v}}{\partial t} = \mathbf{gr\bar{a}d}\left(\frac{\bar{v}^2}{2}\right) + (\mathbf{c\bar{u}rl}\ \bar{v}) \times \bar{v}^2 = -\mathbf{gr\bar{a}d}\,(p + \rho gz) + \mu \bar{\nabla}^2 \bar{v} \quad (2.4)$$

where p is pressure, g is gravity, and μ is viscosity. The topographic roughness, upon which the strength of the friction force depends, therefore provides an additional feedback mechanism via the boundary layer (Fig. 2.4) (Csanady, 1982; Pettigrew & Murray, 1986). Apart from the role friction plays in the momentum balance (far-right term in eqn. (2.3)), it is responsible for cumulative dissipation of energy from flows. Friction is also enhanced as a result of loose-bed roughness (e.g., Grant & Madsen, 1982). This contributes to the feedback between sediment dynamics and boundary layer flows (not shown in Fig. 2.4).

Negative feedback and equilibrium

Negative feedback is a damping mechanism that acts against departures from a morphodynamic state. It is responsible for *self regulation* and is therefore a stabilising process. *Self regulation* applies to the behaviour of a system in response to small perturbations. It is the mechanism by which *equilibrium* is re-established (Wright & Thom, 1977). Although equilibrium tendencies are a manifestation of *self regulation*, morphological hysteresis, together with the stochastic nature of boundary conditions, can prevent steady-state conditions from being attained. Therefore, morphodynamic adjustment is modulated by frequency-response characteristics which reflect the way changes in sediment masses lag behind variations in environmental inputs (see page 58).

An example of *self regulation* due to negative feedback in morphodynamic processes, operating over time scales of minutes to hours, is provided by the interaction between edge waves and beach cusps. Subharmonic edge-wave amplitudes are progressively suppressed by the growth in topographic relief of beach cusps that the edge waves themselves produce (Guza & Bowen, 1981). The plan-form adjustment of shorelines to changes in wave direction and period is an example of *self regulation* occurring over time spans of hours to days (Davies, 1958; Komar, 1973, 1976). *Self regulation* also occurs at *engineering* and *geological* time scales. Examples include maintenance of the equilibrium profile on the upper shoreface over time (Dean, 1991; Kotvojs &

Cowell, 1991), despite episodic perturbations due to storms (Vellinga, 1986), and the infill of the Waddensea (Netherlands), in which the rate of deposition appears to be in dynamic equilibrium with the local sea-level rise (Eysink, 1991).

Equilibrium is the product of negative feedback and represents a morphodynamic state that is stable for a given range of environmental conditions. It is now recognised that equilibrium can take three forms: (i) steady state; (ii) periodic; and (iii) 'chaotic' (Phillips, 1992). Steady-state equilibrium in coastal evolution can either mean that sediment flux averages to zero through time, or that there is no gradient in the sediment flux vector, in which case the topography acts as a sediment pathway rather than a source or sink. Therefore, equilibrium may be expressed in terms of sand-mass continuity (eqn. (2.1)) as

$$\frac{\partial h}{\partial t} = \varepsilon \left(\frac{\partial C}{\partial t} + \frac{\partial Q_x}{\partial x} + \frac{\partial Q_y}{\partial y} \right) - \frac{\partial V}{\partial t} = 0 \qquad (2.5)$$

The spatial terms in eqn. (2.5) define source and sink zones, such that

$$\frac{\partial q_x}{\partial x} + \frac{\partial q_y}{\partial y} \begin{vmatrix} < 0, & \text{sink} \\ = 0, & \text{pathway} \\ > 0, & \text{source} \end{vmatrix} \qquad (2.6)$$

As Wright *et al.* (1991) point out, whether or not equilibrium exists depends upon the length of time over which eqns. (2.5) or (2.6) are integrated (i.e., it is a question of time scale). For example, net cross-shore transport of sediments during a single storm can be cancelled out by successive storms (Vincent, Young, & Swift, 1981). Moreover, morphodynamic equilibrium is unlikely at any instant because of the stochastic nature of boundary conditions and finite morphological response times. Nevertheless, we can assume from the physical imperative expressed in eqn. (2.5), that coastal evolution is driven by a disequilibrium-stress in the continuity of sediment dynamics.

There are two approaches to analysis of equilibrium, entailing (i) *real-time morphodynamics* and (ii) *regime morphodynamics*. The first of these involves dynamic modelling of fluid-flows, sediment-transport fields and their co-evolution with the morphology through time until steady state is attained. Because of the complexities involved, especially in the sediment dynamics (Huntley & Bowen, 1990), simulations have traditionally employed physical modelling (e.g., Dalrymple, 1985; Sunamura, 1989), but increasingly computer-based numerical studies are being undertaken into morphodynamic evolution in shallow water (de Vriend, 1987c), in the surf zone and on the shoreface (Coeffe & Pechon, 1982; Huthnance, 1982; Stive, 1987; Boczar-Karakiewicz & Davidson-Arnott, 1987; Steijn, *et al.*, 1989), as well as on the inner shelf (Boczar-Karakiewicz & Bona, 1986; Boczar-Karakiewicz, Bona & Cohen, 1987; de Vriend, 1987b, 1990; Boczar-Karakiewicz, Bona & Pelchat, 1991), and in estuaries (de Vriend, 1987a; de Vriend *et al.*, 1989). A dynamic steady-state occurs within this approach when net sediment transport becomes either spatially uniform or zero. Under these conditions, the flow field becomes stationary in the statistical sense (i.e., time-average velocities remain constant from one time interval to the next), and time-average morphological change ceases. However, geographic variation in hydrodynamic conditions and geologically inherited effects produce site-specific differences in equilibrium form (Pilkey *et al.*, 1993).

Regime morphodynamics sets out conditions necessary for equilibrium in terms of a bulk representation of processes. Thorne & Swift (1991b) elaborate the approach in relation to the 'regime' concept in hydraulics and the 'grade' concept in geomorphology. Equilibrium morphology is determined directly without consideration of initial or intermediate states and employs related surrogate quantities for sediment transport. A clear example of this approach is the model for shoreline equilibrium (Davies, 1958; Komar, 1973), which involves determination of shoreline alignments that cause a balance in wave-driven alongshore sediment transport or, in the case of deltaic shorelines, a balance between alongshore transport and the discharge rate of sediments at the river mouth (Komar, 1973). *Regime equilibrium* models exist for beach and shoreface profiles based upon force balance (Bowen, 1980; Bailard & Inman, 1981) and wave-energy dissipation rates (Bruun, 1954; Dean, 1977, 1991; Chappell, 1981). In reality, these formulations are guided by physical insights that are generally incomplete (Pilkey *et al.*, 1993) and rely heavily upon empirical tuning (Kotvojs & Cowell, 1991). The result is therefore little different from relying on inductive insights to formulate equilibrium relationships directly from an empirical basis, as in the case of dimensional scalings available for the morphologies of tidal inlets (O'Brien, 1969; Bruun, 1978) and tidal basins (e.g., Eysink, 1991).

Positive feedback, thresholds and self organization

Positive feedback produces *self-forcing* morphodynamic behaviour which is necessarily of limited duration and unstable (Wright & Thom, 1977). Positive feedback is illustrated by the infilling of deep estuaries with marine sediments when a positive tidal asymmetry exists. The rate of infilling grows as the estuary becomes shallower since the tidal asymmetry also increases as the friction and shoaling effects are enhanced by the reduced water depths (Dronkers, 1986). Eventually, the infilling of the estuarine basin produces the development of tidal flats, as exemplified in the formation of the 'Big Swamp' phase of the Holocene history of the South Alligator River (see Fig. 1.5). This marks a reversal in feedback from positive to negative. As the tidal flats become more extensive, and a larger volume of the tidal flow occurs across them, the positive asymmetry of the tide progressively decreases so that the estuarine morphology approaches steady state as sediment imports and exports equilibrate (Chappell & Thom, 1986; Woodroffe *et al.*, 1989). The relatively brief duration of the 'Big Swamp' phase (22% of the stillstand period) is indicative of the transient nature of morphodynamic processes governed by positive feedback. The reversal in feedback in this case itself constitutes a *threshold* (Wright & Thom, 1977).

Thresholds occur when changing environmental inputs, or changes in the mechanical strength of sediment masses, drive morphodynamic state variables toward limiting values for the operational range of existing *self-regulation* mechanisms (Schumm, 1979; Roberts, Suhayda & Coleman, 1980). Beyond these critical values, transformation of state occurs such that a new set of processes operates with a new set of state variables. Examples of coastal-process thresholds at *instantaneous* scales include the often-used illustration of breaking-wave transformations (e.g., Thom, 1975; Chappell, 1983), the transitions from rippled-bed to flat-bed sand transport under water at a critical shear stress (Nielsen, 1992) and the onset of sediment motion at a critical shear velocity in the sediment-dynamics subsystem (Carstens, Nielsen & Altinbilek, 1969; Nielsen, 1990).

At larger (*event*) scales, changes in beach state between reflective and fully dissipative extremes, or a variety of intermediate states, each with its own characteristic bar morphology and surf-zone hydrodynamics, represents a discontinuous range of morphodynamic processes operating over a continuous range of input wave parameters at a given site (e.g., Wright *et al.*, 1979; Wright & Short, 1984; Wright, Short & Green, 1985). For example, it is well known that barred intermediate and dissipative beaches respond to rising wave conditions through an increasing width of the surf zone associated with a shifting of bars offshore (Shepard, 1950; Bascom 1954; Sunamura, 1989). However, reflective beaches accommodate such conditions through an increase in height of the beach step so that waves can propagate to the shoreline without breaking (Hughes & Cowell, 1986). The transition between the two modes occurs at a threshold wave height when waves begin to break seaward of the beach. Other examples of thresholds at *event* scales include the breaching of barriers by tidal-inlet formation during extreme storms (Coates & Vitek, 1980), lateral switching of over-extended deltaic distributary channels, triggered by floods when hydraulic gradients become too low, and various types of mass movement, including diapiric intrusions (Roberts *et al.*, 1980). Examples of state thresholds at *engineering* to *geological* time scales include transformations between barrier, tidal-delta and innershelf sand bodies, especially under conditions of rising sea level (Swift, Phillips & Thorne, 1991b; Roy *et al.*, 1992), and the alternation between mud progradation and coarse, carbonate chenier-ridge deposition on mangrove coasts in tropical Australia, where threshold conditions involve the occurrence of major storms with return intervals exceeding that required for critical levels of shell production on the shoreface (Rhodes, 1982; Chappell & Grindrod, 1984).

Where input variables (environmental conditions in Fig. 2.4) are non-stationary, as in the case of climate and sea-level change, morphodynamic processes may go beyond their stable range. This is exemplified in coastal evolution by barriers overstepped during rising sea levels. Under certain circumstances, a critical point is reached at which the landward, upslope migration of a barrier fails to keep pace with the rate of sea-level rise (Penland, Suter & Boyd, 1985). The process, shown in Fig. 2.5, has been observed on the Nova Scotian coast (see Chapters 1, 4 and 10) where rising sea level caused

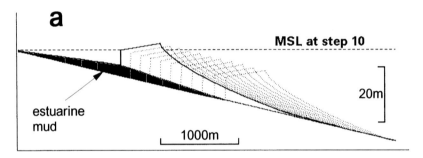

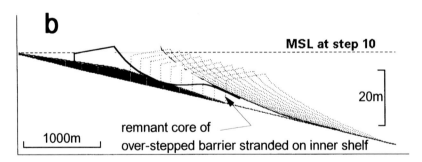

Figure 2.5. Schematic representation of barrier translation in response to a 10 metre sea-level rise, in increments of 1 metre, which corresponds to c.1500 years for the last post-glacial marine transgression. Two modes of coastal evolution are shown: (*a*) continuous *roll-over*, and (*b*) the occurrence of a threshold resulting in barrier *overstepping*. MSL represents mean sea level.

overstepping and remobilisation of a barrier superstructure landwards (Forbes *et al.*, 1991; Orford, Carter & Forbes, 1991). A sudden landward shift in barrier position occurred and the barrier core was left offshore to form a shoreface ridge in the manner proposed (contentiously) for such ridges on the US New Jersey coast (Stubblefield, McGrail & Kersey, 1984; Swift, McKinney and Stahl, 1984; Rine *et al.*, 1991).

Self organisation, which is the subject of recent developments in non-linear dynamics, is driven initially by positive feedback (Waldrop, 1992; Phillips, 1992). It involves the evolution of morphology toward a new equilibrium regime following creation of a new system or an extensive shift in process boundary conditions. Large glacio-eustatic sea-level changes and major engineering interventions, for example, often have these consequences (at *geological* and *engineering* scales respectively). *Self organisation* is, however, an inevitable consequence of morphodynamic feedback at all scales. It has been simulated recently at *instantaneous scales* with the spontaneous development of beach cusps (Werner & Fink, 1993). Emergence of cusps as a global pattern on the beach face resulted from morphodynamic feedback involving the interaction of swash, sediment transport and time-dependent topographic changes on the beach face. The simulated feedback initially was positive, causing growth in small topographic irregularities, but then switched spontaneously to negative feedback as cusps became well developed, providing a stabilising mechanism for cusp survival (Werner & Fink, 1993). *Self organisation* at *event* scales is exemplified in nature by the reworking of dune-sand masses into beach profiles of emergent equilibrium form during storm surges (Wang, 1985; Vellinga, 1986). A heuristic example at *engineering* time scales (months to years) is provided by the partial, artificial closure of the Rhine/Meuse and Eastern Scheldt estuaries in The Netherlands (Steijn *et al.*, 1989; de Vriend *et al.*, 1989), where longitudinal, shore-normal ebb-tide shoals were reorientated to form sand bars with shore-parallel alignment due to the increased importance of waves among the constituents in the shore-normal velocity field. These changes involved the *self organisation* of antecedent, estuarine-tidal dominated morphologies into shoreface morphologies tending toward dynamic equilibrium with waves.

Self organisation at *geological* time scales is illustrated in Fig. 2.6 by the Holocene evolution of broad estuarine sediment sinks into channelised pathways due to progressive infilling involving morphodynamic adjustment to stillstand conditions (Roy, Thom & Wright, 1980). The infilling results (in accordance with eqn. (2.6)) from sharp, negative gradients in transport fluxes for marine and fluvial sediments entering the estuary. The transport gradients are associated with expansion of tidal and stream flows within the estuarine

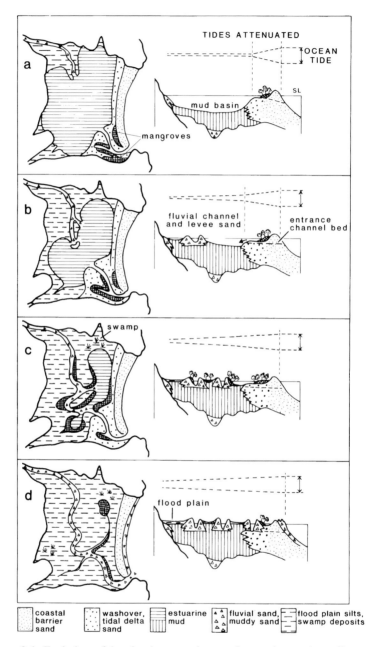

Figure 2.6. Evolution of bay-barrier estuaries on the south-east Australian coast involves *self organisation* in which progressive infilling over the Holocene ultimately leads to sediment bypassing corresponding to a steady state in the estuarine sediment volume (from Roy, 1984).

basin. The scale of the expansion diminishes as the basin fills until transport gradients become zero and sediment bypassing is established, signifying overall steady state conditions. The evolutionary sequence is similar to that described for South Alligator River, despite vast differences in environment as well as detail. Both examples illustrate how *self-organising* behaviour involves a sequence of positive feedback, which drives morphodynamic change toward a new state, followed by negative-feedback processes, which stabilise the sedimentation regime.

Cumulative evolution and state dependence

Coastal evolution is a cumulative process because the morphological outputs are included amongst the inputs for the next cycle of change (Fig. 2.1). This section outlines the principles involved in cumulative, state-dependent evolution, which fundamentally entail *non linearity* in the process–form interaction (Fig. 2.1), *Markovian inheritance* and *response-time* effects.

Although cumulative evolution operates at all scales, its significance in the development of major depositional landforms at longer, *geological* scales is paramount. An example of such state-dependent evolution in the Quaternary history of the south-east Australian coast is given in Fig. 2.7 with respect to the

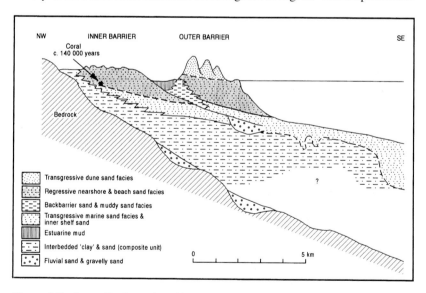

Figure 2.7. Generalised stratigraphic section of the Newcastle Bight embayment, 100 km north of Sydney on the south-east coast of Australia, based upon core logs and radiocarbon dates. The inner and outer barriers are of Last-Interglacial and Holocene ages respectively. (After Thom *et al.*, 1992.)

120 ka and Holocene high sea levels and the post-glacial marine transgressions that preceded them. These episodes both caused deposition of transgressive and regressive sediment sequences, each of which presents a different substrate to their successors. The stratigraphy shows that the earliest deposits occur over substrates which are broader and shallower (see Chapter 4). Shorefaces associated with these deposits were also more protected from wave energy since they generally lie within the landward recesses of embayments on this coast. Later shoreface sediments were deposited under altered morphodynamic conditions since substrates became progressively steeper, deeper and more exposed. The geometric changes produced an increasing accommodation volume per unit width of shoreface as high-stand shorelines were relocated progressively further seaward (Fig. 2.7). Recent thermo-luminescence dating of deposits in another embayment, 100 km further north, indicates that there may have been at least four barrier sequences with ages of 240 ka, 120 ka, 90 ka and Holocene (Roy *et al.*, 1992). Each sequence remains largely intact, each represents intermediate states in the evolution of the coast, and each has presented a different substrate during successive transgressions (Fig. 4.12).

Non linearity

Non linearity is an inevitable consequence of the feedback effects (Phillips, 1992) that make coastal evolution cumulative (Figs. 2.1 and 2.4). Non linearity 'means that the act of playing the game has a way of changing the rules' (Gleick, 1988, p. 24). The evolving morphological outputs (eqn. (2.1)), which comprise the boundary conditions for fluid dynamics (via eqns. (2.3) and (2.4)), are included amongst the inputs for the next cycle of change. Morphological change in coastal evolution therefore 'cannot be explained simply by studying the 'inputs' individually' (Wright & Thom, 1977, p. 418), since morphodynamic processes themselves vary according to the antecedent system state.

Such non-linear processes are neither additive nor homogeneous (Bendat & Piersol, 1971, p.37), which translates respectively to: (i) the old expression that 'the whole is more [or less] than the sum of the parts'; and, more strictly, (ii) the output of the system is not in constant proportion to the input. Thus, a system in which the output, Y is a function of its input, X such that

$$Y = G(X) \qquad (2.7)$$

is non linear if

$$G = f(X) \neq \text{constant} \qquad (2.8)$$

in which the gain G, resulting from operation of the system (i.e., the process), is itself dependent upon the system state. Applying eqn. (2.7) to coastal evolution, Y might be topography and X the flows forcing sediment transport. Morphodynamic feedback (Fig. 2.1) prescribes that X changes as Y evolves, resulting in modification of sedimentation processes (eqn. (2.8)). Changes in G may be continuous for non-linear systems but thresholds can also produce discrete changes (e.g., at the onset of motion in sediment dynamics). In general, non linearity arises if representative relationships involve terms containing powers, products of the output, and/or dependent variables (Bennett & Chorley, 1978, p. 37).

Non linearity of morphodynamic processes is inherent in the overall structure of coastal systems since the interactions between the morphodynamic components (Figs. 2.1 and 2.4) are not weak (von Bertalanffy, 1968). For example, the evolving substrate in the Newcastle Bight (Fig. 2.7) resulted in shorefaces becoming progressively more exposed to waves and other open coast processes as successive shorelines were displaced seaward within the coastal embayment. Apart from consequent changes in cross-shore flow and sedimentation regimes, alongshore-transport budgets changed profoundly as the coast evolved from being laterally bounded toward an open coastline configuration. Chronostratigraphic evidence from other parts of south-east Australia indicates that such changes provided the primary controls over the Quaternary evolution of this coast (Roy *et al.*, 1992), implying strong variations in G (eqn. (2.7)) and thus marked non linearity in the overall evolution of the morphodynamics (see Chapter 4).

Moreover, many of the subsystems themselves are strongly non linear by nature (de Vriend, 1991a). This is especially the case for hydrodynamic processes, which in general behave like wave motions, with different flow categories occupying distinctive frequency bands (e.g., Mooers, 1976). These motions usually become increasingly non linear as water depth diminishes toward the shore, provided flows have horizontal length scales smaller than the continental-shelf width. These flows undergo shoaling toward the coastline, resulting in a steepening of water surface slopes, which causes increasing interaction between fluid-dynamic processes (Ursell, 1953). The effect of water-surface slope also becomes important for flows with very long wave-lengths, such as tides and shelf waves, if they are constricted within straits, bays or estuaries.

The significance of non-linear hydrodynamics in producing sediment transport in a coastal morphodynamic system is that the behaviour of particular flows changes as their non linearity grows. In addition, they gain the capacity to interact with other non-linear flows to produce mutual changes involving a transfer of energy between flows, often at different frequencies (Guza & Thornton, 1981; de Vriend, 1991a). This is an important attribute for the selective growth of resonant motions. Such motions and interactions are known to characterise the shallow-water hydrodynamics of surf zones leading to complex bar-trough topographies (Bowen & Guza, 1978; Huntley & Bowen, 1975; Chappell & Wright, 1979; Wright et al., 1979; Holman, 1981; Huntley, Guza & Thornon, 1981; Wright, Guza & Short, 1982; Mei, 1985).

Non linearity in some hydrodynamic subsystems is fundamental to coastal morphodynamic adjustment. For example, the non-linear properties of 'finite-amplitude' gravity waves are deemed responsible for onshore sand transport in several models for profile morphodynamics of the shoreface and surfzone (e.g., Bowen, 1980; Bailard & Inman, 1981; Roelvink & Stive, 1991). It also provides one explanation for the formation of surfzone bars (Boczar-Karakiewicz & Davidson-Arnott, 1987). Moreover, the non-linear interactions within and between component morphodynamic processes often require the full non-linear equations to be solved so that detail and accuracy in the hydrodynamic model is sufficient to meet the needs of the sediment transport model (Nielsen, 1990; de Vriend, 1991a; GESAMP, 1991, pp. 39–40). At a more general level, self-regulating non-linear systems possess greater resistance to change than do linear systems which, 'given a slight nudge, tend to remain slightly off track' (Gleick, 1988). This inherent stability is attributable to negative feedback loops that epitomise 'control systems' and enable them to return to their former state after a perturbation. An example of this is the recovery of the surfzone profile to its pre-storm position following its upward and landward translation in response to raised water levels during a storm (Vellinga, 1986; Sumamura, 1989; Dean, 1991).

Most of the preceding examples are from processes operating at the shorter time scales in Fig. 2.2. This reflects a greater understanding of details in non-linear dynamics at such scales than is available at present for morphodynamics operating over either *engineering* or *geological* time spans. Nevertheless, the principles can be expected to apply at all scales. Moreover, the very 'interactiveness' responsible for non linearity is likely to cause small-scale effects to be compounded into larger-scale coastal evolution.

Markovian inheritance

The effect of stochastic variation in external inputs (Fig. 2.4) is superimposed upon the state-dependent control of coastal evolution. Coastal evolution thereby proceeds through *Markovian inheritance* (e.g., Fig. 2.8). In general, each morphological state in an evolutionary sequence has a set of states to which it may further evolve during the next step in the succession (e.g., Goodman & Ratti, 1979, p. 196). The set of succeeding states is limited by the position of the existing state within the *Markov chain*. Selection of the new state from the set of possibilities depends upon ensuing conditions created by the stochastic, environmental inputs. In principle therefore, probabilities can be ascribed to state selection based upon: (i) the transfer function representing system processes (eqn. (2.7)); (ii) the state of the system immediately before the change; and (iii) probability distributions for input variables. State transitions in nature may be continuous through time. In addition, an unlimited number of possibilities in the set of outcomes may exist for each transition (e.g., a new bed elevation, $h_j = h_i + \Delta h$). In practice, however, the set of outcomes must be approximated by a finite number of discrete states with transitions occurring on an event basis, such as in the case of Swift, Ludwick & Boehmer's (1972) Markov process model for progressive sorting of sediments across the shoreface. In this model different storm-event histories result in different sediment-dispersal patterns. Nevertheless, there are problems in nature where the potential transitions are both discrete and finite in number. Examples include Sonu & James' (1973) markov chain for the beach face, and models for surfzone bars by Wright & Short (1984) and Sunamura (1989), each of which contain a total of only six or eight states overall, arranged in cycles of erosion and accretion.

Markovian inheritance at larger scale is illustrated in Fig. 2.8 by the evolution of a coastal sand body during a marine transgression. Locally, the sand body may exist in one of three states: a *barrier*, *mainland beach* or *inlet*. There are three potential state transitions at each time step, and transition between any of the states is possible (see state-transition table in Fig. 2.8). The *barrier* may continue to 'roll over' as the sea level rises (Leatherman, 1983) or undergo transformation into a *mainland beach* if overstepping occurs (Penland *et al.*, 1985) or if the substrate becomes too steep. Local transition from *barrier* to *inlet* may occur either through the arrival of an *inlet* that is migrating alongshore, or if the *barrier* is breached during a storm event. Restoration of a *barrier* may occur locally at a subsequent step with further migration of the *inlet* away from the site, or through shoaling and atrophy of the *inlet*. Transition from *mainland beach* to either *barrier* or *inlet* requires flooding of the low area behind the beach as sea level rises, in accordance with Hoyt's (1967) detachment mechanism, provided that the substrate is not too steep (see Chapter 4).

The potential course of coastal evolution entails a myriad of possibilities which increase over time (at rate of s^n, where s is the number of states at each step and n is the number of time steps). In this simplest of Markov trees, each

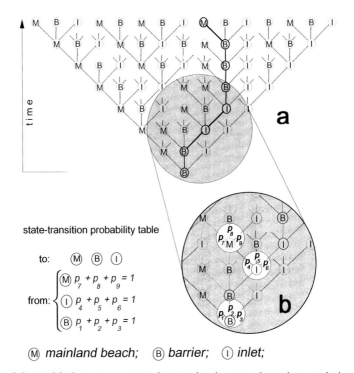

state-transition probability table

to: (M)   (B)   (I)

from: $\begin{cases} \text{(M)}\ p_7 + p_8 + p_9 = 1 \\ \text{(I)}\ p_4 + p_5 + p_6 = 1 \\ \text{(B)}\ p_1 + p_2 + p_3 = 1 \end{cases}$

(M) *mainland beach;*   (B) *barrier;*   (I) *inlet;*

Figure 2.8. (*a*) Markov-tree representing stochastic, state-dependent evolution of a coastal sand body during a marine transgression over 7 discrete time steps. Three possible states exist, signified by the circled letters in the tree: mainland beach (M), barrier (B), and tidal inlet (I). The dotted circles (states) and transition lines map the myriad of possible sequences, of which only one can be realised (hypothetically, the solid circles and lines). Inset (*b*) details the probabilities (*p*) associated with each type of state transition. For any transition $\Sigma P_i = 1$, whereas through time, probabilities multiply such that $P \rightarrow 0$.

state can undergo transition to any other state. Nevertheless, each evolutionary trajectory through Fig. 2.8 is unique. The general geomorphology of *beaches*, *barriers* and *inlets* may be identical at every occurrence, unless such geomorphology is sensitive to antecedent conditions. However, different evolutionary sequences result in unique differences overall. Stratigraphies underlying each of the geomorphological elements are the product of recent antecedent states (e.g., Fig. 2.7). Residual surfaces left by the landward retreat of sand bodies also depend upon the state sequence. For example, a succession of *inlet* states leaves behind *shoal-retreat-massif* and *shelf-valley* morphology on the inner shelf, whereas a *barrier* sequence leaves a *ravinement* surface (Swift *et al.*, 1991a). The sequencing of *barrier* and *inlet* states through time therefore determines the character of the residual innershelf surface.

The probability that any specific evolutionary sequence will occur becomes progressively smaller with the passage of time since successive transition probabilities multiply. Thus, if all three transitions in Fig. 2.8 have equal probabilities ($P = 0.3$), the probability for any sequence after just three time steps is a mere $P = 0.04$. The hypothetical sequence illustrated in Fig. 2.8 never persists for more than three steps as any one sand-body type. Accordingly the residual surface and sand-body stratigraphy would be complex. Coastal evolution is therefore essentially a matter of historical accident. Joint dependence upon intermediate states and stochastic inputs means that evolutionary history is unlikely to repeat itself exactly, even through closed Markov cycles such as those for the beach face and surf zone. For example, in Sonu & James' (1973) closed-cycle Markov simulation of the beach face, there were only six states, four of which could undergo transition to two other states, while a single transition was possible in the case of the two end members. In Sonu & James' computer simulations, the minimum number of six transitions, in a single cycle of erosion and accretion, occurred only 6% of the time. However, one cycle went through 98 transitions before completion. The median number of transitions per cycle was 16, for which there exists 81 permutations in transition sequencing!

The possibility of geomorphological convergence (e.g., Davies, 1980, p. 7) seems to be denied by the diversity of coastal evolutionary potential epitomised by Fig. 2.8. Schumm (1991) suggests that geomorphological convergence is illusory since landform histories may appear convergent upon superficial inspection, but reveal idiosyncratic differences when subjected to closer examination. Nevertheless, the possibility of geomorphological convergence remains if '*absorbing*' *Markov chains* exist. '*Absorbing*' chains possess at least one '*absorbing state*', which can be reached from any other '*non-absorbing state*', but from which transitions to another state do not exist (e.g., Goodman & Ratti, 1979, p. 316). Evolutionary sequences become locked into a single state (or fixed cycle of states) under such circumstances. Residual surfaces and stratigraphies are then predictable, regardless of initial or antecedent conditions. This type of Markov chain possesses *stationarity* as does the 'simple' illustration in Fig. 2.8. In general, however, probabilities ascribed to the state transitions are likely to change, along with the morphodynamic processes, as the coast evolves. The resulting *nonstationarity* is an important cautionary consideration in application of time-series analysis to evolutionary sequences (Jenkins & Watts, 1968; Bennett & Chorley, 1978). Similarly, the interpretation of stratigraphy based on the 'Law of Uniformitarianism' must take *nonstationarity* into account (Kraft & Chrzastowski, 1985).

Finite response-time effects

Morphodynamic response times compound *Markovian* behaviour since the input–event sequencing, upon which this behaviour depends, also affects the strength of hysteresis. Morphodynamic adjustment is constrained by frequency-response characteristics which reflect the finite time required for sediment-volume transfers to occur. The *relaxation time*, T_R needed for the morphological change (Δh) necessary to attain equilibrium, following a shift in boundary conditions, is related to sediment transport rates (Melton, 1958; Hardisty, 1987), through sediment continuity (eqn. (2.1)), by

$$T_R = \frac{\Delta h}{(\partial h / \partial t)} \tag{2.9}$$

Relaxation times of coastal sand bodies in south-east Australia (examined in more detail in Chapter 4) were in the order of 10^3 years (Fig. 2.9) following onset of stillstand conditions c. 6000 years ago (Fig. 2.10). Barriers, which

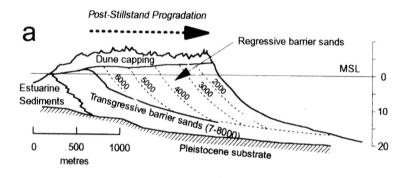

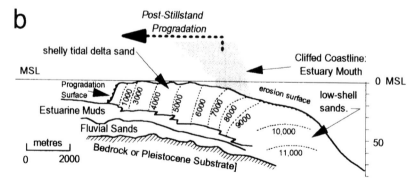

Figure 2.9. Holocene *relaxation times* indicated in chronostratigraphy from south-east Australia for: (*a*) *prograded barriers*; and (*b*) *flood-tide deltas*. The isochrons are labelled in years BP. (Adapted from Nielsen & Roy, 1981.)

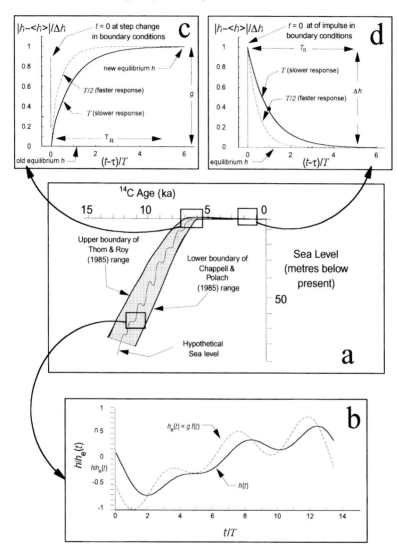

Figure 2.10. Response times for topographic change in a sand barrier during the last phase of the Holocene marine transgression and stillstand. A hypothetical sea-level curve is plotted in (*a*) within an envelope based on radiocarbon dates from south-east Australia (Thom & Roy, 1985) and the Huon Peninsula, Papua New Guinea (Chappell & Polach, 1991). Response times are plotted relative to the time constant (*T*) for : (*b*) continuous variations in rate of sea-level rise; (*c*) a step response (comparing two values of *T*) following a change from transgressive to stillstand conditions; and (*d*) an impulse response, during the stillstand, following a sudden event of short duration, such as severe storm (comparing two values of *T*). The abscissa in (*c*) and (*d*) is time (relative to *T*) after the respective events, and <*h*> denotes time-average bed elevation; *g* is the gain; *f* is the forcing and τ is the time elapsed since the input event.

translated landward whilst the marine transgression was under way, subsequently prograded over a period of at least 4000 years (Fig. 2.9a). Flood-tide deltas presently continue to prograde into flooded paleochannels (Fig. 2.9b).

Relaxation times are scale dependent since the volume (*Vol*) of sediment that must be displaced to affect a given Δh is

$$Vol = \Delta h L_x L_y \tag{2.10}$$

where L_x and L_y are the horizontal dimensions of a geomorphological feature. Furthermore, the total net rate at which this volume is displaced,

$$Q_{total} = \frac{\partial q_x}{\partial x} L_x + \frac{\partial q_y}{\partial y} L_y \tag{2.11}$$

tends over larger features to be more strongly subjected to a limiting effect imposed by the finite range in q_x and q_y across the feature. This effect involves the inverse variation of $\partial q_x/\partial x$ and $\partial q_y/\partial y$ with L_x and L_y in eqn. (2.11).It also results, through eqn. (2.1), in smaller values of $\partial h/\partial t$ in eqn.(2.9), and hence longer T_R. Thus, big features change slowly for given sediment-transport rates, a point that is fundamental to scale relationships in coastal evolution. Scale dependence of response time is illustrated by the evolution of a sand body on the inner shelf off the south Sydney coast, Australia (Field & Roy, 1984). The average thickness of the post stillstand portion of this feature is $\Delta h \approx 4$ m, with $T_R \approx 7500$ years based upon radiocarbon dates (Roy, 1986). Regional measurements of near-bed flows on the inner shelf, and estimates of wave-induced sand mobility, indicate that the entire deposit was readily attributable to q_x, while only 20% of the deposit was potentially attributable to q_y in the time available (Cowell, 1986). The difference in potential contri-butions stems from the aspect ratio, $L_x : L_y$ of the sand body (2000 m : 15 000 m) which makes the feature appear much less extensive to cross-shelf sediment transport than it does to the coast-parallel component. Because of this, $\partial q_x/\partial x$ exceeds $\partial q_y/\partial y$ by more than a factor 5 in eqn.(2.1). These differences in transport gradients produce corresponding differences in respective estimates of $\partial h/\partial t$ from eqn. (2.9).

In reality, $\partial h/\partial t$ in eqn.(2.9) generally diminishes through time as morpho-dynamic equilibrium is approached. If the reduction in $\partial h/\partial t$ is described by an exponential decay, then the morphological response time can be characterised by a *time constant*, T; a similar measure to radioactive half life. More formally, T is the ratio of the energy storage and dissipative parameters of a system (Bennett & Chorley, 1978, p. 45) and is therefore a measure of the filtering present in the system. In morphodynamics, this is predominantly attributable to the cumulative effects of the sediment budget during coastal evolution.

Furthermore, $1/T$ can be thought of as a measure of morphological sensitivity representing the resistance of morphology to change (Brunsden & Thornes, 1979). The overall behaviour of exponential memory-decay can be described as

$$T\frac{dh}{dt} + h(t) = f(t) \tag{2.12}$$

where $f(t)$ is the input process forcing changes in h (Jenkins & Watts, 1968, p. 38). This behaviour is illustrated in Fig. 2.10 which describes hypothetical responses of a *transgressive barrier* to changes in sea level during the post-glacial marine transgression and following Holocene stillstand (Fig. 2.10a). Generally, environmental inputs vary continuously rather than in discrete steps. This is illustrated in Fig. 2.10b which describes the effects on bed elevations across the *barrier* due to fluctuations in the rate of sea-level rise. Changing sea level causes the forcing, $f(t)$, which represents what the geomorphological outcome would be at each instant in the absence of a *time lag* (or if boundary conditions had been constant for an infinite duration). The effect of the *time lag* is evident in the actual morphological variation, $h(t)$, which occurs as a smoothed (or filtered) response to fluctuations in boundary conditions. The sediment-transport budget constitutes the filter controlling the morphological responses to variations in the fluid dynamics. This behaviour is characterised in the frequency domain by the *frequency response function*, $F(\omega)$ (Jenkins & Watts, 1968, pp. 38–9). Working in the frequency domain has two advantages. First, the gain, $G(\omega)$ and phase lag, $\phi(\omega)$, become simple functions of the Fourier coefficients (A and B), which for eqn. (2.12) gives

$$G(\omega) = \sqrt{[A^2(\omega) + B^2(\omega)]} = 1/\sqrt{[1 + (T\omega)^2]} \tag{2.13}$$

and

$$\phi(\omega) = \arctan(T\omega) = \arctan - (B/A) \tag{2.14}$$

(where $\phi = \delta\omega$ in terms of the time lag, δ). Second, the actual response of morphology in the frequency domain is the product

$$h(\omega) = F(\omega)f(\omega) \tag{2.15}$$

which is much simpler than the equivalent convolution in the time domain (Jenkins & Watts, 1968, p. 45). A geomorphological time series incorporating the effects of *frequency response* is provided readily by the inverse Fourier transform of $h(\omega)$.

The behaviour illustrated by Fig. 2.10b occurs whenever T_R is longer than time scales characterising significant changes in boundary conditions. A documented example is provided by Bruun-type analyses of bed-level change and shoreline recession observed with respect to water-level fluctuations in Lake Michigan, USA (Hands, 1983). Although equilibrium may not have time to fully develop in nature, where the stochastic inputs vary continuously, an equilibrium potential (*attractor*) exists at each instant. The equilibrium tendencies inherent in the processes result, through time, in what Wright & Thom (1977) term 'optimal' or 'most-probable' states. Coastal evolution therefore involves dynamic homeostasis (Wright & Thom, 1977). A time series of equilibrium states, although unrealisable, represents the potential that forces evolution along ($h_e(t)$ in Fig. 2.10c). The realised evolution results from modulation of this forcing potential by the response characteristics ($h(t)$ in Fig. 2.10c). The effect is amplified as the resistance of bed material to movement increases. The consequence of morphodynamic hysteresis is that the time lag, δ, permits only a partial morphological response toward transient equilibrium (Chappell, 1983).

A step-like change in environmental conditions occurred at the end of the transgression in south-east Australia, whereupon a sea-level stillstand has ensued for c. 6000 years (Fig. 2.10c). The behaviour of coastal sand bodies under transgressive conditions is markedly different to that during stillstand (see Chapter 4). Although boundary conditions have remained constant during the stillstand, morphodynamic adjustment took at least 4000 years (Fig. 2.9). This adjustment involves a transition from one state of *dynamic equilibrium* to another and is described by a *step-response function* (Fig. 2.10c). The limiting value of this (at $t \to \infty$) is the steady-state gain, g which defines the change in state from the old to the new equilibrium (h_1 to h_2 respectively) as

$$h_2 = gh_1 \qquad (2.16)$$

where

$$g = \int_0^\infty S(\tau)\,d\tau \qquad (2.17)$$

and $S(\tau)$ is the transfer weighting function at time τ following the occurrence of a unit-step change in the input lasting for an infinite duration (Fig. 2.10c).

The *step-response function* is the integral of the *impulse response* which describes morphodynamic behaviour following an event manifesting itself as an impulse in environmental inputs. The *impulse response*, describing the recovery of barrier morphology following a major storm event, is exemplified in Fig. 2.10d (again assuming that eqn. (2.12) is applicable). Facets of such barrier behaviour include recovery of the modal beach-type (Short, 1979; Wright & Short, 1984; Wright et al., 1985) and profile disposition (Vellinga,

1986; Kotvojs & Cowell, 1991; Dean, 1991) after the storm causes perturbation from average wave conditions. Restoration of equilibrium for these morphologies is then given by

$$h(t) = S(t)f(t) \tag{2.18}$$

where $S(t)$ is the *impulse response function*, which for eqn. (2.12) is written as

$$S(t) = \begin{cases} \dfrac{1}{T} e^{-(t-\tau)/\tau} & , \tau \leqslant t \\ 0 & , \tau > t \end{cases} \tag{2.19}$$

Eqn. (2.19) is a transfer function that weights the extent to which morphology remembers the effect of inputs that occurred at time τ in the past. It represents the finite time taken for morphology to respond to a disturbance (Fig. 2.10*d*). Such behaviour is illustrated by the recovery of Moruya Beach, south-east Australia (Thom & Hall, 1991), which took about six years following major storms in 1974 and 1975 (see Fig. 2.11).

The response-time characteristics described by eqns. (2.12) to (2.19) have an important bearing upon effects with respect to input magnitude and

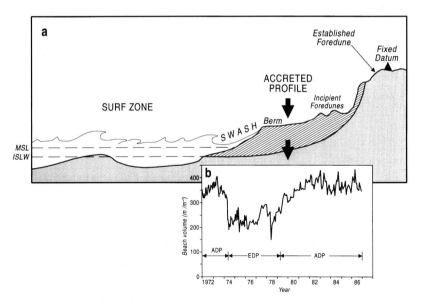

Figure 2.11. Fluctuating LSCB evident in data from a beach-monitoring programme spanning two decades (Thom & Hall, 1991). (*a*) Definition of sand volume monitored. (*b*) Almost all of one long cycle has emerged, incorporating two accretion-dominated periods (ADP) and one erosion dominated period (EDP). ISLW = Indian Spring Low Water.

frequency (Wolman & Miller, 1960). The relationship between $h(t)$ in eqn. (2.18) (or more simply, the recovery time, T_R) and the return interval for events from a given magnitude-frequency distribution determines the potential for different morphological responses to events of the same magnitude (Wolman & Gerson, 1978). For example, coastal storms of similar intensity can cause very different levels of shoreline erosion (Thom, 1978). This further illustrates the Markovian behaviour associated with event sequencing (e.g., Swift, Ludwick & Boehmer, 1972) outlined in the previous section. If input conditions are regarded as changing continuously (Fig. 2.10b), rather than as discrete events, then different morphologies inevitably result whenever identical input magnitudes occur in otherwise dissimilar time series.

Large-scale coastal behaviour

General concepts: LSCB definitions

Computing power, in principle, is now sufficient to apply the precept that coastal evolution over the long term is the time-integrated result of processes operating over the short term (Wright & Thom, 1977). The mathematical physics developed for *instantaneous* morphodynamic processes could then be applied directly to all other scales in Fig. 2.2. Unfortunately, because of the uncertainties outlined above, it has transpired that deterministic principles are not yet directly transferable to the explanation and prediction of long-term coastal evolution (Terwindt & Battjes, 1991; de Vriend, 1991a). This area therefore holds the next big challenge in coastal morphodynamics (de Vriend, 1991a). The problem is now formally termed 'large-scale coastal behaviour' (LSCB) which, by preliminary convention, applies to time scales of decades and distances in the order of 10^1 kilometres (e.g., Stive *et al.*, 1991; Terwindt & Battjes, 1991). Interest in LSCB is driven by practical imperatives of coastal management, including the effects of both engineering interventions and changing boundary conditions (such as sea-level rise and changes in wave climate) and therefore focuses on coastal evolution over *engineering* time scales (Terwindt & Battjes, 1991). Topographic evolution at these LSCB time scales is illustrated in Fig. 2.11 by changes in beach-sediment volumes over more than two decades in south-east Australia. The data indicate a strong cyclical signal at an inter-decadal time scale, corresponding to Stive *et al.*'s (1991) 'fluctuating LSCB'. The amplitude of the cycle is large compared to the geological trend evident in radiometrically calibrated stratigraphic data characterising coastal evolution during the last 2000 years (Thom & Hall, 1991).

LSCB can be defined more generically as topographic evolution over time and space scales that are too long to permit conventional methods of prediction based upon separate analysis of isolated parts of the process. At *engineering* and *geological* scales, predictive uncertainties due to *state-determining* behaviour in coastal evolution (or 'sensitive dependence upon initial conditions') are compounded by problems associated with morphodynamic *response times* and *stochastic* boundary conditions. LSCB steady state is the outcome of time-averaged morphodynamics operating at *instantaneous* and *event* scales driven by magnitude–frequency distributions possessing *stationarity*. Such distributions incorporate natural fluctuations in boundary conditions related to the effects of inter-annual climatic variations such as that associated with ENSO. Change in LSCB steady state is a morphodynamic response to variations in synoptic boundary conditions, which can be natural or artificial. Natural variations are associated with climate change as well as denudational and tectonic processes, whilst artificial changes relate to the impacts of engineering interventions, which may be formal or informal, and intentional or unintentional (Stive *et al.*, 1991).

Sand exchanges between the shelf and surf zone are at conventional LSCB dimensions (Wright, 1987; Wright *et al.*, 1991; Roelvink & Stive, 1991). However, the relaxation times for shoreface-related geomorphological units in Fig. 2.9, estimated from radiocarbon dates, are on the order of 10^3 years. This is consistent with measured trends in shoreface-profile change reported from elsewhere, such as the central Dutch coast (Stive *et al.*, 1991). Therefore, to account for the effects of such long relaxation times, it may be necessary to extend the LSCB class of processes from *engineering* into *geological* time scales, two to three orders of magnitude beyond that of the original definition (Stive *et al.* 1991; Terwindt & Battjes, 1991). Although extension of the LSCB definition into *geological* time scales may seem excessive, the underlying trends during the Holocene evolution of coasts manifest themselves at *engineering* time scales, such as in the case of long-term coastal recession and shoaling of estuaries, and must be factored into coastal management strategies. In that respect, the original definition has already been revised up one order of magnitude (de Vriend, 1992a).

Theoretical approaches to LSCB

Computational and theoretical morphodynamics are emerging to form a new discipline in its own right (de Vriend, 1991a,b; 1992a,b). At the heart of this effort is the objective of modelling long-term behaviour based upon knowledge about short-term processes. One consensus exists among some key coastal scientists that this is an impossible goal since LSCB involves 'unpredictable inputs fed into a non-linear system' (Terwindt & Battjes, 1991). Deterministic prediction of LSCB, which incorporates both Markovian behaviour and sensitive dependence on initial conditions, is not currently achievable and will require hitherto unforeseen advances in the studies of deterministic chaos in general and *theoretical morphodynamics* in particular.

An alternative approach to deterministic modelling is to draw upon the wealth of existing geological knowledge of coastal evolution at a large scale. Paradoxically, this may turn James Hutton's 'Law of Uniformitarianism' on its head, since we now see the possibility that 'the past is the key to the future' in terms of LSCB prediction (e.g., Roy & Thom, 1987; Thom & Roy, 1988). In reality, however, development of knowledge in coastal morphodynamics proceeds most effectively through mutual induction from studies of the present and past undertaken in parallel (Wright & Thom, 1977). Research into the morphodynamics of LSCB thereby becomes a question of scaling up from the laws of physics and scaling down from the geomorphological and geological principles of coasts. This pragmatic approach combines laws obtained from deterministic studies with geostatistical relationships for coastal geomorphologies and their geological evolution (Terwindt & Battjes, 1991; de Vriend, 1992b).

The development of morphodynamic models for LSCB, based upon the scaling up of principles applicable to *instantaneous* and *event* scales, is examined prospectively in detail by de Vriend (1992b). The approaches fall into two groups, input filtering and process filtering, of which empirical relationships and output filtering are both variants (de Vriend, 1992a; Terwindt & Battjes, 1991). Each presupposes that the response-time characteristics are known, especially with respect to the effects of average conditions and episodic events in terms of their relative importance to the range of processes at smaller scales. Similarly, both approaches depend upon the degree to which linearity can be assumed. Scaled-up formulations based upon mathematical physics may involve solving equations for small-scale processes directly for the scale of interest, or aggregating the effects of processes operating at smaller scales into simplified processes at LSCB dimensions, such as in the treatment of smaller-scale residual transports as diffusion at a larger scale (Terwindt & Battjes, 1991). Alternatively, advances in the science of complexity may offer scope for dealing with the effect of the non-linear interactions (de Vriend, 1992b), although it is difficult to see how this approach might deal with the stochastic elements introduced through the Markovian properties (see page 54). This last effect, operating through dependence upon intermediate states under the influence of naturally varying inputs, remains the greatest constraint in the development of deterministic models for LSCB. Implementation at a practicable level may require that such models are formulated to permit dynamic-stochastic simulation so that probabilistic predictions are given. These techniques involve representation of stochastic input variables by their probability distributions. Process subsystems are then coupled in terms of the probabilities of their inputs and outputs. Although these techniques are well established in industrial systems analysis (e.g., Banks & Carson, 1984), their great promise has not been exploited in coastal morphodynamics.

Scaling down from geological models to LSCB dimensions is a possibility that, although acknowledged (Terwindt & Battjes; 1991; and de Vriend, 1992b), has received inadequate attention, given the maturity of rationale for generic geological models. Stratigraphic and topographic data stemming from coastal evolutionary sequences formed during the late Pleistocene in response to extreme changes in boundary conditions provide good control for development and testing LSCB models that incorporate the effects of subsidence, variations in sediment supply and sea-level change (Thom & Roy, 1988). These models are based on the ideas of Curray (1964) concerning transgressive (cut) or regressive (fill) shoreline displacements caused by glacio-eustatic fluctuations and validated by topographic and stratigraphic data from Quaternary coastal deposits. The models have become progressively more formalised over the last two decades in Australia (e.g., Thom, 1978, 1984; Roy & Thom, 1981; Thom, Bowman & Roy, 1981; Thom & Roy, 1985; Roy *et al.*, 1992) and other parts of the world (e.g., Swift, 1976; Kraft & John, 1979; Niedoroda *et al.* 1985; and Swift *et al.*, 1985, 1991b), and have been extended to time scales spanning the Phanerozoic (Nummedal & Swift, 1987; Vail *et al.*, 1991). Scaled-down morphodynamic models are based upon concepts of *regime* sedimentation and equilibrium (Thorne & Swift, 1991b) which do not directly involve process physics for fluid motions and sediment transport (see page 66). Stratigraphic sequences record information on process invariants that have emerged through rigorous interpretation of depositional histories (Thorne & Swift, 1991a). For example, techniques of sequence stratigraphy are well established as the basis of measurement and theoretical methods used world wide at a practical level, notably in the petroleum and mineral exploration industries (Swift *et al.*, 1991a).

Model formulation and process uncertainties

Regardless of the approach, the need for inductive insight and validation of modelling rationale demand a geostatistical analysis of geomorphological changes over historical time scales (e.g., Chappell & Thom, 1986; Terwindt & Battjes, 1991; Dolan, Fenster & Holme, 1991; Thom & Hall, 1991), and sediment budgets at regional scales (Terwindt & Battjes, 1991). Direct measurement of LSCB is illustrated by beach-profile monitoring spanning more than two decades in south-east Australia (Fig. 2.11). Unfortunately, because such data sets are rare, and because morphodynamic attributes are often inherently correlated, it is unlikely that geostatistical analysis can provide a direct means of explanation and prediction in LSCB. Morphodynamic relationships underlying LSCB at *instantaneous* and *event* scales also demand continued investigation since knowledge of these processes in nature is still far from definitive at any scale (de Vriend, 1991b). Further field experimentation is especially called for to improve definition of both input and state variables, their relative importance and the individual process relationships that constrain them. The type of work reported by Wright *et al.* (1991) for the shoreface is an eminent example of such research.

Based upon principles outlined in earlier sections, the structure of a morphodynamic model must be defined in terms of the interrelationships between primary process constituents (Fig. 2.4). Within each of these subsystems is a nested hierarchy of processes, with the lowest level describing greater detail than that required in the solution. Some assessment remains necessary in terms of sensitivity to processes at even lower ('sub-grid') levels (GESAMP, 1991, p. 47). A modular structure of the model is essential (Fig. 2.4) for separate verification and validation of components and independent evaluation of performance (de Vriend, 1991a).

Formulation of a morphodynamic model is first done conceptually and then translated into a set of governing equations for a mathematical (usually numerical) model (GESAMP, 1991). Conceptual development is not trivial and often is as far as the formulation phase can be taken for particularly intractable morphodynamic problems, such as that of long-term coastal evolution. The geological models for evolution of barriers (e.g., Thom, 1978, 1984; Thom *et al.*, 1981; Thom & Roy, 1985; Niedoroda *et al.*, 1985; Swift *et al.*, 1991b; Thorne, and Swift, 1991b) river mouths (Wright, 1985) and estuaries (Roy *et al.*, 1980) are notable examples. In the case of short-term coastal evolution, models based upon generic classifications derived from heuristic knowledge of morphodynamic processes provide practical tools for the understanding and prediction of the characteristic processes and morphology of beach states (Wright & Short, 1984; Sunamura, 1989; Lippmann & Holman, 1990), tidal inlets and estuaries (Hayes, 1975; Davies, 1980), river mouths (Wright, 1985) and dunes (Short & Hesp, 1982; Hesp, 1988; Hesp & Thom, 1990; Carter, Hesp & Nordstrom, 1990). Conceptual models of coastal morphodynamics 'based upon established knowledge of processes and with little or no parameterisation' can be more useful than predictive models that provide little insight into the underlying processes or have a restrictive range of application (GESAMP, 1991, p. 49).

Numerical-model development requires that each of the processes and process interactions are translated into mathematical terms that are necessarily simplified representations of reality. This 'parameterisation' process usually also involves a degree of approximation for practical reasons (GESAMP, 1991, p. 47; Lakhan, & Trenhaile, 1989; Lakhan, 1989). In principle, more refined models involve a greater level of parameterisation, both in terms of sophistication and number of variables and dimensions. This results in greater accuracy (less uncertainty) but only up to a point. Understanding of the processes in coastal morphodynamics is far from definitive, especially with respect to sediment transport (e.g., Huntley & Bowen, 1990; Roelvink & Stive, 1991; Nielsen, 1992). This uncertainty is incorporated into each of the component models and amplified with integration to more complex representations (de Vriend, 1987c, 1991a; GESAMP, 1991) as well as to larger scales in both space and time (Terwindt & Battjes, 1991; de Vriend, 1991*b*). The compounding of this certainty means that there is an optimal level of parameterisation and detail (Fig. 2.12). Model formulation at a practical level involves solution of the governing equations through application of conservation laws to various types of solution procedures. These procedures include analytical methods but, for complex models, finite difference and finite element schemes are called for.

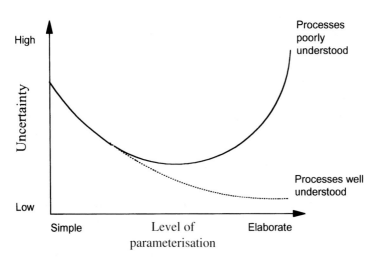

Figure 2.12. The level of uncertainty depends upon the level of detail in the model and the degree to which processes are understood (based on GESAMP, 1991).

Parametric morphological-behaviour models and simulation experiments

Parametric morphological-behaviour models

These formulations offer a way of incorporating theoretical and empirical relationships from both process and geological studies, and therefore represent a compromise between scaling up and scaling down. Such models involve identifying the effects of processes at smaller scales for which bulk parameters are established and incorporated into a gross model for large-scale processes (Terwindt & Battjes, 1991). The inclusion of details of the small-scale processes is therefore implicit only. The gross model entails a set of kinematic decision rules that are based upon concepts about morphological behaviour enshrined in geological models. This approach is therefore flexible with respect to the source of knowledge it draws upon. Furthermore, this type of simple parameterisation makes 'tuning' the model more straightforward for specific problems (GESAMP, 1991).

Examples of this LSCB modelling rationale exist for the evolution of rock coasts (Trenhaile, 1989) and beach shorelines (Hanson, 1989), as well as innershelf and coastal sand bodies (Everts, 1987; Cowell & Roy, 1988; Thorne & Swift, 1991b; Stive *et al.*, 1991; Cowell, Roy & Jones, 1991, 1992). Response of coastal sand bodies to sea-level rise can be modelled using computer simulation of LSCB based only on principles of sand-mass conservation and geometric rules for shoreface and barrier morphology (Cowell *et al.*, 1991, 1992). The geometric rules are derived in terms of bulk parameters from process studies; which may be empirical, analytical and/or numerical (de Vriend, 1991b). Stratigraphic and surface-morphology data are also used in establishing the geometric rules and to check the model's performance. The sand body translates over the pre-existing coastal substrate, which undergoes reworking as a consequence. This produces changes in position of the coastline as well as reconfiguration of the backshore and inner-continental shelf geomorphology and stratigraphy. Fig. 2.13 illustrates how such a computer-based parametric-behaviour model can deal with quite complicated, evolutionary LSCB. The pre-existing substrate (Fig. 2.13, Inset A) includes a flat-topped interstadial barrier complex, 4 km wide, sitting on a 0.3° shelf ramp, corresponding to the morphology assumed to predate the post-glacial marine transgression on the Bermagui shelf, south-east Australia. The initial position of the transgressive barrier is shown on the lower-right end of the ramp in the inset. The simulation includes slow estuarine sedimentation ($1.5\,\mathrm{mm\,a^{-1}}$) at the same time as sea level is rising. The overall outcome is displayed in Fig. 2.13 with more detailed views given for segments seaward of the interstadial barrier in Inset *B* and seaward of the modern shoreface in

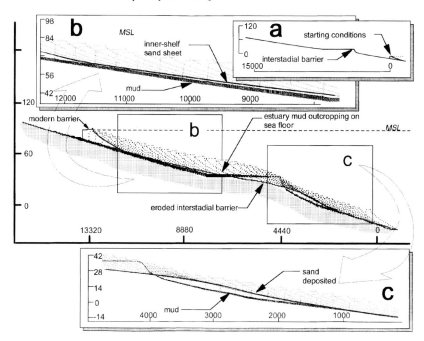

Figure 2.13. Computer simulation showing the result of marine transgression over a pre-existing substrate off the coast, 270 km south of Sydney, Australia (see text for further details). The axes are scaled in metres with the origin at the shoreline of the translating shoreface at time zero. Dotted outlines of the shoreface and transgressive barrier are shown for each of the 2 m increments in sea-level rise, totalling 90 m in all. Estuarine sedimentation is depicted as solid black shading. (From Cowell *et al.*, 1992.)

Inset *C*. The simulated stratigraphy and residual topography of the innershelf agrees well with that revealed in the chronostratigraphy interpreted from vibrocores, radiocarbon dates and seismic records (Cowell *et al.*, 1991, 1992).

A parametric-behaviour approach can also be applied to simulation and assessment of beach nourishment requirements due to sea-level rise (Stive *et al.*, 1991; Cowell & Nelson, 1991; Stive, 1992). This is illustrated by computer experiments exploring the possible responses, to prospective sea-level rise, of a semi-protected barrier beach, located within the mouth of Pittwater, 35 km north of Sydney, Australia (Cowell & Nelson, 1991). The shoreface at Mackerel Beach is heavily affected by sea grass and the alongshore sediment budget is inextricably linked to a larger flood-tide delta complex of which the barrier beach is but a component. The beach and its barrier are smaller in scale than their equivalents on the open coast because of differences in exposure to waves. Sheltered beaches are, in principle, more vulnerable to the effects of sea-level rise because they, and their dunes, are lower. Forecasts were simulated using flexible combinations of topographic dimensions based upon both native morphology and general principles (Fig. 2.14). The range of results show that the diminutive scale of such beaches not only makes them more vulnerable to sea-level rise, it also makes their defence more feasible through artificial nourishment. A total of only 28 000 cubic metres of sand is required to maintain the 600-m-long shoreline in its

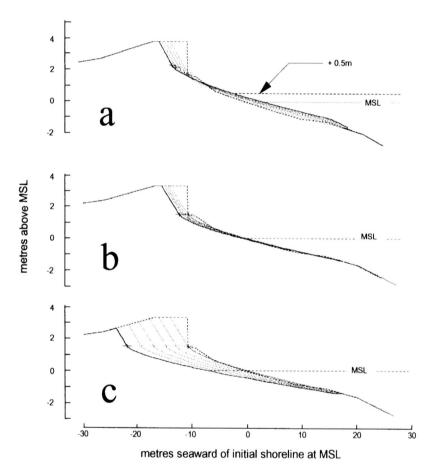

Figure 2.14. Modelled shoreline response to: (*a*) a 0.5 m rise in sea level; (*b*) a negative alongshore transport budget of 0.1% of the total transport; and (*c*) a negative alongshore transport budget of 0.5% of the total transport. The drift-related dune recession in (*b*) and (*c*) is shown over a five-year period. Shoreline morphology (dotted lines) is shown at equal time intervals.

present position as sea level rises by a total of 0.5 metres, which is expected to take at least 50 to 100 years (Fig. 2.14*a*) as a result of the greenhouse effect. More conventional processes were found to be capable of causing much worse shoreline recession. Changes to the alongshore sediment budget, due to variation in sea level, overwhelm any Bruun-type effects. The simulations show that a 0.1% disturbance to this budget can produce almost as much shoreline recession in 5 years (Fig. 2.14*b*) as greenhouse-related problems will cause in more than 50 years (Fig. 2.14*a*). A 0.5% disturbance in this budget, whilst too small to detect directly, causes a dramatic shoreline recession (Fig. 2.14*c*).

Computer simulation experiments

It is apparent from the preceding example that parametric-behaviour formulations provide a methodological rationale at the convergence between upscaling and downscaling approaches. This facilitates a direct way of incorporating multidisciplinary knowledge into development of LSCB models. More specifically, model formulation and implementation can benefit from insights and constraints imposed by principles concerning both short-term process (down to the scale of grain dynamics and turbulence) and the realities of long-term evidence in both the soft and hard rock records covering time scales ranging from the Holocene to the Cainozoic (Fig. 2.2). Observance of the theoretical and empirical constraints for validation of component models and global solutions, at both lower and higher scales, seems to make parametric-behaviour modelling a promising short-term solution to LSCB problems (de Vriend, 1992b), and perhaps even more importantly, an alternative means of experimentation into LSCB.

Forward simulation modelling is the conventional, deterministic approach of employing physical laws to predict the value of observable output variables for given values of the model parameters (Tarantola, 1987). Forward models yield a unique solution based upon processes that can be formulated as properly posed problems in the mathematical sense (Anderssen, 1992). The numerical morphodynamic models of de Vriend, Boczar-Karakiewicz and their colleagues, which are based upon process physics, are '*forward*' models. Parametric-behaviour models, such as that upon which Figs. 2.13 and 2.14 are based, can be used in *forward simulation* if the parameter values are specified deterministically, employing theoretical or empirical relationships, with the purpose of finding the resulting solution.

Inverse simulation modelling uses measurements of observable output variables to infer the value of the model parameters (Tarantola, 1987). It therefore seeks solutions to an *undetermined* problem in the mathematical sense. *Inverse modelling* thus 'involves an examination of alternatives rather than the determination of specific well-defined outcomes' (Anderssen, 1992). When undertaken using a parametric-behaviour model, *inverse simulation* entails knowing the solution in advance, and solving for the parameter values that satisfy the solution, which was the objective underlying Fig. 2.13. This type of 'what-if' modelling is more in keeping with the historical perspective on coastal evolution that pertains to the traditional techniques in geomorphology and geology described by Schumm (1991). The arguments that Schumm presents to justify the historical approach fundamentally derive from the conundrum posed due to the non-linear, Markovian properties of morphodynamics outlined above. The modelling illustrated in Figs. 2.13 and 2.14 clearly provides a way of injecting quantitative rigour into such an approach. *Inverse modelling* thus enables exploratory experimentation to yield an envelope of results rather than the single outcome that is the objective of *forward modelling* (Tarantola, 1987; Anderssen, 1992).

Inverse simulation of coastal evolution is necessary because generally the 'available data and constraints support more than one solution' (Anderssen, 1992). This again is a consequence of the stochastic nature of morphodynamics due to which: (i) future coastal evolution can never be predicted with certainty, and (ii) past evolution can be explained usually in terms of more than one Markov sequence. Therefore, *inverse* morphodynamic modelling is a 'hands-on' experimental tool for understanding morphology and stratigraphic sequences (e.g., Cowell & Roy, 1988; Cowell *et al.*, 1992). However, this type of simulation model is also of value in testing scenarios of future coastal evolution over *engineering* time scales (e.g., Cowell & Nelson, 1991). *Inverse modelling* for this purpose entails a 'trial and error methodology for the solution of problems which are *undetermined*'; i.e., a class of mathematical problems which have more than one outcome and therefore are not properly posed (Anderssen, 1992). It is used in finding combinations and permutations that give particular solutions, such as in the famous intergalactic research of the white mice, who had forgotten the question by the time their *forward model* had given them the solution; i.e., 42 (Adams, 1983). The new, *inverse* problem for the white mice became something like

$$x+y+z+a+b+c+ \dots = 42 \qquad (2.20)$$

There is no one answer; the white mice must explore a *solution space*. Similarly, modelling coastal evolution involves an experimental exploration of a *solution space*, whether for the reconstruction of depositional histories (Schumm, 1991; Cowell *et al.*, 1992) or for the prediction of LSCB.

Simulation of coastal evolutionary sequences presents almost limitless combinations of possibilities, many of which are usually trivial or unlikely scenarios. Fortunately, strong qualitative patterns are recognisable in coastal evolution, and are embodied in a wealth of descriptive models for the geomorphology and geology of coasts. *Inverse modelling* therefore can benefit from the interactive involvement of qualitative professional experience in the decision logic of simulated morphodynamic time series. The model upon which Figs. 2.13 and 2.14 are based permits the option of such intervention between each time step to invoke logic based upon professional expertise, rather than codified decision rules, in evaluation and selection of alternative solution paths. Ultimately, it may be possible to develop these procedures further through the application of neural-network techniques, which provide built-in learning and adaptation capabilities (Holland *et al.*, 1986; Goldberg, 1989).

Conclusions

Computer-based modelling has added a new dimension to morphodynamic research into coastal evolution. It has facilitated the mathematical carpentry envisaged by Longuet-Higgins in 1981. This has been achieved through numerical treatment of the strong non linearity that is inherent both in the time-dependence of coastal evolution and the all-important feedback between different parts of the morphodynamic process. Thus, the conceptual objectives for morphodynamics, put forward by Wright & Thom over one and a half decades ago, have come near to realisation at a practical level (de Vriend, 1987c). However, special difficulties remain in relation to large-scale coastal behaviour due to predictive uncertainties. These uncertainties are conferred through the inimitable morphodynamic property of *Markovian inheritance* in which the problematical, internal dependence of coastal evolution upon antecedent morphological states is compounded by stochastic environmental conditions.

Fortunately, computer-based research also offers an additional avenue in the study of coastal evolution through *inverse modelling*. This method involves morphodynamic-simulation experiments, the results of which are little different from raw data obtained by more conventional methods. The data are of limited use until they have been analysed and interpreted with some clear objective in mind. Computer-based simulation modelling therefore must be viewed as another experimental method for scientific exploration and testing. Whilst this is no real substitute for laboratory and field measurements, it does provide information and insights that cannot be obtained in other ways, or otherwise with great difficulty. ('In a computer experiment data flows like wine from a chalice. In a laboratory experiment you had to fight for every drop.' Gleick, 1988, p. 210.) For this reason, morphodynamic models are not, and probably will never be expert systems, and their application to coastal evolution very much depends upon expertise (GESAMP, 1991, p. 4).

Acknowledgments

We are grateful for helpful comments, on early drafts, from the late Bill Carter (Department of Environmental Studies, University of Ulster) and Gerd Masselink (Coastal Studies Unit, University of Sydney).

References

Adams, D. (1983). *The hitchhiker's guide to the galaxy*. In *Hitchhiker's trilogy*. New York: Harmony Books. xi, 468 pp.

Anderssen, B. (1992). Linking mathematics with applications: the comparative assessment process. *Mathematics and Computers in Simulation*, **33**, 469–76.

Atkinson, L.P. Oka, E., Wu, S.Y., Berger, T.J., Blanton, J.O. & Lee, T.N. (1989). Hydrographic variability of southern United States shelf and slope waters during the genesis of Atlantic lows experiment: winter 1986, *Journal of Geophysical Research*, **C94**, 10, 699–713.

Bailard, J.A. & Inman, D.L. (1981). An energetics total load sediment transport model for a plain sloping beach. *Journal of Geophysical Research*, **C86**, 938–54.

Baines, P.G. (1986), Internal tides, internal waves, and near-internal motions, In *Baroclinic processes on continental shelves*, Coastal & estuarine sciences monograph vol. 3, ed. C.N.K. Mooers, pp. 19–62. Washington, DC: American Geophysical Union.

Bane, J.M. (1983), Initial observations of the subsurface structure and short-term variability of the seaward deflection of the Gulf Stream off Charleston, South Carolina, *Journal of Geophysical Research*, **C88**, 4673–84.

Banks J. & Carson, J.S. (1984). *Discrete-event system simulation*. New Jersey: Prentice-Hall. 514 pp.

Bascom, W.H. (1954). Characteristics of natural beaches. In *Proceedings 4th international conference on coastal engineering*, pp. 163–80. New York: American Society of Civil Engineers.

Bendat, J.S. & Piersol, A.G. (1971). *Random data: analysis and measurement procedures*. New York: Wiley. 407 pp.

Bennett R. J. & Chorley, R. J. (1978). *Environmental systems: philosophy, analysis and control*. London: Methuen. 624 pp.

Bishop, C.T. & Donelan, M.A. (1989). Wave prediction models. In *Applications in coastal modeling,* ed. V.C. Lakhan & A.S. Trenhaile, pp. 75–105. Amsterdam: Elsevier.

Blanton, J.O., Lee, T.N., Atkinson, L.P., Bane, J.M., Riordan, A. & Raman, S. (1987). Oceanographic studies during project GALE. *EOS*, **68**, 1626–38.

Boczar-Karakiewicz, B. & Bona, J.L. (1986). Wave-dominated shelves: a model of sand-ridge formation by progressive infragravity waves. In *Shelf sands and sandstones*, ed. R.J. Knight & J.R. McLean, pp. 163–79. Canadian Society of Petroleum Geology Memoir II.

Boczar-Karakiewicz, B., Bona, J.L. & Cohen, D.L. (1987). Interaction of shallow-water waves and bottom topography, In *Dynamical problems in continuum physics*, ed. J.L Bona, C. Dafermos, J.L. Erikson & D. Dinderlelhrer, pp. 131–76. New York: Springer-Verlag.

Boczar-Karakiewicz, B., Bona, J.L. & Pelchat, B. (1991). Interaction of internal waves with the sea bed on continental shelves. *Continental Shelf Research*, **11**, 1181–97.

Boczar-Karakiewicz, B. & Davidson-Arnott, R.G.D. (1987). Nearshore bar formation by non-linear wave processes: a comparison of model results and field data. *Marine Geology*, **77**, 287–304.

Bowen, A.J. (1980). Simple models of nearshore sedimentation, beach profiles and longshore bars. In *The coastline of Canada*, ed. S.B. McCann, pp. 1–11. Geological Survey of Canada Paper, 80–10.

Bowen, A.J. & Guza, R.T. (1978) Edge waves and surf beat. *Journal of Geophysical Research*, **83**, 1913–20.

Boyd, R., Suter, J. & Penland, S. (1989). Relation of sequence stratigrahy to modern depositional environments. *Geology*, **17**, 926–29.

Braatz, B.V. & Aubrey, D.G. (1987). Recent relative sea-lvel change in eastern North America. In *Sea-level fluctuations and coastal evolution*, Special Publication No. 41, ed. D. Nummedal, H.P. Orring & J.D. Howard, pp. 29–46. Tulsa, OK: Society of Economic Paleontologists and Mineralogists.

Bretshneider, C.L. (1966). Wave generation by wind, deep and shallow water. In *Estuarine and coastal hydrodynamics*, ed. A.T. Ippen, pp. 133–96, New York: McGraw Hill.

Brink, K.H. (1987).Coastal ocean physical processes, *Review of Geohpysics*, **25**, 204–16.

Brunsden, D. & Thornes, J.B. (1979). Landscape sensitivity and change. *Transactions of the Institute of British Geographers,* new series, **4**, 463–84.

Bruun, P. (1954). *Coast erosion and the development of beach profiles.* US Army Corps of Engineers, Beach Erosion Board, Technical Memorandum No. 44, 79 pp.

Bruun, P. (1978). *Stability of tidal inlets.* Amsterdam: Elsevier. 510 pp.

Bryant, E.A. (1983). Regional sea level, Southern Oscillation and beach change, New South Wales, Australia. *Nature*, **305**, 213–16.

Cannon, G.A. & Lagerloef, G.S.E. (1983). Topographic influences on coastal circulation: a review. In *Coastal oceanography*, ed. H.G. Gade, A. Edwards & H. Svendsen, pp. 235–52. New York: Plenum.

Carstens, M.R., Nielsen, F.M. & Altinbilek, H.D. (1969). *Bedforms generated in the laboratory under oscillatory flow: analytical and experimental study.* US Army Corps of Engineers, CERC, Technical Memorandum No. 28. 39 pp.

Carter, R. W. G., Hesp, P. A. & Nordstrom, K. (1990). Geomorphology of erosional dune landscapes. In *Coastal dunes: processes and morphology.* ed. K. N. Nordstrom, N. Psuty & R.W.G. Carter, pp. 217–50. Chichester: J. Wiley & Sons.

Chappell, J. (1981). Inshore–nearshore morphodynamics: a predictive model. In *Proceedings 17th International Conference on Coastal Engineering*, pp. 963–77. New York: American Society of Civil Engineers.

Chappell, J. (1983). Thresholds and lags in geomorphological changes. *Australian Geographer*, **15**, 357–66.

Chappell, J. & Grindrod, J. (1984). Chenier plain formation in northern Australia. In *Coastal geomorphology in Australia,* ed. B.G. Thom, pp. 197–231. Sydney: Academic Press.

Chappell, J. & Polach, H. (1991). Post-glacial sea-level rise from a coral record at Huon Peninsula, Papua New Guinea. *Nature*, **349**, 147–9.

Chappell, J., Rhodes, E.G., Thom, B.G. & Wallensky, E.P. (1982). Hydro-isostasy and the sea level isobase of 5500 BP in north Queensland, Australia. *Marine Geology*, **49**, 81–90.

Chappell, J. & Thom, B.G. (1986). Coastal morphodynamics in North Australia: review and prospect. *Australian Geographical Studies*, **24**, 110–27.

Chappell, J. & Wright, L.D. (1979). Surf zone resonance and coupled morphology. In *Proceedings 16th International Conference on Coastal Engineering*, pp. 1359–77. New York: American Society of Civil Engineers.

Coates, D.R. & Vitek, J.D. (1980). Perspectives on geomorphological thresholds. In *Thresholds in geomorphology*, ed. D.R. Coates & J.D. Vitek, pp. 3–23. London: George Allen & Unwin.

Coeffe, Y. & Pechon, P. (1982). Modeling sea bed evolution under wave action. In *Proceedings 18th International Conference on Coastal Engineering*, pp. 1149–60. New York: American Society of Civil Engineers.

Cowell, P.J. (1982). *Breaker stages and surf structure on beaches*. Technical Report 82/7. Coastal Studies Unit, University of Sydney. 239 pp.

Cowell, P.J. (1986). Wave-induced sand mobility and deposition on the south Sydney inner-continental shelf. *Geological Society of Australia Special Publication*, **2**, 1–28.

Cowell, P.J. & Nelson, H. (1991). Management of beach erosion due to low swell, inlet and greenhouse effects: case study with computer modeling. In *Proceedings 10th Australasian Conference on Coastal and Ocean Engineering*, pp. 329–33.

Cowell, P.J. & Roy P.S. (1988). *Shoreface transgression model: programming guide* (Outline, assumptions and methodology). Unpubl. Report, Coastal Studies Unit, Marine Studies Centre, University of Sydney. 23 pp.

Cowell, P.J., Roy, P. S. & Jones, R. A. (1991). Shoreface translation model: application to management of coastal erosion. In *Applied Quaternary studies*, ed. G. Brierley & J. Chappell, pp. 57–73. Canberra: Biogeography & Geomorphology, Australian National University.

Cowell, P.J., Roy, P. S. & Jones, R. A. (1992). Shoreface translation model: computer simulation of coastal-sand-body response to sea level rise. *Mathematics and Computers in Simulation*, **33**, 603–8.

Cresswell, G.R., Ellyett, C., Legeckis, R. & Pearce, A.F. (1983). Nearshore features of the East Australian Current system. *Australian Journal of Marine and Freshwater Research*, **34**, 105–14.

Csanady, G.T. (1977). The coastal boundary layer. In *Estuaries, geophysics and the environment*, National Research Council, pp. 57–68. Washington DC: National Academy of Sciences.

Csanady, G.T. (1982). *Circulation in the coastal ocean*. Dortrecht: Reidel: 279 pp.

Curray, J.R. (1960). Sediments and history of Holocene transgression, continental shelf, northwest Gulf of Mexico. In *Recent sediments, Northwest Gulf of Mexico*. ed. F.P. Shepard, F. Phleger & Tj Van Andel, pp. 221–26. Tulsa, OK: American Association of Petroleum Geologists.

Curray, J.R. (1964). Transgressions and regressions, In *Papers in marine geology*, ed. R.C. Miller, pp. 175–203. New York: McMillan.

Dalrymple, R.A., ed. (1985). *Physical modeling in coastal engineering*, Rotterdam: Balkema. 276 pp.

Davies, J.L. (1958). Wave refraction and evolution of shoreline curves. *Geographical Studies*, **5**, 1–14.

Davies, J.L. (1980). *Geographical variation in coastal development*, 2nd edn. New York: Longman. 212 pp.

Davies, J.L. (1986). The coast. In *Australia: a geography*, vol. 1, ed. D.N. Jeans, pp. 203–22. Sydney: Sydney University Press.

Dean, R.G. (1977). *Equilibrium beach profiles. U.S. Atlantic and Gulf Coasts.* Department of Civil Engineering, Ocean Engineering Report 12, Newark, DE: University of Delaware. 45 pp.

Dean, R.G. (1991). Equilibrium beach profiles: characteristics and applications. *Journal of Coastal Research*, **7**, 53–84.

Deser, C. & Wallace, J.M. (1987). El Niño events and their relation to the Southern Oscillation, *Journal of Geophysical Research*, **C92**, 14 189–96.

de Vriend, H.J. (1987a). *Two and three dimensional mathematical modeling of coastal morphology*, Delft Hydraulics Communication No. 377. 37 pp.

de Vriend, H.J. (1987b). Analysis of horizontally two-dimensional morphological evolutions in shallow water. *Journal of Geophysical Research*, **C92**, 3877–93.

de Vriend, H.J. (1987c). 2DH mathematical modeling of morphological evolutions in shallow water. *Coastal Engineering*, **11**, 1–27.

de Vriend, H.J. (1990). Morphological processes in shallow seas. In *Residual currents and long-term transport*, Coastal and estuarine studies 38, ed. R.T. Cheng, pp. 276–301. New York: Springer-Verlag.

de Vriend, H.J. (1991a). Coastal morphodynamics. In *Coastal sediments '91*, pp. 356–70. New York: American Society of Civil Engineers.

de Vriend, H.J. (1991b). Modeling in marine morphodynamics. In *Computer modeling in ocean engineering '91*, ed. A.S. Arcilla, M. Pastor, O.C. Zienkiewicz & B.A. Schrefler, pp. 247–60. Rotterdam: Balkema.

de Vriend, H.J. (1992a). Mathematical modeling and large-scale coastal behaviour. Part 1. Physical processes. *Journal of Hydraulic Research, Special Issue, Maritime Hydraulics*, 727–40.

de Vriend, H.J. (1992b). Mathematical modeling and large-scale coastal behaviour. Part 2. Predictive models. *Journal of Hydraulic Research, Special Issue, Maritime Hydraulics*, 741–53.

de Vriend, H.J., Louters, T., Berben, F. & Steijn, R.C. (1989). Hybrid prediction of sandy shoal evolution in a mesotidal estuary. In *Hydraulic and environmental modeling of coastal, estuarine and river waters,* ed. R.A. Falconer, P. Goodwin & R.G.S. Matthew, pp. 145–56. Aldershot: Gower Technical.

Dolan, R., Fenster, M.S. & Holme, S.J. (1991). Temporal analysis of shoreline recession and accretion, *Journal of Coastal Research*, **7**, 723–44.

Dronkers, J. (1986). Tidal asymmetry and estuarine morphology. *Netherlands Journal of Sea Research*, **20**, 117–31.

Everts, C.H. (1987). Continental shelf evolution in response to a rise in sea level. In *Sea-level fluctuations and coastal evolution*, Special publication no. 41, ed. D. Nummedal, H.P. Orring & J.D. Howard, pp. 49–57. Tulsa, OK: Society of Economic Paleontologists and Mineralogists.

Eysink, W.D. (1991). Morphological response of tidal basins to changes. In *Proceedings 22nd International Conference on Coastal Engineering*, pp. 1948–61. New York: American Society of Civil Engineers.

Field, M.E. & Roy, P.S. (1984). Offshore transport and sand body formation: evidence from a steep, high energy shoreface, southeastern Australia. *Journal of Sedimentary Petrology*, **54**, 1292–302.

Flemming, B.W. (1988). Pseudo-tidal sedimentation in a non tidal shelf environment (southeastern African margin). In *Tide-influenced sedimentary environments and facies*, ed. P.L. Deboer, A. van der Gelde & S.D. Nio, pp. 167–80. Dordrecht: D. Reidel.

Forbes, D.L., Taylor, R.B., Orford, J.D., Carter, R.W.G. & Shaw, J. (1991). Gravel barrier migration and overstepping. *Marine Geology*, **97**, 305–13.

Frey, R. W. & Basan, P. B. (1985). Coastal salt marshes. In *Coastal sedimentary environments*, 2nd edn, ed. R.A. Davis, pp. 225–301. New York: Springer-Verlag.

GESAMP (IMO/FAO/UNESCO/WMO/WHO/IAEA/UN/UNEP Joint Group of Experts on the Scientific Aspects of Marine Pollution) (1991). *Coastal modeling*, GESAMP Reports and Studies No. 43. Vienna: International Atomic Energy Agency. 192 pp.

Gleick, J. (1988). *Chaos: making a new science*. London: Cardinal. 352 pp.

Goldberg, D.E. (1989). *Genetic algorithms in search, optimization and machine learning*. Reading, MA: Adison-Weslely. 412 pp.

Goodman, A.W. & Ratti, J.S. (1979). *Finite mathematics with applications*, 3rd edn. New York: Macmillan. 584 pp.

Grant, W.D. & Madsen, O.S. (1982). Moveable bed roughness in unsteady oscillatory flow. *Journal of Geophysical Research,* **87**, 469–81.

Griffin, D.A. & Middleton, J.H. (1992). Upwelling and internal tides over the inner New South Wales continental shelf. *Journal of Geophysical Research*, **C97**, 14 389–405.

Guza, R.T. & Bowen, A.J. (1981). On the amplitude of beach cusps. *Journal of Geophysical Research*, **C86**, 4125–32.

Guza, R.T. & Thornton, E.B. (1981). Local and shoaled comparisons of sea surface elevations pressures and velocities. *Journal of Geophysical Research*, **C85**, 1524–30.

Hamon, B.V. (1966). Continental shelf waves and effects of atmospheric pressure and wind stress on sea level. *Journal of Geophysical Research*, **71**, 2883–93.

Hands, E.B. (1983). The Great Lakes as a test model for profile response to sea level changes. In *Handbook of coastal processes and erosion*, ed. P.D. Komar, pp. 176–89. Boca Raton, FL: CRC Press.

Hanson, H. (1989). Genesis: A generalised shoreline change numerical model. *Journal of Coastal Research*, **5**, 1–27.

Haq, B.U., Hardenbol, J. & Vail, P.R. (1987). Chronology of fluctuating sea levels since the Triassic. *Science*, **235**, 1156–67.

Hardisty, J. (1987). The transport response function and relaxation time in geomorphic modeling. *Catena*, Supplement **10**, 171–9.

Hayes, M.O. (1975). Morphology of sand accumulation in estuaries. In *Estuarine research, geology and engineering*, vol. 2, ed. L.E. Cronin, pp. 3–22. New York: Academic Press.

Hays, J.D., Imbrie, J. & Shackleton, N.J. (1976). Variations in the Earth's orbit: pacemaker of the ice ages. *Science*, **194**, 2212–32.

Hesp, P. A. (1988). Surfzone, beach, and foredune interactions on the Australian south east coast. *Journal of Coastal Research*, Special Issue No. **3**, 15–25.

Hesp, P. A. & Thom, B. G. (1990). Geomorphology and evolution of erosional dunefields. In *Coastal dunes: processes and morphology*. ed. K.F. Nordstrom, N.P. Psuty & R.W.G. Carter, pp. 253–88. Chichester: J. Wiley & Sons.

Holland, J.H., Holyoak, K.J., Nisbett, R.E. & Thagard, P.R. (1986). *Induction: processes of inference, learning and discovery*. Cambridge, MA: MIT Press. 398 pp.

Holman, R.A. (1981). Infragravity energy in the surf zone, *Journal of Geophysical Research*, **C86**, 6442–540.

Hoyt, J.H. (1967). Barrier island formation. *Geological Society of America Bulletin*, **78**, 1123–36.

Hsu, S.A. (1987). Structure of airflow over sand dunes and its effect on eolian sand transport in coastal regions. In *Proceedings of coastal sediments '87*, pp. 188–201. American Society of Civil Engineers.

Hughes, M.G. & Cowell, P.J. (1986). Adjustment of reflective beaches to waves. *Journal of Coastal Research*, **3**, 153–67.

Huntley, D.A. & Bowen, A.J. (1975). Field observations of edge waves and a discussion of their effect on beach material. *Journal Geological Society of London*, **131**, 69–81.

Huntley, D.A. & Bowen, A.J. (1990). Modeling sand transport on continental shelves. In *Modeling marine systems*, ed. A.M Davies, pp. 221–54. Boca Raton, FL: CRC Press.

Huntley, D.A., Guza, R.T. & Thornton, E.B. (1981). Field observations of surf beat. I. Progressive edge waves. *Journal Geophysical Research*, **C86**, 6451–66.

Huthnance, J.M. (1982). On the formation of sand banks of finite extent. *Estuarine, Coastal and Shelf Science*, **15**, 277–99.

Huthnance, J.M., Mysak, L.A. & Wang, D.P. (1986). Coastal trapped waves. In *Baroclinic processes on continental shelves*, Coastal & estuarine sciences monograph vol. 3, ed. C.N.K. Mooers, pp. 1–18. Washington, DC: American Geophysical Union.

Huyer, A., Smith, R.L., Stabeno, P.J., Church, J.A. & White, N.J. (1988). Currents off south-east Australia: results from the Australian Coastal Experiment. *Australian Journal of Marine and Freshwater Research*, **39**, 245–88.

Inman, D.L. & Brush, B.M. (1973). The coastal challenge. *Science*, **181**, 20–32.

Inman, D.L. & Nordstrom, C.E. (1971). On the tectonic and morphological classification of coasts. *Journal of Geology*, **709**, 1–21.

Jenkins, G.M. & Watts, D.G. (1968). *Spectral analysis and its applications*. Holden-Day: San Francisco. 525 pp.

Johnson, J.A. & Rockcliff, N. (1986). Shelf break circulation processes. In *Barolcinic processes on continental shelves*, ed. N.K. Moores, pp. 33–62. American Geophysical Union.

Komar, P.D. (1973). Computer models for delta growth due to sediment input from rivers and longshore transport. *Geological Society of America Bulletin*, **84**, 2217–26.

Komar, P.D. (1976). *Beach processes and sedimentation*. Englewood Cliffs, NJ: Prentice Hall. 333 pp.

Komar, P.D. & Enfield, D.B. (1987). Short-term sea-level changes and coastal erosion. In *Sea-level fluctuations and coastal evolution*, Special publication no. 41, ed. D. Nummedal, O.H. Pilkey & J.D. Howard, pp. 17–27. Tulsa, OK: Society of Economic Paleontologists and Mineralogists.

Kotvojs, F.J. & Cowell, P.J. (1991). Refinement of the Dean profile model for beach design. *Australian Civil Engineering Transactions*, **CE33**, 9–15.

Kraft, J.C. & Chrzastowski, M.J. (1985). Coastal stratigraphic sequences. In *Coastal sedimentary environments*, 2nd edn, ed. R.A. Davis, pp. 625–63. New York: Springer-Verlag.

Kraft, J.C. & John, C.J. (1979). Lateral and vertical facies relations of transgressive barriers. *American Association of Petroleum Geologists*, **63**, 2145–63.

Lakhan, V.C.(1989). Modeling and simulation of the coastal system. In *Applications in coastal modeling*, ed. V.C. Lakhan & A.S. Trenhaile, pp. 17–41. Amsterdam: Elsevier.

Lakhan, V.C. & Trenhaile, A.S. (1989). Models and the coastal system. In *Applications in coastal modeling*, ed. V.C. Lakhan & A.S. Trenhaile, pp. 1–16. Amsterdam: Elsevier.

Leatherman, S.P. (1983). Barrier dynamics and landward migration with Holocene sea-level rise. *Nature*, **301**, 415–18.

Lee, T.N., Williams, E., Wang, J. & Evans, R. (1989). Response of South Carolina continental shelf waters to wind and Gulf Stream forcing during winter of 1986. *Journal of Geophysical Research*, **C94**, 10 715–54.

Lippmann, T.C. & Holman, R.A. (1990). The spatial and temporal variability of sand bar morphology. *Journal of Geophysical Research*, **C95**, 11 575–90.

Longuet-Higgins, M.S., (1981). The unsolved problem of breaking waves. In *Proceedings 17th International Conference on Coastal Engineering*, pp. 1–28. New York: American Society of Civil Engineers.

Mei, C.C. (1985). Resonant reflections of surface waves by periodic sand bars. *Journal of Fluid Mechanics*, **152**, 315–35.

Melton, M.A. (1958). Correlation structure and morphometric properties of drainage systems and their controlling agents. *Journal of Geology*, **66**, 442–60.

Mooers, C.N.K. (1976). Introduction to the physical oceanography and fluid dynamics of continental margins. In *Marine sediment transport and environmental management*, ed. D.J. Stanley & D.J.P. Swift, pp. 7–21. New York: Wiley.

Nichols, M. M. & Biggs, R. B. (1985). Estuaries. In *Coastal sedimentary environments*, 2nd edn, ed. R.A. Davis, pp. 77–186. New York: Springer-Verlag.

Niedoroda A.W., Swift, D.J.P. & Hopkins, T.S. (1985). The shoreface. In *Coastal sedimentary environments*, 2nd edn, ed. R.A. Davis, pp. 533–624. New York: Springer-Verlag.

Nielsen, P. (1990). Coastal bottom boundary layers and sediment transport. In *Port engineering*, 4th edn, vol. 2, ed. P. Bruun, pp. 550–85. Houston, TX: Gulf Pub. Co.

Nielsen, P. (1992). *Coastal bottom boundary layers and sediment transport.* Advanced series on ocean engineering, vol. 4, Singapore: World Scientific Pub. Co. 324 pp.

Nielsen, A.F. & Roy, P.S. (1981). Age contamination of radiocarbon dates in shell hash from coastal sand deposits: SE Australian examples. In *Proceedings 5th Conference on Coastal and Ocean Engineering*, pp. 1177–82. Institute of Engineers Australia.

Nordstrom, K. N., Psuty, N. & Carter, R.W.G., eds. (1990). *Coastal dunes: processes and morphology.* Chichester: J. Wiley & Sons. 392 pp.

Nummedal, D. & Swift, D.J.P (1987). Transgressive stratigraphy at sequence-bounding unconformities: some principles derived from Holocene and Cretaceous examples. In *Sea-level fluctuations and coastal evolution*, Special publication no. 41, ed. D. Nummedal, O.H. Pilkey & J.D. Howard, pp. 241–60. Tulsa, OK: Society of Economic Paleontologists and Mineralogists.

O'Brien, M.P. (1969). Equilibrium flow areas and inlets on sandy coasts. *Journal of Waterways, Harbours and Coastal Engineering*, **15**(WW1), 43–52.

Orford, J.D., Carter, R.W.G. & Forbes, D.L. (1991). Gravel barrier migration and sea level rise: some observations from Storey Head, Nova Scotia, Canada. *Journal of Coastal Research*, **7**, 477–88.

Penland, S., Suter, J.R. & Boyd, R. (1985). Barrier island arcs along abandoned Mississippi river deltas. *Marine Geology*, **63**, 197–223.

Pettigrew, N.R. & Murray, A.P. (1986). The coastal boundary layer and inner shelf. In *Baroclinic processes on continental shelves*, Coastal & estuarine sciences monograph vol. 3, ed. C.N.K. Mooers, pp. 95–108. Washington, DC: American Geophysical Union.

Phillips, J.D. (1992). Nonlinear dynamical systems in geomorphology: revolution or evolution. *Geomorphology*, **5**, 219–29.

Pilkey, O.H., Young, R.S., Riggs, S.R., Smith, A.W., Wu, H. & Pilkey, W.D. (1993). The concept of shoreface profile equilibrium: a critical review. *Journal of Coastal Research*, **9**, 255–78.

Pugh, D.T. (1987). *Tides, surges, and mean sea level*, Wiley. 472 pp.

Rhodes, E.G. (1982). Depositional model for a chenier plain, Gulf of Carpentaria, Australia. *Sedimentology*, **29**, 201–21.

Rine, J.M. Tillman, R.W., Culver, S.J. & Swift, D.J.P. (1991). Generation of late Holocene sand ridges on the middle continental shelf of New Jersey, USA: evidence for formation in a mid-shelf setting based on comparisons with a nearshore ridge. In *Shelf sand and sandstone bodies: geometry, facies and sequence stratigraphy*, Special publication no. 14 of the International Association of Sedimentologists, ed. D.J.P. Swift, G.F. Oertel, R.W. Tillman & J.A. Thorne, pp. 395–423. Oxford: Blackwell Scientific Publications.

Roberts, H. H., Suhayda, J.N. & Coleman, J.M. (1980). Sediment deformation and transport on low-angle slopes: Mississippi River delta. In *Thresholds in geomorphology*, ed. D.R. Coates & J.D. Vitek, pp. 131–67. London: George Allen & Unwin.

Roelvink, J.A. & Stive, M.J.F. (1991). Sand transport in the shoreface of the Holland coast. In *Proceedings 22nd International Conference on Coastal Engineering*, pp. 1909–21. New York: American Society of Civil Engineers.

Roy, P.S. (1984). New South Wales estuaries: their origin and evolution. In *Coastal geomorphology in Australia*, ed. B.G. Thom, pp. 99–121. Sydney: Academic Press.

Roy, P.S. (1986). *The geology of marine sediments on the south Sydney inner shelf, S.E. Australia*. Geological Survey Report GS1984/158, Department of Mineral Resources, NSW. 128pp.

Roy, P.S. & Thom. B.G. (1981). Late Quaternary marine deposition in New South Wales and southern Queensland: an evolutionary model. *Journal of the Geological Society of Australia*, **28**, 471–89.

Roy, P.S. & Thom, B.G. (1987). Sea-level rise: an eminent threat. In *Proceedings 8th Australasian Conference on Coastal and Ocean Engineering*, pp. 186–90. Institute of Engineers Australia.

Roy, P.S., Thom, B.G. & Wright, L.D. (1980). Holocene sequences on an embayed high-energy coast: an evolutionary model. *Sedimentary Geology*, **16**, 1–9.

Roy, P.S., Zhuang, W-Y, Birch, G.F. & Cowell, P.J. (1992). *Quaternary geology and placer mineral ptotential of the Forster–Tuncurry shelf, southeastern Australia*. Geological Survey Report: GS 1992/201, Department of Mineral Resources, NSW. 164 pp.

Schumm, S.A. (1979). Geomorphic thresholds: the concept and its applications. *Transactions of the Institute of British Geographers*, new series, **4**, 485–515.

Schumm, S.A. (1991). *To interpret the Earth: ten ways to be wrong*. Cambridge University Press. 133 pp.

Shackleton, N.J. (1987). Oxygen isotopes, ice volume and sea level. *Quaternary Science Reviews*, **6**, 183–90.

Shepard, F.P. (1950). *Beach cycles in Southern California*. US Army Corps of Engineers, Beach Erosion Board Technical Memo. no. 20. 26 pp.

Short, A.D. (1979). Wave power and beach stages: a global model. In *Proceedings 16th International Conference on Coastal Engineering*, pp. 1145–62. New York: American Society of Civil Engineers.

Short, A.D. & Hesp, P.A. (1982). Wave, beach and dune interactions in S.E. Australia. *Marine Geology*, **48**, 259–84.

Sonu, C.J. & James, W.R. (1973). A Markov model for beach profile change. *Journal of Geophysical Research*, **78**, 1462–71.

Steijn, R.C., Louters, T., van der Spek, A.J.F. & de Vriend, H.J. (1989). Numerical model hindcast of the ebb tidal delta evolution in front of the delta works. In *Hydraulic and environmental modeling of coastal, estuarine and river waters,* ed. R.A. Falconer, P. Goodwin & R.G.S. Matthew, pp. 255–64. Aldershot: Gower Technical.

Stive, M.J.F. (1987). A model for cross shore sediment transport. In *Proceedings 20th International Conference on Coastal Engineering*, pp. 1551–64. New York: American Society of Civil Engineers.

Stive, M.J.F. (1992). Sea-level and shore nourishment: a discussion. *Coastal Engineering*, **16**, 147–63.

Stive, M.J.F., Roelvink, D.J.A. & de Vriend, H.J. (1991). Large-scale coastal evolution concept. In *Proceedings 22nd International Conference on Coastal Engineering*, pp. 1962–74. New York: American Society of Civil Engineers.

Stubblefield, W.L., McGrail, D.W. & Kersey, D.G. (1984). Recognition of transgressive and post-transgressive sand ridges on the New Jersey continental shelf. In *Siliclastic shelf sediments*, Special publication no. 34, ed. R.W. Tillman & C.T. Siemers, pp. 1–23. Tulsa, OK: Society of Economic Paleontologists and Mineralogists.

Sunamura, T. (1989). Sandy beach geomorphology elucidated by laboratory modeling. In *Applications in coastal modeling,* ed. V.C. Lakhan & A.S. Trenhaile, pp. 159–213. Amsterdam: Elsevier.

Svendsen, I.A. & Buhr Hansen, J. (1978). On the deformation of long period waves over a gently sloping bottom. *Journal of Fluid Mechanics*, **87**, 433–48.

Swift, D.J.P. (1976). Coastal sedimentation. In *Marine sediment transport and environmental management*, ed. D.J. Stanley & D.J.P. Swift, pp. 255–310. New York: Wiley.

Swift, D.J.P., Ludwick, J.C. & Boehmer, W.R. (1972). Shelf sediment transport, a probability model. In *Shelf sediment transport: process and pattern*, ed. D.J.P. Swift, D.B. Duane & O.H. Pilkey, pp. 195–223. Stroudsburg: Dowden Hutchinson & Ross.

Swift D.J.P., McKinney, T.F. & Stahl, L. (1984). Recognition of transgressive and post-transgressive sand ridges on the New Jersey continental shelf: Discussion. In *Siliclastic shelf sediments*, Special publication no. 34, ed. R.W. Tillman & C.T. Siemers, pp. 25–36. Tulsa, OK: Society of Economic Paleontologists and Mineralogists.

Swift, D.J.P., Niedoroda, A.W., Vincent, C.E. & Hopkins, T.S.(1985). Barrier island evolution, Middle Atlantic Shelf, USA. Part I. Shoreface dynamics. *Marine Geology*, **63**, 331–61.

Swift, D.J.P., Phillips, S. & Thorne, J.A. (1991a). Sedimentation on continental margins. V. Parasequences. In *Shelf sand and sandstone bodies: geometry, facies and sequence stratigraphy*, Special publication no. 14 of the International Association of Sedimentologists, ed. D.J.P. Swift, G.F. Oertel, R.W. Tillman & J.A. Thorne, pp. 153–87. Oxford: Blackwell Scientific Publications.

Swift, D.J.P., Phillips, S. & Thorne, J.A. (1991b). Sedimentation on continental margins. IV. Lithofacies and depositional systems. In *Shelf sand and sandstone bodies: geometry, facies and sequence stratigraphy*, Special publication no. 14 of the International Association of Sedimentologists, ed. D.J.P. Swift, G.F. Oertel, R.W. Tillman & J.A. Thorne, pp. 89–152. Oxford: Blackwell Scientific Publications.

Swift, D.J.P. & Thorne, J.A. (1991). Sedimentation on continental margins. I. A general model for shelf sedimentation. In *Shelf sand and sandstone bodies: geometry, facies and sequence stratigraphy*, Special publication no. 14 of the

International Association of Sedimentologists, ed. D.J.P. Swift, G.F. Oertel, R.W. Tillman & J.A. Thorne, pp. 3–31. Oxford: Blackwell Scientific Publications.

Tarantola, A. (1987). *Inverse problem theory: methods for data fitting and model parameter estimation.* Amsterdam: Elsevier. 613 pp.

Terwindt, J.H.J. & Battjes, J.A. (1991). Research on large-scale coastal behaviour. In *Proceedings 22nd International Conference on Coastal Engineering*, pp. 1975–83. New York: American Society of Civil Engineers.

Thom, B. G. (1978). Coastal sand deposition in southeast Australia during the Holocene. In *Landform evolution in Australasia*, ed. J.L. Davies & M.A.G. Williams, pp. 197–214. Canberra: ANU Press.

Thom, B. G. (1984). Transgressive and regressive stratigraphies of coastal sand barriers in eastern Australia. *Marine Geology*, **7**, 161–8.

Thom, B.G., Bowman, G.M. & Roy, P.S. (1981). Late Quaternary evolution of coastal sand barriers, Port Stephens: Myall Lakes area, central New South Wales, Australia. *Quaternary Research*, **15**, 345–64.

Thom, B. G. & Hall, W. (1991). Behaviour of beach profiles during accretion and erosion dominated periods. *Earth Surface Processes and Landforms*, **16**, 113–27.

Thom, B.G. & Roy, P.S. (1985). Relative sea levels and coastal sedimentation in southeast Australia in the Holocene. *Journal of Sedimentary Petrology*, **55**, 257–64.

Thom, B.G. & Roy, P.S. (1988). Sea-level rise and climate: lessons from the Holocene. In *Greenhouse: planning for climatic change*, ed. G.I. Pearman, pp. 177–88. Melbourne: CSIRO Division of Atmospheric Research

Thom, B.G., Shepherd, M., Ly, C.K., Roy, P.S., Bowman, G.M. & Hesp, P.A. (1992). *Coastal Geomorphology and Quaternary Geology of the Port Stephens–Myall Lakes Area.* Canberra: Department of Biogeography and Geomorphology, ANU. 407 pp.

Thom, R. (1975) *Structural stability and morphogenesis.* Reading, MA: Benjamin. 384 pp.

Thorne, J.A. & Swift, D.J.P. (1991a). Sedimentation on continental margins. VI. A regime model for depositional sequences, their component system tracts, and bounding surfaces. In *Shelf sand and sandstone bodies: geometry, facies and sequence stratigraphy*, Special publication no. 14 of the International Association of Sedimentologists, ed. D.J.P. Swift, G.F. Oertel, R.W. Tillman & J.A. Thorne, pp.189–255. Oxford: Blackwell Scientific Publications.

Thorne, J.A. & Swift, D.J.P. (1991b). Sedimentation on continental margins. II. Application of the regime concept. In *Shelf sand and sandstone bodies: geometry, facies and sequence stratigraphy*, Special publication no. 14 of the International Association of Sedimentologists, ed. D.J.P. Swift, G.F. Oertel, R.W. Tillman & J.A. Thorne, pp. 33–58. Oxford: Blackwell Scientific Publications.

Trenhaile, A. S. (1989). Sea level oscillations and the development of rock coasts. In *Applications in coastal modeling*, ed. V.C. Lakhan & A.S. Trenhaile, pp. 271–95. Amsterdam: Elsevier.

Ursell, F. (1953). The long wave paradox in the theory of gravity waves. *Proceedings Cambridge Philosophical Society*, **49**, 685–92.

Vail, P.R., Audemard, F., Bowman, S.A., Eisner, P.N. & Perez-Cruz, C. (1991). The stratigraphic signatures of tectonics, eustasy and sedimentology: an overview. In *Cycles and events in stratigraphy*, ed. G. Einsele, W. Ricken & A. Seilacher, pp. 617–59. Berlin: Springer-Verlag.

van Rijn, L.C. (1990). *Principles of fluid flow and surface waves in rivers, estuaries, seas, and oceans.* Amsterdam: Aqua Publications. 335 pp.

Vellinga, P. (1986). *Beach and dune erosion during storm surges.* Delft Hydraulics Communications No. 372. 169 pp.

Vincent, C.E., Young, R.A. & Swift, D.J.P. (1981). Bedload transport under waves and currents. *Marine Geology*, **39**, 71–80.

von Bertalanffy, L. (1968). *General systems theory: foundations developments applications.* London: Allen Lane Penguin Press. 311 pp.

Waldrop, M.M. (1992). *Complexity: the emerging science at the edge of order and chaos.* Viking. 380 pp.

Wang, H. (1985). Beach profile modeling. In *Physical modeling in coastal engineering.* ed. R.A. Dalrymple, pp. 237–70. Rotterdam: Balkema.

Werner, B.T. & Fink, T.M. (1993). Beach cusps as self-organized patterns. *Science*, **260**, 968–70.

Wolman, M.G. & Gerson, R. (1978). Relative scales of time and effectiveness of climate in watershed geomorphology. *Earth Surface Processes*, **3**, 198–208.

Wolman, M.G. & Miller, J.P. (1960). Magnitude and frequency of forces in geomorphic processes. *Journal of Geology*, **68**, 54–74.

Woodroffe, C.D., Chappell, J., Thom, B.G. & Wallensky, E. (1989). Depositional model of a macrotidal estuary and floodplain, South Alligator River, Northern Australia. *Sedimentology*, **36**, 737–56.

Wright, L.D. (1976). Nearshore wave power dissipation and the coastal energy regime of the Sydney-Jervis Bay region, New South Wales: a comparison. *Australian Journal of Marine and Freshwater Research*, **27**, 633–40.

Wright, L.D. (1985). River deltas. In *Coastal sedimentary environments* (2nd edn), ed. R.A. Davis, pp. 1–76. New York: Springer-Verlag.

Wright, L.D. (1987). Shelf-surfzone coupling: diabathic shoreface transport. *Coastal Sediments '87*, 25–40.

Wright, L.D. (1993). Micromorphodynamics of the inner continental shelf: a middle Atlantic Bight case study. *Journal of Coastal Research*, SI **15**, 93–124.

Wright, L.D., Boon, J.D., Kim, S.C. & List, J.H. (1991): Modes of cross-shore sediment transport on the shoreface of the Middle Atlantic Bight. *Marine Geology*, **96**, 19–51.

Wright, L.D., Chappell, J., Thom, B.G., Bradshaw, M.P. & Cowell, P.J. (1979). Morphodynamics of reflective and dissipative beach and inshore systems: southeastern Australia. *Marine Geology*, **32**, 105–40.

Wright, L.D., Guza, R.T. & Short, A.D. (1982). Dynamics of a high-energy dissipative surf zone, *Marine Geology*, **45**, 41–62.

Wright, L.D. & Short, A.D. (1984). Morphodynamic variability of surf zones and beaches: a synthesis. *Marine Geology*, **56**, 93–118.

Wright, L.D., Short, A.D. & Green, M.O. (1985). Short-term changes in the morphodynamic states of beaches and surf zones: an empirical model. *Marine Geology*, **62**, 339–64.

Wright, L.D. & Thom, B.G. (1977). Coastal depositional landforms: a morphodynamic approach. *Progress in Physical Geography*, **1**, 412–59.

3

Deltaic coasts

J.R. SUTER

Introduction

Deltas are broadly defined as coastal accumulations of sediment extending both above and below sea level, formed where a river enters an ocean or other large body of water (Fig. 3.1) . The key element in this definition is the presence of a fluvially derived point source of sediment (Boyd, Dalrymple & Zaitlin, 1992). This chapter concentrates on the Late Quaternary evolution of marine deltas.

The term 'delta' was first applied about 450 BC by the Greek historian Herodotus, who noted the similarity of the Greek letter to the shape of the accumulation of sediment around the mouth of the Nile River. Despite the fact that many of the world's deltas do not show this particular morphology, the term has remained.

Deltas are extremely important depositional systems, both ecologically and economically. They often contain extensive wetlands, whose high biological productivity makes them vital nursery grounds for fisheries. Just as significant are the agricultural activities supported by the fertile soils of the world's deltas. Many of the world's largest ports are located on distributaries of major deltas. Huge amounts of coal, oil, and natural gas, still the principal fuels of the world economy, are derived from subsurface deltaic deposits.

The purpose of this chapter is to examine the geomorphology and sedimentology of modern deltas and provide a synthesis of the evolutionary elements. This will be done in several steps: 1. Examination of the processes responsible for deltaic deposition; 2. discussion of basic deltaic environments, 3. development of idealized, end-member models of deltaic sedimentation; 4. discussion of the evolutionary implication of these concepts. Although these will be wholly derived from examination of modern deltas, this approach can be usefully applied to ancient deltas in a sequence-stratigraphic context. Finally, the chapter examines interactions of human activities and deltaic

J.R. Suter

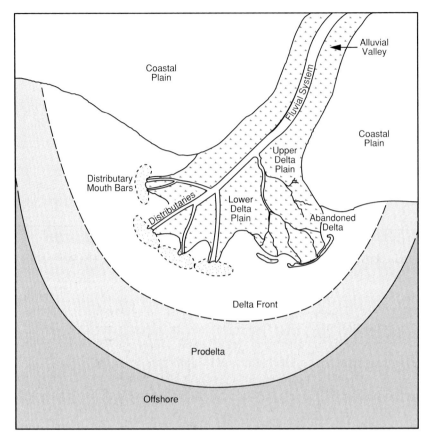

Figure 3.1. Basic environments of a fluvially dominated delta.

processes to show how an understanding of such processes can contribute to a
more enlightened approach to coastal management.

Deltaic processes

Marine deltas form where a fluvial system delivers sediment faster than
marine processes can rework it. Continued sediment supply results in
progradation of the shoreline and the characteristic protuberance of the delta
(Fig. 3.1). Deltaic sediments accumulate in three main environments: the delta
plain, dominated by fluvial processes; the delta front, reflecting river–marine
interaction, and the prodelta, which is fully marine. The expression of these
environments is the result of the interaction of many forces (Fig. 3.2). These

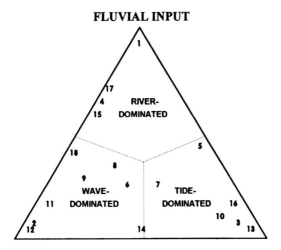

FLUVIAL INPUT

WAVE ENERGY   TIDAL ENERGY

1	MISSISSIPPI	10	COLORADO
2	BRAZOS	11	RHONE
3	GANGES-BRAHMAPUTRA	12	SÃO FRANCISCO
4	ATCHAFALAYA	13	KLANG-LAGAT
5	MAHAKAM	14	COPPER
6	NIGER	15	DANUBE
7	MEKONG	16	FLY
8	ORINOCO	17	PO
9	NILE	18	EBRO

Figure 3.2. The process-based deltaic classification. Positions of specific deltas are plotted on the basis of general delta-front morphology, which reflects the dominant processes acting upon the system (after Galloway, 1975).

include the nature of the fluvial system (drainage basin, climate, discharge, sediment calibre and load), river-mouth processes, basin morphology and tectonics, marine reworking, and evolutionary factors such as autocyclic delta switching or eustatic fluctuations.

Fluvial input

Fluvial input to a delta is largely a function of the rate and seasonality of discharge, and the amount and calibre of the sediment load. These are controlled by the drainage basin climate, tectonics, and lithology.

Climate determines the system's discharge, and affects the rate of physical and chemical weathering, governing the amount and calibre of the sediment load supplied. The seasonality of rainfall also affects the uniformity of discharge, with important implications for the type of fluvial systems delivering water and sediment to the river mouth. Climatic factors also greatly influence the deposition and preservation of organic material on the delta plain.

Sediment load and calibre are determined by the nature of the drainage basin. An active tectonic upland leads to a coarser-grained fluvial system, such as in the Ganges-Brahmaputra River; a more quiescent drainage area contributes to a finer sediment supply, as in the Mississippi River. Some aspects of deltaic deposition are a direct function of the type of sediment supplied to the marine receiving basin.

River-mouth factors

Fluvial–marine interaction at the river mouth is a primary control on deltaic deposition. As the river waters mix with those of the receiving basin, the inflow expands, diffuses and loses competence, resulting in deposition. The sedimentation pattern is influenced by the nature of the river waters, amount and calibre of the sediment load, and many basinal factors, such as water depth, bathymetry, tidal range (and thus the strength of tidal reworking),wave energy, and oceanic currents. These interactions produce three primary forces controlling deposition at the river mouth: 1. the inertia of issuing river water and associated turbulent diffusion; 2. friction between the effluent and the sea bed, and 3. buoyancy resulting from density contrasts (Fig. 3.3). Most considerations of river-mouth processes draw heavily on the work of Bates (1953). This 'rational theory' of delta formation (Fig. 3.3), was a pioneering application of hydrodynamic principles to geological problems. Wright (1977, 1985) provides excellent reviews of river-mouth processes.

Where fluvial input is large, density contrasts near the river mouth are usually small, and flow is referred to as homopycnal (Fig. 3.3*a*). The primary forces acting on the jet flow are inertial, causing the effluent to spread as an axial turbulent jet (Fig. 3.3). Coarse-grained sediment is deposited radially near the river mouth as bars with steeply dipping foreset beds, separating lower-angle topset and bottomset beds (Fig. 3.4*a*). Such deposits were first described in lacustrine settings by Gilbert (1885), and are now commonly termed Gilbert-type deltas. This behaviour is not common in marine deltas, although some homopycnal behaviour may occur during periods of high discharge.

Hyperpycnal conditions occur when the inflow is more dense than the receiving basin water (Fig. 3.3*b*). Incoming waters flow along the bottom as

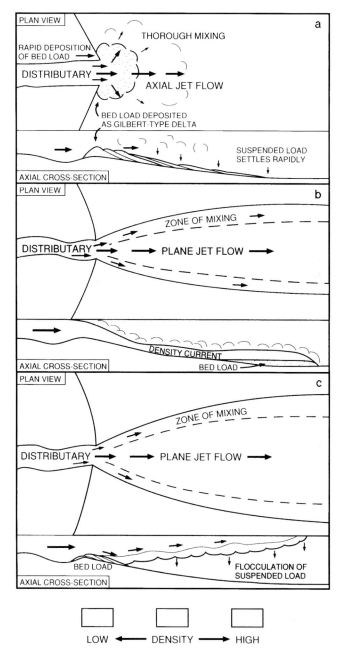

Figure 3.3. Modes of interaction between fluvial and marine waters at distributary mouths: (*a*) homopycnal flow, (*b*) hyperpycnal flow, (*c*) hypopycnal flow (after Bates, 1953, and Fisher *et al.*, 1969).

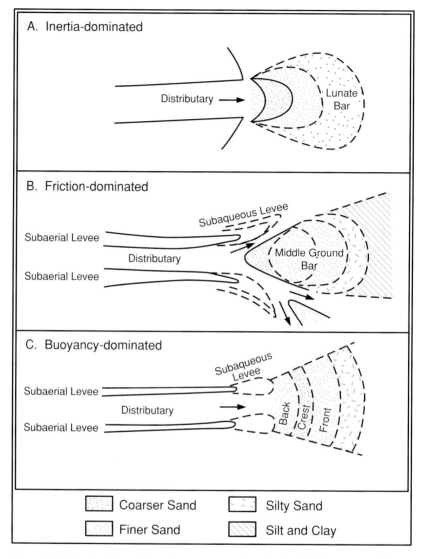

Figure 3.4. Idealized depositional patterns associated with distributary-mouth bars: (*a*) inertia-dominated, (*b*) friction-dominated, and (*c*) buoyancy-dominated (after Wright, 1977).

turbidity currents, bypassing the shoreline. Such behaviour is seen in the silt-rich Huanghe delta of China (Wright *et al.*, 1988), and is often associated with coarser-grained systems characterized by flash floods and sediment gravity flows (Orton & Reading, 1993).

Most marine deltas form under hypopycnal conditions, in which the river effluent is less dense than the saline receiving basin (Fig. 3.3c). Turbulent mixing results in deposition of the sediment load, with the coarser material being deposited near the river mouth. Finer sediments are transported further offshore and deposited from suspension as the plume disperses. When outflow is into relatively shallow water, friction between the sea bed and the base of the outflow causes deposition of a 'midground' distributary-mouth bar bordered by bifurcating channels (Fig. 3.4b). Shallower delta platforms and coarser-grained sediment loads favour this friction-dominated river-mouth style. Outflow into deeper water results in lessened bottom friction, and subsequent dominance of buoyant forces. Under these conditions, fresh water spreads as a narrow expanding plume above a salt-water wedge, which may extend inland up the distributary channel for a considerable distance. This process results in elongate distributary-mouth bars which extend a considerable distance into the basin, as shown by the Balize complex of the Mississippi delta (Figs. 3.4c, 3.5).

The above processes are idealized end members of the river-mouth spectrum. Real-world deposits display a combination of inertial, frictional, and buoyant processes, influenced by variations in discharge and characteristics of the receiving basin.

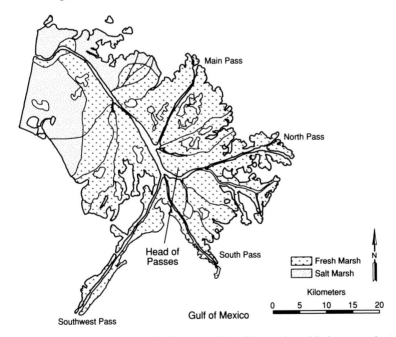

Figure 3.5. The river-dominated Balize lobe of the Plaquemines–Modern complex of the Mississippi delta, showing the elongate 'birdfoot' morphology.

Marine reworking

Once a fluvial system has delivered sediment into a marine basin, waves, tides, and oceanic currents begin to rework the sediment. Basin tectonics and morphology greatly influence marine processes. An open-ocean basin has a greater fetch, and thus the potential for greater wave energy, making a wave-dominated delta (Figs. 3.2, 3.6) more likely. An elongate, enclosed basin can

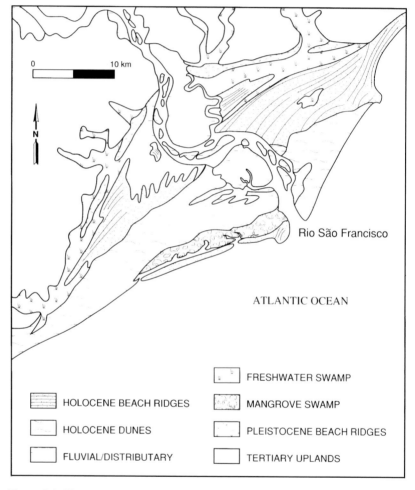

Figure 3.6. The wave-dominated mouth of the São Francisco River of Brazil. This feature is often cited as a type-example of wave-dominated deltas, although beach ridges on either side of the river may have different provenances. Despite such complications, this may still be considered a delta because of the protruding shoreline, fluvial sediment source, and coeval delta plain (after Dominguez *et al.*, 1992).

amplify tidal currents, resulting in tide-dominance (Figs. 3.2, 3.7). River-dominated deltas, such as the Mississippi (Figs. 3.2, 3.5), tend to occur in tectonically quiet basins with more limited marine energy. Rivers entering open-ocean settings are subject to the full effects of oceanic processes,

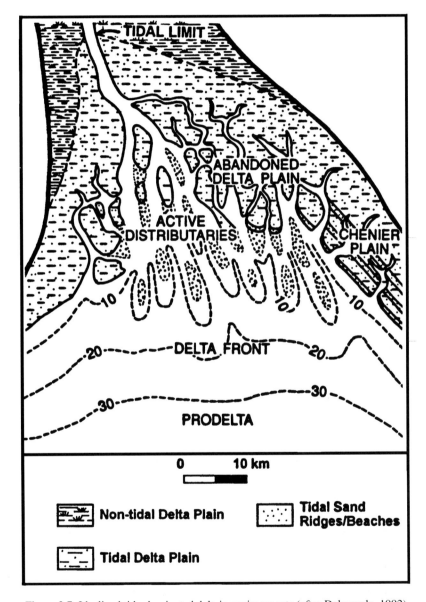

Figure 3.7. Idealized tide-dominated deltaic environments (after Dalrymple, 1992).

including waves, tides, storms, and semi-permanent ocean currents. These rework the sediment to a greater or lesser degree. The Niger delta (Allen, 1970; Oomkens, 1974) is an excellent 'mixed-energy' example. The world's largest river, the Amazon, which carries an extensive sediment load, is unable to establish a subaerial delta, although its subaqueous platform extends for several hundred kilometres offshore and alongshore (Nittrouer *et al.*, 1986).

In wave-dominated settings, breaking waves cause immediate mixing of fresh and salt water (see Chapter 4). Typically, freshwater flow velocity decelerates rapidly. A bar may form in the immediate vicinity of the distributary mouth, often supplemented by landward migrating swash bars. These bars are constantly reworked by wave action, and sediment is moved alongshore, usually being deposited as a series of beach ridges centred around distributary mouths (Fig. 3.6). Favoured directions of wave approach can result in asymmetric beach ridges, and may cause the progradation of a spit across the river mouth, resulting in channel flow oblique or parallel to shore. Finer-grained sediments are carried offshore to form the subaqueous portions of the deltaic platform, which is usually areally restricted as compared with more fluvially dominated deltas. Some well-documented examples of wave-dominated deltas include the Brazos, Nile, Senegal, Rhone, and São Francisco (Figs. 3.2, 3.6).

Tidal currents at the river mouth cause mixing of the basinal and incoming waters and thus lessen density stratification. Tidal currents may be stronger than the fluvial input, dominating sediment dispersal and moving the land–sea and river–marine interfaces horizontally and vertically (Wright, 1985). Distributary mouths of tide-dominated deltas (Figs. 3.2, 3.7) are usually funnel shaped (see Chapter 5). Under such conditions, sediments tend to be deposited as linear sand ridges, flanked by different channels for ebb and flood tides. These tidal ridges replace the normal distributary-mouth deposits. Tidal influence can reach a considerable distance up the river itself, resulting in landward transport of coarse-grained material. Finer-grained sediments are carried either offshore by ebb flows, or upriver by flood tides. These may accrete as marginal or interchannel flats. Many of the tide-dominated or tide-influenced deltas are found in tropical areas, such as the Mekong, Klang–Lagat, and Mahakam deltas of southeast Asia. These commonly have delta plains dominated by mangroves or palms in the areas of tidal influence, which grade into more typical delta plains landward (Fig. 3.7).

Prevailing oceanic currents or geostrophic storm flows may also distribute deltaic sediments. The general effect of such currents is to align deposition more alongshore (Coleman & Prior, 1982). Sediments may be transported great distances away from the river mouth, and not be considered part of the deltaic system at all. Numerous rivers, including the Mississippi, Amazon–Orinoco and Yangtze, have deposited extensive marginal chenier plains (Augustinus, 1989).

Syndepositional deformation

Sediment redistribution and modification of deltaic sediments may occur by contemporaneous mass movement. This is the result of high rates of deposition, producing elevated water contents and excess pore pressures in sediments. Biogenic gas produced by decomposing organic matter adds fluid pressure and decreases internal strength. If sufficient stress is applied by waves, currents, or sediment loading, failure may occur. Modern deltas which show such contemporaneous slumping include the Mississippi, Huanghe, Niger, Amazon, Nile, Orinoco, and Magdalena (Coleman & Prior, 1982; Lu *et al.*, 1991; Wright *et al.*, 1988). Such mass movement may take numerous forms, including small-scale mudflows and gullies, diapiric intrusions ('mudlumps'), growth faults, and shelf-edge failures (Coleman & Prior, 1982; Coleman, Prior & Lindsay, 1983). Syndepositional deformation is responsible for much downslope transport of sediments in modern deltas, and was common in Late Pleistocene lowstand deltas (Lehner, 1969; Winker & Edwards, 1983; Suter & Berryhill, 1985).

Autocyclicity and allocyclicity

Evolutionary factors affecting deltaic systems take two major forms. Autocyclic processes are those that are a part of the system itself, while allocyclic processes, such as climatic changes or eustatic fluctuations, act independently of deltaic deposition.

The classic autocyclic process is delta switching, in which an alluvial system or distributary is abandoned or avulsed for a more hydraulically favoured route. This occurs in some form in virtually all deltas, but the best known examples of this process lie in the Mississippi delta of the Gulf of Mexico (Figs. 3.5, 3.8). Here, delta switching occurs at two scales. During a period in which the main trunk of the Mississippi River occupies a particular position, a number of delta lobes prograde by a series of distributary channels, which are successively abandoned. Abandonment initiates the transgressive phase of the delta cycle (Scruton, 1960). The abandoned delta subsides, and coastal processes rework the seaward margin (see Penland, Boyd, & Suter, 1988). New lobes form nearby. These lobes stack together to form a delta complex (Frazier, 1967). Seven major deltaic complexes during

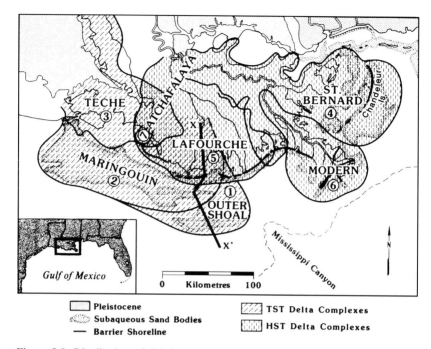

Figure 3.8. Distribution of deltaic complexes and associated transgressive deposits built during the last 7000 years in the Mississippi River delta plain. Line X–X' shows location of cross-section shown in Fig. 3.12. TST = transgressive systems tract; HST = highstand systems tract. (From Boyd *et al.*, 1989; after Frazier, 1967.)

the Holocene mark trunk stream avulsions of the Mississippi River. One is ongoing as the river tries to switch to a new course down the Atchafalaya (Fig. 3.8). Lobe switching is typical of river-dominated deltas which have a single favoured distributary path, and results in the accumulation of a large, complex delta plain (Fig. 3.8).

Delta switching may also occur by channel extension, in which several distributaries originate from a common point, taking turns at being the active conduit. Series of beach ridges oriented about the various river mouths are a typical result. The Rhone delta is an excellent example of this process (Oomkens, 1970). Avulsions far updip on the delta plain may also initiate switching, as in the Ganges–Brahmaputra delta.

Allocyclic processes dramatically affect deltaic deposition. Climatic variations may alter discharge or sediment supply as discussed above. Eustatic fluctuations also have major impacts. Falling sea level forces the deltaic system to prograde rapidly across the emerging continental shelf, and may result in entrenchment of a particular distributive network. Rising sea level creates greater accommodation space, favouring aggradation or retrogradation over progradation. In a later section it will be shown that different deltaic styles characterize each portion of a eustatic cycle.

Constituent facies: deltaic successions

Broadly, deltas can be divided into subaerial, transitional, and subaqueous components. Within these major subdivisions, individual environments occur (Fig. 3.1). The delta plain is entirely subaerial, the delta front is transitional between fluvial and marine environments, while the prodelta is entirely subaqueous. Each of these major environments can be recognized in some form in all delta types, although a host of specific differences are present which are caused by the dominant forcing factors operating on the particular system.

Delta plain

The delta plain is a mosaic of distributary channel and interchannel environments. Interchannel areas may consist of a variety of bays, lagoons, estuaries, flood plains, lakes, tidal flats and creeks, marshes, swamps, dune fields, or salt flats, depending on the type of delta and the climate. In large deltaic systems like the Mississippi, the active delta plain area is small in comparison to the area of abandoned deltaic plain, which continues to evolve under transgressive conditions (cf. Penland *et al.*, 1988; Fig. 3.9).

The delta plain is subdivided into the upper and lower delta plains. The upper delta plain is essentially a fluvial environment. Deposition is by fluvial processes, both within and between channels. Vegetation can be quite variable, but a simple distinction is that the upper delta plain is mostly a freshwater environment and usually has more diverse vegetation.

The lower delta plain is typically a brackish to saline environment and is the scene of more active deposition, mostly by crevassing or overbanking from the deltaic distributaries, modified by input of marine sediments from storms. Because the natural levées are usually more poorly developed in the lower delta plain, such crevassing is more frequent, as witness the subdeltas of the Balize complex of the Mississippi delta (Coleman & Prior, 1982). Vegetation is necessarily salt tolerant, and is confined to those plants that can withstand prolonged inundation, such as *Spartina* grasses or mangroves, depending on climate. In arid settings, the lower delta plain can be a zone of salt flats or active dune fields, especially on wave-dominated deltas with their extensive beach-ridge plains (Fig. 3.6).

The coarsest-grained deposits within delta plains are associated with

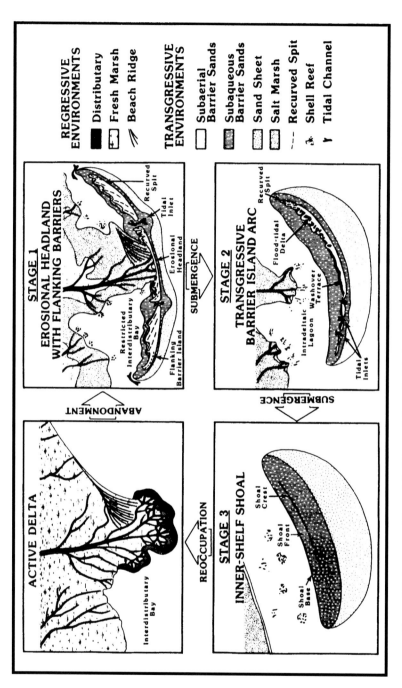

Figure 3.9. A geomorphic model of the evolution of a lobe/complex in the Mississippi delta (from Penland *et al.*, 1988).

the distributary channels. Processes within upper delta-plain channels are essentially fluvial and facies consist of fining-upward successions with erosional bases. On the lower delta plain, distributaries are more influenced by marine processes, such as tidal fluctuations and salt-water wedge intrusion (Fig. 3.3). Abandoned distributary channels may develop into estuaries, and be filled by marginal marine deposits. For example, Bayou Lafourche, an active distributary of the Mississippi River until about 300 years ago (Frazier, 1967), remains largely open and is essentially a brackish estuary (Fig. 3.8). The proportion of distributary channel fill to other deltaic facies is largely a function of the degree of fluvial dominance of the delta. Wave-dominated deltas usually have fewer distributaries than their river-dominated counterparts and a correspondingly larger proportion of interchannel and marginal marine facies. The delta plains of tide-dominated deltas are complexes of swamps, marshes, tidal flats and channels, reflecting the mix of fluvial and tidal processes (Fig. 3.7).

Another aspect of deltaic plain sedimentation is the deposition of organic lithofacies, often used as analogues for coal deposits in the rock record. Many ancient coal deposits are considered to have formed in deltaic settings (e.g., Horne *et al.*, 1978, Ryer; 1981), although this view is undergoing some re-evaluation (McCabe, 1984, 1991; McCabe *et al.*, 1989; Haszeldine, 1989; Kosters, Chmura & Bailey, 1987; Kosters, 1989; Kosters & Suter; 1993, Scott, 1989). However, organic lithofacies can form an important component of deltaic deposits in modern settings. Peats can form throughout the upper and lower delta plain. Because peat accumulation depends upon organic productivity, subsidence, and lack of clastic sediment supply, higher quality peats have typically been associated with the upper delta plain or in abandoned portions of deltaic complexes. Brackish and salt marshes on the lower delta plain tend to produce organic-rich muds or high-sulphur peats, due to admixture with salt water and frequent clastic influx. Climatic conditions are of paramount importance. Substantial differences in plant type and water table level are found from high to low latitudes and from arid to humid regions. Some peats may form during active deltaic deposition as raised mires (e.g., Coleman, Gagliano & Smith, 1970; Styan & Bustin, 1983). These deposits are elevated above levels of river flooding and clastic influx, and are not coupled to the groundwater table (Diessel, 1992).

Delta front

Most of the active deposition in a delta takes place in the transition zone from the fluvial to the marine environment, referred to as the delta front. Usually, the coarsest material is deposited at the mouths of the distributaries as bars, variously called distributary-mouth, channel-mouth, stream-mouth and middle-ground bars (Fig. 3.4; see discussion above). In river-dominated deltas, finer-grained materials travel further offshore in suspension, or by traction and gravity-currents. Progradation gives the deltaic deposit its characteristic upward-coarsening signature, in grain size, bed thickness, and scale of sedimentary structures.

Offshore of the distributary-mouth bars is the marine portion of the delta front. Wave-dominated delta fronts differ very little from standard shoreface environments, although the latter would have a more diverse microfauna and more bioturbation. In the immediate area of the distributary mouth there may be a subdued distributary-mouth bar. However, these deposits can be completely reworked by wave energy following abandonment of the distributary. Good documentation of stratigraphic successions of some wave-dominated deltas can be found in Bernard *et al.* (1970), Oomkens (1970), Dominguez & Wanless (1991), Chen, Warne, & Stanley (1992), and Stanley & Warne (1993).

The tide-dominated delta fronts are complex mazes of tidal sand ridges and channels, grading offshore into open shelf environments. Less information is available about these stratigraphic successions than for other deltaic types. Allen, Laurier & Thouvenin (1979) and Gastaldo & Huc (1992) offer some stratigraphic information about the Mahakam delta, and Meckel (1975) provides some on the Colorado delta of the Sea of Cortez. Coleman & Wright (1975) showed idealized vertical successions for tide-dominated deltas.

Prodelta

Entirely subaqueous, the prodelta is the finest-grained portion of a delta. Sediments here are deposited mostly from suspension, or from dilute turbidity current flows. In most cases, the prodelta merges imperceptibly landward into the delta front, and seaward into the open shelf environment (Fig. 3.1). Prodelta deposits are usually laterally continuous, and show less lithological variation than the delta front. The finest sediments are found at the greatest depths, and usually a coarsening-upward signature is present. Parallel laminations are the dominant sedimentary structure. Beds are thinnest at the base, thickening upward.

Relatively slow or intermittent deposition can permit marine organisms to colonize the sediments of the prodelta. These communities are most abundant on the seaward margins and in the basal sections of prodelta deposits. Landward and upward, greater stress by increasing sedimentation rates results in less bioturbation.

Deltaic models

Models are commonly used in geology as idealized simplifications of complex natural environments and processes. Models may be scale representations, mathematical constructs, visualizations, inductive, or actualistic (Reading, 1986a). Facies models are general summaries of a depositional system, which can be used to interpret ancient deposits, guide future observations, and predict facies distributions and process-responses (Reading, 1986b; Walker, 1992). Deltaic facies models have long been used in the energy and minerals industries, where decisions of great economic consequence must be made routinely with limited data. Today, such concepts are finding increasing use in hydrogeology, environmental studies and in coastal management.

Deltaic models and classifications have undergone a considerable refinement since the days in which the Balize complex of modern Mississippi delta (Fig. 3.5), the familiar 'birdfoot' delta, was used as an exploration model for virtually all deltaic deposits. As more deltas have been investigated and understanding of sedimentary processes has evolved, it has become increasingly difficult to synthesize the existing database into the simple, idealized forms alluded to above. A multitude of processes and variables interact in the deposition of deltas and a very large number of types are possible. Although every delta is different to some extent, there are common characteristics which can be used to define useful classification schemes which have some predictive capability. The challenge is to prepare models that both capture this variety and synthesize the available information into a useful form.

Deltas have been classified and modelled on the basis of their alluvial feeder system (e.g., Holmes, 1965; McPherson, Shanmugan, & Moiola, 1987), general morphology (Galloway, 1975), tectonic and physiographic settings (Ethridge & Wescott, 1984), position in a eustatic cycle (Suter & Berryhill, 1985; Suter, 1991), thickness distribution patterns (Coleman & Wright, 1975; Wright, 1985), dominant grain size (McPherson *et al.*, 1987; Orton, 1988; Orton & Reading, 1993), and numerous combinations of all of these factors (e.g., Postma, 1990). The references cited above are by no means a complete listing. Excellent recent reviews of this subject can be found in Elliot (1986, 1989), Nemec (1990), and Orton & Reading (1993). To capture the variety of deltaic systems, many of these classifications are quite detailed and complex, requiring large amounts of information for accurate 'pigeon-holing' of a particular deposit. The predictive capability of such detailed models is quite high, but applicability is limited to those deposits which share many similar characteristics and are dominated by the same forcing factors. In other words, in order to use the models as a predictor of facies or an interpretive guide, one must already know the facies distributions and forcing factors involved. If a specific detailed model is chosen improperly it can do more harm than good. Such detailed comparisons are more properly referred to as analogues, applicable to single systems rather than the more general term facies model.

Amongst sedimentologists, the most widely accepted deltaic facies models have been process based. Building upon earlier efforts (e.g., Fisher *et al.*, 1969; Morgan, 1970; Wright & Coleman, 1973), Galloway (1975) proposed a simple, elegant ternary classification which differentiated modern deltas into end members based upon delta-front morphology (Fig. 3.2). Deltas are considered to be fluvially or river dominated if they have sufficient sediment input to prograde the shoreline significantly and show little modification by marine basinal processes. If marine reworking is sufficient to redistribute sediments away from the river mouth, deltas are classified as wave or tide dominated, depending upon the morphology. This classification allows the recognition of dominant processes acting upon a given deposit, and predicts within a broad framework what the resulting style of deltaic deposition will be.

This simple scheme does have some drawbacks. Classifications dependent upon delta-front morphologies tell us little about the deltaic plain. The classification ignores the grain size or calibre of sediment input, a critical factor in determining deltaic type. Orton & Reading (1993) have modified the delta triangle into a prism, with its long axis being various dominant grain-size populations. Interested readers are referred to this paper, which contains a thorough treatment of many of the factors alluded to above. However, despite its greater complexity and useful modifications and extensions of the original idea, this approach ultimately classifies deltas as river, wave, or tide dominated.

At present there is much debate about the wave- and tide-dominated apices of the delta triangle (Bhattacharya & Walker, 1992; Dominguez, Martin & Bittencourt, 1987; Walker, 1992). Many wave-dominated 'deltas' are actually composite features formed by longshore drift of sediments from other sources. The distributary system provides some sediment input, and also acts as a barrier or 'groyne' to the longshore drift. Excellent examples of this type of depositional system are the beach-ridge plains of the Brazilian coast, such as the São Francisco delta, in which different provenances can be demonstrated for beach ridges on either side of 'distributary' channels (Dominguez *et al.*, 1987; Fig. 3.6). Numerous other deltas, e.g., the Orinoco (van Andel, 1967), the Brazos (Bernard *et al.*, 1970; Suter & Morton, 1989) and the Rhone (Oomkens, 1970) derive some of their delta-front sediments by longshore drift. These deposits might be considered as prograding strandplains (Curray, Emmel, & Crampton, 1969; Dominguez & Wanless, 1991; Boyd *et al.*, 1992).

However, similar beach-ridge plains are also seen in some lobes of the Mississippi delta, which are overall river dominated (Penland *et al.*, 1988). Strandplains are strike-fed depositional systems which are welded onto a pre-existing coastal plain, whereas a true delta has a fluvial point source and a coeval delta plain (cf. Boyd *et al.*, 1992).

Tide-dominated deltas present special problems due to their possible confusion with tide-dominated estuaries (Bhattacharya & Walker, 1992; Dalrymple, 1992; Walker, 1992; see Chapter 5). The critical point of distinction lies in the dominant sediment source, which in the estuarine case may be non-fluvial (Boyd *et al.*, 1992), and in the morphology of the shoreline. In the Bay of Fundy, perhaps the type example of a tide-dominated estuary, the updip fluvial source (the Schubenacadie and Salmon rivers) is small in comparison to the sediment being derived from erosion of the valley walls (Dalrymple, pers. commun., 1989). However, examples such as the Ganges-Brahmaputra (Wright, 1985; Umitsu, 1993) in the Indian Ocean, the Mahakam (Allen *et al.*, 1979; Gastaldo & Huc, 1992) and Mekong (Kolb & Dornbusch, 1975) deltas in the South China Sea and the Colorado (Meckel, 1975) delta in the Sea of Cortez show that it is entirely possible for the shoreline to be prograded by fluvially derived sediment, whilst showing a tidally modified or dominated morphology. Such systems are properly considered deltas.

A final shortcoming of process–response models relates to deltaic evolution. Deltas evolve throughout their existence, and reflect processes extant at a given time, which may not be representative of the entire deltaic cycle. A lobe or complex of the Mississippi delta evolves from fluvial dominance during its deposition, to wave dominance during its transgressive phases (Fig. 3.9) (Russell, 1936; Fisk, 1944; Scruton, 1960). Penland *et al.* (1988) discussed this process in detail, showing how the reworking of deltaic deposits during the abandonment phase changes the nature of even the deltaic plain. Use of models as predictive tools, either for subsurface exploration or coastal management, requires an understanding of these processes.

Deltaic style also varies throughout a eustatic cycle (Winker & Edwards, 1983; Suter & Berryhill, 1985; Elliot, 1989; Suter, 1991). Shelf-phase or shoal-water deltas, the typical modern variety, can differ dramatically from their shelf-margin, or deeper water counterparts, which are more commonly deposited during eustatic lowstands (Figs. 3.10 and 3.11). Quaternary deltaic facies models which incorporate their evolution throughout complete eustatic cycles are more apt to be applicable to the ancient record than those which concentrate solely on the current highstand version. Most ancient deltaic deposits are really composite systems which record the evolution of depositional systems through a variety of phases of basin subsidence, eustasy, and sediment supply.

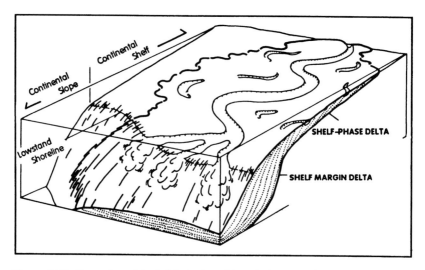

Figure 3.10. General model of deltaic deposition during eustatic fall in the Gulf of Mexico. Rates of progradation are enhanced by absolute sea-level fall, and ensuing fluvial incision removes much of the delta plain (after Suter & Berryhill, 1985). Shelf-phase delta is characterized by low accommodation, rapidly eustatically enhanced progradation, thin, widespread, low-angle clinoforms, and extensive fluvial incision. Shelf-margin delta is characterized by high accommodation and subsidence, aggradation, thick localized wedge, and steeper clinoforms.

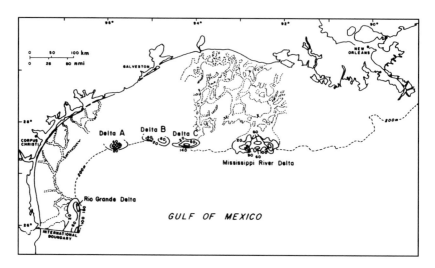

Figure 3.11. Location of Late Quaternary incised fluvial systems and associated lowstand deltaic systems in the northwestern Gulf of Mexico (after Suter & Berryhill, 1985).

Quaternary deltaic evolution

Early deltaic models were constructed largely on the basis of the subaerial geomorphology of a few well-studied deposits around the world. The advent of high-resolution seismic reflection profiling coupled with the movement of the petroleum industry into the offshore made available a new database on deltaic deposits on the continental shelf and shelf margin (Moore & Curray, 1964; Lehner, 1969; Winker & Edwards, 1983; Suter & Berryhill, 1985). These data allow the consideration of deltaic evolution during a complete eustatic cycle, which can be used to construct models in the context of the new field of sequence stratigraphy.

Sequence stratigraphy is the study of rock relations within a chronostratigraphic framework of genetically related strata, bounded by surfaces of erosion or nondeposition, or their correlative conformities (van Wagoner *et al.*, 1990). These successions are termed depositional sequences, and contain smaller genetic units. Systems tracts are linkages of contemporaneous depositional systems (Posamentier & Vail, 1988; van Wagoner *et al.*, 1990). These are defined objectively on the basis of types of bounding surface, position within a sequence, and parasequence or parasequence set stacking patterns (van Wagoner *et al.*, 1990). Lowstand, transgressive and highstand systems tracts accumulate during various phases of a eustatic cycle.

Parasequences are relatively conformable successions of genetically related beds or bedsets bounded by marine-flooding surfaces or their correlative surfaces (van Wagoner *et al.*, 1990). Flooding surfaces are boundaries between individual parasequences. Progradation and abandonment of a single delta lobe creates a parasequence (Boyd, Suter & Penland, 1989; Kosters & Suter, 1993; van Wagoner *et al.*, 1990), and related groups of deltaic lobes, i.e. a delta complex (Frazier, 1967), can be considered a parasequence set (Kosters & Suter, 1993). The stacking patterns of these parasequences and parasequence sets can be related to glacio-eustatic fluctuations.

Quaternary deltaic evolution in the northwestern Gulf of Mexico has followed a predictable if complicated pattern, controlled by relative sea level, nature of the receiving basin, and the rate of sediment influx (Suter, 1991). During a glacio-eustatic cycle, sea level fluctuates in a roughly sinusoidal fashion, divisible into falling, lowstand, rising, and highstand phases. Each of these divisions shows a particular style of deposition (Suter & Berryhill, 1985; Suter, 1991). During falling stage, deltaic systems prograde rapidly across the previously submerged shelf. Accommodation space is relatively low, and these shelf-phase deltas (Suter & Berryhill, 1985) are relatively thin and stacked progradationally (Fig. 3.10). Much of the present continental shelf is covered by these deposits. Because of eustatic fall, these deposits do not undergo the full deltaic cycle, lacking a fully developed transgressive phase. Such systems become progressively more wave dominated during the overall fall, as frictional attenuation of wave energy is lessened by the decreasing shelf width. As eustatic fall continues, fluvial systems incise into the deltas, removing much of the pre-existing deltaic sediments (Fig. 3.11). Fluvial incision and subaerial weathering processes create an erosional unconformity on the Pleistocene surface. This incised lowstand surface is a Type-1 sequence boundary in the nomenclature of sequence stratigraphy (van Wagoner *et al.*, 1990).

At eustatic lowstand, deltaic systems are fixed at the heads of the incised valleys, at or near the shelf margin. Greater accommodation space from increased water depth, higher subsidence and sea-floor gradients creates a distinctly different deltaic style (Suter & Berryhill, 1985; Berryhill & Suter, 1986). These shelf-margin deltas (Fig. 3.11) reach thicknesses of over 150 m during a single glacio-eustatic cycle, and tend to stack aggradationally, comprising the lowstand systems tract. Syndepositional deformation processes are common, and serve as a primary mechanism for downslope sedimentation. Shelf-margin deltas are a primary mechanism of basin-filling, and are preferred hydrocarbon exploration targets. Numerous petroliferous deposits in the Tertiary of the Gulf of Mexico have been interpreted as shelf-margin deltas (Edwards, 1981; Morton *et al.*, 1991).

Eustatic rise causes retrogradation into the incised valleys. Fill within these laterally restricted containers is largely controlled by sediment supply. High sediment supply systems choke their valleys with fluvial sediments, while lower sediment supply systems have more complicated fill, reflecting the creation of estuarine conditions. Thin (<10 m) lacustrine and bayhead deltas are deposited under these conditions. These deposits are characteristic of the transgressive systems tract. In the Gulf of Mexico these deposits are mostly fluvially dominated. In areas with greater tidal ranges, tide-dominated or influenced deltas may be deposited due to amplification of tidal energy by the embayments (Allen, 1991; Dalrymple, Zaitlin, & Boyd, 1992).

Once the valley has been filled, deltaic style changes. Sufficient sediment supply results in the progradation of shelf-phase deltas during stillstands in the overall rise. These backstep, or stack retrogradationally, with each successively younger delta being thinner and occurring further landward (Fig. 3.12). This reflects a diminishing sediment supply as well as relative sea-level rise. This retrogradational stacking pattern defines the transgressive systems tract (Boyd *et al.*, 1989).

Completion of a eustatic cycle results in highstand conditions similar to those now extant. Larger fluvial systems, such as the Rio Grande (Fig. 3.11),

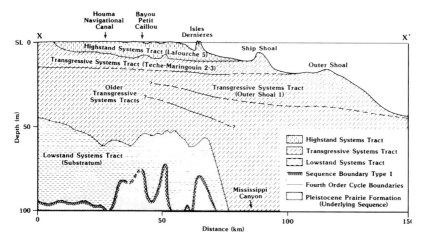

Figure 3.12. The Holocene Mississippi delta can be subdivided into a number of eustatic systems tracts. Location of cross-section shown in Fig. 3.8 (after Boyd *et al.*, 1989).

Colorado and Brazos maintain relatively small, wave-dominated shelf-phase deltas. These deltas are currently in transgressive phases. Most other streams debouch into partially filled valleys, i.e., the estuaries of the modern coast, as bayhead deltas. Such deposits are mostly fluvially dominated (Fig. 3.13). Much of the area remains in the transgressive systems tract.

The Mississippi delta itself presents a somewhat different picture. During the Wisconsinan lowstand, the Mississippi River incised a deep alluvial valley across the continental shelf (Fisk, 1944), which ultimately breached the shelf margin as a submarine canyon (Bouma & Coleman, 1985; Feeley *et al.*, 1990; Weimer, 1990). The youngest lobe of the Mississippi Fan received sediment through this canyon, largely by mass-movement processes (Coleman, Prior, & Lindsay, 1983; Goodwin & Prior, 1989).

The melting of the Wisconsinan glaciers brought relative sea level from about 125 m below present (Berryhill, 1986) to about 20 m below present by about 9000 years BP (Frazier, 1974), and brought huge volumes of meltwater and sediment down the river (Emiliani, Rooth & Stipp, 1978; Trainor & Williams, 1991; Kolla & Perlmutter, 1993). The river filled its incised valley and began depositing a number of progressively shallower shelf-phase deltas (Boyd *et al.*, 1989; Fig. 3.12). Backstepping of these deposits indicates that sediment volume supplied by the river was unable to keep pace with the overall eustatically driven transgression, and the deltas were probably developed during stillstands. As rates of relative sea-level rise increased these earlier

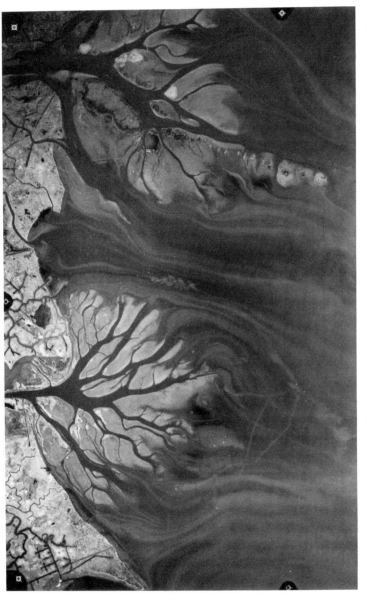

Figure 3.13. Two new bayhead deltas of the Mississippi River are forming in Atchafalaya Bay: the Atchafalaya (right) and the Wax Lake (left) lobes. The Atchafalaya, subaerial since 1973, shows substantial modification by the ship channel dredged down its main distributary. The Wax Lake lobe, formed by an artificial outlet of the Atchafalaya river, displays a classic fluvially dominated morphology (photo courtesy of O. Huh, taken December 1990).

shelf-phase deltas were themselves transgressed and reworked (Frazier, 1967, 1974; Penland *et al.*, 1988). Shelf-phase delta complexes onlap the Pleistocene Type 1 unconformity with Holocene coastal sedimentary facies, comprising the transgressive systems tract.

The culmination of eustatic rise occurred around 3–4000 years BP in the Mississippi delta and resulted in the retreat of the coastline to the mouth of the Mississippi alluvial valley. The exact timing of the eustatic stillstand is debatable and may not have begun until about 2500 years BP (Penland, Suter & McBride, 1987). The subsequent development of the Holocene Mississippi delta consists of a series of prograding delta complexes that began at the early St. Bernard shoreline and advanced more than 150 km southeast (Frazier, 1967) to the position of the currently active Balize lobe of the Modern delta. Determination of the highstand systems tract of the Mississippi is based on the change from retrogradational to progradational stacking. Three new delta complexes have formed during the highstand, the Lafourche (active 3500–400 years BP), the Plaquemines–Modern (active 1000 years BP to present), and the Atchafalaya (active c. 1950–present). The central region of the Mississippi River delta plain contains the transgressive systems tract of the Teche–Maringouin delta overlain by the highstand systems tract of the Lafourche delta. The transgressive systems tract is separated from the highstand systems tract by a regional ravinement surface (Penland *et al.*, 1987). During the highstand, sediment supply has easily outstripped regional subsidence and any minor eustatic rise. Stacking of the St. Bernard, Lafourche and Modern delta complexes has produced a progradational parasequence set, making up the highstand systems tracts still in the process of formation. The Atchafalaya complex is in the incipient phase of prograding over the older Teche–Maringouin deltas.

Each of these complexes has evolved in a similar fashion, though the process has been carried to different extents. During the initial phase of new delta creation, the incipient lobe debouches onto foundering deposits of the pre-existing delta. This primarily results in lacustrine delta deposition (cf. Tye & Coleman, 1989). Once the lakes on the upper delta plain are filled, the new bayhead deltas are deposited in the bays and estuaries created by subsidence and transgression of the abandoned delta plain. The present Atchafalaya and Wax Lake deltas (Fig. 3.13) typify this phase (Roberts *et al.*, 1980; van Heerden & Roberts, 1988). Continued progradation fills the bay and extends seaward as a shelf-phase delta, typified by the currently abandoned Lafourche complex (Fig. 3.8). Further sediment supply and progradation can then create the circumstance which exits today in which the highstand delta has prograded to the shelf margin (Figs. 3.5, 3.8).

Although the evolution of individual deltaic systems throughout the Late Quaternary differs in detail, the above model provides a reasonable general synthesis for many Late Quaternary deltas. Broadly similar evolutionary histories are shown by the Tiber (Bellotti, Tortora, & Valeri, 1989), Ebro (Farran & Maldonado, 1990), Niger (Oomkens, 1974), Nile (Chen, Warne, & Stanley, 1992; Stanley & Warne, 1993), Rhone (Tesson *et al.*, 1990), Saloum (Ausseil-Badie *et al.*, 1991), and Mahakam deltas (D. Nummedal, pers. commun., 1993).

Deltas and coastal management

Human impacts upon coastal zones have increased dramatically through the last half of the twentieth century. This is partly a result of increasing population pressure and utilization of coastal regions, and partly due to the ongoing sea-level rise These impacts are keenly felt in deltaic areas. A full discussion of the complex issues of coastal management of deltaic areas is beyond the scope of this chapter, but we can consider a few examples of how deltaic processes directly impact human affairs.

Deltas form and evolve in a process-response mode. The natural geomorphology of a given delta is the result of complex interactions of sediment supply, relative sea-level changes, and marine reworking. Human interference with any of these factors inevitably alters the form and evolution of the delta. For example, widespread forest clearing and agricultural utilization has led to dramatically increased sediment supply in some deltas. It is estimated that heavy farming has increased the sediment load of the Huanghe River of China by an order of magnitude over its early Holocene level (Milliman *et al.*, 1987). Such variations in sediment supply will obviously have dramatic impacts on deltaic development.

The Nile delta has changed markedly in the Holocene and continues to evolve rapidly. This evolution has been controlled by eustatic fluctuations, climatic variations, wave reworking, neotectonics, and human activities (Stanley, 1988). For millennia, Egyptian civilizations have depended on the Nile River floods for irrigation and nutrient supply for agriculture and for fisheries. Construction of the Aswan dams, commencing in 1902 and culminating in the closure of the High Aswan dam in 1964, has completely shut off the fluvial sediment supply to the coast. As a result, the Nile delta has now entered an abandonment phase and is essentially a wave-dominated coastal plain (Stanley & Warne, 1993; Sestini, 1989).

Stanley & Warne (1993) extrapolated coastal changes affecting the Nile delta to predict its future evolution. Current deltaic degradation will continue well into the twenty-first century, with potentially alarming consequences for the population of Egypt. Deterioration of the delta could be slowed by massive responses, such as construction of artificial wetlands, freshwater management, sewage recycling, pollution control, and emplacement of coastal protection structures (Stanley & Warne, 1993). Funding for such heroic undertakings is an open question and, at current levels of population growth, perhaps even such measures would be insufficient.

Several environmental geological problems of huge economic importance confront residents of the Mississippi delta area. European settlement began along the river in the early eighteenth century. Early settlers were aware of the dangers of flooding, but took an active approach to make the delta plain more conducive to their lifestyles, building levées and communities along the course of the main distributary. Successive floods over the next several centuries resulted in even higher and more extensive levées. A devastating flood in 1927 resulted in the completion of an extensive artificial levée system to keep the river in its banks and prevent the overflow of waters onto the communities of the delta plain.

However, through construction of the levées, the natural processes of the delta were drastically altered. Overbank flooding and sediment supply to the delta plain were eliminated. Subsidence of the abandoned lobes of the Mississippi delta is quite rapid, up to almost $2\,\mathrm{cm\,a^{-1}}$. (Penland *et al.*, 1988). Depleted sediment supply from overbanking has accelerated the long-term degradation of the deltaic plain, and the very existence of many communities is threatened. Political struggles are underway to determine the response. Initial plans focused on building artificial structures to maintain shorelines and protect communities. More recently, solutions are being proposed which attempt to work with the natural systems, ranging from small scale artificial breaches in the levée systems to wholesale diversion of the Mississippi River to a new route. These attempt to ameliorate land loss by providing new sediment to subsiding areas.

Another major coastal management problem in the Mississippi delta relates to the process of delta switching. Heavy settlement and industrial development has occurred along the Mississippi River, including the cities of Baton Rouge and New Orleans and one of the world's largest petrochemical complexes. However, the current hydraulically favoured distributary is not the Lower Mississippi itself, but the Atchafalaya River, located some 200 km to the west (Fig. 3.8). Avulsion to this more efficient course is only prevented by two massive River Control structures built by the US Army Corps of Engineers above Baton Rouge, some 300 km north of the mouth of the Mississippi River. At present, the control structures restrict flow down the Atchafalaya to 30%, funnelling the remainder to the Lower Mississippi.

Should the control structures ever prove inadequate, dire economic and human consequence would ensue, most of which probably cannot be accurately predicted or appreciated. The city of New Orleans would be deprived of its fresh water supply as the salt water wedge migrated up the then abandoned lower Mississippi River. The industrial complex located along the river would be deprived of fresh water for its operation. Land loss in the eastern delta plain would accelerate to an even greater extent, increasing the threats to coastal residents. Valuable wildlife habitat would be lost. Navigation along the current Mississippi River would suffer. Various coastal communities in the path of the 'new' distributary would require protection or relocation. Transportation and other infrastructure would have to be repaired or replaced. At the same time, renewed deltaic sedimentation and progradation in western Louisiana would create a new landscape, a process already underway (Penland & Suter, 1989).

These examples serve to illustrate the effects of deltaic processes on human affairs. As people continue to migrate towards the coastal zone, it is inevitable that further population increases and economic development will increase demands upon deltaic areas and resources. Understanding of deltaic processes is of paramount importance to minimizing deleterious impacts, and managing the use of the shorelines.

Acknowledgements

I thank Exxon Production Research for permission to publish this paper. A review by and discussions with Kevin Bohacs were very helpful. Numerous ideas contained within this paper, particularly those dealing with Quaternary sequence stratigraphy and deltaic evolution, have been developed through collaborations and discussions with Henry Berryhill, Ron Boyd, Elisabeth Kosters, Robert Morton, Dag Nummedal and Shea Penland. I also thank the editors of this book, the late Bill Carter and Colin Woodroffe, for their patience throughout the production of the manuscript.

References

Allen, G., Laurier, D. & Thouvenin, J. (1979). *Etude sedimentologique du delta de la Mahakam*. Paris: TOTAL, Compagnie Francaise des Petroles, Notes et Memoires 15. 156 pp.
Allen, G.P. (1991). Sedimentary processes and facies in the Gironde estuary: a recent model for macrotidal estuarine systems. In *Clastic tidal sedimentology*, ed. D.G. Smith, G.E. Reinson, B.A. Zaitlin & R.A. Rahmani, pp. 29–39. Canadian Society of Petroleum Geologists Memoir 16.

Allen, J. R. L. (1970). Sediments of the modern Niger delta, a summary and review. In *Deltaic sedimentation modern and ancient*, ed. J. P. Morgan, pp. 138–51. Society of Economic Paleontologists and Mineralogists Special Publication 15.

Augustinus, P.G.E.G. (1989). Cheniers and chenier plains: a general introduction. *Marine Geology*, **90**, 219–30.

Ausseil-Badie, J., Barusseau, J. , Descamps, C., Salif Diop, E. H., Giresse, P. & Pazdur, M. (1991). Holocene deltaic sequence in the Saloum Estuary, Senegal. *Quaternary Research*, **36**, 178–94.

Bates, C. C. (1953). Rational theory of delta formation. *American Association of Petroleum Geologists Bulletin*, **37**, 2119–62.

Bellotti, P., Tortora, P. & Valeri, P. (1989). Sedimentological and morphological features of the Tiber Delta. *Sedimentology*, **36**, supplement.

Bernard, H. A., Major, C. F., Parrott, B. S. & LeBlanc, R. J., Sr. (1970). *Recent sediments of Southeast Texas*. Texas Bureau of Economic Geology Guidebook, 11.

Berryhill, H. L. (1986). Sea level lowstand features off southwest Louisiana. In *Late Quaternary facies and structure, northern Gulf of Mexico*, ed. H. L. Berryhill, pp. 225–40. American Association of Petroleum Geologists Studies in Geology 23.

Berryhill, H. L. & Suter, J. R. (1986). Deltas. In *Late Quaternary facies and structure, northern Gulf of Mexico*, ed. H. L. Berryhill, American Association of Petroleum Geologists Studies in Geology 23.

Bhattacharya, J. & Walker, R.G. (1992). Deltas. In *Facies models: response to sea level change*, ed. R. G. Walker & N. P. James, pp. 157–77. Geological Association of Canada.

Boyd, R.L., Suter, J.R. & Penland, S. (1989). Relationship of sequence stratigraphy to modern sedimentary environments. *Geology*, **17**, 926–9.

Boyd, R. L., Dalrymple, R. W. & Zaitlin, B. A. (1992). Classification of clastic coastal depositional environments. *Sedimentary Geology*, **80**, 139–50.

Bouma, A. H. & Coleman, J. M. (1985). Mississippi Fan: Leg 96 program and principal results. In *Submarine fans and related turbidite systems*, ed. A. H. Bouma, W. R. Normark, & N. E. Barnes, pp. 247–52. New York: Springer-Verlag.

Chen, Z., Warne, A. & Stanley, D. J. (1992). Late Quaternary evolution of the northwestern Nile delta between the Rosetta Promontory and Alexandria, Egypt: *Journal of Coastal Research*, **8**, 527–61.

Coleman, J. M., Gagliano, S. M. & Smith, W. G. (1970). Sedimentation in a Malaysian high-tide tropical delta. In *Deltaic sedimentation: modern & ancient*, ed. J. P. Morgan & R. H. Shaver, pp. 185–97. Society of Economic Paleontologists and Mineralogists Special Publication 15.

Coleman, J. M. & Prior, D. B. (1982). Deltaic environments. In *Sandstone depositional environments*, ed. P. A. Scholle, & D. Spearing, pp. 129–78. American Association of Petroleum Geologists.

Coleman, J. M., Prior, D. B. & Lindsay, J. F. (1983). Deltaic influences on shelf edge instability processes. In *The shelfbreak: critical interface on continental margins*, ed. D. J. Stanley & G. T. Moore, pp. 121–37. Society of Economic Paleontologists and Mineralogists Special Publication 33.

Coleman, J. M. & Wright, L. D. (1975). Modern river deltas: variability of processes and sand bodies. In *Deltas: models for exploration*, ed. M. L. Broussard, pp. 87–98. Houston, TX: Houston Geological Society.

Curray, J. R., Emmel, F. J. & Crampton, P. J. S. (1969). Holocene history of a strandplain, lagoonal coast, Nayarit, Mexico. In *Coastal lagoons: a symposium*, ed. A. A. Castanares & F. B. Phleger, pp. 63–100. Mexico: Universidad National Autonoma.

Dalrymple, R.W. (1992). Tidal depositional systems. In *Facies models: response to sea level change*, ed. R. G. Walker & N. P. James, pp. 195–218. Geological Association of Canada.

Dalrymple, R.W., Zaitlin, B.A. & Boyd, R. (1992). Estuarine facies models: conceptual basis and stratigraphic implications, *Journal of Sedimentary Petrology*, **62**, 1030–43.

Diessel, C.F.K. (1992). *Coal-bearing depositional systems*. New York: Springer-Verlag. 721 pp.

Dominguez, J. M. L., Bittencourt, A. C. S. P. & Martin, L. (1992). Controls on Quaternary coastal evolution of the east-northeastern coast of Brazil: roles of sea-level history, trade winds and climate. *Sedimentary Geology*, **80**, 213–32.

Dominguez, J. M. L., Martin, L. & Bittencourt, A. C. S. (1987). Sea-level history and Quaternary evolution of river mouth-associated beach-ridge plains along the east-southeast Brazilian coast: a summary. In *Sea-level fluctuations & coastal evolution*, ed. D. Nummedal, O. H. Pilkey & J. D. Howard, pp. 115–27. Society of Economic Paleontologists and Mineralogists Special Publication 41.

Dominguez, J. M. L. & Wanless, H. R. (1991). Facies architecture of a falling sea-level strandplain, Doce River coast, Brazil. In *Shelf sand and sandstone bodies: facies and sequence stratigraphy*, ed. D. J. P. Swift, G. F. Oertel, R. W. Tillman & J. A. Thorne, pp. 259–81. International Association of Sedimentologists Special Publication 14.

Edwards, M. B. (1981). Upper Wilcox Rosita delta system of south Texas: growth-faulted shelf edge deltas, *American Association of Petroleum Geologists Bulletin*, **65**, 1574–85.

Emiliani, C., Rooth, C. & Stipp, J.J. (1978). The late Wisconsinan flood into the Gulf of Mexico. *Earth and Planetary Science Letters*, **41**, 159–62.

Ethridge, F. G. & Wescott, W. A. (1984). Tectonic setting, recognition and hydrocarbon reservoir potential of fan-delta deposits. In *Sedimentology of gravels and conglomerates*, ed. E. H. Koster & R. J. Steel, pp. 217–35. Memoir Canadian Society of Petroleum Geologists 10.

Elliot, T. (1986). Deltas. In *Sedimentary environments and facies*, ed. H. G. Reading, pp. 113–54. Oxford: Blackwell Scientific Publications.

Elliot, T. (1989). Deltaic systems and their contribution to an understanding of basin-fill successions. In *Deltas: sites and traps for fossil fuels*, ed. M. K. G. Whateley & K. T. Pickering, pp. 3–10. Oxford, Blackwell Scientific Publications. Geological Society Special Publication 41.

Farran, M. & Maldonado, A. (1990). The Ebro continental shelf: Quaternary seismic stratigraphy and growth patterns. *Marine Geology*, **95**, 289–312.

Feeley, M. H., Moore, T.C., Jr., Loutit, T.S. & Bryant, W.R. (1990). Sequence stratigraphy of Mississippi Fan related to oxygen isotope sea level index. *American Association of Petroleum Geologists*, **74**, 407–24.

Fisher, W. L., Brown, L. F., Jr., Scott, A. J. & McGowen, J. H. (1969). *Delta systems in the exploration for oil and gas: a research colloquium*, The University of Texas at Austin, Bureau of Economic Geology. 212 pp.

Fisk, H. N. (1944). *Geological investigation of the alluvial valley of the lower Mississippi River*. Vicksburg: Mississippi River commission. 78 pp.

Frazier, D. E. (1967). Recent deltaic deposits of the Mississippi delta: their development and chronology. *Gulf Coast Association of Geological Societies, Transactions*, **17**, 287–315.

Frazier, D. E. (1974). Depositional episodes: their relationship to the *Quaternary stratigraphic framework in the northwestern portion of the Gulf basin*, Austin, Texas, Bureau of Economic Geology, Geologic Circular 74–1. 28 pp.

Galloway, W.E. (1975). Process framework for describing the morphologic & stratigraphic evolution of deltaic depositional systems. In *Deltas, models for exploration*, ed. M. L. Broussard, pp. 87–98. Houston, TX: Houston Geological Society.

Gastaldo, R. A. & Huc, A. (1992). Sediment facies, depositional environments, and distribution of phytoclasts in the Recent Mahakam River delta, Kalimantan, Indonesia. *Paliaos*, **7**, 574–90.

Gilbert, G.K. (1885). The topographic features of lake shores. *United States Geological Survey, Annual Report*, **5**, 69–123.

Goodwin, R. & Prior, D. B. (1989). Geometry and depositional sequences of the Mississippi Canyon, Gulf of Mexico. *Journal of Sedimentary Petrology*, **59**, 318–29.

Haszeldine, R. S. (1989). Coal reviewed: depositional controls, modern analogs, and ancient climates. In *Deltas: sites and traps for fossil fuels*, ed. M. K. G. Whateley & K. T. Pickering, pp. 3–10. Oxford: Blackwell Scientific Publications. Geological Society Special Publication 41.

Holmes, A. (1965). *Principles of physical geology*, 2nd edn, London: Thomas Nelson and Sons. 1288 pp.

Horne, J. C., Ferm, J. C., Caruccio, F. T. & Baganz, B. P. (1978). Depositional models in coal exploration and mine planning in Appalachian region. *American Association of Petroleum Geologists Bulletin*, **62**, 2379–411.

Kolb, C. R. & Dornbusch, W. K. (1975). The Mississippi and Mekong Deltas: a comparison. In *Deltas, models for exploration*, ed. M. L. Broussard, pp. 193–208. Houston, TX: Houston Geological Society.

Kolla, V. & Perlmutter, M. A. (1993). Timing of turbidite sedimentation on the Mississippi Fan. *American Association of Petroleum Geologists Bulletin*, **77**, 1129–41.

Kosters, E. C. (1989). Organic-clastic facies relationships and chronostratigraphy of the Barataria interlobe basin, Mississippi delta plain. *Journal of Sedimentary Petrology*, **59**, 98–113.

Kosters, E. C., Chmura, G. L. & Bailey, A. (1987). Sedimentary and botanical factors influencing peat accumulation in the Mississippi Delta. *Journal of Sedimentary Petrology*, **144**, 423–34.

Kosters, E. C. & Suter, J. R. (1993). Facies relationships and systems tracts in the Late Holocene Mississippi delta plain. *Journal of Sedimentary Petrology*, **63**, 727–33.

Lehner, P. (1969). Salt tectonics and Pleistocene stratigraphy on continental slope of northern Gulf of Mexico. *American Association of Petroleum Geologists Bulletin*, **53**, 2431–79.

Lu, Z., Suhayda, J., Prior, D. B., Bornhold, B. D., Keller, G. H., Wiseman, W. J., Wright, L. D. & Yang, Z. S. (1991). Sediment thixotropy and submarine mass movement, Huanghe delta, China. *Geomarine Letters*, **11**, 9–15.

McCabe, P. J. (1984). Depositional environments of coal and coal-bearing sequences. In *Sedimentology of coal and coal bearing sequences*, ed. R. A. Rahmani, & R. M. Flores, pp. 13–42. International Association of Sedimentologists Special Publication 7.

McCabe, P.J. (1991). Geology of coal: environments of deposition. In *Economic geology, U.S., The geology of North America volume P-2,* ed. H. J. Gluskoter, D. D. Rice, & R. B. Taylor, pp. 469–82. Geological Society of America.

McCabe, P. J., Breyer, J. A., Kosters, E. C., Chmura, G. L. & Bailey, A. (1989). Discussion of sedimentary and botanical factors influencing peat accumulation in the Mississippi Delta. *Geological Society of London Journal*, **146**, 877–80.

McPherson, J. G., Shanmugan, G. & Moiola, R. J. (1987). Fan-deltas and braid deltas: varieties of coarse-grained deltas. *Geological Society of America Bulletin*, **99**, 331–40.

Meckel, L. D. (1975). Holocene sand bodies in the Colorado delta, Salton sea, Imperial County, California. In *Deltas, models for exploration*, ed. M. L. Broussard, pp. 239–66. Houston, TX: Houston Geological Society.

Milliman, J. D., Qin Y., Ren, M. & Saito, Y. (1987). Man's influence on the erosion and transport of sediment by Asian rivers: the Yellow River (Huanghe) example. *Journal of Geology*, **95**, 751–62.

Moore, D. G. & Curray, J. R. (1964). Sedimentary framework of the drowned Pleistocene delta of Rio Grande de Santiago, Nayarit, Mexico. In *Developments in sedimentology, I, Deltaic and shallow marine deposits*, ed. L. M. J. U. van Straaten, pp. 277–81. Amsterdam: Elsevier.

Morgan, J. P. (1970), Depositional processes and products in the delta environment, In *Deltaic sedimentation modern & ancient*, ed. J. P. Morgan, pp. 31–47. Society of Economic Paleontologists and Mineralogists Special Publication 15.

Morton, R. A., Sams, R. H. & Jirik, L. A. (1991). Plio-Pleistocene depositional sequences of the southeastern Texas continental shelf: geologic framework, sedimentary facies and hydrocarbon distribution. *Texas Bureau of Economic Geology Report of Investigations* 200. 90 pp.

Nemec, W. (1990). Deltas: remarks on terminology and classification. In *Coarse-grained deltas*, ed. A. Colella & D. B. Prior, pp. 3–12. International Association of Sedimentologists Special Publication 10.

Nittrouer, C. A., Kuehl, S.A., DeMaster, D.J. & Kowsmann, R.O. (1986). The deltaic nature of Amazon shelf sedimentation. *Geological Society of America Bulletin*, **97**, 444–58.

Oomkens, E. (1970). Depositional sequences and sand distribution in the postglacial Rhone delta complex. In *Deltaic sedimentation modern & ancient*, ed. J. P. Morgan, pp. 198–212. Society of Economic Paleontologists & Mineralogists Special Publication 15.

Oomkens, E. (1974). Lithofacies relations in the Late Quaternary Niger Delta complex. *Sedimentology*, **21**, 195–221.

Orton, G. J. (1988). A spectrum of Middle Ordovician fan deltas and braidplain deltas, North Wales: a consequence of varying fluvial clastic input. In *Fan deltas: sedimentology and tectonic settings*, ed. W. Nemec & R. J. Steel, pp. 3–13. London: Blackie & Son.

Orton, G. J. & Reading, H. G. (1993). Variability of deltaic processes in terms of sediment supply, with particular emphasis on grain size. *Sedimentology*, **40**, 475–512.

Penland, S., Boyd, R.L & Suter, J. R. (1988). Transgressive depositional systems of the Mississippi delta plain: a model for barrier shoreline and shelf sand development. *Journal of Sedimentary Petrology*, **58**, 932–49.

Penland, S. & Suter, J. R. (1989). The geomorphology of the Mississippi River Chenier Plain. *Marine Geology*, **90**, 231–58.

Penland, S., Suter, J. R. & McBride, R. A. (1987). Delta plain development and sea level history in the Terrebonne coastal region, Louisiana. In *Coastal sediments '87*, pp. 1689–705. New Orleans, LA: Water Ways Division, American Society of Coastal Engineers.

Posamentier, H.W. & Vail, P.R. (1988). Eustatic controls on clastic deposition. II. Sequence and systems tract models. In *Sea-level changes: an integrated approach.* ed. C.K. Wilgus, B.S. Hastings, C.G. St C. Kendall, H.W. Posamentier, C.A. Ross & J.V. van Wagoner, pp. 125–154. Society of Economic Paleontologists and Mineralogists, Special Publication 42.

Postma, G. (1990). Depositional architecture and facies of river and fan deltas: a synthesis. In *Coarse-grained deltas*, ed. A. Colella, & D. B. Prior, pp. 13–27. International Association of Sedimentologists Special Publication 10.

Reading, H. G. (1986a). Facies. In *Sedimentary environments and facies* (2nd edn), ed. H. G. Reading, pp. 4–19. London: Blackwell Scientific Publications.

Reading, H. G., ed. (1986b). *Sedimentary Environment and Facies* (2nd edn). London: Blackwell Scientific Publications. 615 pp.

Roberts, H. H., Adams, R. D. & Cunningham, R. H. W. (1980). Evolution of sand-dominant subaerial phase, Atchafalaya delta, Louisiana. *American Association of Petroleum Geologists Bulletin*, **64**, 264–79.

Russell, R. J. (1936). Physiography of lower Mississippi River delta. In *Reports on the geology of Plaquemine and St. Bernard Parishes, Louisiana.* pp. 3–199. Louisiana Department of Conservation, Geological Bulletin 8.

Ryer, T.A. (1981). Deltaic coals of the Ferron sandstone member of Mancos shale: predictive model for Cretaceous coal-bearing strata of Western Interior *American Association of Petroleum Geologists Bulletin*, **65**, 2223–340.

Scruton, C. (1960). Delta building and the deltaic sequence, In *Recent sediments, Northwest Gulf of Mexico, 1951–1958*, ed. F. Shepard *et al.*, pp. 82–102. American Association of Petroleum Geologists Special Publication.

Scott, A. C. (1989). Deltaic coals: an ecological and palaeobotanical perspective. In *Deltas: sites and traps for fossil fuels*, ed. M. K. G. Whateley & K. T. Pickering, pp. 3–10. Oxford: Blackwell Scientific Publications. Geological Society Special Publication 41.

Sestini, G. (1989). Nile delta: a review of depositional environments and geological history. In *Deltas: sites & traps for fossil fuels*, ed. M.K.G. Whateley & K.T. Pickering, pp 99–127. Oxford Blackwell Scientific Publications, Geological Society Special Publication 41.

Stanley, D.J. (1988). Subsidence in the northeastern Nile delta: rapid rates, possible causes and consequences. *Science*, **240**, 497–500.

Stanley, D.J. & Warne, A.G. (1993). Nile delta: recent geological evolution and human impact. *Science*, **260**, 628–34.

Styan, W.B. & Bustin, R.M. (1983). Sedimentology of Fraser River Delta peat deposits: a modern analogue for some deltaic coals. *International Journal of Coal Geology*, **3**, 101–43.

Suter, J. R. (1991). Late Quaternary chronostratigraphic framework, northern Gulf of Mexico. In 12th Annual Research Conference Proceedings. Gulf Coast Section, Society of Economic Paleontologists and Mineralogists.

Suter, J. R. & Berryhill, H.L. (1985). Late Quaternary shelf-margin deltas, northwest Gulf of Mexico. *American Association of Petroleum Geologists Bulletin*, **69**, 77–91.

Suter, J. R. & Morton, R. A. (1989). Coastal depositional systems, northwest Gulf of Mexico. *28th International Geological Congress Field Trip Guidebook T370.* 52 pp.

Tesson, M., Gensous, B., Allen, G.P. & Ravenne, C. (1990). Late Quaternary deltaic lowstand wedges on the Rhone continental shelf, France. *Marine Geology*, **91**, 325–32.

J.R. Suter

Trainor, D. M. & Williams, D. F. (1991). Quantitative analysis and correlation of oxygen isotope records from planktonic and benthonic foraminifera and well log records from OCS Well G1267 No. A-1, South Timbalier Block 198, north-central Gulf of Mexico. *Gulf Coast Section–SEPM 12th Annual Research Conference Proceedings*, pp. 363–77.

Tye, R. S. & Coleman, J. M. (1989). Depositional processes and stratigraphy of fluvially dominated lacustrine deltas: Mississippi delta plain. *Journal of Sedimentary Petrology*, **59**, 973–96.

Umitsu, M. (1993). Late Quatenary sedimentary environments and landforms in the Ganges Delta. *Sedimentary Geology*, **83**, 177–86.

van Andel, T.J. (1967). The Orinoco delta, *Journal of Sedimentary Petrology*, **37**, 297–310.

van Heerden, I. L. & Roberts, H. H. (1988). Facies development of Atchafalaya Delta, Louisiana: a modern bayhead delta. *American Association of Petroleum Geologists Bulletin*, **72**, 439–53.

van Wagoner, J.C., Mitchum, R.M., Campion, K.M. & Rahmanian, V.D. (1990). *Siliciclastic sequence stratigraphy in well-logs, cores, and outcrops.* American Association of Petroleum Geologists, Methods in Exploration Series **7**, 55 pp.

Walker, R. G. (1992). Facies, facies models, and modern stratigraphic concepts. In *Facies models: responses to sea level change*, ed. R. G. Walker & N. P. James, pp. 1–14. Geological Association of Canada.

Weimer, P. (1990). Sequence stratigraphy, facies geometries, and depositional history of the Mississippi Fan, Gulf of Mexico. *American Association of Petroleum Geologists Bulletin*, **74**, 425–53.

Winker, C. D. & Edwards, M. B. (1983). Unstable progradational clastic shelf margins. In *The shelfbreak: critical interface on continental margins*, ed. D. J. Stanley, & G. T. Moore, pp. 139–57. Society of Economic Paleontologists and Mineralogists Special Publication 33.

Wright, L. D. (1977). Sediment transport and deposition at river mouths: a synthesis. *Geological Society of America Bulletin*, **88**, 857–68.

Wright, L. D. (1985). River deltas. In *Coastal sedimentary environments*, ed. R. A. Davis, Jr., pp. 1–76. New York: Springer-Verlag.

Wright, L. D., Wiseman, W. J., Bornhold, B. D., Prior, D. B., Suhayda, J., Keller, G. H., Yang, Z. S. & Fan, Y. B. (1988). Marine dispersal and deposition of Yellow River silts by gravity-driven underflows. *Nature*, **332**, 629–32.

4

Wave-dominated coasts

P.S. ROY, P.J. COWELL, M.A. FERLAND AND B.G. THOM

Introduction

Wave-dominated sedimentary coasts comprise accumulations of detrital sand and gravel-sized material which undergo high levels of physical reworking, interspersed with periods of burial before finally being deposited as the coastal deposits we see today (Davis & Hayes, 1984). Quite commonly sediments tend to be of clean sand and gravel, often quite well sorted and abraded, containing relatively high proportions of more resistant minerals and rock types such as quartz, chert and heavy minerals. Waves and wave-induced currents are the dominant mechanisms for moving and depositing sand on shorefaces and beaches of the open coast, although winds, river discharge, tidal currents and Ekman flows variously act as transporting agents landward of the beach, in estuaries and seaward of the shoreface. In relation to the shoreface and beach, open coastal types are determined by four factors: (i) substrate gradient, (ii) wave energy versus tidal range; (iii) sediment supply versus accommodation volume (Swift & Thorne, 1991); and (iv) rates of sea-level change. At one extreme are steep, high-energy, sediment-deficient coasts that have bedrock cropping out as headlands, with negligible sand at their base and relatively deep water offshore (autochthonous, accommodation-dominated coast of Swift & Thorne, 1991). At the other extreme are low-gradient, low-energy coasts that are typically muddy with a coastal fringe of wetland vegetation. Here, incident wave action is dissipated over very shallow offshore gradients such as those associated with deltaic environments at river mouths (allochthonous, sediment-supply dominated coast of Swift & Thorne, 1991; see Chapter 3). But even here, rare high-energy events such as cyclones can cause episodes of wave reworking leading to the formation of cheniers.

The objectives of this chapter are to examine the evolution of wave-dominated coasts through the Late Quaternary, especially the Holocene, and to demonstrate how present coastal geomorphologies and stratigraphies are predicated upon this evolution. The focus therefore is on principles involving the four controls enumerated above, and on how, as indicated by Pilkey *et al.* (1993), geological inheritance plays a much more important role than suggested in some previous reviews. Nonetheless, reviews such as Swift, Phillips & Thorne (1991a) provide a comprehensive basis for the concepts presented here, which are drawn from case studies in southeast Australia (24.5° to 38° S). The variety of substrate slopes that have existed on this shelf and coast at different Quaternary sea levels furnish backgrounds against which different types of wave-dominated coastal development can be evaluated. The southeast Australian shelf presents a substrate ranging from very low angle (0.1–0.2°) on the outer shelf to relatively steep (>1.0°) immediately seaward of the present shoreline (Wright, 1976), with very steep slopes (>5°) occurring locally. In contrast, gradients on other well-studied coasts, such as those of the US Atlantic and Gulf of Mexico, are very low angle (<0.1°), immediately seaward of the present coast. The natural variability of the southeast Australian coast and shelf therefore affords valuable experimental control. At lower sea levels in the Late Quaternary, a low-gradient, open coast prevailed. At sea levels approaching those of today, the coast is much steeper with a wide range of open and embayed coastal sectors (Roy & Thom, 1981).

Two approaches are applied in order to develop the concepts in this chapter. The first is based upon extensive field investigations of the Quaternary geology on the southeast Australian coast and continental shelf, involving surface sediment sampling, vibrocores, side-scan sonar, and high-resolution seismic reflection data. Granulometric analyses and radiocarbon dates of numerous vibrocore samples provide the necessary ground-truthing for interpreting the seismic records. These investigations include morphostratigraphic surveys of present-day barriers (Roy & Stephens, 1980; Roy & Thom, 1981; Roy & Crawford, 1981; Thom, 1984; Thom *et al.*, 1992), shoreface (Gordon & Roy, 1977; Roy 1985), and inner shelf (Roy & Crawford, 1980; Roy & Thom, 1981; von Stackelberg, 1982; Field & Roy, 1984; Cowell & Nielsen, 1984; Hudson, 1985; Roy, 1985; Cowell, 1986; Hudson & Ferland, 1987; Roy & Hudson, 1987; Ferland, 1986, 1987, 1988, 1990; Roy & Ferland, 1987), as well as former environments at lower sea levels that are partly preserved on the present inner and mid shelf areas (Kudrass, 1982; Roy 1985; Roy & Hudson, 1987; Roy *et al.*, 1992; Bickford *et al.*, 1993; Roy & Keene, 1993). One of the problems of investigating coastal evolution is that much of the field evidence is not preserved, as deposits have been reworked and modified through the action of evolutionary processes. Computer modelling therefore provides the second approach underpinning the principles examined in this chapter. The fundamental rationale for this computer modelling is based on principles of large-scale coastal behaviour (Terwindt & Wijnberg, 1991) and is outlined in Chapter 2 (also see Cowell, Roy & Jones, 1991, 1992).

The scope of this chapter does not extend to consideration of the role of tectonics and continental-margin setting in basin evolution (see Chapter 12). Relationships at this scale are assumed to be implicit in variables such as substrate slope and the degree to which coasts are laterally bounded by headlands or drift divides. Similarly, climatic variables are ignored as are the influences of tides and wind-driven currents. Furthermore, coastal evolution is primarily examined in terms of its two-dimensional behaviour in the vertical plane perpendicular to the coastline using a computer simulation model (Cowell *et al.*, 1992). The model expresses the dynamic complexity of the coast in simple shape terms employing the principle of conservation of sand mass. Barrier shape is empirically defined by active surfaces (shoreface and barrier top) that reflect ambient conditions and thus can act as a surrogate for a site-specific set of hydrodynamic processes and sediment properties. The role of coastline configuration in plan form is of first-order importance to the evolution of wave-dominated coasts, especially given the significance of littoral sand transport in the surf zone (Wright *et al.*, 1991) and its effects on shoreline adjustment (Komar, 1976; Gordon, 1988). The effects of shoreline configuration are examined through their implicit control over coastal-profile response, via the alongshore sediment budget which, in turn, primarily depends upon coastal plan form and the direction of incident wave energy (Davies, 1958; Komar, 1976).

Description of the lithofacies found on wave-dominated coasts is provided only to the extent required in considering evolutionary processes. Such descriptions can be found elsewhere (e.g., Swift, 1976a; Greenwood & Davis, 1984; Davis, 1985, 1992; Niedoroda, Swift & Hopkins, 1985; Swift *et al.*, 1991a). Present-day studies of barrier beaches and tidal inlets are invariably dominated by process measurements and historical reconstructions. While these studies are invaluable in documenting the mechanics of short-term coastal change, they do not address the broad questions of barrier genesis and evolution over geological time spans of one or more glacial-interglacial cycles.

The principles presented here are, for the most part, grossly simplified. Notably in this respect, coastal evolution in response to marine transgression is presented in relation to the high rates of postglacial sea-level change that occurred 20 to 7 ka BP. While it may be possible to draw implications for the evolution of wave-dominated coasts subject to slower rates of sea-level rise, such as that occurring along the present US Atlantic coast, substantial differences exist due to local variations in sediment budgets. It is probably helpful to keep these differences in mind when comparing concepts developed in this chapter with models presented by others such as D.J.P. Swift (e.g., Swift & Thorne, 1991; Thorne & Swift, 1991a, 1991b) and J.C. Kraft (e.g., Kraft, 1971; Kraft & John, 1979; Belknap & Kraft, 1985). The question of time scale underpinning different models for evolution of wave-dominated coasts is also important. Many aspects of previous ideas are based upon the long-term and large-scale perspective of basin infill studies, which incorporate modern-analogue processes into geological time scales (e.g., Haq, Hardenbol & Vail, 1987; Vail *et al.*, 1991; Swift, Phillips & Thorne, 1991b). The critical question here is whether pre-Quaternary coasts experienced glacial-interglacial sea-level oscillations of anything like the magnitude and frequency of those of the last 700 ka. Its answer will determine the extent to which lessons from the Holocene can be extrapolated into the geological past. The perspective of this chapter, however, concerns the much more rapid sea-level changes characteristic of Late Pleistocene and early Holocene conditions that are more immediately relevant to the antecedents of present-day wave-dominated coasts.

Background

In the past, wave-dominated coasts have been described in terms of beaches (Zenkovich, 1967; King, 1972), the nature of which was seen to be solely the product of contemporary wave processes (Pilkey *et al.*, 1993). In contrast, and in keeping with the evolutionary focus of this book, our emphasis is geological: we propose to use the term 'coastal sand barrier' as the basic depositional element for wave-dominated coasts. The concepts of geological inheritance, as well as morphodynamics, are used to describe their nature. Within this framework, the beach is seen as only part of a larger entity. Following the classifications of Curray (1969) and Boyd, Dalrymple, & Zaitlin (1992), barriers are defined as elongated, shore-parallel sand bodies, extending above sea level, and consisting of a number of sandy lithofacies including beach, dune, shoreface, tidal delta, inlet and washovers. Together they form a continuum of barrier morphologies (Fig. 4.1). Barrier islands with wide lagoons are the end member on low-gradient, open coasts (Davis, 1992), but as substrates progressively steepen, the lagoon becomes narrower so that the special case of a mainland beach constitutes the other end member, with negligible backbarrier morphologies (Fig. 4.1). Between these end members lie bay barriers (King, 1972, p. 502) which reflect the degree to which the coast is laterally bounded (Fig. 4.1). Embayedness generally relates to the extent of bedrock outcrop. This can be expected to become more frequent where regional substrates are steeper, as is the case in southeast Australia where the

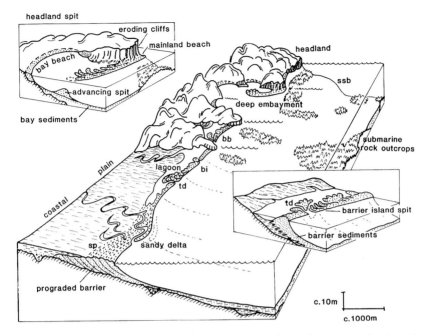

Figure 4.1. Both sediment-rich and sediment-starved settings exist on wave-dominated coasts. These range from low-lying coastal strand plains and barrier island/lagoon systems to rocky coasts with bay barriers and estuaries in embayments and rock reefs on the sea bed off headlands. The insets show in more detail the settings and stratigraphies of headland and barrier island spits. (ssb: shelf sand body; bb: bay barrier; bi: barrier island; td: tidal delta; sp: strand plain)

sea intersects erosional bedrock landforms comprising drowned valleys and interfluves that form headlands (Fig. 4.1) (Roy & Stephens, 1980; Roy & Thom, 1981). Submarine deposits of shelf sand seaward of the shoreface zone are residuals of coastal deposits formed during lower phases of sea level and are reworked to varying degrees by waves and shelf flows (Everts, 1987; Niedoroda *et al.*, 1985; Rine *et al.*, 1991).

Wave-dominated coasts correspond mainly to Swift & Thorne's (1991) 'accommodation-dominated' coastal settings in which basins receiving sediments are volumetrically large compared with the rate at which sediment is supplied to them. Nevertheless, many 'supply-dominated' (allochthonous) coasts are presently building sandy deltas with wave-dominated littoral zones; the São Francisco in Brazil (see Fig. 3.6) is an example (Wright, 1985). The main distinction between barriers in accomodation- as opposed to supply-dominated coastal settings is the thickness of their lithofacies, especially on the shelf (Fig. 4.2). The former are associated with autochthonous shelves that have sandy sediments and experience extensive reworking (Fig. 4.2*a* and *b*). In

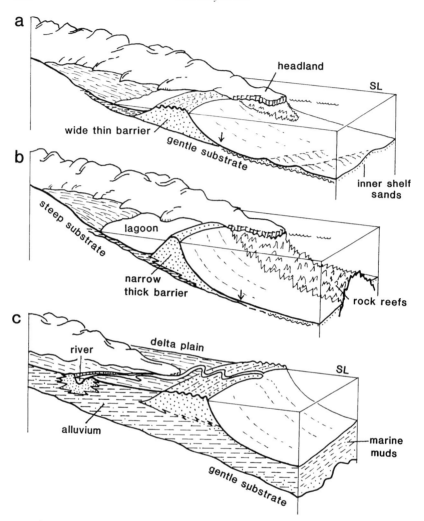

Figure 4.2. Barrier shape is influenced by embayment geometry and accommodation space. Even where barrier sand volumes are the same, very different surface morphologies are produced in embayments with shallow, gently sloping substrates (*a*) compared with embayments with deeper and steeper slopes (*b*), that probably experience higher energy waves than (*a*). In supply-dominated (*allochthonous*) settings (*c*) with the same wave regime as (*b*), a similar-sized barrier sand body shows markedly different stratigraphic relationships. The small arrows indicate the toe of the shoreface which, because the barrier sand bodies are the same size but energy levels are different, occurs in deeper water, further offshore in (*b*) compared to (*a*); in (*c*) the toe of the shoreface is indeterminate. The width of each section are in the order of 5–10 km and the maximum water depths are 10–30 m. SL = sea level.

contrast, supply-dominated regimes produce allochthonous shelves characterised by more rapid sedimentation and thick muddy sequences in which barrier sand bodies may be encased (Fig. 4.2*c*). However, Swift & Thorne's (1991) reliance on contemporary processes to discriminate between different types of coastal deposits takes insufficient account of the influence of inherited geological phenomena or antecedent topography (Belknap & Kraft, 1985; Pilkey *et al.*, 1993; Riggs & Cleary, 1993). As Fig. 4.2*a* and *b* shows, even with the same sediment supply, variations in accommodation space (specifically, in substrate slope) can produce different coastal geometries and stratigraphies.

Table 4.1 is an alternative framework based on inherited landscape, but without implying a specific tectonic setting (see also Short, 1988). This tripartite division is a way of grouping coastal deposits in plan form; features on coastal plain (unbounded) coasts generally have the longest dimensions coast-parallel while those on embayed (bounded) or cliffed coasts are smaller and more variable lithologically. There are also cross-sectional differences, between the three classes of coasts, in the slope or gradient of the pre-existing land surface on which contemporary coastal deposits have formed. In general terms, slopes are steepest on cliffed or protruding coastal sectors and flattest on coastal plain coasts; slopes are most variable on embayed coasts.

For more than a century the origin of barriers and barrier islands has been debated (e.g., Hoyt, 1967; Schwartz, 1973; Otvos 1979). Emergent shoals, submergent bars and spit progradation were some of the early hypotheses. Some workers recognised the importance of the postglacial marine transgression (e.g., Curray, 1964; Kraft, 1971; Leatherman, 1983), and now it is widely acknowledged that, rather than being solely the product of interglacial highstands of the sea, barriers have also existed at lower sea levels (Swift, 1975a; Hovland & Dukefoss, 1981; Field & Trincardi, 1991; Roy *et al.*, 1992). Vibrocores that encountered relict estuarine sediments in present-day shelf environments provide evidence for the existence of barrier lagoons when

Table 4.1. *Regional coastal depositional morphologies*

Coastal plain coasts (unbounded)	Embayed coasts (bounded)	Protruding or cliffed coasts
Linear barriers	Bay barriers	Cliff-top dunes
Barrier islands	Headland spits	Mainland beaches
Shelf ridge fields and inner shelf sand sheets	Inner shelf sand sheets	Headland – attached shelf sand bodies

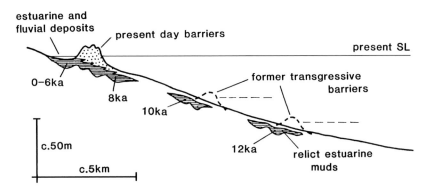

Figure 4.3. Continental shelf coring investigations such as those in southeast Australia encounter relict estuarine sediments in the subsurface with ages which indicate they formed as the shelf was inundated by the postglacial marine transgression. Estuarine and lagoonal environments only exist in the protection of barriers which means that these latter features must also have existed at times of lower sea levels. The lack of open marine deposits behind barriers at the present coast accords with their onshore migration as sea level rose, rather than by *in situ* accretion when sea level stabilised.

sea level was lower (Sanders & Kumar, 1975; Field & Duane, 1976; Williams, 1976; Kudrass, 1982; Colwell & Roy, 1983) (Fig. 4.3). Furthermore, the absence of open marine sediments on the mainland shore behind barriers indicates that, rather than emerging in place as offshore bars, most barriers migrated shorewards during the postglacial marine transgression from an initial position on the outer shelf (Hoyt, 1967; Swift, 1975a; Field & Duane, 1976). Thus, modern models of barrier formation must accommodate sea-level oscillations of the order of 100 m every 100 000 years or so over the last 400 000 years (Chappell & Shackleton, 1986). For non-tectonic coasts, this means that shorelines were lower than present levels for more than 90% of Quaternary time (and occupied a modal zone between –25 and –75 m; Roy & Thom, 1981) and that two of the major variables in barrier formation – shelf morphology and sediment supply – are essentially imposed on contemporary barriers through inheritance.

There is an important distinction between the flux of sediment on wave-dominated coasts and the movement of barrier sand masses during sea-level transgressions, stillstands and regressions. As distinct from headland spits, which grow alongshore (Fig. 4.1, inset), barriers translate across the shelf in a shore-normal sense as sea levels change, although both alongshore and shore-normal sediment fluxes are involved in their translation. The emplacement of barriers is seen as a complex response to changes in the rate of sea-level rise, substrate gradient, sediment supply and energy conditions. However, the geological continuity (and longevity) of barrier shorelines may be interrupted in particular cases: high latitude coasts that were permanently ice covered during glacials are one example (see Chapter 9) and shallow, land-locked seas (e.g., the North Sea and Gulf of Thailand) that were exposed as dry land during sea-level lowstands are another. But even in these cases, wave action re-established its dominance as the ice retreated and sea levels rose with the postglacial marine transgression. Overall, however, wave-dominated deposits of various types have existed either continuously or for substantial periods of time on most of the world's coasts throughout the Quaternary. It is not surprising, therefore, that the rapid eustatic changes that have characterised this geological period have had an important influence on the style of present-day coastal development.

More is known about coastal development under stable or slowly changing sea levels than when sea level is changing rapidly, because we are presently in an interglacial period. Periods of changing sea level were more common in the past, but our understanding of the events accompanying them is more speculative and relies on indirect techniques such as simulation modelling supported by limited field evidence at specific localities. In Fig. 4.4*a*, various types of wave-dominated deposits are arranged according to: 1. whether relative sea level (RSL) is rising, falling or stable, and 2. the morphology of the land surface over which the sea is moving. However, because many continental shelves tend to be more rugged on their inner parts and planar elsewhere, low-gradient coastal morphologies were much more common at times of lower sea level than they are today. This general relationship is shown schematically in Fig. 4.4*b* for southeast Australia and indicates that, for most of the Quaternary, coastal settings during both marine transgressions and regressions have been low-gradient and planar with linear barrier shorelines. Cliffed and embayed coasts probably represented only a minor proportion of the region's coastal settings over this time period.

The following sections use field studies from southeast Australia, together with the results of computer modelling, to describe in morphodynamic terms the evolution of wave-formed coastal features under conditions of marine transgression, marine regression and prolonged stillstand. The section on the evolution of wave-dominated coasts under stillstand conditions has the most direct bearing on the present-day coasts, which mostly have been subjected to the relatively stable sea levels of the late Holocene compared with those of the preceding 100 000 years. This section is therefore presented in more detail. Sections on sea-level transgression and regression are presented first, however, because these are the antecedents of the Holocene stillstand. Again, because the marine transgression is the more recent of the two antecedents, more is

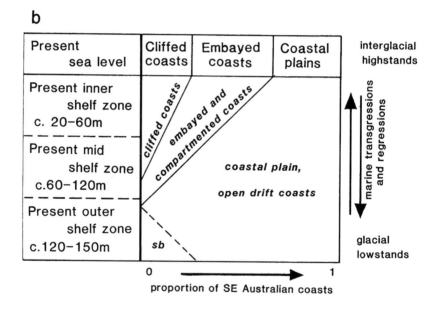

Figure 4.4. Field diagrams indicating (*a*) relationships between regional morphologies, changing sea levels and the occurrence of various types of wave-formed coastal deposits, and (*b*) the relative proportion of various coastal morphologies as functions of relative sea level and of geological time (glacial-interglacial cycles). The "sb" in the bottom left corner denotes the situation during lowstands on shallow shelves (e.g., the US Atlantic shelf) when the sea drops below the shelf break and encounters the relatively steep upper slope.

known about coastal evolution during rising sea levels. These two sections are therefore presented in the reverse order of their chronology in nature, to convey a firmer foundation for principles of coastal evolution driven by changing sea levels.

Marine transgression

This section focuses on coastal evolution during a rapid marine transgression, such as that of the postglacial marine transgression or that predicted under some scenarios for Greenhouse-induced sea-level rise (Roy & Thom, 1987; Thom & Roy, 1988). The effects of changing substrate slopes, sediment budgets and backbarrier accommodation volumes are also examined.

Low-gradient coasts, whether laterally unbounded or embayed, that are exposed to wave action at times of rising sea level are characterised by transgressive barriers (Fig. 4.5a). These correspond to Swift's 'erosional shoreface retreat' model and are primarily made up of tidal inlet and washover deposits. Transgressive barriers are essentially transitory features that maintain themselves in dynamic equilibrium with rising sea level by the landward transfer of sand, eroded from the shoreface, to backbarrier settings (Fig. 4.5b). The shape and dimensions of such transgressive barriers have been reconstructed from investigations of coasts presently experiencing slow relative sea-level rise (e.g., US and Canadian Atlantic coasts – Kraft & John, 1979; Boyd, Bowen & Hall, 1987) and from more stable coasts where drilling and dating of prograded barriers have delineated their transgressive facies that were deposited just before sea level stabilised (Roy & Thom, 1981; Thom, 1984; Roy *et al.*, 1992). On high-energy coasts, transgressive barriers have dimensions of 15–25 m thick and 2-3 km wide and extend 2–4 m above mean sea level depending on the storm surge level (Thom, 1984). Southeast Australian studies suggest that their cross-sectional areas are in the order of 10 000–30 000 m^2 (Roy & Thom, 1981; Roy *et al.*, 1992).

Calculations based on modelling a constant morphology during barrier translation while sea level is rising at around 15 mm a^{-2} (its average rate during the postglacial marine transgression in southeast Australia) involves eroding as much as 200 m^3 m^{-1} a^{-1} of sand from the front of the barrier and transporting it to the rear by tidal inlet and overwash processes. The volumes are much less than this on low-wave-energy coasts (Cowell & Nelson, 1991). Volumetric changes far in excess of this are thought to occur on the shoreface during single storms on this coast (Thom & Hall, 1991). Thus, there would seem to be ample energy available to shift barriers landward even while sea

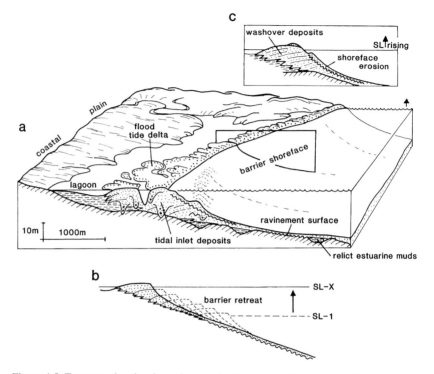

Figure 4.5. Transgressive barriers characterised most coasts when sea levels were rising on low-gradient substrates. They were composed almost entirely of tidal delta (*a*) and washover (*c*) deposits that retrograded into estuarine/lagoonal environments as sea level rose. Landward migration was through a process of erosional shoreface retreat (*b*) as the barrier adjusted to changing sea level. The sea bed exposed by the retreating barrier is an erosion or ravinement surface.

level is rising relatively rapidly. As much, or even more, time is available for sediment transfers when the rate of marine transgression slows down. However, another factor – foredune development – may alter barrier dynamics under these conditions. While sea level is rising rapidly, aeolian processes only have time to build low, discontinuous foredunes on the barrier surface that provide little impediment to storm washovers (Fig. 4.5*c*). As the rate of sea-level rise slows, more dunes build above storm surge levels and the frequency of washovers is reduced thus retarding barrier translation. The development of a continuous foredune leads to *in situ* submergence as sea level continues to rise, and to erosion of the barrier superstructure which, in planform, shows a progressive narrowing from both the ocean and the lagoon side. Leatherman (1983) identified this historical pattern on Fire Island, New York. Sanders & Kumar (1975) present controversial evidence from the Long Island shelf that this process has led to in-place drowning and overstepping of barriers under conditions of slow sea-level rise. However, it is more likely that landward barrier translation recommences when the sea rises to a level that effectively narrows the barrier and once again allows storm washovers (Leatherman, 1979; 1983). It follows from this discussion that barrier translation is fairly continuous while sea level is rising relatively rapidly but may become intermittent, with hiatuses of many hundreds of years, when it slows down.

While external conditions of substrate slope, wave regime, sediment budget and rate of RSL rise are maintained, it seems reasonable to assume that parameters such as shoreface profile shape, inlet geometry, backbarrier width and relief on the barrier surface remain approximately constant. Under these conditions, the rate of shoreface translation is simply a function of the substrate slope and the speed at which RSL is rising; computer modelling suggests that landward translation of the transgressive barrier is accomplished without eroding the substrate (Fig. 4.5*b*) and that its basal contact corresponds to the former land surface. Later, an erosional ravinement surface (the product of wave reworking at the toe of the shoreface) is developed on the shelf after the transgressive barrier has passed but contemporary estuarine sediments may be preserved in topographically low areas related to antecedent topographies (Belknap & Kraft, 1985) (Fig. 4.5*a*).

The role of substrate slope

Changes to the external parameters influence transgressive barriers in ways that simulation modelling suggests are subtle and, in some cases, unexpected. The most sensitive parameter seems to be substrate slope – a geologically inherited property. Changes to the coastal sand budget that mostly arise from local variations in littoral transport rate are generally less important when sea level is changing rapidly.

The types of behaviour of a transgressive barrier translating over a range of substrates are listed in Fig. 4.6, based on simulation modelling of a typical southeast Australian barrier (Cowell *et al.*, 1991). The modelling experiments show that for a constant RSL rise, the rate of shoreface retreat, the amount of sand recycled each year and the cross-sectional area of the transgressive barrier all decrease as the substrate steepens from very gentle gradients of 0.1° to about 0.8° (Fig. 4.6*a*, *b* and *c*). On slopes steeper than approximately 0.8°, for barrier geometries applicable to southeast Australia, the model predicts a Bruun-like response involving a net offshore sand transfer as RSL rises (Fig. 4.6*d* and *e*). Shoreface changes on 'steep' gradients involve a continuing input of sand either from coastal erosion (standard Bruun Rule; Dean & Maurmeyer,

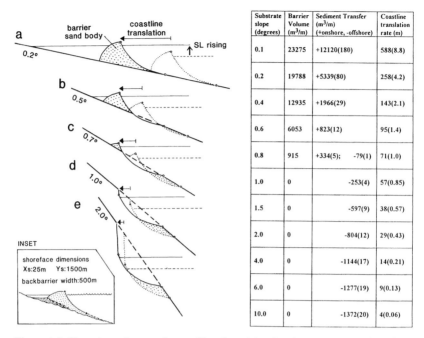

Substrate slope (degrees)	Barrier Volume (m³/m)	Sediment Transfer (m³/m) (+onshore, -offshore)	Coastline translation rate (m)
0.1	23275	+12120(180)	588(8.8)
0.2	19788	+5339(80)	258(4.2)
0.4	12935	+1966(29)	143(2.1)
0.6	6053	+823(12)	95(1.4)
0.8	915	+334(5); -79(1)	71(1.0)
1.0	0	-253(4)	57(0.85)
1.5	0	-597(9)	38(0.57)
2.0	0	-804(12)	29(0.43)
4.0	0	-1144(17)	14(0.21)
6.0	0	-1277(19)	9(0.13)
10.0	0	-1372(20)	4(0.06)

Figure 4.6. Changing substrate slopes affect depositional styles on coasts undergoing a marine transgression. On gentle slopes (*a* and *b*), rates of coastal recession are high and there is a mass onshore transfer of sand with the migration of transgressive barriers. In contrast, on relatively steep substrates (*d* and *e*) the coast is in encroachment mode and rates of coastal recession are slow. Sediment moves offshore to the lower shoreface, and is lost to the barrier which consequently loses volume. In (*c*) onshore and offshore sand movements are approximately balanced. (Each illustration shows equal amounts of erosion and deposition). The table documents simulation results of the shoreface translation model (inset) for a range of substrate slopes. Sand transfer and coastline translation rates are for a 1.0 m rise in sea level; those in brackets are annual rates if the sea was rising at 1.5 m/century.

1983) or by littoral transport (i.e. a sediment budget imbalance). The amount of sand deposited offshore increases as the substrate slope steepens but if the external sand supply is starved or cut off, the inshore sea bed becomes a drowned, scoured surface with exposed bedrock – a situation that occurs off many southeast Australian headlands (Ferland, 1990).

Changes in the rate at which RSL rises not only control shoreface translation but also influence the amount of work that waves expend on the shoreface and beach. They also influence the shape of the barrier, especially the shoreface profile, although little is known about this (Pilkey *et al.*, 1993). On very gentle gradients (<0.1°) bottom-friction effects reduce incident wave energy levels (Wright, 1976) such that equilibrium shoreface profiles and barrier widths (and volumes) become smaller thus reducing the quantity of

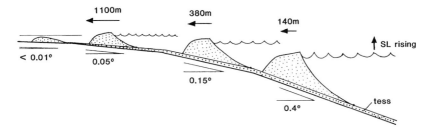

Figure 4.7. A marine transgression on a substrate that progressively flattens in a landward direction (0.4° to <0.05°) inevitably leads to an increase in the rate of coastal recession (from 140 to 1100 mm m^{-1} rise in sea level). Frictional effects of the shoaling sea bed tend to diminish incident wave energy, and also shoreface and barrier dimensions. On extremely shallow shelves, this culminates in submerged sand banks. Reduction in barrier size is accomplished by shedding sand which, if not dispersed by littoral currents, forms a trailing-edge sand sheet (tess) on the shelf surface.

sand transferred landward per unit time. This trend is illustrated in Fig. 4.7 on a flattening substrate; a similar effect is produced on a constant slope by an acceleration in the rate of RSL rise (rates as high as c. 20–25 mm a^{-1} were probably achieved during the postglacial marine transgression – Fairbanks, 1989). As indicated in Fig. 4.6a, the quantities of sand that are redistributed on very low-gradient substrates can be very large and may exceed the transporting capacity of the incident waves. In response, the barrier probably reduces size to accommodate a more rapid shoreface translation, leaving behind a trailing-edge sand sheet (Fig. 4.7). With slow rates of RSL change, more time exists for waves to mould the barrier shoreface into an 'equilibrium' configuration and to increase the width of the backbarrier (tidal delta and washover) deposits. Under quasi-stillstand conditions the shoreface profile becomes wider and extends to deeper water depths than when RSL is changing rapidly. Therefore, barriers attain their maximum dimensions under prolonged stillstand conditions, as discussed in the section on stable sea levels.

In many respects, the pattern produced by a transgression over a relatively steep slope (Fig. 4.8) corresponds to Swift's (1976a) trailing-edge sand sheet that he associates with low-gradient shelves. In southeast Australia, transgressive sand sheets of the type shown in Fig. 4.8b occur on relatively steep substrates and are composed of uniformly fine sand, typical of the lower shoreface. They are found off headlands and steep sectors of coast, and form the lower unit of shelf sand bodies (SSBs) (Field & Roy, 1984; Roy, 1985; Ferland, 1986, 1990) (Fig. 4.8a). The age of these transgressive deposits in the Sydney region is shown in Fig. 4.8c; the overlying unit was deposited under stillstand conditions and is discussed in a later section.

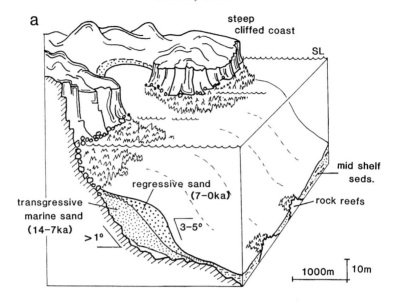

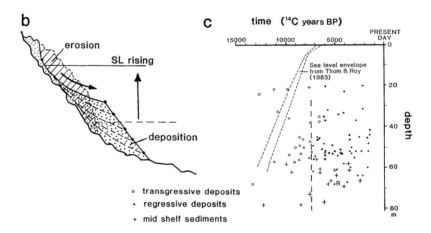

Figure 4.8. Shelf sand bodies that occur along steep and deep sections of coast in southeast Australia are made up of transgressive and regressive units (*a*) deposited subaqueously during the postglacial marine transgression and the following stillstand. At all stages of their formation, the direction of net sand movement was seaward (*b*). Radiocarbon dates (environmentally corrected) on shell fragments from the two depositional units in the Sydney shelf sand body, and from mid shelf sediments lying immediately seaward of it, are plotted in (*c*).

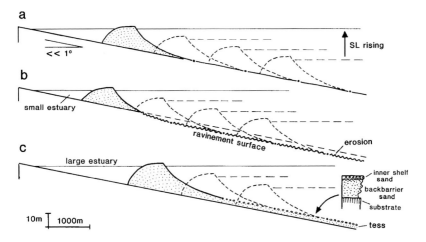

Figure 4.9. Simulation modelling predicts that variations in the marine sand budget will cause a transgressive barrier to erode the sea bed if there is a net loss of sand (case *b*) and to cause deposition of a trailing-edge sand sheet (tess) if there is a net addition of sand (case c) (The sediment budget is balanced in case *a*). The trailing-edge sand sheet in case *c* is composed of backbarrier sand, but subsequent marine reworking may change its character to a marine deposit and re-arrange its sheet-like morphology into sand ridges.

The role of the nearshore sediment budget

The predominance of 'coastal plain, open drift' coastal settings in Fig. 4.4*b* indicates that transgressive barriers existed on low-gradient continental shelves for much of the Quaternary, at times of lower sea levels. On these shelves, alongshore sediment movement was essentially unimpeded and operated largely independently of onshore sand transfers. However, as Boyd *et al.* (1987) recognised on the highly indented, glaciated coast of Nova Scotia, local changes in sediment supply during transgressions strongly influence barrier behaviour.

　　Fig. 4.9 exemplifies a number of situations involving sediment budget imbalances during a marine transgression on a relatively gently sloping substrate. In cases *a*, *b* and *c*, the littoral-transport differential (the difference between littoral drift rates at various places along the coast) is important, not the total drift rate. In the case of a balanced budget (Fig. 4.9*a*) the dimensions of the transgressive barrier and its rate of landward translation remain constant with time provided that the substrate gradient and rate of RSL rise are unchanged. A ravinement surface is formed by waves reworking the inner shelf surface, but with a balanced sediment budget the ravinement surface corresponds closely to the original land surface. With a negative sand imbalance (Fig. 4.9*b*), and the same boundary conditions as Fig. 4.9*a*, landward barrier translations occur at a similar rate but the toe of the shoreface erodes into the underlying substrate and, in so doing, generates additional sand to maintain transgressive barrier volume. However, the cross-sectional dimensions of this barrier are significantly smaller than in Fig. 4.9*a*. The reverse is true for a positive sand imbalance (Fig. 4.9*c*). Excess sand, originally deposited in backbarrier environments, is left as a sand sheet on the shelf surface after the transgression passes and the transgressive barrier itself is considerably larger than in the case of Fig. 4.9*a*. For example, a drift imbalance of $5\,m^3\,m^{-1}$ length of coast/year will produce about 3.5 m of erosion/ deposition on a low angle shelf ($0.4°$ slope) undergoing transgression. This drift imbalance is equivalent to a difference between littoral drift inputs and outputs of $50\,000\,m^3\,a^{-1}$ on a 10 km long sector of coast experiencing a RSL rise of $10\,mm\,a^{-1}$. The negative imbalance causes a 40% decrease in barrier size; the positive imbalance produces a 55% increase in its size. Despite this, rates of coastline retreat are dominated by the marine transgression and are only slightly modified ($\pm 10\%$) by imbalances in the sand budget. These results are for steady-state conditions, which the simulation experiments show are only achieved after 30 to 50 time steps, the equivalent of many thousands of years in nature. They also show that similar coastal impacts can be produced by changing the rate of RSL rise as by varying the sand budget.

Lagoons, estuaries and accommodation space

The preceding discussion focused on shelf sand budgets. Fig. 4.10, on the other hand, illustrates the situation where an estuary or lagoon, located behind a transgressive barrier, is receiving sediment at the same time as sea level is rising. In Fig. 4.10*a*, the estuary is infilling at a much slower rate than sea level is rising (the result of a small sediment supply or very rapid sea-level rise), which gives rise to a thin layer of backbarrier sediments on the sea bed after the barrier has translated landward. With estuarine sedimentation occurring at almost the same rate as sea level is rising (Fig. 4.10*b*), the estuary becomes wider as the barrier translates somewhat more slowly over the estuarine deposits which form a relatively thick deposit on the inner shelf. Compared with Fig. 4.10*a*, the barrier in Fig. 4.10*b* becomes smaller in the process of attaining steady-state conditions. The sand it loses to the inner shelf veneers the muds to some extent, although the sea bed undoubtedly experiences erosion during severe storm events. Slower translation of the barrier shoreface, due to rapid infilling of the estuary (equivalent to a delta-margin environment), leads to a protruding coastline which may be exposed to marine erosion. This

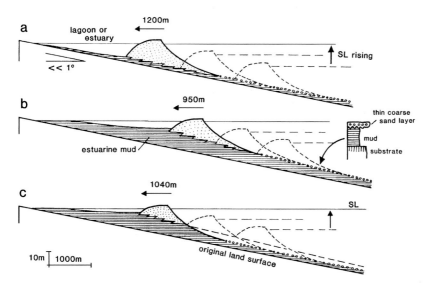

Figure 4.10. The model simulates the effect of lagoonal mud sedimentation (i.e., a fluvial sediment budget gain) on the same transgressive barrier as shown in Fig. 4.9. A small amount of mud deposition (case *a*) has little effect on barrier size or coastal behaviour, but with mud accumulating almost as fast as sea level is rising (case *b*), the barrier decreases in size and recession slows down. In case *c*, a deficit in the marine sand budget is superimposed on case *b*. In all cases lagoonal muds conformably blanket the substrate beneath and to seaward of the barrier; after the barrier passes, wave reworking ensures that the sea bed is soon veneered by a coarse lag (the inner shelf sand sheet). (The arrows and distances indicate comparative translations for the same increment of sea-level rise).

is modelled in Fig. 4.10*c* as a shelf sand loss that causes the shoreface to erode into the mud substratum, the barrier to decrease in volume and the rate of shoreline translation to increase. In this scenario, the estuarine sediments are muds, which are dispersed by wave action; if they contain sand it would be added to the shoreface and incorporated in the transgressive barrier which would, again, slow the rate of shoreface retreat.

Lagoons and estuaries also act as long-term sediment sinks for marine sand (Swift & Thorne, 1991). The importance of accomodation volume to the rear of the barrier is demonstrated by Fig. 4.10*b* (where the estuary is infilling rapidly), which shows a reduction in the translation rate as the estuary becomes shallower. The overall behaviour is indicative of the state-dependent evolution outlined in Chapter 2, with the accommodation volume providing a self-regulating feedback mechanism. In nature, the physical character of the estuary mouth is controlled by a dynamic balance between littoral and tidal sand movements on the open coast and in the tidal inlet, respectively (Roy,

1984; Nichols & Biggs, 1985; Dalrymple, Zaitlin, & Boyd, 1992). Usually inlets in bay barriers are fixed against bedrock outcrops or headlands and do not form extensive tidal inlet deposits (see section on stable sea levels). However, in the case of barrier island chains, local imbalances in the littoral sand budget may lead to inlet migration and the formation of recurved spits at the ends of the adjacent barrier. Under these conditions large quantities of sand become stored in abandoned channel and flood tide delta deposits. Where the channels are deep (>10 m) these deposits may make up the bulk of the barrier sand mass (Hoyt & Henry, 1967; Moslow & Heron, 1978).

Marine regression

All barrier sand bodies that prograde in a seaward direction are regressive in nature *sensu stricto*; however, in this chapter, we restrict usage of the term 'regressive barrier' to features deposited by waves under conditions of falling

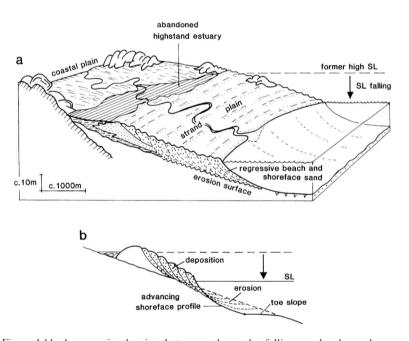

Figure 4.11. A regressive barrier that progrades under falling sea levels produces a tabular, gently seaward-inclined sand deposit 10–20 + m thick with an erosional base (*a*). As sea level falls, the inner shelf surface erodes and sand moves onto the shoreface (*b*). The surface of the regressive barrier forms a wide strand plain but without estuaries comparable to those formed under conditions of stable or rising sea level.

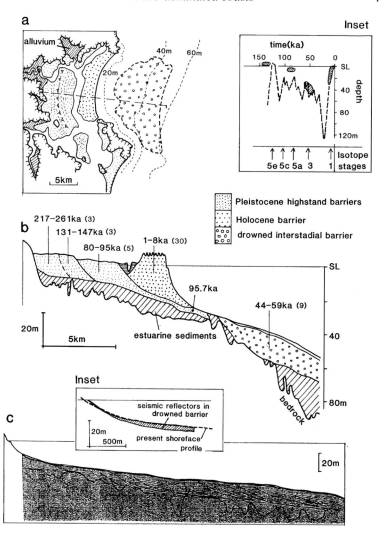

Figure 4.12. (*a*) The Tuncurry embayment in central NSW (lat. 32°S) showing the distribution of barriers onshore and on the inner shelf. (*b*) A shore-normal cross-section (dashed line in *a*) through the various barriers showing their dimensions and age groupings. The pre-Holocene ages are based on thermoluminescence measurements by the University of Wollongong documented in Roy *et al.* (1992); the numbers in brackets indicate the numbers of samples dated. The inset shows the group ages of the various barrier systems (shaded) and the estimated position of the sea when they formed in relation to the sea-level curve (dashed line) of Chappell & Shackleton (1986). (*c*) Shallow marine seismic 'Uniboom' profile (Line 12 in Roy *et al.*, 1992) through the axis of the Tuncurry embayment. The interpretation of a drowned (regressive) barrier system is based in part on the concordance between the attitude of its dipping reflectors and the present shoreface profile (inset).

RSL – the so-called 'forced or eustatic regression' (Fig. 4.11). Similar deposits when formed at times of stable, or even slowly rising, RSL are termed 'prograded barriers' and are discussed in the next section.

Very little is known about the effects of marine regression, particularly at the scale associated with the onset of glacial periods. The little that is known comes from field evidence on contemporary coasts that are undergoing slow uplift due to tectonic or isostatic processes (see Chapter 12). Regressive barriers appear to be features of low-gradient, relatively sediment-rich shelves that have not experienced excessive erosion during the postglacial marine transgression. The principles presented in this section are based upon field data from a mid shelf barrier complex drowned in 30 to 50 m of water north of Sydney, Australia (Roy *et al.*, 1992) (Fig. 4.12), supplemented by the results of computer modelling. Regressive barriers preserved on present-day shelves are not widely reported but where they have been identified in southeast Australia, they seem to be as wide or wider than their stillstand counterparts (Fig. 4.12*a*) and contain almost as much sand as the onshore barriers. In the Forster-Tuncurry area, the regressive (drowned) barrier shown in Fig. 4.12*a* and *b* is up to 7 km wide and 15 m thick. Highstand barriers onshore are almost 10 km wide and range in thickness from 12 to 23 m (Roy *et al.* 1992). Thermoluminescence dating of both the highstand and shelf barriers indicates a history of intermittent progradation spanning the last 260 ka with at least two regressive barrier systems forming during Late Pleistocene interstadials (isotope stages 5a and 3) (Fig. 4.12, inset). The drowned barrier on the inner shelf prograded during the period 60–40 ka when RSL was falling slowly at about 0.5 m per thousand years (see the inset in Fig. 4.12). Its base is erosional, in places intersecting bedrock (Fig. 4.12*c*); the upper surface is thought to have been modified during passage of the postglacial marine transgression which caused 3 to 5 m of erosion (Roy *et al.*, 1992).

The role of substrate slope

Results of a simulated regression are illustrated in Fig. 4.13 for both gentle and steeper-gradient shelves. In the former case (Fig. 4.13*a* and *b*), as RSL falls, the highstand barrier experiences shoreface progradation with sand eroded from further offshore. Erosion occurs on the lower shoreface and in a zone extending seawards of its toe (the 'toe slope' zone shown in Figs. 4.11*b* and 4.13*b*). With continued RSL fall, the regressive barrier builds seawards into the excavated zone and reconstitutes the former sea-bed morphology (Fig. 4.13*b*). Shoreface reflectors in the seismic profile of the drowned barrier shown in Fig. 4.12*c* have very similar geometry to the onshore barriers and to

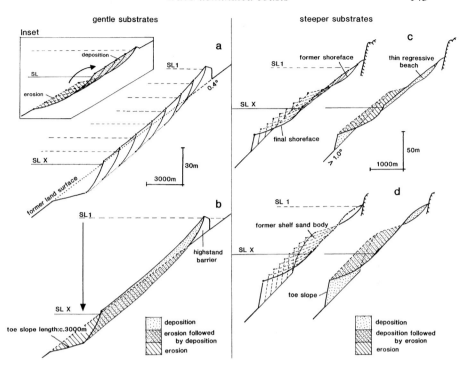

Figure 4.13. Mechanisms for regressive barrier formation, as sea levels fall on gently sloping shelf surfaces, involve erosion of the sea bed and the onshore movement of sand to prograde the shoreface. The inset shows a closed system with no net gains or losses of sand to the barrier. The simulation model (*a*) suggests that the sea bed experiences erosion for distances of many kilometres seaward of the barrier in high-energy, open-coast environments such as southeast Australia. In (*b*), the regressive barrier lies on the newly-formed erosion surface and represents a zone of massive *in situ* reworking equivalent to the barrier's thickness. The model also shows that steady-state conditions are only achieved very slowly during a marine regression. On a relatively steep, sandy substrate (*c*) falling sea level causes extensive reworking of the sea bed and a progressive offshore movement of sand. On slopes that are not too steep, a thin veneer of coarse beach sand may be left stranded as the sea recedes. A marine regression completely reworks a shelf sand body (*d*) transporting much larger quantities of sand offshore than is the case in (*c*).

the present shoreface profile (Fig. 4.12*c*, inset). Wellner, Ashley & Sheridan (1993) report similar seismic patterns from the New Jersey inner shelf (USA) and also attribute them to interstadial barriers. Furthermore, the erosional nature of the base of the drowned barrier in the Forster-Tuncurry area is explained by the initial phase of offshore erosion that precedes barrier progradation in the simulation model (Fig. 4.13). Initially, progradation of regressive barriers was slow but gradually increased until it reached a steady state when the surface of the regressive barrier tracked the former sea bed. As shown previously, steady-state conditions take thousands of years to achieve in the model; in nature they probably are never reached before environmental parameters (such as sea level or substrate slope) change.

As is the case for marine transgressions, substrate slope plays a critical role in controlling barrier size, as well as the rate of coastal recession, as RSL falls. For the same wave conditions, the largest regressive barriers are predicted to occur on the most gently sloping surfaces. In reality, however, bed friction tends to reduce incident wave energy and also barrier size in such cases. The best-developed regressive barriers should occur on moderately low-gradient, but high-energy, shelves when RSL is falling slowly, thus allowing maximum time for the shoreface profile to fully develop. The southeast Australian case study illustrated in Fig. 4.12 seems to exemplify these conditions. Like transgressive barriers, barriers formed during regressions recycle their local sand budget as RSL falls (Fig. 13a and b) and, while they are affected by littoral sand supplies, they are not dependent on them for their existence. Regressive barriers should therefore be widespread globally on relatively energetic, low-gradient and sand-rich shelves.

In general terms, regressive barriers can be expected to form on slopes of <1°, but simulation studies depicted in Fig. 4.13c and d suggest that as substrates steepen the sand flux reverses and becomes directed dominantly offshore. As shelf gradients steepen, increasingly large amounts of sand are deposited beyond the toe of the shoreface. Fig. 4.13c and d which exemplifies a marine regression on a relatively steep slope, illustrates two cases: a planar substrate sloping at >1° (Fig. 4.13c) and the same substrate with a convex shelf sand body formed under highstand conditions (Fig. 4.13d). In the former case, however, a thin beach deposit, possibly composed of coarse sand and enriched in heavy minerals, is left behind as the shoreface retreats; a larger amount of sand is transferred offshore in the latter case (Fig. 4.13d) and the SSB is completely reworked. Initially, the regression causes massive erosion of the existing substrate but, eventually, steady-state conditions are approached at which time reworking is focused on the body of fine sand deposited at the toe of the shoreface (Fig. 4.13d). Because, in nature, steeper substrate slopes are found off headlands or protruding sectors of coast with high wave energy and littoral transport rates, it is unlikely that the offshore sand accumulations will remain intact during a regression. Rather, they will be reworked and added to the downdrift coastal sediment budget. A case study where sand eroded from a SSB is incorporated in regressive barriers forming contemporaneously in an adjacent embayment is discussed in the conclusions to this chapter.

Lagoons and estuaries during marine regressions

It is clear from Figs. 4.11 and 4.13 that lagoons and estuaries are spatially restricted during marine regressions. The combined effects of a massive onshore movement of sand, falling water levels and fluvial deposition reduced estuary size to a minimum and changed their morphologies from the common 'barrier lagoons' of today to the more constricted 'river estuary' type at the mouths of the larger and more active rivers (Roy, 1984; Wright, 1985) (Fig. 4.11*a*). Furthermore, there was little opportunity for the formation of the coast-parallel topographic depressions landward of the shoreface that characterise coastal evolution during transgressions over gentle substrates. Estuarine waters therefore would have been restricted mainly to the seaward extremities of distributary channels and estuarine habitats would thus be greatly curtailed during the long periods of the Quaternary that RSL was falling. This has special implications for estuarine habitats and species: perhaps only those estuarine species that were able to adapt to more marine conditions on the open coast would survive (see Woodroffe & Grindrod, 1991 for a similar discussion about mangroves).

Stable sea-level conditions

Cessation of barrier transgression and barrier stabilisation is a function of sand supply in relation to RSL. In principle, a large sand influx while sea level is still rising could cause a barrier to stabilise and grow upwards in place. However, these conditions probably occur rarely and then only near active river mouths (Swift, 1976a). While sea level is rising, most river mouths are constantly being inundated and the offshore sea bed is deepening, thus cutting off a direct supply of river sand to the coast and reducing onshore movement of sand from the adjacent sea bed. Conditions favourable to barrier progradation occur only when rates of sea-level rise slow or stabilise. In the case of some southeast Australian barriers, coastal progradation was triggered when sea level stabilised at around 6.5 ka. At this time, the offshore sea bed began slowly to equilibrate, thus providing an influx of shelf sand; a process that has continued for thousands of years on this coast (see Fig. 2.9) (Roy, Thom & Wright, 1980; Thom, Bowman & Roy, 1981; Thom & Roy, 1985).

Changing RSL, substrate slopes and sediment budgets all contribute to coastal evolution under rising sea levels, but as the marine transgression slows, and RSL eventually stabilises, the role played by imbalances in sediment budgets becomes dominant. This is illustrated in Fig. 4.14, which shows the simulated translation of a transgressive barrier experiencing constant drift imbalances (negative and positive) during a marine transgression that slows

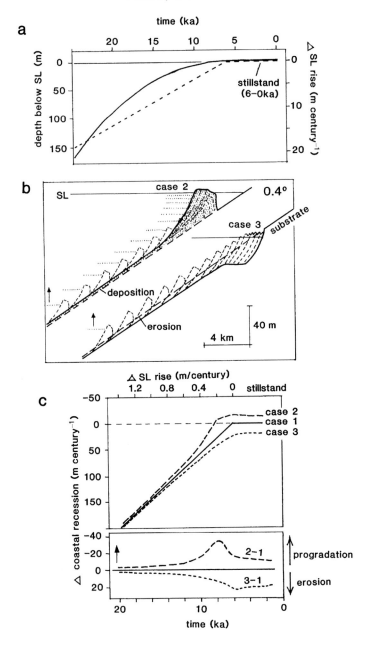

Figure 4.14. As the rate of RSL rise slows, sediment budget differentials play an increasingly important role in coastal change. (*a*) Shows a theoretical marine transgression from 25 ka to 6 ka BP (solid line) with the rate of SL rise slowing down

down and then stabilises (stillstand). The sea-level history is shown in Fig. 4.14*a*. The sand imbalance modelled in Fig. 4.14b ($\pm\,5\,m^3\,m^{-1}\,$RSL rise a^{-1}) is equivalent to a 10 km long embayment with a differential of $50\,000\,m^3\,a^{-1}$ between alongshore sand inputs and outputs. During most of the marine transgression (in this case from 25 ka to 10 ka) coastal changes are dominated by rising RSL and the sediment budget has a very minor impact on rates of coastal recession (Fig. 4.14*c*). However, when rates of sea-level rise slowed to less than about 0.5 m per century, sediment budget effects become more noticeable and under stillstand conditions they eventually dominate coastal change. With a positive sediment budget, coastal recession slows rapidly and is reversed before sea level stabilises (Fig. 4.14*b*, case 2). Barriers prograde during the stillstand but the rate of progradation slowly declines as the shoreface builds into deeper water. Fig. 4.14*b* (case 3) shows that, with a negative sediment imbalance, the rate of coastal recession decreases to a minimum when RSL stabilises (around 20 m per century in Fig. 4.14*c*) and remains just below this value as the shoreface erodes into the substrate (Fig. 4.14*b*, case 3).

The following sections examine stillstand coastal deposits ranging from sediment-rich to sediment-deficient types. The models are based on the wide variety of bay barriers recognised along the mainly embayed coast of southeast Australia, but they have their counterpart on coastal plain coasts elsewhere in the world (Dickinson, Berryhill & Holmes, 1972; Thom, 1984; Short, 1988). Each has characteristic lithofacies architecture reflecting sediment budget imbalances that arose either during the stillstand or were inherited at the end of the postglacial marine transgression. The various types of stillstand barriers and inner shelf sand deposits are listed in Fig. 4.15 which also indicates their inferred sediment budgets/fluxes. In southeast Australia most barriers are composed of 'marine' or shelf sand – a mixture of relatively mature clastic grains, with lesser amounts of calcareous detritus, reworked and transported

Caption for Figure 4.14 (*cont.*).
(dotted line) and a stillstand of the sea after 6 ka BP. (*b*) Simulations of two transgressive barriers responding to the SL changes in (*a*) and experiencing positive (case 2) and negative (case 3) sediment budget imbalances. (The case of a balanced sediment budget is not shown.) (*c*) The behaviour of transgressive barriers, in terms of coastal recession, is plotted against time and rate of sea-level rise for the case of a balanced sand budget (case 1), for a barrier gaining sand (case 2), and for a barrier losing sand (case 3). The lower graph compares cases 2 and 3 to case 1. It shows that, in the case of a barrier gaining sand, the sediment budget factor had the greatest effect on coastal behaviour around 8 ka BP, well before sea level stabilised. With the barrier loosing sand, the effect gradually increased as the rate of sea-level rise slowed and reached a maximum when sea level stabilised at 6 ka BP, after which it stayed relatively constant.

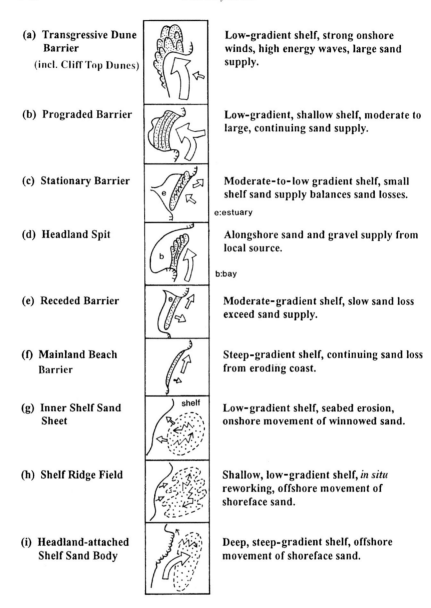

(a) Transgressive Dune Barrier (incl. Cliff Top Dunes)		Low-gradient shelf, strong onshore winds, high energy waves, large sand supply.
(b) Prograded Barrier		Low-gradient, shallow shelf, moderate to large, continuing sand supply.
(c) Stationary Barrier	e	Moderate-to-low gradient shelf, small shelf sand supply balances sand losses.
		e:estuary
(d) Headland Spit	b	Alongshore sand and gravel supply from local source.
		b:bay
(e) Receded Barrier	e	Moderate-gradient shelf, slow sand loss exceed sand supply.
(f) Mainland Beach Barrier		Steep-gradient shelf, continuing sand loss from eroding coast.
(g) Inner Shelf Sand Sheet	shelf	Low-gradient shelf, seabed erosion, onshore movement of winnowed sand.
(h) Shelf Ridge Field		Shallow, low-gradient shelf, *in situ* reworking, offshore movement of shoreface sand.
(i) Headland-attached Shelf Sand Body		Deep, steep-gradient shelf, offshore movement of shoreface sand.

Figure 4.15. List of wave-dominated coastal deposits formed under quasi-stillstand conditions and the main factors responsible for their formation. Relative magnitudes of long-term sand budgets and net directions of sediment fluxes on the shoreface are indicated by arrows. The jagged arrows in (*g*), (*h*) and (*i*) indicate irregular transport on the inner shelf.

landward from the continental shelf. Except in a few cases, rivers on this coast are still infilling their estuaries or are delivering sand to the coast in quantities that are small in comparison to the littoral sediment flux (Roy & Crawford, 1977; Roy & Thom, 1981). Elsewhere in the world, more active rivers than those that occur in Australia have delivered large quantities of sediment to the coast throughout the Holocene. As a result, coasts in these areas are characterised by barriers at the sediment-rich end of the spectrum of coastal types listed in Fig. 4.15 (transgressive dune and prograded barriers) and are composed of immature fluvial sand. The various types of stillstand coastal deposit are discussed below.

Transgressive dune barriers

The geomorphology and evolution of transgressive dunes are described by Hesp & Thom (1990) who, following others, identify three largely independent factors as being important in their development: rate of sand supply, wind and wave energy and the effectiveness of sand-binding plants. Many transgressive dune barriers occur where sand is supplied by waves in such large quantities to an exposed, windy coast that the local pioneer plants are inundated and killed before they can stabilise the sand in incipient foredunes (Short & Hesp, 1982; Hesp & Thom, 1990). They may also result from the remobilisation of a previously stabilised barrier surface (Davies, 1980); the transgressive dunes at Newcastle Bight, 150 km north of Sydney, are an example (Thom *et al.*, 1992).

Transgressive dune barriers are transitional between barriers with complex foredunes and non-barrier (desert) coastlines with aeolian dune sheets that, in migrating inland, have become detached from the shoreline. The cliff-top dunes described by Short (1988) from the southern Australian coast are detached remnants of what were formerly much larger transgressive dune fields. Originally these incorporated sand ramps a hundred metres or more high connecting coastal sand sources to long-walled parabolic dunes on the cliff tops (Jennings, 1967). Subsequently, when the marine sand supply was cut off, the sand ramps and beaches were destroyed – either blown inland or reworked by waves, or a combination of both. Many cliff-top dunes appear to have formed in the early Holocene while RSL was still rising (Pye & Bowman, 1984), while others formed at various times in the Pleistocene (Bryant *et al.*, in press).

Transgressive dunes contain larger quantities of sand per unit length of coast than other barrier types and occur in the most exposed locations on the southern Australian coast (Short, 1987). In southeast Australia, where they

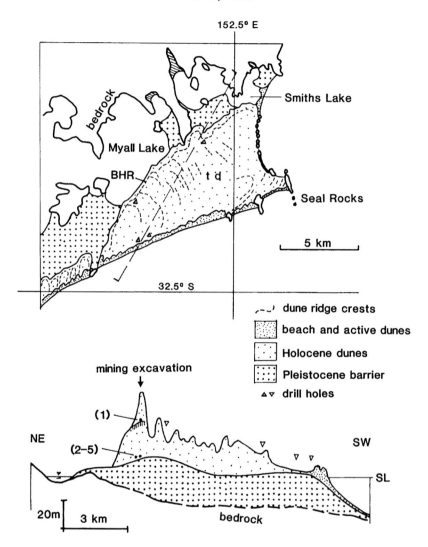

Figure 4.16. Generalised morphology and stratigraphy of the Seal Rocks transgressive dune field (td), Myall Lakes, southeast Australia. Dashed line shows transect in b. (b) Topography along the transect. The landward margin of the dune field is marked by Bridge Hill Ridge (BHR), a large precipitation ridge that reaches elevations of 100 m in places. Radiocarbon dates on charcoal pieces sampled from buried soil surfaces in a sand-mining excavations (1. 1265 ± 135 years BP; 2. 3025 ± 135 years BP; 3. 2485 ± 135 years BP; 4. 1975 ± 110 years BP; and 5. 2600 ± 110 years BP) suggest late Holocene phases of transgressive dune construction for the bulk of the large sand ridge (based on Thom et al., 1992).

occur interspersed with other barrier types, their occurrence can be attributed to locally high rates of sand supply (and trapping) at the downdrift terminus of a littoral drift system (Chapman *et al.*, 1982) rather than to variations in wind energy or vegetation (Fig. 4.15*a*). Here they form long-walled transgressive dunes similar to Cooper's (1958) 'precipitation ridges'. The largest examples are the sand islands of southeast Queensland: Fraser, Moreton and Stradbroke Islands (24.5–28°S, Roy & Thom, 1981; Stephens, 1982). These are composite features comprising beach as well as aeolian deposits that formed episodically throughout much of the Quaternary. A NSW example is Myall Lakes (Fig. 4.16) where the Bridge Hill Ridge reaches elevations of 100 m (Thom *et al.*, 1992). Here, radiocarbon dates from buried swamp deposits and soil horizons on transgressed surfaces are mostly less than 3000 years BP and probably reflect the migration time for the dunes to travel inland (in this case,

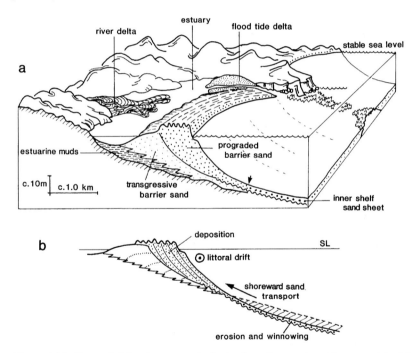

Figure 4.17. Prograded barriers occur in bedrock embayments and on coastal plain coasts. (*a*) The transgressive part was deposited in backbarrier environments (tidal delta and washover facies) as sea level was rising (see Fig. 4.5) (the small arrow indicates the toe of the shoreface). In autochthonous settings such as southeast Australia, the regressive (prograded) part of the barrier is composed of sand winnowed from the adjacent shelf and/or carried alongshore by littoral currents (*b*). A very different situation occurs on allochthonous shelves as shown in Fig. 4.2*c*.

a distance of 2–5 km). Both at Myall Lakes and Evans Head (29.2°S) it is possible that large transgressive dune fields were initiated by sudden changes in coastal alignment following the failure (erosion) of isolated rock outcrops anchoring one end of a barrier system (Roy, 1982). Elsewhere, drilling in episodic transgressive dunes has encountered older soil horizons with ages of 8000–10 000 years BP (Roy & Crawford, 1981; Pye & Bowman, 1984; Cook, 1986), indicating that transgressive dune development commenced in some areas during the postglacial marine transgression well before barriers began prograding. In many cases, these dunes mantle bedrock headlands and may have been associated with former barriers ('proto barriers') and sand ramps that formed at lower sea level (Roy & Crawford, 1981).

Prograded barriers

Prograded barriers or strand plain complexes (Clifton & Hunter, 1982) (Fig. 4.17) are characterised by multiple, coast-parallel beach or foredune ridges, the genesis of which has been discussed by Davies (1957), Bird (1976) and Hesp (1984a,b, 1988). Foredune ridges have an aeolian (foredune) cap on the accreted beach berm which reflects the occurrence of winds sufficiently strong to transport beach sand and a climate able to support sand-binding plants to trap sand on the upward accreting dune surface (Hesp, 1984a). Beach ridges are low-relief, wave-formed berms that rarely rise more than 3 m above mean sea level but foredune ridges and swales have greater amplitude (3–5 m) and crest elevations of 7–10 m above sea level (Thom, Polach & Bowman, 1978).

Beach and foredune ridge plains represent coastal progradation that, in Australia, spans the last 6.5 ka (Fig. 4.18), but in other countries with different RSL histories, progradation began somewhat later (e.g., Galveston Island, Texas, USA, Bernard & LeBlanc, 1965, and the Dutch coast, van Straaten, 1965). Prograded barriers are usually well developed near large, active river mouths on wave-dominated coasts (e.g., Western Africa, Brazil, the Gulf coast of the USA), but in southeast Australia the sand source, in all but a few cases, is the offshore sea bed (Fig. 4.17b) (Thom, 1984). In cross-section, regressive sequences tend to thicken seaward, due to embayment geometry, and become progressively younger towards the present coast (Thom, 1984). Dimensions and shapes of typical prograded barriers have been determined by detailed drilling and dating as exemplified by the Tuncurry Holocene barrier (Roy et al., 1992) (Fig. 4.19a and b); ground-penetrating radar shows its internal structure (Fig. 4.19c). Southeast Australian prograded barriers tend to completely infill bedrock embayments leaving space only for small wetlands

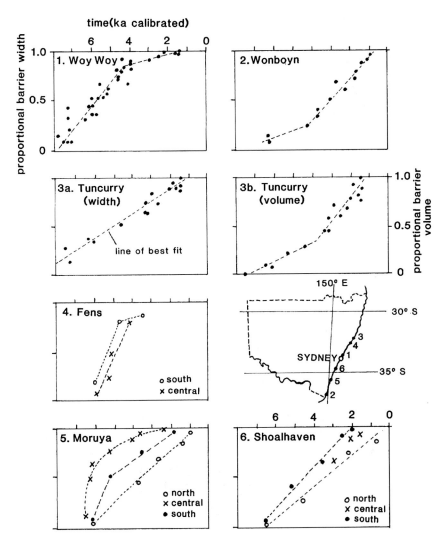

Figure 4.18. Age structure of selected prograded barriers in NSW based on drilling and radiocarbon dating, with sample locations plotted as a proportion of total barrier width (the Tuncurry data is also plotted as a proportion of total barrier sand volume). The dates are on shell fragments and the problems of mixing are discussed in Roy (1991). Sites range from moderately low- energy bays (Sites 1 and 2) to exposed, high-energy beaches on the open coast (Sites 5 and 6); histories of progradation show no consistent trends. Dating of barriers with multiple drilling transects show that shore-normal progradation occurred roughly simultaneously along the length of the barrier, and not by alongshore spit progradation. Average rates of barrier progradation (m a^{-1}) are: Woy Woy 0.57; Wonboyn 0.38; Tuncurry 0.26; Fens 0.35; Moruya 0.34; Shoalhaven 0.24 (based on Chapman *et al.*, 1982 and Thom *et al.*, 1978).

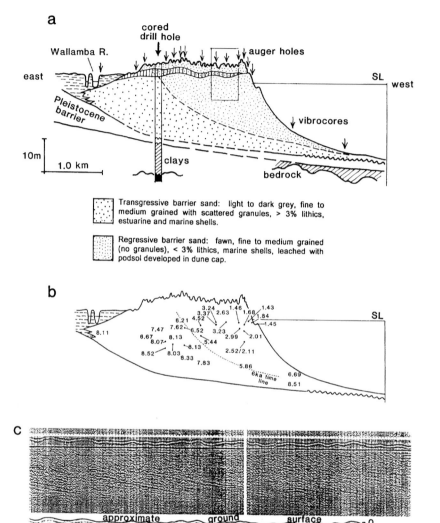

Figure 4.19. The Tuncurry barrier (see Fig. 4.12) is a Holocene prograded bay barrier with multiple foredune ridges. Its subdivision into transgressive and regressive (prograded) components in (*a*) is based on sedimentological studies of drillhole samples and dating of shell fragments. Radiocarbon dates in (*b*) indicate that the transgressive sands are more than 6 ka old and were transported onshore during the closing stages of the postglacial marine transgression. The regressive deposits are less

at their landward margin. This contrasts with stationary barriers that more commonly enclose substantial estuary water bodies.

Within bedrock embayments where inlets are relatively stable, as in southeast Australia, ridge patterns shown on air photos are quite regular and are aligned sub-parallel to the present coast. However, on coastal plain coasts in other parts of the world, complex ridge morphologies arise where some parts of a barrier are eroding while other parts adjacent to inlets are accreting. 'Drumstick' barrier islands are characterised by recurved spits and flared ridge patterns that occur where large ebb tide deltas locally cause wave refraction and drift convergence at the mouths of tidal inlets (Hayes, 1979). Prograded barrier widths vary from a few hundred metres to 10 km or more and reflect different rates of sand supply as well as substrate slope. Ridge spacing typically is in the order of 20–40 m and some prograded barriers have up to 80 ridges. On air photos these appear continuous over distances of kilometres but, on the ground, individual foredunes are quite variable in height, width and continuity.

The generally accepted view that depositional surfaces dip seawards parallel to the present beach and shoreface (Kraft & John, 1979) has been substantiated by recent ground-penetrating radar results (Baker, 1991; Roy *et al.*, 1992) from the Forster/Tuncurry area in southeast Australia (Fig. 4.19c). These penetrate up to 20 m into clean sands in the regressive parts of the barriers, and show complex internal structures dominated by seaward dipping reflectors. Cored drill holes show the barrier sands to be uniformly fine to medium grained and moderately well sorted - subtle changes in sorting or packing are presumably responsible for the discontinuities that produce the reflectors.

Seaward progradation during the stillstand has been confirmed by ^{14}C dating of numerous barriers (Thom *et al.,* 1978, 1981; Thom, 1984). Some barriers show an early phase of rapid growth at the beginning of the stillstand (6500–3000 years BP) – possibly due to initially abundant sand reserves – while others appear to have prograded relatively constantly (Fig. 4.18). The dates suggest that barriers ceased growing in the last 1000–2000 years but, in most cases where new barrier sand is derived through reworking of older

Caption for Figure 4.19 (*cont.*).
than 6 ka and young in a seaward direction indicating shoreface progradation during the stillstand under environmental conditions generally similar to today (see Fig. 4.18). Ground-penetrating radar results (*c*) are from a shore-normal transect through the seaward part of the barrier (see box in *a* for location). Reflectors suggest that beach bedding occurs above present mean sea level and dips seaward at 3° to 4° flattening to about 1° below sea level, corresponding to the upper shoreface zone. Irregularities at the contact of the two zones probably represent relict surf zone morphologies.

deposits on the continental shelf, this is an artefact caused by mixing of shell fragments of different ages (Nielsen & Roy, 1982; Roy, 1991). In fact, most barriers on this coast are probably still prograding, albeit slowly. Average rates of horizontal shoreline progradation vary considerably but, in southeast Australia, have rarely exceeded $1.0 \, \mathrm{m \, a^{-1}}$ during the stillstand period, and most are less than $0.5 \, \mathrm{m \, a^{-1}}$ (Fig. 4.18). Since growth involves not only the beach and foredune, but also shoreface/nearshore sands that often extend down to water depths of 20–25 m, the quantity of sand involved is in the order of 15–$30 \, \mathrm{m^3 \, m^{-1} \, a^{-1}}$. Multiple transects dated from the same bay barriers at Moruya and Shoalhaven (Fig. 4.18) show shoreface accretion to have occurred almost simultaneously along the entire barrier.

Stationary barriers

Stationary barriers are recognised on the basis of two criteria: firstly, the absence of significant evidence of progradation over the past 6000 years; and secondly, the presence of complex foredune structures (Thom, 1974). These types of barrier are intermediate between eroded and prograded coasts and possibly fluctuate between these two conditions to give the impression of relative stability over long periods of time (Shepherd, 1991). In southeast Australia they tend to occur in compartmented embayments that, at the end of the postglacial marine transgression, apparently had limited reserves of sand available for barrier building (Roy & Thom, 1981; Thom, 1984). They often enclose substantial estuary water bodies, and in places appear to be anchored on remnants of earlier Pleistocene bay barriers (Roy & Thom, 1981).

The stationary barrier superstructure is usually narrow (0.5 to 1.0 km) with beach and dune facies fronting a well-developed backbarrier sand flat (Thom *et al.*, 1978). On tectonically stable coasts such as southeast Australia, the backbarrier sand flats are relict features that ceased forming once sea level stabilised and the development of a foredune ridge prevented overwash from occurring; their radiocarbon ages in Fig. 4.19 are older than 7 ka (Thom, 1978, 1984). Their preservation behind stationary barriers is generally considered to be evidence of shoreline stability (on eroding coasts, backbarrier deposits are destroyed by the landward translation of the shoreface). However, some stationary barriers may represent a transitional stage in the evolutionary progression from prograded barrier to receded barrier/mainland beach (Fig. 4.20). This follows an hypothesis by Roy *et al.* (1980), that barrier coasts experienced a growth phase in the mid Holocene as the sea bed re-equilibrated and sand moved onshore, followed by an erosional phase in the late Holocene as shelf sand reserves were depleted and coastal sand budget losses exceeded gains.

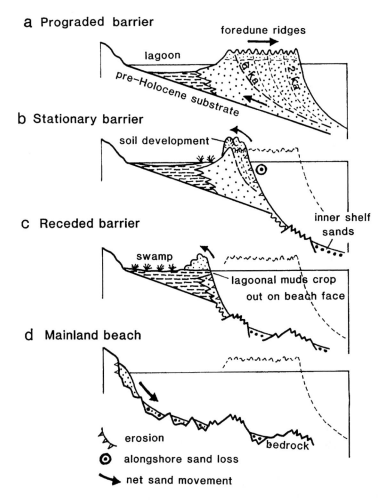

a **Prograded barrier**

foredune ridges

lagoon

pre-Holocene substrate

b **Stationary barrier**

soil development

inner shelf
sands

c **Receded barrier**

swamp

lagoonal muds crop
out on beach face

d **Mainland beach**

erosion

bedrock

⊙ alongshore sand loss

net sand movement

Figure 4.20. The succession of stillstand barriers, from prograded to mainland beach (*a* to *d*), reflect diminishing quantities of sand at the coast. The store of available sand may be an inherited feature of the inner shelf or, as is indicated here, arise through progressive erosion of an initial prograded barrier (dashed line) throughout the stillstand. The erosion states shown by (*b*), (*c*) and (*d*) could represent progressive time steps (durations of erosion) or increased intensities of erosion.

Dunes on stationary barriers may rise to heights of 30–40 m and their morphology is typically hummocky. Discontinuous buried podsols indicate alternating periods of stability (soil formation) and localised sand mobility in blowouts. Commonly, dune development is asymmetrical alongshore with the largest dunes at the downdrift end of the embayment where the beach is most exposed to onshore winds. Inland dune migration at the northern end of some bay barriers causes the barrier surface to widen to the north (Hesp & Thom, 1990).

Beneath the dune cap, stationary-barrier stratigraphy is characterised by a relatively thick unit (10–15 m) of backbarrier/transgressive sands dating 5–7+ ka old that merge landward into estuarine sediments, and into beach and shoreface sands to seaward. Since, by definition, stillstand progradation of stationary barriers has been limited, beach and shoreface sands are not well developed. For example, a stationary barrier at Merimbula (36.9° S) on the New South Wales south coast has only two or three foredune ridges. Thom (1984) suggested that stationary barriers could have stabilised before the end of the postglacial marine transgression and, for a time, grew upwards in response to rising sea level as may have occurred at Padre Island in Texas (Fisk, 1959). This is difficult to document as the available data on environments and times of deposition are not sufficiently precise. Alternatively, it is as likely that most barriers stabilised when sea level stopped rising, and foredunes began growing upwards when washovers ceased.

Headland spits

Headland spits are stillstand barriers that have principally grown alongshore in response to a dominant littoral drift (Allen, 1982). Like all barriers they rely on a supply of sediment and sufficient accommodation space, in the form of an open embayment or shelf, in their direction of growth (Fig. 4.1, inset). They are found on rocky, embayed coasts (e.g. Britain; King, 1972), glaciated coasts (Leatherman, 1987) or adjacent to sandy river mouths on drowned coastal plain coasts (Senegal delta; Wright, 1985). Spit morphologies and stratigraphies range from prograded, stationary to receded barrier types depending on the balance between onshore and alongshore sediment supplies. Criteria for distinguishing between headland spits and barrier island spits include the nature of their underlying lithofacies (Fig. 4.1, insets). Since headland spits have grown from some point of attachment across an embayment, the barrier sands overlie bay or shallow shelf facies. Barrier island spits, on the other hand, lie on an eroded old substrate (where they are associated with a deep scouring inlet) or with estuarine/lagoonal or related backbarrier facies. The progressive alongshore growth of headland spits is characterised by the curvature of prograded beach ridge ends. Although tidal currents are involved in building the subtidal platform, new ridge progradation occurs episodically during storms that erode the seaward face and overwash and recurve their exposed ends. The recurved ridges enclose intertidal deposits that provide a record of progressive spit growth (Schwartz, 1972). The change from updrift erosion to downdrift accretion is usually accompanied by a progressive decrease in the mean size of the beach sediment.

Receded barriers

Receded barriers occur on coasts undergoing a marine transgression and long-term erosion either because RS is rising (e.g., US Atlantic coast) or because of a negative imbalance in the sediment budget (parts of the New South Wales coast). They are distinguished by outcrops on the beach of tidal flat muds or freshwater peats, which indicate that the shoreface has retreated landwards to completely erode the former barrier (Fig. 4.20c). Typically, beach and foredune deposits form a relatively narrow and low ridge on top of relict backbarrier sediments over which they have translated. Dates reported by Thom *et al.* (1978) from southeast Australia indicate that here recession has occurred over the past 3000 years. Barriers with long recession histories probably indicate embayments with very limited sediment input throughout the Holocene (Fig. 4.15e); Fig. 4.20c suggests another scenario involving loss of sediment from a compartment to create a negative sediment imbalance and coastal recession. Under stable sea-level conditions as in southeast Australia, erosional retreat of the barrier superstructure involves slow encroachment of the foredune into estuary/lagoon systems to the rear of the barrier – washovers are virtually non existent under these conditions. Thicknesses of receded barrier deposits are less than 10 m, much less than other types of barrier on the same coast. In contrast, transgressive barriers, which are also associated with coastal recession, typically are thicker than receded barriers and contemporary storm washovers and tidal inlet deposits form their dominant lithofacies. The US Atlantic coast contains examples of barriers that are receding for both reasons (sea-level rise and sand deficit). Gravel barriers move in a similar fashion (Orford, Carter & Forbes, 1991; Forbes *et al.*, 1991), and under conditions of shoreface recession, relict gravel deposits may be locally exposed and reworked into the barrier (see also Chapters 1 and 10).

Mainland beach barriers

In the spectrum of barrier types, mainland beaches are an end-member. They comprise thin veneers of beach and shoreface sand – usually less than 5 m thick – mantling a pre-Holocene erosional substrate (Figs. 4.15f, 4.20d), which may be bedrock. On embayed coasts, rock outcrops are common as cliffs at the rear of the beach. These beaches often contain gravel layers derived from cliff erosion and submarine rock outcrops on the nearshore sea bed where the sand cover is particularly thin. Alternatively, the substrate may be made up of semi-consolidated, Pleistocene coastal sand deposits impregnated with humate as in northern New South Wales (Thom, 1984), cemented calcareous sands as in southern Australia (Short, 1988) or old piedmont fan deposits as in the Gulf of Thailand. In these cases, the mainland beach is composed mainly of weathered products eroded from the adjacent coast.

The main stratigraphic characteristics of mainland beaches is that they represent a zone of contemporary reworking. During particularly violent storms, beach and nearshore sands are substantially reworked to expose the substrate; after a storm the beach reforms, often with newly eroded parent material. Conditions are essentially erosional over the long term (Fig. 4.15*f*). On littoral drift-dominated coasts, mainland beaches represent a moving carpet of sand eroded from the backshore. In both southeast and southwest Australia, many of the economic heavy mineral deposits occur within these types of deposit. In pocket beach embayments, they reflect a sediment deficit which is usually associated with steep offshore slopes. In the most extreme case, this leads to the formation of gravel or boulder beaches at the base of eroding bedrock slopes (similar to those formed from eroding drumlins in Nova Scotia, Boyd *et al.*, 1987; Carter *et al.*, 1987; Carter & Orford, 1993, see Chapters 1 and 10).

Inner shelf sand deposits

Of all the various morphologies of wave-dominated deposits, inner shelf sand sheets such as those from southeast Australia (Kudrass, 1982, Colwell & Roy 1983; Bergs, 1993) and shelf ridge fields of the type described from the US Atlantic shelf (e.g., Duane *et al.*, 1972; Swift *et al.*, 1978; Field, 1980; Swift & Field, 1981) are the most extensive in terms of area and probably, on some shelves, in overall volume as well. Despite their remoteness from the present-day coast, these features are included with other wave-formed coastal deposits because of the substantial, if not dominant, role played by waves and associated currents in their formation.

Sediment dispersal on storm-dominated shelves is described by Wright *et al.* (1991) as driven by both regional and local weather systems and occurring in a hierarchy of spatial scales. At a scale of tens to hundreds of kilometres, regional variations in shelf width and depth define along-shelf zones with increasing and decreasing flow velocity at compartment boundaries. Smaller dispersal subsystems that operate within compartments are superimposed on these systems. These involve both alongshore and across-shelf flows that lead

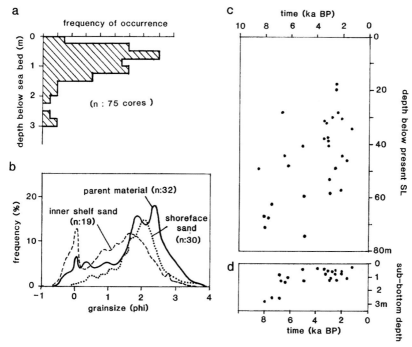

Figure 4.21. (*a*) Thickness of the inner shelf sand sheet, which is mainly less than 1.5 m thick, based on vibrocores from the shelf in various parts of southeast Australia. A detailed study of vibrocores from the Tuncurry shelf (Bergs, 1993) found evidence of winnowing of the inner part of the inner shelf sands in water depths of 20–35 m. The three curves in (*b*) represent envelopes encompassing multiple grain size frequency curves from the inner shelf sand sheet, the underlying parent material and barrier sand from the adjacent shoreface. The shoreface sands are much finer than the inner shelf sands and slightly finer than the parent material. Radiocarbon dates on shells from the inner shelf sand unit are plotted against water depth (*c*) and sub-bottom depth (*d*). The ages mostly correspond to the stillstand period and the absence of significant trends in both plots probably reflects contemporary reworking of the deposits.

to shoreface accretion/erosion, especially where the shoreface interfaces with estuarine and shelf regimes.

Inner shelf sand sheets

In southeast Australia, inner shelf sands form a surficial sheet-like unit that generally is between 0.25 and 1.5 m thick (Fig. 4.21*a*) and occupies a coast-parallel zone 5–10 km wide in water depths of 20-70 m (Fig. 4.21*c*) (Roy & Stephens, 1980). Off headlands, the continuity of the zone may be interrupted by exposed bedrock outcrops or shelf sand bodies (see below) (von

Stackelberg, 1982; Roy & Stephens, 1980). While the dominant sediment type is clean, medium sand, variations include shell gravels near reefs and irregular patches of coarse gravelly sand that are particularly common near the toe of the shoreface (Swift *et al.*, 1991a; Roy & Cowell, 1991). Shelf sands may be blanketed by mid shelf fine sands and muds along its seaward margin (Davies, 1979; Marshall, 1980; Colwell & Roy, 1983). The sand sheet rests on a ravinement surface marked by a discontinuous and generally thin gravel lag that contains robust estuarine shell valves, reworked during the postglacial marine transgression from former estuarine/lagoonal deposits on the shelf. Elsewhere on the inner shelf, shells are predominantly marine species and are usually abraded by prolonged wave reworking. Their ^{14}C ages range from 1–8 ka (Fig. 4.21*c*) indicating a stillstand age of formation in southeast Australia, but show no trends related to present water depth or sub-bottom depth (Fig. 4.21*d*). Ages of estuarine shells from the basal lag tend to be somewhat older than the marine shells and to become younger in a shoreward direction (Roy *et al.*, 1992).

The inner shelf surface is planar and slopes gently seaward (0.2–0.3°). Bedforms range from megaripples in patches of coarse sand to low-amplitude (1-3 m), long-wavelength (300–1000 m) sand ridges. The latter are generally ill-defined but in some areas appear to be aligned obliquely to the coast (Roy, 1985). Comparisons of grain-size frequency curves from inner shelf sands, the underlying parent material and barrier sands on the adjacent shoreface (Fig. 4.21*b*), suggest that episodic storm reworking is accompanied by winnowing and, in water depths less than 35 m, there is a tendency for fine sand to migrate shoreward (Bergs, 1993). However, there seems to be no preferred direction of mass transport for the sand sheet as a whole (see Fig. 4.15*g*). Certainly, there is far less tendency to reorder the southeast Australian inner shelf into the ridge and swale morphologies as found on other shelves (Swift *et al.*, 1991a).

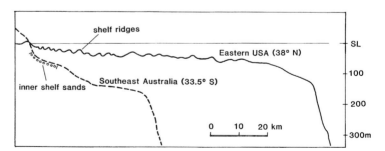

Figure 4.22. Representative shelf cross-sections from eastern USA and southeast Australia. The extent of the inner shelf sand sheet is indicated on the Australian section; in the US example, the same zone probably incorporates the ridge fields and covers virtually the entire shelf surface.

Shelf ridge fields

Linear ridges composed of reworked shelf sands reportedly occur on many temperate latitude shelves. The best-studied examples of wave-formed ridges are described from the mid-Atlantic shelf of the USA (e.g., Duane *et al.*, 1972; Swift *et al.*, 1978; Field, 1980; Swift & Field, 1981; Swift, McKinney & Stahl, 1984), and are distinct from the tidally formed sand ridges of the North Sea (Stride, 1982). Shelf ridge fields appear to be absent from the southeast Australian shelf, which extends to twice the water depth of the Atlantic shelf (Fig. 4.22) and is therefore more likely to experience Csanady's (1977) baroclinic coastal boundary layer, in which the bed is protected from storm-driven currents by the thermocline. On the Atlantic shelf, fields of sand ridges mainly occupy the broad interfluves or shelf segments between shoal retreat massifs and shelf valley complexes that represent retreat paths during the postglacial marine transgression of cuspate headlands and large estuary mouths respectively (Swift, 1975b). The ridges are up to 10 m high (mean height 3–4 m), 2–4 km apart and make angles of 25–45° with the coastline, opening into the prevailing southwest storm flow (Duane *et al.*, 1972; Swift & Field, 1981). Some of the largest ridges reportedly occur on the front of Sable Island on the exposed outer shelf off Halifax, Canada (R. Dalrymple, pers. commun., 1993). Here storm wave heights regularly exceed 10 m suggesting that ridge development is linked to storm intensity. The ridges are initiated on the barrier shoreface by down-welling flows that carry sand obliquely seaward by advection. Swift *et al.* (1991a) believe them to be an autochthonous response to transgressive conditions on a low-gradient, storm-dominated shelf; they reflect variations in the littoral dispersal system at the leading edge of the transgression.

A model for the formation of the ridge fields involves erosion of the shoreface with newly formed ridges being let down onto the inner shelf surface where they continue to evolve through erosion and ridge crest accretion. As a result, the shelf sand sheet becomes discontinuous as Holocene backbarrier deposits or older Pleistocene or Tertiary substrates are exposed in erosional windows between ridges (Belderson & Stride, 1966; Swift, 1976b; Swift *et al.*. 1985). Relative sea level along the Atlantic Bight coast has been slowly rising (c. 2–5 mm a^{-1}; Belknap & Kraft, 1977) since the mid Holocene and, as the water column deepens over the shelf surface and the shoreline recedes, ridges undergo continuing evolution; ridge heights and spacings increase, side slopes decrease and with time the ridges become senescent or moribund in depths of 50–60 m on the mid shelf.

The morphological differences between the planar sand sheets of the southeast Australia shelf and the more irregular ridge morphologies on the US

Atlantic shelf reflect different environmental conditions: both are storm-dominated environments, but compared with Australia the US Atlantic shelf is shallower and wider (Fig. 4.22; see also Wright, 1976) and presently is experiencing a slow marine transgression. As a result, its water mass is less stratified and bottom sediments experience more effective reworking associated with downwelling storm currents that carry shoreface sand offshore. The resulting long-term deficit in the coastal sand budget created by this mechanism ensures coastal recession at rates exceeding that attributable to the marine transgression alone. In contrast, the somewhat deeper inner-shelf region off embayments in southeast Australia is characterised by more-frequent occurrences of swell during intervals between storms, which give predominance to the onshore component of cross-shelf transport (Wright *et al.*, 1991). The occurrence of prograded barriers in many embayments indicates that downwelling-induced offshore transport tends to be more than offset by onshore transport by shoaling swell waves in the intervals between storms on this coast. This situation is not ubiquitous, as is discussed in the following section dealing with shelf sand bodies found attached or adjacent to prominent headlands on the southeast Australian coast.

Headland-attached shelf sand bodies

Headland-attached shelf sand bodies (SSBs) occur on the inner continental shelf off prominent headlands in southeast Australia. They form convex-up accumulations of shelf sand which are generally 20 to 30 m thick, 2 to 4 km wide and extend along-coast for distances of 5 to 30 km in water depths of 25 to 80 m (Ferland, 1986, 1990) (Fig. 4.8). SSBs are the product of waves and currents and have accumulated subaqueously both during the postglacial marine transgresssion and the subsequent stillstand on relatively steep substrates in the presence of an adequate sand supply (Field & Roy, 1984; Roy, 1985). In terms of volume they contain between 0.5 and 1.0 km^3 of sand and are comparable to medium-sized bay barriers (Ferland, 1990). Research in southeast Australia indicates that SSBs were initiated as transgressive sand sheets in areas where variations in continental shelf relief in the vicinity of prominant bedrock headlands interrupted the continuity of coast-parallel sand transport at times of lower sea level. The inner shelf adjacent to these prominent headlands is extremely narrow, with steep shelf gradients (>1°) and water depths of 50 to 70 m within 2–3 km of the coast; the headlands themselves are often capped by cliff-top dunes.

Initially, a few SSBs were recognised as anomalously thick deposits of clean, quartzose sand on the inner shelf (von Stackelberg, 1982; Field & Roy,

1984). More-detailed analyses of their location and morphology have shown that the linear, shore-parallel features occur adjacent to almost every prominent headland in southeast Australia (Ferland, 1990). SSB morphology consists of a low-gradient planar surface (<0.5°) and a steep, convex seaward slope (1–5°) (Fig. 4.8). Seismic profiles show that SSBs are characterised by discontinuous, seaward-dipping internal reflectors, which are particularly apparent in the convex upper portion and indicate offshore sediment transport.

Vibrocoring and reverse-circulation drilling in the top 5–9 m of a number of SSBs have encountered fine to medium grained, moderately to well sorted, quartz sand with 10–15% biogenic carbonate and usually less than 1% mud (Kudrass, 1982; Colwell & Roy, 1983; Roy, 1985; Hudson, 1985). The mud increases slightly in the toe of the SSB where energy levels are lowest. Seaward of this the shelf sediment is fine to very fine carbonate-rich muddy sand (the mid-shelf muddy sand facies of Roy & Stephens, 1980; Roy & Thom, 1981).

The macro- and micro-fauna of the SSB sand indicates deposition under relatively shallow-water, high-energy conditions such as those found on the inner shelf today. Extensive radiocarbon dating defines two stratigraphic units (Fig. 4.8c): (i) A lower, fine-grained sand unit, similar to the lower shoreface sand in the adjacent barriers, which is more than 7 ka old and was deposited during the postglacial marine transgression. (ii) An upper unit that is often coarser grained, contains barnacles and rock fragments from the adjacent cliffs, and is less than 7 ka old (Hudson, 1985; Roy, 1985). 'Geologically averaged' rates of accumulation of the upper unit are in the order of 1 m/1000 years (Ferland, 1990).

A model of SSB development is based on detailed studies of a large SSB at Sydney (Field & Roy, 1984; Roy, 1985; Ferland, 1990; J. Hudson, pers. commun., 1992). Sedimentation was initiated during the postglacial marine transgression when the rising sea encountered steeper than normal substrate slopes off some sectors of coast. The depositional environment was essentially part of the lower shoreface (Fig. 4.8b) and the result was a transgressive sand sheet whose thickness was a function of sand supply and substrate slope. While composed in part of barrier sand, these features are not thought to be drowned or overstepped barriers. The oversteepened seaward face of the transgressive deposits ensures that, under stillstand conditions, offshore sand transport continues under the influence of storm downwelling currents. However, progradation rates have varied depending on sediment supply and regional setting. SSBs in northern New South Wales are connected to contemporary littoral drift systems and occur in shallower water depths than SSBs on the south coast, which is compartmented (Roy & Thom, 1981; Ferland, 1988, 1990). Fig. 4.23 illustrates a range of stillstand stratigraphies that reflect

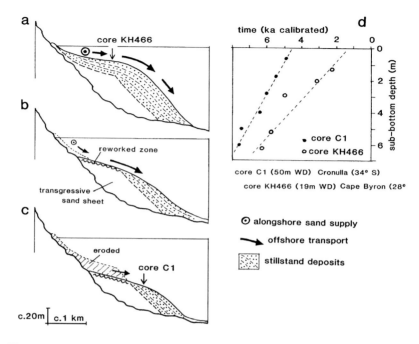

Figure 4.23. As shown in Fig. 4.8, shelf sand bodies (SSBs) are composite features with a transgressive core and a regressive, seaward part formed under stillstand conditions. The development of the latter deposits involves a net offshore sand transfer and the resulting SSB morphology thus depends on the local sediment budget. With a large littoral sand supply (*a*) the SSB aggrades, as well as progrades. The lesser supply shown in (*b*) maintains the upper surface of the SSB, which acts as a bypass zone for sand moving offshore. If the sand supply is cut off (*c*), the upper surface of the SSB slowly degrades as downwelling currents carry sand seaward. Dated vibrocores in (*d*) show the age structure of two regressive SSB deposits corresponding to cases (*a*) and (*c*). The time-lag effect of intermixing old shell is more evident in core C1 from the eroded SSB than from core KH466 from the aggraded SSB.

different rates of sand supply. Fig. 4.23*a* represents the Cape Byron SSB where a littoral sand supply of 100s of thousands of cubic metres to the north annually has resulted in stillstand aggradation of at least 6 m of sediment, as well as rapid seaward progradation (Roy & Stephens, 1980; Kudrass, 1982). Fig. 4.23*b* refers to the southern part of the Sydney SSB (lat. 33.7° S). Here, alongshore sand transport rates were less than at Cape Byron and have declined over the stillstand, but were sufficient to prograde the front of the SSB without causing severe erosion of its upper surface. Fig. 4.23*c* exemplifies a SSB that has had its along-shelf sand supply cut off during the stillstand. It is modelled on the northern part of the Sydney SSB, which occurs in deep water and is compartmented by rock reefs. In this case, progradation on the SSB's

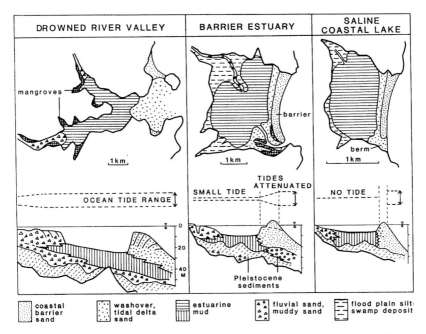

Figure 4.24. Three main estuary types in New South Wales showing idealised sediment distributions in plan and in section (dimensions are only approximate). Tidal ranges in the estuaries are shown in relation to that in the open ocean. (From Roy, 1984.)

seaward face was supplied by erosion of its upper surface, together with contributions from the coastal cliffs and submarine rock reefs. Some of the very deep SSBs on the far south coast may be essentially moribund in terms of sediment supply, although their convex morphology is apparently maintained by contemporary processes (Ferland, 1990).

Estuarine evolution during stable sea levels

Southeast Australian estuaries occur behind wave-built sand deposits and have evolved under stillstand conditions as one of three primary estuary types (Roy *et al.*, 1980; Roy, 1984, in press) (Fig. 4.24). Here, the coast has 1.5–2.0 m tides and is exposed to moderately high-energy waves that tend to suppress the development of ebb tide deltas except at the mouths of larger rivers (e.g., the Clarence River in northern NSW at 29.4°S; Floyd & Druery, 1976, and Chapter 5). 'Drowned valley estuaries' occupy deeply incised palaeo-valleys. Present-day estuary mouths are wide and deep and open into semi-enclosed bays where wave energy levels are considerably reduced. These estuaries tend to be tide-dominated and to contain large flood-tide delta sand bodies that have grown landward for distances of 5–10 km during much of the Holocene. Intermediate-sized flood-tide delta deposits are found in 'barrier estuaries' with constricted inlet channels that reduce tidal prisms but remain permanently open to the sea. These estuaries occur in broad, shallow palaeo-valleys with bay barriers across their mouths. Radiocarbon dating of these flood deltas suggests that the bulk of their sand was deposited in the closing stage of the postglacial marine transgression and only a small amount has accumulated during the stillstand (Roy, 1984; Nichol, 1992). The smallest inlet deposits – formed mainly of storm washover lobes – occur in the mouths of 'saline coastal lagoons' whose inlets are mostly closed and only open briefly when flood waters breach the beach berm. The catchments of these estuaries are so small that the accretionary action of waves on the beach face far exceeds the effect of stream discharge at the estuary mouth (see also Chapter 6). Thus, geologically inherited factors such as bedrock geometry and river discharge not only determine estuary type along the southeast Australian coast, but also the rate of estuary infilling with sediment (Roy *et al.*, 1980; Roy, 1984).

Discussion and conclusions

The large-scale behaviour of wave-dominated coasts can be understood in terms of two concepts. The first relates to geological inheritance, or the imprint of various land-forming processes that have operated over very long periods of geological time to create the regional landscapes in the vicinity of the present-day coast. Regional substrate slopes and the extent to which the coast is laterally bounded by headlands influence characteristics of wave-formed coastal deposits, such as their spatial extent, while climate and the lithologies in the catchments control their sediment composition. Three general coastal types can be recognised on this basis: low-lying coastal plain coasts, more irregular embayed coasts and relatively steep, cliffed or protruding sectors of coast (see Fig. 4.4*a*). Also, the pre-existing substrate beneath the present-day coastal deposits is an inherited property, and its morphology and material properties strongly influence the nature of coastal development.

The second concept involves the evolution of wave-formed coastal deposits, operating over shorter periods of time than the first (usually hundreds or thousands of years) and concerns the large-scale morphodynamics of the deposits themselves. Local reworking of sediments by waves and currents under changing environmental conditions of RSL and sediment supply create particular situations in which different coastal morphologies can evolve (see Chapter 2). Resulting coastal deposits are characterised by their cross-sectional size, shape and the stratigraphic architecture of their lithofacies (Fig.

4.25). On this basis, three main groups of wave-formed coastal deposits are recognised: those formed during marine transgressions, those formed during relative stillstands of the sea, and those formed during marine regressions (Fig. 4.25). Most of these deposits are coastal sand barriers of various types, but a number of subaqueous sand deposits on the inner shelf are included, whose origin is due to wave action working in concert with currents. The character of the transgressive and regressive barriers principally reflects changing sea levels, but under stillstand conditions imbalances in the shelf sand budget are responsible for producing a spectrum of different size barriers, some of which may be composite features (e.g., prograded barriers that have been destabilised and are now blanketed by transgressive dunes; Thom, 1978). The dichotomy between the roles of changing sea level on the one hand and sand budget imbalances on the other, which forms the basis for understanding coastal evolution, becomes blurred when both factors are small and act together. For example, the eastern seaboard of the USA is presently experiencing a slow marine transgression and in many areas this phenomenon is clearly the over-riding control on coastal development. In other areas, however, shoreface retreat (or even progradation) is occurring due to local imbalances in the marine sediment budget which either reinforces or counterbalances the effect of rising sea level. The balance between these forcing factors changes subtly along sectors of coast and is responsible for the diversity of coastal landforms and their histories of evolution.

It follows from this analysis that the nature of a particular barrier provides a measure of its long-term sediment budget and of the coastal changes that have occurred at the site. However, such interpretations need to be substantiated by quantifying specific mechanisms, such as littoral drift imbalances, that are responsible for sand losses or gains. For example, it has long been assumed that measured rates of coastal recession of 0.5–$2.0\,\mathrm{m\,a}^{-1}$ in parts of southeast Australia (especially the north-coast region) are attributable to naturally occurring imbalances in the littoral transport system (Stephens, Roy & Jones, 1981; Gordon, 1988). Over a distance of about $600\,\mathrm{km}$, northward drift rates progressively increase from zero to perhaps $10^6\,\mathrm{m}^3\,\mathrm{a}^{-1}$ at Fraser Island ($25°\mathrm{S}$). Long-term deficits between inputs and outputs are thought to have been made-up with sand eroded from the beach and shoreface – a natural phenomenon presumed to have operated throughout the late Holocene. The unlikely implication of this is that the coastline in mid-Holocene times would have been located several kilometres further to the east, which, in most areas, puts it seaward of the present bedrock headlands (Chapman *et al.*, 1982). An alternative hypothesis is that the littoral-transport system, rather than being fed by coastal erosion, is sustained by sand that moves onshore from the inner

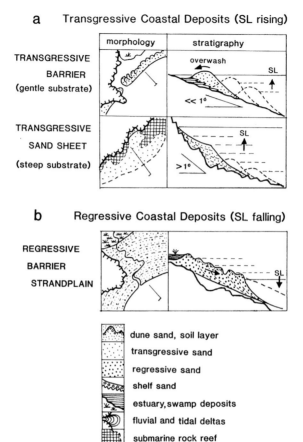

a   Transgressive Coastal Deposits (SL rising)

b   Regressive Coastal Deposits (SL falling)

dune sand, soil layer
transgressive sand
regressive sand
shelf sand
estuary, swamp deposits
fluvial and tidal deltas
submarine rock reef

Figure 4.25. (Caption on p. 171.)

shelf – a process that has operated over thousands of years to form prograded barriers in compartmented embayments elsewhere on this coast (Roy & Thom, 1981). This hypothesis is not directly testable since small annual contributions of a few cubic metres of sand per metre length of coast are unmeasurable in an open drift situation, especially when they are overprinted by massive onshore/offshore sand movements during storms (Thom & Hall, 1991). Nevertheless, it is more consistent with the principles of large-scale coastal behaviour and geological sediment budgets on which ideas of coastal evolution are presently based. Within this framework, historical erosion trends in particular areas are more likely to be due to the sand-trapping effects of a century of coastal engineering works, such as breakwaters, than to natural phenomenon spanning millennia.

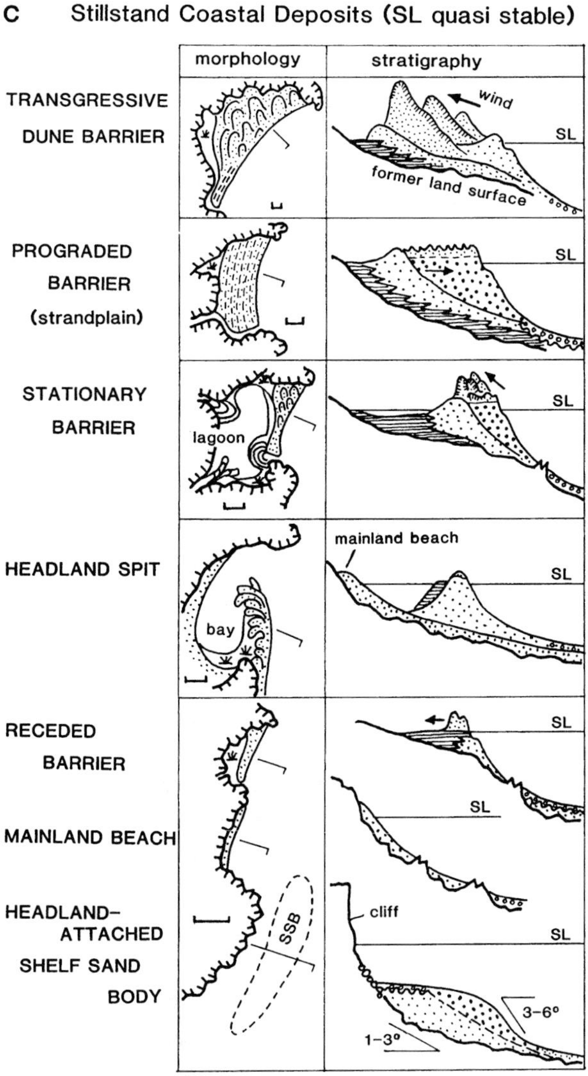

C   Stillstand Coastal Deposits (SL quasi stable)

Figure 4.25. Generalised geometries and stratigraphies of wave-formed coastal features (barriers and inner shelf deposits), based on southeast Australian examples.

Fig. 4.26 is an estimate of the relative sizes and lithofacies make-up of coastal deposits found on tectonically stable and moderately high-energy coasts such as those of southeast Australia, where sediment inputs are small. While limited data exist at this time, remnants of transgressive barriers are rare on this shelf and mainly occur in the core of stillstand barriers at the present coast.

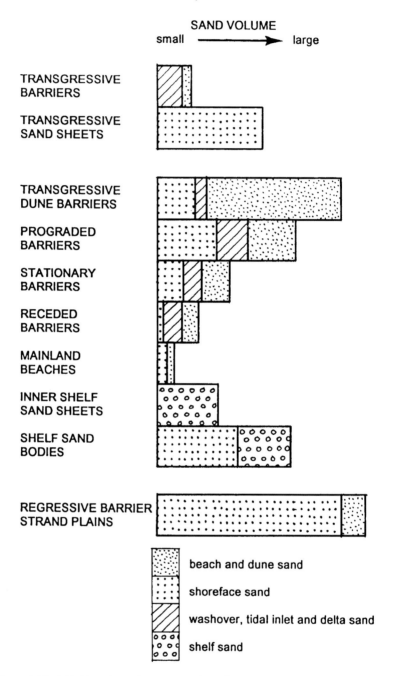

Figure 4.26. A highly schematic representation of the relative size and compositional make-up of typical examples of the various types of coastal deposits formed under both changing and stable sea-level conditions.

Transgressive sand sheets, on the other hand, are thought to be more common on steep sectors of the southern and central shelf where they occur seaward of, and in association with, SSBs. In very general terms, the stillstand barriers on the present coast belong to a size-continuum, ranging from large transgressive dune barriers to small mainland beaches, which reflects variations not only in sediment supply, but also in local wave- and wind-energy regimes. The little-understood interplay of these factors is further complicated by the geometry and accomodation space of individual compartments, which can disrupt the size-relationships between the different barrier types shown in Fig. 4.26. For example, some prograded barriers in long embayments far exceed small transgressive dune barriers in size. Compared with the barriers, stillstand deposits on the shelf are relatively thin; however, the inner shelf sands in particular extend over thousands of square kilometres and may be more important than is shown in Fig. 4.26. Their widespread occurrence reflects not only the intensity of storm wave-reworking, but also the autochthonous, sediment-starved nature of the southeast Australian shelf. These relatively coarse-grained shelf sands are relatively under-represented on allochthonous shelves (e.g., the Oregon-Washington shelf, USA), even where energy levels are quite high. Deposits of shoreface sands in drowned, interstadial barriers (regressive strand plains) appear to be particularly widespread on the southeast Australian shelf, especially on its northern part. Here, stacking of successive strand plains has apparently led to a regional shoaling of the shelf surface by as much as 40 m (Roy & Thom, 1981; Schluter, 1982).

The geological preservation of the highstand coastal deposits schematised in Fig. 4.25c depends on positive tectonic movements of the type described by Cook *et al.* (1977) for the Murray Basin coast, which elevated the deposits above sea level. Alternatively, under subsiding conditions their incorporation in the rock record requires a very large and prolonged influx of sediment, such as occurs in depositional basins. However, on sediment-starved coasts undergoing slow tectonic subsidence, earlier highstand deposits tend to be eroded during successive marine transgressions. Thus, the absence of a succession of Quaternary highstand barriers in southeast Australia probably indicates that the region is stable or is very slowly subsiding (Bryant, Roy & Thom, 1988). The effects of more rapid glacio-eustatic changes in sea level on barrier preservation are more problematic. Compared with large, prograded barrier systems, small highstand deposits (stationary and receded barriers, and mainland beaches) are less likely to be preserved in the geological record because of their erosional setting, especially under renewed transgression. On the other hand, deposits such as transgressive sand sheets and shelf sand bodies that were formed under transgressive conditions (Fig. 4.25a) are unlikely to survive a marine regression. The same applies to flood-tide delta sand bodies in drowned-valley estuaries, which are susceptable to fluvial erosion during sea-level lowstands. In contrast, regressive barrier strandplains (Fig. 4.25*b*) seem to experience little erosion during marine transgressions, except where sand deficits are very large. Considering the long periods of the Late Quaternary during which sea levels have been slowly falling, it is reasonable to expect that regressive barriers are more widespread on low-gradient shelves than hitherto reported in the literature.

In the context of large-scale coastal behaviour (Chapter 2), the geological sediment budget concept refers to the net product of the various processes that cause erosion, transport and deposition of sediment within a sector of coast and continental shelf over geological time spans of hundreds to thousands of years. At the largest scale, Roy & Thom (1991) and Roy & Keene (1993) describe the geological sediment budget of the Tasman Sea margin of southeast Australia as involving the net transfer of sand from south to north throughout the Quaternary and the growth of barriers and transgressive dune islands in the northern region at the expense of the southern region, which is sediment starved. However, on a somewhat smaller scale, it is becoming apparent that alongshelf sediment movements here and elsewhere in the world do not occur continuously but have been interrupted by discrete episodes of deposition and erosion orchestrated by changes in Quaternary sea levels. These changes in large-scale coastal behaviour are initiated at physical discontinuities such as headlands, drowned valleys and changes in coastal alignment (Ferland, 1990; Swift *et al.*, 1991a; Roy *et al.*, 1992). On wave-dominated coasts such as those of southeast Australia, their impact is especially marked near the present coast where topographic relief is greatest, and at times when the sea is at or above the modal zone (c. -70 m WD and above) during interstadials and interglacials. In contrast, during full glacial and stadial periods when the sea was below the modal level, alongshelf sediment movements on the generally low-relief middle and outer shelf were relatively unimpeded (Roy & Thom, 1981).

The intricacies of coastal sediment dynamics, operating locally over geological time scales are exemplified by the relationship between coastal sand barriers and SSBs – features that are associated with very different physical settings that impose contrasting styles of sedimentation. Where they are juxtaposed, as in the Seal Rocks-Tuncurry area (lat. 32.3° S) (Fig. 4.27*a*), they both constitute large depocentres of sand with similar lithologies but with different morphologies and preservation potentials. In the latter part of the Quaternary SSBs in this area have grown and decayed a number of times while the Tuncurry embayment has been a more permanent repository for a succession of barrier sand bodies (Fig. 4.12). In both cases, processes of

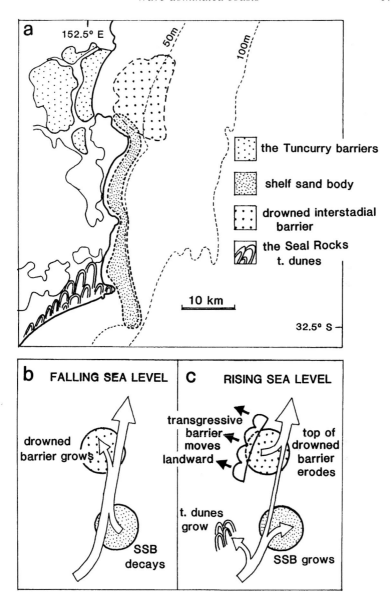

Figure 4.27. Conceptual model based on the Forster–Tuncurry case study (Roy *et al.*, 1992) (*a*) showing sediment budget relationships between a shelf sand body, a drowned (regressive) barrier, and a transgressive barrier during times of falling sea level (*b*) and rising sea level (*c*). Direction of small arrows indicate gains and losses of sand from the various features; the large arrow symbolises the dominant northward littoral transport.

sedimentation and erosion are focused in the active shoreface zone and the differently timed responses of SSBs and coastal sand barriers to marine transgressions and regressions have had an overriding influence on both regional and local sediment budgets (Roy, Ferland & Cowell, 1992; Roy *et al.*, 1992). One of the main impacts during rising sea level is for littoral sand to be

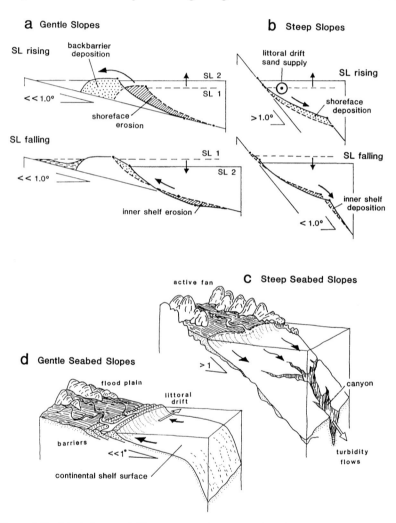

Figure 4.28. Simulation modelling suggests that on gentle substrates (*a*) sand moves onshore – and on steeper slopes (*b*), it moves offshore, despite the direction of sea-level change. Thus, on submarine slopes steeper than 1.0° (*c*) wave action tends to move sediment offshore into canyons and thence to the deep sea in turbidity flows. On continental shelf surfaces with slopes much less than 1.0° (*d*) waves move sand onshore and alongshore, and mud is dispersed by ocean currents.

diverted into growing SSBs at the expense of coastal barriers further downdrift. During falling sea levels, because SSBs are located on exposed sectors of coast, they erode thus supplying sand to prograde barriers downdrift. These sediment fluxes are illustrated diagramatically in Fig. 4.27. With the onset of the last glaciation, large quantities of SSB sand were reworked and released into the littoral transport system, carried downdrift (to the north) and deposited in receptive embayments as sea level continued to fall (Fig. 4.27*b*). The growth of the drowned, interstadial barrier on the Tuncurry shelf 40–60 ka ago was a result of this process. It also led to the almost complete destruction of the SSB. When sea levels began to rise again the cycle was reversed due to the persistence of a strong northward littoral transport throughout the postglacial marine transgression. Sand supplied from updrift (from the south) was added to the SSB creating a downdrift deficit and erosional conditions in the Tuncurry embayment. Here, in response to rising sea levels, a transgressive barrier was migrating landward over the surface of the pre-existing drowned barrier (Fig. 4.27*c*). Because of the deficit in the alongshore sand supply, recycling of the transgressive barrier involved erosion into the substrate – in this case the upper 3–5 m of the drowned barrier – and its partial incorporation into the transgressive barrier (Roy *et al.*, 1992).

The above example mainly concerns along-coast sediment exchanges in response to rapid sea-level fluctuations. However, simulation modelling has yielded important implications for across-shelf sand dispersal over much longer geological time spans. The results suggests that, irrespective of changing sea level, the steepness of the regional substrate is the primary influence on coastal sediment budgets and fluxes (Fig. 4.28*a*). The critical slope is slightly less than 1.0°; on slopes steeper than this, sandy sediment moves offshore and on flatter slopes, it tends to move onshore. The shoreface zone thus acts as a gateway controlling sediment dispersal between land and sea and appears to have played a disproportionately important role in long-term sedimentation trends on passive margins such as that of southeast Australia and the Tasman Sea basin (Roy & Keene, 1993). This margin's main morphological elements are a gently sloping craton surface, a relatively steep, faulted margin edge and an ocean floor that subsided throughout the Cainozoic. Initially, during the Paleogene, when the shoreline was in contact with the steep terrain (regional slopes 5° to 10°), erosion products from the land bypassed the shoreface zone and were carried directly to the ocean basin via river systems that probably connected to canyon heads (Fig. 4.28*b*). True continental shelf conditions did not exist at this time and coastal sediment budgets were directed offshore. This initial phase of infilling of the Tasman Sea basin is represented by turbidites and distal hemipelagic muds derived from the Eastern Highlands.

They vary in thickness from at least 500 to 1500 m and at DSDP site #283 in the central part of the Tasman Sea basin (Kennett *et al.*, 1975), the muds range in age from 80 to around 40 Ma, at which time the terrigenous sediment supply was apparently cut off. Today, the continental slope is blanketed by relatively thin calcareous muds that extend over the basin floor.

At about 30–40 Ma ago thermal subsidence and tilting of the margin brought the more gently sloping upper surface of the craton into contact with the sea and for the first time created a shallow continental shelf environment with regional slopes much less than 1.0°. Instead of bypassing clastic sediment to deep water, terrigenous sand transported by shoaling waves was directed onshore leading to the progressive accumulation of sediment on the shelf surface (Fig. 4.28*b*). The outer shelf along the entire length of the southeast Australian margin comprises a 500 m-thick and 20–30 km-wide sediment wedge that extends to about 1000 m water depth on the upper slope (Davies, 1979). The age of the base of the wedge where it was drilled off southern Queensland (24.7° S) is 25–30 Ma (Grimes, 1980; Roy & Thom, 1991). In the Late Quaternary, clastic sediment has been confined to the present coast and inner shelf in a zone dominated by strong northward littoral transport and its transfer to the ocean basin has virtually ceased. Lowstand deposits of calcareous sands and shell gravel occur on the outer shelf (Bickford *et al.*, 1993; Ferland & Roy, 1993). Except possibly in southern Queensland, the submarine canyons appear to be inactive. Along this coast there is growing evidence of a nexus between large-scale coastal behaviour and the development of inherited geological characteristics. Here, the cumulative result of along-shelf sediment movements over distances of thousands of kilometres throughout the Cainozoic is responsible for the dichotomy between southern and northern regions: steep, rocky and sediment deficient in the south and more gently sloping, shallower and sediment-rich in the north. In turn, these regional morphologies have influenced the styles of coastal deposition in more recent times when sea levels have been changing rapidly (Roy & Thom, 1981, 1991; Ferland, 1990).

The behaviour of sandy sediments under the influence of shoaling waves on shelves with different slopes leads not only to the development of a range of different types of coastal deposits on a local scale, but can also be responsible for basin-wide patterns of deposition and dispersal. The concept of large-scale coastal behaviour thus provides a framework encompassing both inherited and dynamic factors responsible for coastal evolution over the complete spectrum of spatial and temporal scales.

Acknowledgements

The authors have welcomed the opportunity, in contributing to this book, to synthesise a large body of data collected in mainly eastern Australia over more than two decades. The attempt to devise a broad theory of coastal development at this time, and from this perspective, is undoubtedly premature and somewhat unbalanced. Nevertheless, hopefully it will lead others to revise and refine the ideas presented here. Major aspects of the work were supported by MST and ARC grants No. 83/1142, A18600985 and A38700985, and additional research funding was provided by the New South Wales Government and Industry groups. The authors wish to thank their colleagues for their ongoing debate and the many graduate students that have helped with the work. In particular, we are indebted to Bob Jones for his contribution to the development of the computer simulation model; the skilful drilling by Peter Johnston and Carl Shaw is also appreciated. Sue Ferris provided invaluable help with the word processing. The work is published with the approval of the Director-General, New South Wales Department of Mineral Resources.

References

Allen, J.R.L. (1982). Spits. In *The encyclopedia of beaches and coastal environments*, ed. M.L. Schwartz, pp. 789–92. Stroudsberg, PA: Hutchinson Ross Publishing.

Baker, P.L. (1991). Response of ground–penetrating radar to bounding surfaces and lithofacies variations in sand barrier sequences. *Exploration Geophysics*, **22**, 19–22.

Belderson, R.H. & Stride, A.H. (1966). Tidal current fashioning of a basal bed. *Marine Geology*, **4**, 237–57.

Belknap, D.F. & Kraft, J.C. (1977). Holocene relative sea-level changes and coastal stratigraphic units on the northwest flank of the Baltimore Canyon Trough geosyncline. *Journal of Sedimentary Petrology*, **47**, 610–29.

Belknap, D.F. & Kraft, J.C. (1985). Influence of antecedent geology on stratigraphic preservation potential and evolution of Delaware's barrier systems. *Marine Geology*, **63**, 235–62.

Bergs, M.A. (1993). Origin and evolution of the inner continental shelf sand sheet in Southeast Australia, The Forster Region, NSW Central Coast: a case study. Unpubl. MSc thesis, Department of Geography, The University of Sydney. 175 pp. (plus appendices).

Bernard, H.A. & LeBlanc, R.J. (1965). Resume of the Quaternary geology of the northwestern Gulf of Mexico Province. In *The Quaternary of the United States*, ed. H.F. Wright & D.G. Frey, pp. 137–85. Princeton, NJ: Princeton University Press.

Bickford, G., Heggie, D., Birch, G.F., Ferland, M.A., Jenkins, C., Keene, J.B. & Roy, P.S. (1993). Preliminary results of AGSO *RV Rig Seismic* Survey 112 Leg B: offshore Sydney Basin continental shelf and slope geochemistry, sedimentology and geology. *Australian Geological Survey Organisation Record* 1993/5, 121 pp. (plus appendices). Canberra, Australia: Australian Geological Survey Organisation.

Bird, E.C.F. (1976). *Coasts*, 2nd edn. Canberra: Australian National University Press. 282 pp.

Boyd, R., Bowen, A.J. & Hall, R.K. (1987). An evolutionary model for transgressive sedimentation on the Eastern Shore of Nova Scotia. In *Glaciated coasts*, ed. D.M. Fitzgerald & P.S. Rosen, pp. 88–114. San Diego: Academic Press.

Boyd, R., Dalrymple, R.W. & Zaitlin, B.A. (1992). Classification of clastic coastal depositional environments. *Sedimentary Geology*, **80**, 139–50.

Bryant, E.A., Roy, P.S. & Thom, B.G. (1988). Australia – an unstable platform for tide-guage measurements of changing sea level: a discussion. *Journal of Geology*, **96**, 635–40.

Bryant, E.A., Young, R.W., Price, D.M. & Short, S.A. (in press). Late Pleistocene dune chronology: near-coastal New South Wales and Eastern Australia. *Quaternary Science Reviews*.

Carter, R.W.G. & Orford, J.D. (1993). The morphodynamics of coarse clastic beaches and barriers: a short- and long-term perspective. *Journal of Coastal Research*, SI **15**, 158–79.

Carter, R.W.G., Orford, J.D., Forbes, D.L. & Taylor, R.B. (1987). Gravel barriers, headlands and lagoons: an evolutionary model. In *Coastal sediments '87*, vol. 2, pp. 1776–92. New Orleans: American Society of Civil Engineers.

Chapman, D.M., Geary, M., Roy, P.S. & Thom, B.G. (1982). *Coastal evolution and coastal erosion in New South Wales*. Sydney: Coastal Council of New South Wales. 341 pp.

Chappell, J. & Shackleton, N.J. (1986). Oxygen isotopes and sea level. *Nature*, **324**, 137–40.

Clifton, H.E. & Hunter, R. (1982). Coastal sedimentary facies. In *The encyclopedia of beaches and coastal environments*, ed. M.L. Schwartz, pp. 314–22. Stroudsburg, PA: Hutchinson Ross Publishing.

Colwell, J.B. & Roy, P.S. (1983). Description of subsurface sediments from the east Australian continental shelf (SONNE Cruise SO-15). *Bureau of Mineral Resources Record* 83/21. Canberra, Australia: Bureau of Mineral Resources. 66 pp.

Cook, P.G. (1986). A review of coastal dune building in eastern Australia. *Australian Geographer*, **17**, 133–43.

Cook, P.J., Colwell, J.B., Firman, J.B., Lindsay, J.M., Schwebel, D.A. & von der Borch, C.C. (1977). The late Cainozoic sequence of southeast South Australia and Pleistocene sea level changes. *Bureau of Mineral Resources Australian Journal of Geology and Geophysics*, **2**, 81–8.

Cooper, W.S. (1958). *Coastal sand dunes of Oregon and Washington*. Geological Society of America Memoir, **72**. 168 pp.

Cowell, P.J. (1986). Wave-induced sand mobility and deposition on the south Sydney inner-continental shelf. *Geological Society of Australia Special Publication*, **2**, 1–28.

Cowell, P.J. & Nelson, H. (1991). Management of beach erosion due to low swell, inlet and greenhouse effects: case study with computer modelling. In *Proceedings 10th Australian Conference on Coastal and Ocean Engineering*, ed. R.G. Bell, T.M. Hume & T.R. Healy, pp. 329–33. Hamilton, New Zealand: Water Quality Centre, Department of Scientific and Industrial Research.

Cowell, P.J. & Nielsen, P. (1984). Predictions of sand movement on the South Sydney inner-continental shelf, south east Australia. *Coastal Studies Unit Technical Report* 84/2. Sydney: University of Sydney. 151 pp.

Cowell, P.J. & Roy P.S. (1988). *Shoreface transgression model: programming guide (outline, assumptions and methodology)*. Unpubl. Report, Coastal Studies Unit, Marine Studies Centre, University of Sydney. 23 pp.

Cowell, P.J., Roy, P. S. & Jones, R. A. (1991). Shoreface translation model: application to management of coastal erosion. In *Applied Quaternary studies*, ed. G. Brierley & J. Chappell, pp. 57–73. Canberra: Department of Biogeography and Geomorphology, Australian National University.

Cowell, P.J., Roy, P. S. & Jones, R. A. (1992). Shoreface translation model: computer simulation of coastal-sand-body response to sea level rise. *Mathematics and Computers in Simulation*, **33**, 603–8.

Csanady, G.T. (1977). The coastal boundary layer. In *Estuaries, geophysics and the environment*, National Research Council, pp. 57–68. Washington, DC: National Academy of Sciences.

Curray, J.R. (1964). Transgressions and regressions. In *Papers in marine geology*, Shepard commemorative volume, ed. R.L. Miller, pp. 175–203. New York: Macmillan.

Curray, J.R. (1969). History of continental shelves. In *New concepts of continental margin sedimentation*, ed. D.J.Stanley, pp. JC–6–1 to JC–6–7. Washington, DC: American Geological Institute.

Dalrymple, R.W., Zaitlin, B.A. & Boyd, R. (1992). Estuarine facies models: conceptual basis and stratigraphic implications. *Journal of Sedimentary Petrology*, **62**, 1030–43.

Davies, J.L. (1957). The importance of cut and fill in the development of sand beach ridges. *Australian Journal of Science*, **20**, 105–11.

Davies, J.L. (1958). Wave refraction and evolution of shoreline curves. *Geographical Studies*, **5**, 1–14.

Davies, J.L. (1980). *Geographical variation in coastal development*, 2nd edn. New York: Longman. 212 pp.

Davies, P.J. (1979). *Marine geology of the continental shelf off southeast Australia*. Bureau of Mineral Resources Bulletin 195. Canberra, Australia: Bureau of Mineral Resources. 51 pp.

Davis, R.A. & Hayes, M.O. (1984). What is a wave-dominated coast? *Marine Geology*, **60**, 313–30.

Davis, R.A., Jr. (1985). *Coastal sedimentary environments*, 2nd edn. New York: Springer-Verlag. 716 pp.

Davis, R.A., Jr. (1992). *Depositional systems: an introduction to sedimentology and stratigraphy*. Englewood Cliffs, NJ: Prentice Hall. 604 pp.

Dean, R.G. & Maurmeyer, E. M. (1983). Models for beach profile responses. In *Handbook of coastal processes and erosion*, ed. P.D. Komar, pp. 151–66. Boca Raton, FL: CRC Press.

Dickinson, K.A., Berryhill, H.L. & Holmes, C.W. (1972). Criteria for recognizing ancient barrier coastlines. In *Recognition of ancient sedimentary environments*, Special Publication No.16, ed. J.K. Rigby & W.K. Hamblin, pp. 192–214. Tulsa, OK: Society of Economic Mineralogists and Paleontologists.

Duane, D.B., Field, M.E., Meisburger, E.P., Swift, D.J.P. & Williams, S.J. (1972). Linear shoals on the Atlantic inner continental shelf, Florida to Long Island. In *Shelf sediment transport, process and pattern*, ed. D.J.P. Swift, D.B. Duane & O.H. Pilkey, pp. 447–98. Pennsylvania: Dowden, Hutchinson and Ross.

Everts, C.H. (1987). Continental shelf evolution in response to a rise in sea level. In *Sea-level fluctuations and coastal evolution*, Special Publication No. 41, ed. D. Nummedal, O.H.Pilkey & J.D. Howard, pp. 49–57. Tulsa, OK: Society of Economic Paleontologists and Mineralogists.

Fairbanks, R.G. (1989). A 17 000 year glacio-eustatic sea level record: influence of glacial melting rates on Younger Dryas event and deep-ocean circulation. *Nature*, **342**, 637–42.

Ferland, M.A. (1986). Identification, description and evolution of shelf sand bodies, southeast Australia. In *Abstracts, 12th International Sedimentology Congress*, p. 105. Canberra, Australia: International Association of Sedimentologists.

Ferland, M.A. (1987). *Inner continental shelf sand bodies in southeastern Australia: surface sediments in the Jervis Bay region*. Geological Survey of New South Wales Report No. 1987/092. Sydney, Australia: Geological Survey of New South Wales. 20 pp.

Ferland, M.A. (1988). Similarities in the location of onshore and offshore deposits along the NSW Coast. In *Proceedings of the Australian Marine Science Association Conference*, pp. 67–71. Sydney, Australia: Australian Marine Science Association.

Ferland, M.A. (1990). Shelf sand bodies in Southeastern Australia. University of Sydney, Unpublished PhD thesis. Sydney, Australia. 527 pp.

Ferland, M.A. & Roy, P.S. (1993). Late Quaternary stratigraphy of the mid and inner shelf: central New South Wales. *2nd. Australian Marine Geoscience Workshop, 14–16 July 1993, Department of Geology and Geophysics, University of Sydney, Program and Abstracts*, pp. 20–1.

Field, M.E. (1980). Sand bodies on coastal plain shelves: Holocene record of the U.S. Atlantic inner shelf off Maryland. *Journal of Sedimentary Petrology*, **50**, 505–28.

Field, M.E. & Duane, D.B. (1976). Post-Pleistocene history of the United States inner continental shelf: significance to the origin of barrier islands. *Geological Society of America Bulletin*, **87**, 691–702.

Field, M.E. & Roy, P.S. (1984). Offshore transport and sand body formation: evidence from a steep, high energy shoreface, southeastern Australia. *Journal of Sedimentary Petrology*, **54**, 1292–302.

Field, M.E. & Trincardi, F. (1991). Regressive coastal deposits on Quaternary continental shelves: preservation and legacy. In *From shoreline to abyss*, Special Publication No. 46. pp. 107–22. Tulsa, OK: Society of Economic Mineralogists and Paleonotologists.

Fisk, H.N. (1959). Padre Island and Laguna Madre mud flats, south coastal Texas. In *Coastal Studies Institute, 2nd Coastal Geography Conference*, pp. 103–51. Baton Rouge, LA: Louisiana State University.

Floyd, C.D. & Druery, B.M. (1976). Results of river mouth training on the Clarence River bar,

New South Wales, Australia. In *Proceedings of the 15th International Conference on Coastal Engineering.* New York: American Society of Civil Engineers.

Forbes, D.L., Taylor, R.B., Orford, J.D., Carter, R.W.G. & Shaw, J. (1991). Gravel barrier migration and overstepping. *Marine Geology*, **97**, 305–13.

Gordon, A.D. (1988). A tentative but tantalizing link between sea-level rise and coastal recession in New South Wales, Australia. In *Greenhouse: planning for climatic change*, ed. G.I. Pearman, pp. 121–34. Melbourne: CSIRO Division of Atmospheric Research

Gordon, A.D. & Roy, P.S. (1977). Sand movements in Newcastle Bight. In *Proceedings of the 3rd Australian Conference on Coastal and Ocean Engineering*, Melbourne, pp. 64–9.

Greenwood, B.G. & Davis, R.A., Jr., ed. (1984). *Hydrodynamics and sedimentation in wave-dominated coastal environments. Marine Geology*, SI **60**.

Grimes, K.G. (1980). Stratigraphic drilling report: GSQ Sandy Cape 1 -3R. *Queensland Government Mining Journal*, **83**, 224–33.

Haq, B.U., Hardenbol, J. & Vail, P.R. (1987). Chronology of fluctuating sea levels since the Triassic. *Science*, **235**, 1156–67.

Hayes, M.O. (1979). Barrier island morphology as a function of tidal and wave regime. In *Barrier islands: from the Gulf of St. Lawrence to the Gulf of Mexico*, ed., S.P. Leatherman, pp. 1–27. New York: Academic Press.

Hesp, P. A. (1984a). Foredune formation in southeast Australia. In *Coastal geomorphology in Australia*, ed. B.G. Thom, pp. 69–97. Sydney: Academic Press.

Hesp, P.A. (1984b). The formation of sand 'beach ridges' and foredunes. *Search*, **15**, 289–91.

Hesp, P. A. (1988). Surfzone, beach, and foredune interactions on the Australian south east coast. *Journal of Coastal Research*, SI **3**, 15–25.

Hesp, P. A. & Thom, B. G. (1990). Geomorphology and evolution of erosional dunefields. In *Coastal dunes: processes and morphology.* ed. K.F. Nordstrom, N.P. Psuty & R.W.G. Carter, pp. 253–88. Chichester: J. Wiley & Sons.

Hovland, M. & Dukefoss, K.M. (1981). A submerged beach between Norway and Ekofisk in the North Sea. *Marine Geology*, **43**, M19–M28.

Hoyt, J.H. (1967). Barrier island formation. *Geological Society of America Bulletin*, **78**, 1123–36.

Hoyt, J.H. & Henry, V.J. (1967). Influence of inlet migration on barrier island sedimentation. *Geological Society of America Bulletin*, **78**, 77–86.

Hudson, J.P. (1985). Geology and depositional history of late Quaternary marine sediments on the South Sydney inner continental shelf, southeastern Australia. *Coastal Studies Unit Technical Report* 85/2. Sydney: University of Sydney. 126 pp.

Hudson, J.P. & Ferland, M.A. (1987). Seismic results from the inner continental shelf of the Twofold Bay/Disaster Bay region, Southern NSW: a progress report. *Geological Survey of New South Wales Report GS* 1987/093. Sydney, Australia: Department of Mineral Resources.

Jennings, J.N. (1967). Cliff-top dunes. *Australian Geographical Studies*, **5**, 40–9.

Kennett, J.P., Houtz, R.E., Andrews, P.B., Edwards, A.R., Gostin, V.A., Hajos, M., Hampton, M.A., Jenkins, D.G., Margolis, S.V., Ovenshine, A.T., Perch-Nelson, K., Webb, P.N. & Wilson, G.J. (1975). Site 283. In *Initial reports of the deep sea drilling project, vol. XXIX*, pp. 365–377. Washington, DC: US Government Printing Office.

King, C.A.M. (1972). *Beaches and coasts*, 2nd edn. London: Edward Arnold. 570 pp.

Komar, P.D. (1976). The transport of cohesionless sediments on continental shelves. In *Marine sediment transport and environmental management*, ed. D.J. Stanley & D.J.P. Swift, pp.107–25. New York: John Wiley.

Kraft, J.C. (1971). Sedimentary facies patterns and geologic history of a Holocene marine transgression. *Geological Society of America Bulletin*, **82**, 2131–58.

Kraft, J.C. & John, C.J. (1979). Lateral and vertical facies relations of transgressive barriers. *American Association of Petroleum Geologists Bulletin*, **63**, 2145–63.

Kudrass, H-R. (1982). Cores of Holocene and Pleistocene sediments from the east Australian continental shelf (SO-15 Cruise 1980). In *Heavy mineral exploration of the East Australian Shelf 'SONNE' Cruise SO-15 1980, Geologisches Jahrbuch*, Reihe D, Heft 56, pp. 137–63.

Leatherman, S.P. (1979). Migration of Assateague Island, Maryland, by inlet and overwash processes. *Geology*, **7**, 104–7.

Leatherman, S.P. (1983). Barrier dynamics and landward migration with Holocene sea-level rise. *Nature*, **301**, 415–18.

Leatherman, S.P. (1987). Reworking of glacial outwash sediments along Outer Cape Cod: development of Provincetown Spit. In *Glaciated coasts*, ed. D.M. Fitzgerald & P.S. Rosen, pp. 307–25. San Diego, CA: Academic Press.

Marshall, J.F. (1980). *Continental shelf sediments: southern Queensland and northern New South Wales. Bureau of Mineral Resources Bulletin* 207. Canberra, Australia: Bureau of Mineral Resources. 39 pp.

Moslow, T.F. & Heron, S.D. (1978). Relict inlets: preservation and occurrence in the Holocene stratigraphy of southern Core Banks, North Carolina. *Journal of Sedimentary Petrology*, **48**, 1275–86.

Nichol, S. (1992). A partially preserved last interglacial estuarine fill: Narrawallee Inlet, New South Wales. *Australian Journal of Earth Sciences*, **39**, 545–53.

Nichols, M. M. & Biggs, R. B. (1985). Estuaries. In *Coastal sedimentary environments*, 2nd edn. ed. R.A. Davis, pp. 77–186. New York: Springer-Verlag.

Niedoroda A.W., Swift, D.J.P., & Hopkins, T.S. (1985). The shoreface. In *Coastal sedimentary environments*, 2nd edn. ed. R.A. Davis, pp. 533–624. New York: Springer-Verlag.

Nielsen, A.F. & Roy, P.S. (1982). Age contamination of radiocarbon dates in shell hash from coastal sand deposits: SE Australian examples. In *Proceedings 5th Australian Conference on Coastal and Ocean Engineering*, pp. 177–82. Institute of Engineers Australia.

Orford, J.D., Carter, R.W.G. & Forbes, D.L. (1991). Gravel barrier migration and sea level rise: some observations from Storey Head, Nova Scotia, Canada. *Journal of Coastal Research*, **7**, 477–88.

Otvos, E.G., Jr. (1979). Barrier island evolution and history of migration, north central Gulf coast. In *Barrier islands from the Gulf of St. Lawrence to the Gulf of Mexico*, ed. S.P. Leatherman, pp. 291–319. New York: Academic Press.

Pilkey, O.H., Young, R.S., Riggs, S.R., Smith, A.W., Wu, H. & Pilkey, W.D. (1993). The concept of shoreface profile equilibrium: a critical review. *Journal of Coastal Research*, **9**, 255–78.

Pye, K. & Bowman, G.M. (1984). The Holocene marine transgression as a forcing function in episodic dune activity on the eastern Australian coast. In *Coastal geomorphology in Australia*, ed. B.G. Thom, pp. 179–96. Sydney: Academic Press.

Riggs, S.R. & Cleary, W.J. (1993). Influence of inherited geological framework upon barrier island morphology and shoreface dynamics. In *Large-scale coastal behavior '93*, ed. J.H. List, pp. 173–6. US Geological Survey Open File Report 93-381. St.Petersburg, FL: US Geological Survey.

Rine, J.M. Tillman, R.W., Culver, S.J. & Swift, D.J.P. (1991). Generation of late Holocene sand ridges on the middle continental shelf of New Jersey, USA: evidence for formation in a mid-shelf setting based on comparisons with a nearshore ridge. In *Shelf sand and sandstone bodies: geometry, facies and sequence stratigraphy*. Special Publication No. 14 of the International Association of Sedimentologists, ed. D.J.P. Swift, G.F. Oertel, R.W. Tillman & J.A. Thorne, pp. 395–423. Oxford: Blackwell Scientific Publications.

Roy, P.S. (1982). Regional geology of the central and northern New South Wales coast. In *Heavy mineral exploration of the East Australian Shelf 'SONNE' Cruise SO-15 1980. Geologisches Jahrbuch*, Reihe D, Heft 56, pp. 25–35.

Roy, P.S. (1984). New South Wales estuaries: their origin and evolution. In *Coastal geomorphology in Australia*, ed. B.G. Thom, pp. 99–121. Sydney: Academic Press.

Roy, P.S. (1985). Marine sand bodies on the south Sydney shelf, S.E. Australia. *Coastal Studies Technical Report* 85/1. Sydney: University of Sydney. 180 pp..

Roy, P.S. (1991). Shell hash dating and mixing models for palimpsest marine sediments. *Radiocarbon*, **33**, 283–9.

Roy, P.S. (1992). Coastal Quaternary geology. In *The geology of the Camberwell, Dungog, Bulahdelah and Forster 1:100,000 sheets 9133, 9233, 9333 and 9433*, pp. 244–251. Sydney, NSW: New South Wales Geological Survey.

Roy, P.S. (in press). Holocene estuary evolution: stratigraphic studies from southeastern Australia. In *Incised valley systems: origin and sedimentary facies*, ed. R. Dalrymple & R. Boyd, Special Publicatioin No. 51. Tulsa, OK: Society of Economic Mineralogists and Paleontologists.

Roy, P.S. & Cowell, P.J. (1991). Geological inheritance effects in the response of the shoreface to coastal storms. In *Coastal development: policy and management*. Geographical Society of New South Wales, Conference Papers, 9, pp. 74–82.

Roy, P.S. & Crawford, E.A. (1977). Significance of sediment distribution in major coastal rivers, northern N.S.W. In *Proceedings 3rd Australian Conference on Coastal and Ocean Engineering*, Melbourne, pp. 177–84.

Roy, P.S. & Crawford, E.A. (1980). Quaternary geology of Newcastle Bight inner continental shelf, New South Wales, Australia. *New South Wales Geological Survey Records*, **19**, 145–88.

Roy, P.S. & Crawford, E.A. (1981). Holocene geological evolution of the southern Botany Bay–Kurnell region, central NSW coast. *New South Wales Geological Survey Records*, **20**, 159–250.

Roy, P.S. & Ferland, M.A. (1987). *Seismic results from the inner continental shelf of the Shoalhaven/Jervis Bay, Ulladulla region, southern NSW: a progress report*. Geological Survey of New South Wales Report No. 1987/091. 33 pp.

Roy, P.S., Ferland, M.A. & Cowell, P.J. (1992). Headland-attached shelf sand bodies and drowned barriers: their growth and decay. In *Abstracts, 5th Meeting Australia/New Zealand Geomorphology Research Group, Port Macquarie*, p. 22. Canberra: Australian National University.

Roy, P.S. & Hudson, J.P. (1987). *The marine placer minerals project – seismic results from the continental shelf, southern NSW: Narooma – Montague Island – Bermagui area*. Geological Survey of New South Wales Report No. 1987/094, 65 pp.

Roy, P.S. & Keene, J.B. (1993). Coastal morphodynamics: a control on Tasman Sea basin sedimentation. In *Second Australian Marine Geoscience Workshop*, pp. 52–4. Sydney: Department of Geology & Geophysics, University of Sydney.

Roy, P.S. & Stephens, A.W. (1980). Geological controls on process-response, S.E. Australia. In *Proceedings of the 17th International Coastal Engineering Conference*, pp. 913–33. Sydney, Australia: American Society of Civil Engineers.

Roy P.S. & Thom. B.G. (1981). Late Quaternary marine deposition in New South Wales and southern Queensland: an evolutionary model. *Journal of Geology Society of Australia*, **28**, 471–89.

Roy, P.S. & Thom, B.G. (1987). Sea-level rise: An eminent threat. In *Proceedings 8th Australasian Conference on Coastal and Ocean Engineering*, pp.186–90. Institute of Engineers, Australia.

Roy P.S. & Thom. B.G. (1991). Cainozoic shelf sedimentation model for the Tasman Sea margin of southeastern Australia. In *The Cainozoic in Australia: a reappraisal of the evidence*, ed. M.A.J. Williams, P. DeDeckker & A.P. Kershaw, pp. 119–36, Geological Society of Australia Special Publication No. 18.

Roy, P.S., Thom, B.G. & Wright, L.D. (1980). Holocene sequences on an embayed high-energy coast: an evolutionary model. *Sedimentary Geology*, **16**, 1–9.

Roy, P.S., Zhuang, W-Y., Birch, G.F. & Cowell, P.J. (1992). *Quaternary geology and placer mineral potential of the Forster–Tuncurry shelf, southeastern Australia*. Geological Survey Report: GS 1992/201, Department of Mineral Resources, NSW. 164 pp.

Sanders, J.E. & Kumar, N. (1975). Evidence of shoreface retreat and in-place 'drowning' during Holocene submergence of barriers, shelf off Fire Island, New York. *Geological Society of America Bulletin*, **86**, 65–76.

Schluter, H-L. (1982). Results of a reflection seismic survey in shallow water areas off east Australia, Yamba to Tweed Heads. In *Heavy mineral exploration of the East Australian Shelf 'SONNE' Cruise SO-15 1980. Geologisches Jahrbuch*, Reihe D, Heft 56, pp. 77–95.

Schwartz, M.L., ed. (1972). *Spits and bars*. Stroudsburg, PA: Dowden, Hutchinson and Ross. 452 pp.

Schwartz, M.L., ed. (1973). *Barrier islands*. Stroudsburg, PA: Dowden, Hutchinson and Ross. 451 pp.

Shepherd, M.J. (1991). Relict and contemporary foredunes as indicators of coastal processes. In *Applied Quaternary studies*, ed. G. Brierley & J. Chappell, pp. 17–24. Canberra: Department of Biogeography and Geomorphology, Australian National University.

Short, A.D. (1987). Modes, timing and volume of Holocene cross-shore and aeolian sediment transport, southern Australia. In *Coastal sediments '87*. pp. 1925–37. New York: American Society of Civil Engineers.

Short, A.D. (1988). The South Australia coast and Holocene sea-level transgression. *Geographical Review*, **78**, 119–36.

Short, A.D. & Hesp, P.A. (1982). Wave, beach and dune interactions in S.E. Australia. *Marine Geology*, **48**, 259–84.

Stephens, A.W. (1982). Quaternary coastal sediments of southeast Queensland. In *Heavy mineral exploration of the East Australian Shelf 'SONNE' Cruise SO-15 1980. Geologisches Jahrbuch*, Reihe D, Heft 56, pp. 37–47.

Stephens, A.W., Roy, P.S. & Jones, M.R. (1981). Model of erosion on a littoral drift coast. In *Proceedings of the Fifth Australian Conference on Coastal and Ocean Engineering, Perth, WA*. pp. 171–6. Sydney: The Institute of Engineers, Australia.

Stride, A.H., ed. (1982). *Offshore tidal sands: processes and deposits*. London: Chapman and Hall. 222 pp.

Swift, D.J.P. (1975a). Barrier island genesis: evidence from the central Atlantic shelf, eastern U.S.A. *Sedimentary Geology*, **14**, 1–43.

Swift, D.J.P. (1975b). Tidal sand ridges and shoal retreat massifs. *Marine Geology*, **18**, 105–34.

Swift, D.J.P. (1976a). Coastal sedimentation. In *Marine sediment transport and environmental management*, ed. D.J. Stanley & D.J.P. Swift, pp. 255–310. New York: Wiley.

Swift, D.J.P. (1976b). Continental shelf sedimentation. In *Marine sediment transport and environmental management*, ed. D.J. Stanley & D.J.P. Swift, pp. 311–50. New York: Wiley.

Swift, D.J.P. & Field, M.E. (1981). Evolution of a classic sand ridge field; Maryland sector, North American inner shelf. *Sedimentology*, **28**, 461–82.

Swift D.J.P., McKinney, T.F. & Stahl, L. (1984). Recognition of transgressive and post-transgressive sand ridges on the New Jersey continental shelf: discussion. In *Siliclastic shelf sediments*, Special Publication No. 34. ed. R.W. Tillman & C.T. Siemers, pp. 25–36. Tulsa, OK: Society of Economic Palaeontologists and Mineralogists.

Swift, D.J.P., Niedoroda, A.W., Vincent, C.E. & Hopkins, T.S.(1985). Barrier island evolution, Middle Atlantic Shelf, USA. Part I: Shoreface dynamics. *Marine Geology*, **63**, 331–61.

Swift, D.J.P., Parker, G., Lanfredi, N.W., Perillo, G. & Figge, K. (1978). Shoreface-connected sand ridges on American and European shelves: a comparison. *Estuarine and Coastal Marine Science*, **7**, 257–73.

Swift, D.J.P., Phillips, S. & Thorne, J.A. (1991a). Sedimentation on continental margins. IV. Lithofacies and depositional systems. In *Shelf sand and sandstone bodies: geometry, facies and sequence stratigraphy*, Special Publication No. 14 of the International Association of Sedimentologists, ed. D.J.P. Swift, G.F. Oertel, R.W. Tillman & J.A. Thorne, pp. 89–152. Oxford: Blackwell Scientific Publications.

Swift, D.J.P., Phillips, S. & Thorne, J.A. (1991b). Sedimentation on continental margins. V. Parasequences. In *Shelf sand and sandstone bodies: geometry, facies and sequence stratigraphy*, Special Publication No. 14 of the International Association of Sedimentologists, ed. D.J.P. Swift, G.F. Oertel, R.W. Tillman & J.A. Thorne, pp. 153–87. Oxford: Blackwell Scientific Publications.

Swift, D.J.P. & Thorne, J.A. (1991). Sedimentation on continental margins. I. A general model for shelf sedimentation. In *Shelf sand and sandstone bodies: geometry, facies and sequence stratigraphy*, Special Publication No. 14 of the International Association of Sedimentologists, ed. D.J.P. Swift, G.F. Oertel, R.W. Tillman & J.A. Thorne, pp. 3–31. Oxford: Blackwell Scientific Publications.

Terwindt, J.H.J. & Wijnberg, K.M. (1991). Thoughts on large scale coastal behaviour. In *Coastal Sediments '91*, ed. N.C. Kraus, K.J. Gingerich & D.L. Kriebel, pp. 1476–87. New York: American Society of Civil Engineers.

Thom, B. G. (1974). Coastal erosion in eastern Australia. *Search*, **5**, 198–209.

Thom, B. G. (1978). Coastal sand deposition in southeast Australia during the Holocene. In *Landform evolution in Australasia*, ed. J.L. Davies & M.A.G. Williams, pp. 197–214. Canberra: A.N.U. Press.

Thom, B. G. (1984). Transgressive and regressive stratigraphies of coastal sand barriers in eastern Australia. *Marine Geology*, **7**, 161–8.

Thom, B.G., Bowman, G.M. & Roy, P.S. (1981). Late Quaternary evolution of coastal sand barriers, Port Stephens – Myall Lakes area, central New South Wales, Australia. *Quaternary Research*, **15**, 345–64.

Thom, B.G. & Hall, W. (1991). Behaviour of beach profiles during accretion and erosion dominated periods. *Earth Surface Processes and Landforms*, **16**, 113–27.

Thom, B.G., Polach, H.A. & Bowman, G.M. (1978). *Holocene age structure of coastal sand barriers in New South Wales, Australia.* Geography Department Report. Canberra, Australia: Duntroon, University of New South Wales. 86 pp.

Thom, B.G. & Roy, P.S. (1983). Sea level change in New South Wales over the past 15 000 years. In *Australian sea levels in the last 15 000 years: a review*, ed. D. Hopley, pp. 64–84. Geography Department, James Cook University of North Queensland, Townsville, Monograph Series, Occasional Paper 3.

Thom, B.G. & Roy, P.S. (1985). Relative sea levels and coastal sedimentation in southeast Australia in the Holocene. *Journal of Sedimentary Petrology*, **55**, 257–64.

Thom, B.G. & Roy, P.S. (1988). Sea-level rise and climate: lessons from the Holocene. In *Greenhouse: planning for climatic change*, ed. G.I. Pearman, pp. 177–88. Melbourne: CSIRO Division of Atmospheric Research.

Thom, B.G., Shepherd, M., Ly, C.K., Roy, P.S., Bowman, G.M. & Hesp, P.A. (1992). *Coastal geomorphology and Quaternary geology of the Port Stephens-Myall Lakes Area.* Canberra: Department of Biogeography and Geomorphology, The Australian National University. 407 pp.

Thorne, J.A. & Swift, D.J.P. (1991a). Sedimentation on continental margins. VI. A regime model for depositional sequences, their component system tracts, and bounding surfaces. In *Shelf sand and sandstone bodies: geometry, facies and sequence stratigraphy*, Special Publication No. 14 of the International Association of Sedimentologists, ed. D.J.P. Swift, G.F. Oertel, R.W. Tillman & J.A. Thorne, pp.189–255. Oxford: Blackwell Scientific Publications.

Thorne, J.A. & Swift, D.J.P. (1991b). Sedimentation on continental margins. II. Application of the regime concept. In *Shelf sand and sandstone bodies: geometry, facies and sequence stratigraphy*, Special Publication No. 14 of the International Association of Sedimentologists, ed. D.J.P. Swift, G.F. Oertel, R.W. Tillman & J.A. Thorne, pp. 33–58. Oxford: Blackwell Scientific Publications.

Vail, P.R., Audemard, F., Bowman, S.A., Eisner, P.N. & Perez-Cruz, C. (1991). The stratigraphic signatures of tectonics, eustasy and sedimentology: an overview. In *Cycles and events in stratigraphy*, ed. G. Einsele, W. Ricken & A. Seilacher, pp. 617–59. Berlin: Springer-Verlag.

van Straaten, L.M.J.U. (1965). Coastal barrier deposits in south- and north-Holland; in particular in the areas around Scheveningen and Ijmuiden. *Mededelingen Geologische Stichting*, NS, **17**, 41–75.

von Stackelberg, U. (1982). *Heavy mineral exploration of the East Australian Shelf, 'SONNE' Cruise SO-15, 1980.* Geologisches Jahrbuch, Reihe D, Heft 56, 215 pp.

Wellner, R.W., Ashley, G.M. & Sheridan, R.E. (1993). Seismic stratigraphic evidence for a submerged middle Wisconsin barrier: implications for sea-level history. *Geology*, **21**, 109–112.

Williams, S.J. (1976). *Geomorphology, shallow subbottom structure, and sediments of the Atlantic inner continental shelf off Long Island, New York.* US Army Coastal Engineering Research Center Technical Paper 76–2. 123 pp.

Woodroffe, C.D. & Grindrod, J. (1991). Mangrove biogeography: the role of Quaternary environmental and sea level change. *Journal of Biogeography*, **18**, 479–92.

Wright, L.D. (1976). Nearshore wave power dissipation and the coastal energy regime of the Sydney-Jervis Bay region, New South Wales: a comparison. *Australian Journal of Marine and Freshwater Research*, **27**, 633–40.

Wright, L.D. (1985). River deltas. In *Coastal sedimentary environments*, 2nd edn. ed. R.A. Davis, pp. 1–76. New York: Springer-Verlag.

Wright, L.D., Boon, J.D., Kim, S.C. & List, J.H. (1991). Modes of cross-shore sediment transport on the shoreface of the Middle Atlantic Bight. *Marine Geology*, **96**, 19–51.

Zenkovich, V.P. (1967). *Processes of coastal development.* Edinburgh: Oliver and Boyd. 738 pp.

5

Macrotidal estuaries

J. CHAPPELL AND C.D. WOODROFFE

Introduction

Macrotidal estuaries can be viewed within a continuum of deltaic-estuarine coastal depositional settings, influenced by riverine processes, wave regime and tidal energy (Wright & Coleman, 1973; Wright, 1985; Boyd, Dalrymple & Zaitlin, 1992; see Fig. 1.6). The morphodynamics of individual macrotidal estuaries are a function of sea-level changes and prior or inherited topography. The response of estuaries to sea-level changes, past and future, is affected by tidal range, nearshore wave climate, river inflow and the nature and supply of sediment. All estuaries assumed their present form during the rise of sea level that followed the last glacial maximum, about 18 000 years ago. In areas that are relatively stable, such as northern Australia, estuaries have had similar sea-level histories, whereas in areas of very rapid crustal uplift or glacioisostatic response, estuaries are more likely to have experienced highly individual relative sea-level histories. The variety of estuaries reflects the range of submerged prior landforms, from relatively straight, steep coastlines through valleys in differing stages of infill to rock-barred basins. These different types of prior topography, which are manifest in coasts such as rias and dendritic drowned valley harbours, influence the interplay of processes which redistribute sediment to produce estuarine channels, tidal basins, backwaters and floodplains. Vegetation can modify estuarine morphodynamics through its effects upon sediment trapping and through its influence on the shear strength of channel banks.

This chapter is concerned with macrotidal estuaries, mostly in northern Australia which tectonically is a very stable region, with low sediment yield (about $5-15\,t\,km^{-2}\,a^{-1}$) and broad continental shelves. Rising postglacial sea level invaded coastal valleys relatively late, usually within the last 7000–9000 years (Thom & Roy, 1985), reaching a peak around 5500–6000 ^{14}C years BP, about 1 to 2 m above present sea level at most locations on the Australian

187

mainland coast (Chappell *et al.*, 1982, 1983). Since then, relative sea level has fallen slowly and apparently smoothly to its present position (Chappell, 1982, 1987; Nakada & Lambeck, 1989; Lambeck & Nakada, 1990). In this respect, north Australian estuaries, in common with much of the Old World tropics, and northern South America, have a mid to late Holocene sea-level history which differs from north Atlantic and some other regions where a slowly-rising trend persisted until the present (Pirazzoli, 1991).

Although this chapter is focussed on macrotidal estuaries in north Australia, other systems are discussed which differ in terms of sediment yield from their fluvial catchments, nearshore wave climate or tidal regimes, in order to examine the effects of boundary conditions. Locations of major north Australian systems which are examined are shown in Fig. 5.1 and their characteristics are listed in Table 5.1. Morphodynamics, and sedimentary effects of Holocene sea-level changes, within macrotidal and selected other systems will be examined and possible effects of future sea-level changes are considered.

Principal boundary conditions

Most of the macrotidal estuaries in tropical northern Australia in the region near Darwin are tidal rivers which pass through extensive sedge- and grass-covered plains, seasonally flooded with fresh water during the monsoonal wet season. Tidal limits of the longer estuaries lie 90–140 km inland. They have wide entrances typically set between prograded coastal plains, adopting the funnel-like estuarine form typical of macrotidal rivers (Wright, Coleman & Thom, 1973). Several distinct meandering channel forms, examined below, are recognised. Infilled palaeochannels are visible on the neighbouring flood-plains, and backwater swamps, dominated by paperbark trees (*Melaleuca* spp.), lie near the margins of the plains. Most of the fluvial rivers which feed into these estuaries flow only during the wet season, and discharges are approximately proportional to catchment sizes. On a global scale the sediment yield of their catchments is low to very low. To appreciate similarities with and differences from other types of estuaries, some general effects of boundary conditions are briefly reviewed.

The fluvial–marine spectrum

The relative effects of tides, wave climate and fluvial dominance upon delta morphology were clearly delineated by Wright & Coleman (1973). Estuarine systems, which typically have elements similar to those of deltas without the same degree of seaward progradation, vary across a comparable spectrum (Dalrymple, Zaitlin & Boyd, 1992). The effect of wave climate is

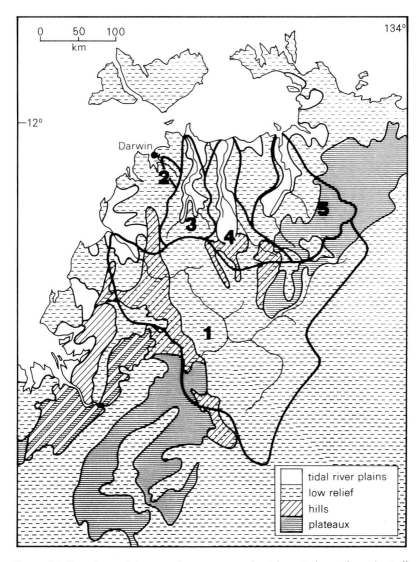

Figure 5.1. Locations of the estuarine systems and catchments in northern Australia which are referred to in this chapter. 1: Daly River; 2: Elizabeth River; 3: Adelaide River; 4: Mary River; 5: South Alligator River.

demonstrated effectively by the mesotidal Clarence estuary in New South Wales (NSW), eastern Australia, which enters the high-energy wave regime of the southwestern Pacific Ocean. The estuary is separated into two fluvio-estuarine basins by a bedrock barrier and the outer basin is impounded by a coastal barrier of marine sands which were driven shoreward during postglacial sea-level rise (see Chapters 2 and 4); in this respect the outer

Table 5.1. *Parameters of selected estuarine systems*

River	TR (m)	A_c (km$^2 \times 10^3$)	Q_t (m^3/s $\times 10^3$)	Q_b (m^3/s $\times 10^3$)	S (t y^{-1})	V_e (m^3)	t_e (ka)	k (m^{-1})	E_w
Daly	6	50	25	1.5	2×10^5	2×10^{10}	3	6×10^{-5}	low
S. Alligator	5	9	30	0.8	4×10^4	3×10^9	2	6×10^{-5}	low
Adelaide	3.5	5	10	0.4	2×10^4	2×10^9	2	4×10^{-5}	low
McArthur	3.5 (d)	12	2	0.6	4×10^4	10^9	6	2×10^{-5}	v. low
Clarence (I)*	1.5	22	0.1	0.6	2×10^5	2×10^9	5	(n)	v. low
Clarence (O)	1.7	23	0.6	0.6	10^5	2×10^9	3	10^{-5}	v. low
Sepik (I)*	1 (d)	77	1	6	4×10^7	6×10^{10}	4	(n)	high
Sepik (O)	1 (d)	77	<1	6	4×10^7	$>10^{10}$	2	10^{-6}	v. low
Yangtse	4	1800	200	30	10^9	5×10^{11}	9	3×10^{-5}	int

TR = Tidal Range (= $2N$; see eqn. (4.1)); A_c = catchment area; Q_t = mean tidal flux at entrance; Q_b = bankfull discharge; and S = fluvial sediment input. V_e = volume of Holocene sediment in estuarine system; t_e = time in which 90% of sediment V_e accumulated, based on stratigraphic and radiocarbon data; k = funnelling coefficient (see eqn. 4.2); E_w = wave energy at coast; I = inner; O = outer; and d = diurnal tides. All values are estimated for mid-Holocene inner basins of Clarence and Sepik systems; (n), funnelling coefficients k are not known.

Clarence basin resembles other estuaries and barrier-basins in southeastern Australia (Roy, 1984). The inner basin of the Clarence, protected by the bedrock barrier, has no high-energy Holocene coastal features and, during its development, the basin divided into extensive low-energy backwater swamps which appear to have been separated by a prograding birdfoot delta (Fig. 5.2).

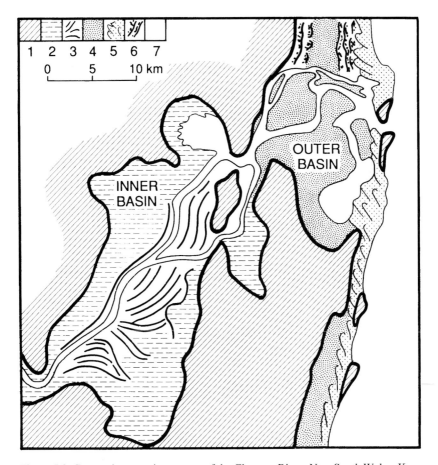

Figure 5.2. Composite estuarine system of the Clarence River, New South Wales. Key: 1. Pre-Holocene land surface, 2. Holocene inner basin sediments: deltaic sands, brackish estuarine and swamp muds beneath thin floodplain silts, 3. Fluvial levée and chute-channel sands, 4. Holocene outer basin estuarine and marine sands, 5. Coastal dune barrier complex, 6. Relict Pleistocene barrier, 7. Estuarine channels and lakes. A bedrock ridge separates inner and outer Holocene sedimentary basins. Estuarine sediment of the inner basin is dominantly of fluvial origin with deltaic and brackish backwater lake and swamp sediments of a low wave-energy environment, capped by fluvial levée, chute-channel and overbank sediments. The outer basin is estuarine sands and muds dominantly of marine origin and is impounded by a high-wave-energy sand barrier at its seaward margin.

The Clarence estuary is fed by the largest of the rivers draining to the NSW coast and represents the fluvial end of a spectrum of NSW sand-barrier estuaries, which includes saline coastal lakes and mud basins at the other extreme (Roy, 1984). A comparable spectrum is recognisable amongst macrotidal estuaries of northern Australia, although two fundamental differences exist. Coastal chenier and beach-ridge plains of the north are not exactly equivalent to the high-wave-energy, sand-barrier coasts of southeastern Australia, although comparable longshore transport processes do occur in the northern, lower-wave-energy regime (Chappell & Thom, 1986). More significantly, the macrotidal, funnelling estuaries are actively formed and maintained by tidal flows, whereas the NSW mesotidal estuary entrances are strongly affected by wave-driven littoral drift and shoal development, and their dimensions are not solely determined by tidal flows (Nielsen & Gordon, 1981; Roy, 1984).

Significance of the fluvial factor in macrotidal northern Australia is expressed by differences between the Daly, South Alligator and the Mary rivers systems, all of which formed within elongate, shallow prior valleys and have prograded coastal plains (Chappell & Thom, 1986; Woodroffe *et al.*, 1986, 1989; Woodroffe, Mulrennan & Chappell, 1993; Chappell, 1993). Catchment and fluvial discharges of the Daly are largest of these three and the Mary is smallest (Table 5.1). The Daly, which is shown in Fig. 5.3*a*, has a very actively meandering estuarine channel and scroll plain, the South Alligator estuary (Fig. 5.5*a*) is characterised by relatively stable, cuspate estuarine meanders, and the formerly large estuary of the Mary is virtually infilled with marine sediment. As shown later, active meandering in the Daly estuary is forced by its large fluvial floods, stable estuarine meanders in the South Alligator reflect statistical equilibrium between fluvial and tidal flows, while extinction of the Mary system is partly due to its relatively small fluvial input.

Prior topography

The range of forms of mesotidal estuaries identified by Roy (1984) in eastern Australia, from drowned valley harbours through barrier estuaries to saline coastal lake, reflects diversity of prior landscapes which were inundated by postglacial rising sea level. Contrasting morphologies and Holocene sediments in the two basins of the Clarence River illustrate the effect of prior topography (Fig. 5.2). The inner basin, protected by a bedrock barrier, contains estuarine sediments of dominantly fluvial origin and channel features are fluvially dominated, while the outer basin contains mostly reworked marine sands and its anastomosing estuary is impounded by a high-wave-energy sand barrier. A comparable case, where a bedrock barrier divides inner and outer

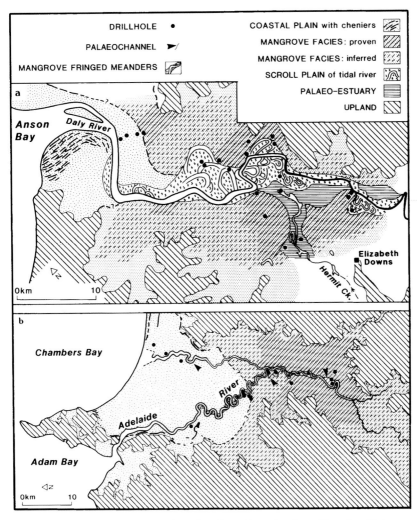

Figure 5.3. Coastal plains, channels, palaeochannels and major mid- to late-Holocene sedimentary facies which underlie the veneer of floodplain sediment, in (*a*) the Lower Daly River estuary (after Chappell, 1993) and (*b*) the Adelaide River estuary (after Woodroffe *et al.*, 1993).

Holocene fluvio-estuarine sedimentary tracts with contrasting facies and morphodynamic styles, is the microtidal Sepik system in Papua New Guinea described by Chappell (1993).

Macrotidal estuaries in northern Australia also vary with prior topography, ranging from small tributaries of drowned valleys (e.g., Darwin Harbour, Semeniuk, 1985a, 1985b) through large, funnelling tidal rivers such as the

Daly to the unusual case of the Adelaide River estuary, which has a bedrock barrier at its entrance and a long reach of sinuous meanders which are inherited from a prior fluvial river (Fig. 5.3*b*). While some morphodynamic differences are due to their differing fluvial inputs, the effects of geometry and depths of their prior valleys are significant.

Fluvial sediment influx

Holocene sedimentation in estuarine systems is strongly influenced by sediment yield of their fluvial catchments. Processes in north Australian macrotidal estuaries with large catchments, such as the Daly, resemble processes in the macrotidal Fly River delta of southern Papua New Guinea, in the dynamics of mud recirculation, funnel morphodynamics and formation of elongate mouth-bar islands. However, the sediment yield of north Australian catchments is very low (about $10 \, t \, km^{-2} \, a^{-1}$) while the yield from tectonically active New Guinea reaches more than $1000 \, t \, km^{-2} \, a^{-1}$. As a consequence, large areas of estuarine plain in the north Australian systems are underlain by vertically accreted mangrove sediments derived from sources to seaward, while Holocene sediments of the New Guinea systems dominantly are of fluvial origin (Chappell, 1993). Fluvially derived sand and gravel also contribute to the infill of macrotidal systems along the higher-relief coast of Queensland (Hacker, 1988). Morphodynamic effects and responses to sea-level changes arising from these differences are examined below.

Classification of tidal estuaries

End-points in the morphologic spectrum of deltas reflect relative dominance of tidal, wave-climate and fluvial regimes (see Chapter 3). Estuaries may be classified in similar terms (Dalrymple *et al.*, 1992), with sediment yield of the fluvial catchment as a fourth variable. There is probably no unambiguous way of expressing form and dynamics in terms of these boundary conditions, because of the effects of prior topography. However, a classification scheme is presented here, which takes several of the process factors into account and separates estuaries with different suites of morphodynamics and sedimentary facies models. The frame is shown in Fig. 5.4. The ratio Q_t/Q_f of tidal discharge defines the horizontal axis; the vertical axis is defined by the ratio V_f/V_e of gross fluvial sediment input, to sediment volume of the estuarine system. Q_t is taken as the mean flood-tide inflow across the entrance, Q_f is bankfull flood discharge entering the estuary and V_e is the volume of Holocene sediment within the estuarine basin. Gross fluvial sediment volume (V_f) is

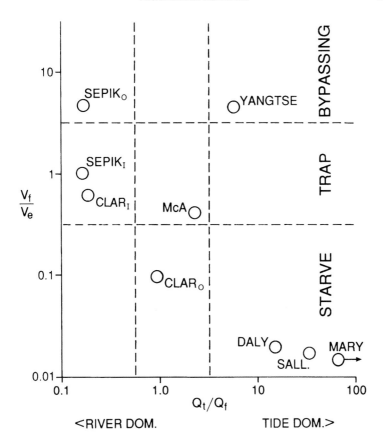

Figure 5.4. Classification diagram for estuaries which are developed in Holocene sediment bodies. The horizontal axis is Q_t/Q_f where Q_t = mean tidal discharge at entrance and Q_f = bankfull fluvial flood discharge at tidal limit. The vertical axis is V_f/V_e where V_f = total volume of fluvial sediment which entered the system during the period in which 90% of the Holocene sediment accumulated (including sediment which has bypassed the estuarine system) and V_e = volume of the Holocene estuarine sediment body. Abbreviations: SALL = South Alligator, McA = McArthur, CLAR = Clarence; subscripts $_o$ = outer basin, $_I$ = inner basin. The figure is compiled from data in Table 5.1. Mary River values were calculated using data for the similar-sized Adelaide River.

estimated by multiplying the mean annual sediment yield by the interval of time (in years) during which 90% of the Holocene sediment body accumulated. This interval (t_e in Table 5.1) is used because in many estuarine systems the bulk of the sediment accumulated within a few thousand years centred on the culmination of postglacial sea-level rise, with little sedimentation in the past several thousand years.

Tidal dominance increases with Q_t/Q_f; vertical dashed lines in Fig. 5.4 separate river-dominated, intermediate and tide-dominated regimes. A value of $V_f/V_e = 1$ separates those systems where most of the input fluvial sediment is retained in the estuarine basin ($V_f/V_e < 1$) from those where some fluvial sediment bypasses the basin and is deposited offshore ($V_f/V_e > 1$). When V_f/V_e is significantly less than one, fluvial sources are insufficient to account for sediment in the basin and it is likely that some sediment is derived from offshore. Fig. 5.4 shows tropical Australian macrotidal systems considered in this chapter (plus several other estuaries and deltas for comparison) – both inner and outer basins of the mesotidal Clarence River estuary, the Daly, Mary and South Alligator macrotidal estuaries, the mesotidal McArthur, the inner and outer basins of the microtidal Sepik system described by Chappell (1993) and the macrotidal Yangtse River delta.

Morphodynamics

Morphodynamics of macrotidal estuaries in northern Australia are forced by strong tidal flows throughout the year and fluvial floodwater flows for a relatively short part of the monsoonal wet season. Tides tend to be symmetrical offshore but become progressively asymmetrical with distance up the tidal rivers. Flood tides have shorter duration and higher flow velocities, including tidal bores in more extreme cases, than the ebb tides (Woodroffe *et al.*, 1986; Vertessy, 1990). This flow asymmetry affects sediment movement. Bedload transport formulae relate transport to a current velocity term raised to some power z, with z in the range from 3 to 5 in most formulae (Bagnold, 1966; TAPSM, 1971). Where highest mean tidal flows are directed upstream the net bedload sediment movement also tends to be upstream, unless tidal asymmetry is offset by downstream fluvial flood discharges. Vertical profiles of suspended sediment concentration (SSC) also vary through the tide cycle. High velocities and shallow water depths early in the flood tide generate high turbulent mixing through the water column, whereas sediment settling at high-tide slack water followed by lower ebb velocities leads to formation of a lutocline with lesser velocities in the underlying, high-SSC water than in the upper, low-SSC water (Wolanski *et al.*, 1988). Vertically integrated velocity–density products differ between flood and ebb flows, which may cause upstream drift of suspended sediment. Salt-wedge structures, which develop when there is freshwater outflow, also generate upstream sediment transport in the underlying saltwater inflow. Finally, evaporative water loss from the estuary in regions with low relative humidity can produce a net water and suspended sediment flux into the tidal system (Wolanski, 1986; Chappell,

1990). Interaction between these various processes is strongly seasonal in northern Australia.

In those north Australian macrotidal estuaries, such as the Daly and South Alligator, which are little affected by bedrock topography and lie completely within bodies of Holocene sediment, competing upstream- and downstream-directed flows and sediment movements generate a variety of morphodynamic elements which are reviewed in the following paragraphs. Examples of the elements described here can be found in the following systems: the Daly (Fig. 5.3*a*), the South Alligator (Fig. 5.5), the Adelaide (Fig. 5.3*b*) and the Mary (Fig. 5.7).

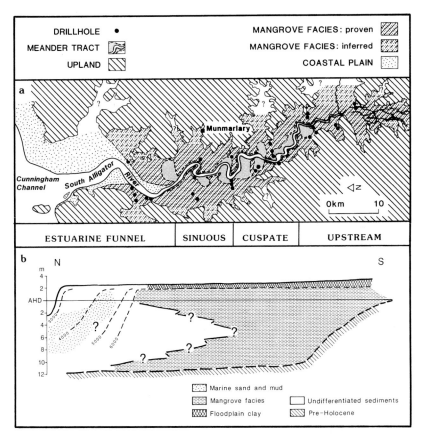

Figure 5.5. (*a*) South Alligator estuary, showing funnel, sinuous, cuspate and anabranching upstream channel segments, plus (*b*) Holocene sedimentary facies. Tentative isochrons indicating coastal progradation are shown. AHD = Australian Height Datum, which approximates mean sea level. (After Woodroffe *et al.*, 1986, 1993.)

The estuarine funnel

Long macrotidal estuaries typically are funnel-shaped in planform, tapering negative-exponentially upstream (Chappell & Woodroffe, 1985), often with a few large, dog-leg bends in their lower reaches (Figs. 5.3a, 5.5 and 5.10a). Radiocarbon dating of sediment cores adjacent to these funnels shows that they can migrate slowly, sometimes accompanied by progradation of the adjoining coast, but they appear to maintain their overall form through time (Woodroffe et al., 1986; Chappell, 1993). Estuarine funnels are tidally generated, but there are different morphodynamic forms of estuarine funnel – flood-tide dominated, ebb-dominated and progradational. North Australian macrotidal funnels are flood-tide dominated and we focus on these, although the other forms are discussed at the end of this section.

The cross-sectional area of the mouth of most estuaries is closely related to magnitude of the tidal prism (Byrne, Gammisch & Thomas, 1981). The cross-section at the point where tidally reversing flow becomes negligible is related to bankfull floods which enter the tidal river (in this chapter, the point where reversing flow is negligible at low fluvial stages is referred to as the tidal limit, despite that elevation of the water surface can vary tidally upstream of this). The following simple analysis indicates that rate of change of channel width with distance is determined by shear strength of the channel banks, for a given tidal regime. Total discharge through a given cross-section during a tidal half-cycle equals tidal prism upstream of the section. For a simple funnel parallel to the x-axis, tidal limit at $x = L$, this is approximately represented for a section at x by

$$W{\cdot}HU{\cdot}t \approx \int_{x}^{L} (4NW) \cdot dx \qquad (5.1)$$

where W is width, HU is the mean product of depth (H) and depth-averaged velocity U, N is tidal amplitude and t is tidal period. Tidal asymmetry and progression of the tidal wave along the channel are neglected in the approximation in eqn (5.1) and can be significant. However, within limits of these assumptions and provided that N, U and H are constant along the length of the funnel, which is approximately true in the estuarine funnels of north Australian tidal rivers such as the South Alligator and Daly (Woodroffe et al., 1986; Vertessy, 1990), then the solution of eqn. (5.1) is

$$W = W_0 \exp(-kx), \quad k = 4N/tHU \qquad (5.2)$$

where W_0 is width at the entrance, $x = 0$. The funnelling coefficient k defines the rate of change of channel width with distance and eqn. (5.2) shows that this

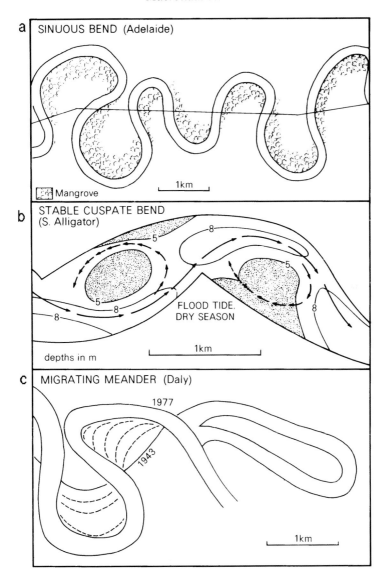

Figure 5.6. (*a*) Inherited sinuous meanders of the Adelaide River estuary. The pre-Holocene palaeochannel is buried beneath 8 m of Holocene 'big swamp' mangrove and estuarine channel sediment. Mangroves on insides of bends indicate extent of bend migration of the last several thousand years. (*b*) Cuspate estuarine meanders of South Alligator river, showing flood-tide course of maximum flow velocities path (solid arrows). Note large eddies (dashed arrows) around mid-channel shoals. (*c*) Rapidly migrating sinuous estuarine meander on Daly River. Dashed lines indicate scroll bars. (After Vertessy, 1990.)

is related to flow velocity as well as tidal amplitude and channel depth. Constancy of flow velocity throughout a funnel implies that channel banks are adjusted to a critical shear stress induced by the tidal current. Actually, the peak shear force which the channel banks can withstand is likely to be directly related to peak current velocity; tide-cycle current measurements in the South Alligator and Daly rivers (Woodroffe *et al.*, 1986, Vertessy, 1990, and unpublished data) indicate that peak velocity in these systems occurs during flood tides and is proportional to U.

Macrotidal estuarine funnels in northern Australia conform closely to exponential eqn. (5.2). Data in Table 5.1 show that the funnels of the South Alligator and Daly have k-values around $5 \times 10^{-5} \, \mathrm{m}^{-1}$ and other macrotidal estuaries in the same region, which lie within similar bodies of Holocene sediment (East Alligator, West Alligator, and Wildman), have similar values of k. Mesotidal systems in tropical Australia have lower k-values and the long reach of inherited sinuous meanders in the Adelaide River estuary has a low k-value. Macrotidal reaches upstream of the funnel in the South Alligator and some other systems also have smaller k-values, reflecting their different morphodynamics.

Not all funnelling estuaries are flood-tide dominated. Tidal creeks which drain intertidal flats typically have highest flow velocities during ebb tide (due to efficient tidal runoff from saturated intertidal flats). Provided that the assumptions underlying eqn. (5.1) apply and that N, U and H are constant, an ebb-dominated funnel will be exponential. This distinction between flood- and ebb-dominated funnels is relevant to the Holocene evolution of funnels, outlined later. Finally, we note that many funnelling tidal creeks, such as occur in mangrove swamps, show flow velocity diminishing upstream from a high value at the mouth to zero near the upstream end. These usually occur in prograded coastal swamps, and funnelling is presumed to reflect continued morphodynamic adjustment of the mouth cross-section to the tidal prism, which enlarges with progradation of the swamp system. In these cases the channel upstream of the mouth is relict rather than in dynamic equilibrium with tidal flow.

Estuarine meanders

River bends in macrotidal channels quite commonly are different from the sinuous bends of meandering fluvial rivers in that, seen in planform, the inner bank forms a sharp cusp opposite the apex of the bend (Fig. 5.6*b*). Mid-channel shoals often occur in the broad reaches which intervene between cuspate bends. Ahnert (1960) termed these 'estuarine meanders' and noted that ebb- and flood-tide flow paths occur on opposite sides of the mid-channel shoals. Through measurements in the monsoonal South Alligator River, Vertessy (1990) showed that morphodynamics of cuspate, estuarine meanders are a little more complex. When they occur, mid-channel shoals tend to be centred beneath large eddies which form in the lee of cusps (Figs. 5.6*b*, 5.10*b*), but shoals sometimes occur at the channel side and some estuarine meanders are not sharply and symmetrically cuspate. Separation of flood and ebb flow paths within a tide cycle occurs in some but not all cuspate meanders although, in monsoonal north Australia, separation occurs between the flood-tide path in the dry season and the ebb-tide path, enhanced by fluvial floodwater, in the wet season. Shoals tend to migrate to and fro with the dominant seasonal flow and cuspate estuarine meanders are likely to be statistically stationary. Aerial photographic runs of estuarine meanders in the South Alligator, spanning 40 years, show almost no detectable changes. This contrasts with very active sinuous meanders observed in the Daly macrotidal estuary, described below.

Abandoned sinuous meanders, pinching out opposite cusps of some estuarine meanders in northern Australia, indicate that some cuspate reaches formed after cutoff (Vertessy, 1990). Other estuarine meanders are not associated with cutoff palaeochannels, however, and appear to be freely developed where sediment entrainment and transport under reversing flows are statistically equal and opposite. Once formed, stationary cuspate meanders show no tendency to revert to the sinuous form, which implies that prior formation of a 'parent' sinuous meander may involve a different fluvio-tidal regime from that which exists today.

Active meanders and scroll plains

The general morphodynamic equilibrium which exists in cuspate estuarine meanders is not sustained when there is strong net drift of bedload sediment in one direction. In this case the cusp, which forms by cutoff of a sinuous meander, acts as a focus for deposition of a point-bar or point-shoal, and the meander redevelops a sinuous form by continued accretion which generates a scroll bar (Fig. 5.6*c*). For example, a 30 km reach within the strongly funnelled Daly River estuary in northern Australia is characterised by active meanders, with migration rates up to $50 \, \mathrm{m \, a^{-1}}$, set into a late Holocene scroll plain with nests of cutoff meanders (Fig. 5.3*a*). Scroll-bar ridges show that active meanders of the Daly prograded in the downstream direction, evidently during monsoonal floods.

The funnelling coefficient k is relatively constant throughout the estuarine funnel and active sinuous meandering reaches of the Daly, which suggests that tidal flows dominate over fluvial discharge in determining channel width despite the fact that fluvial floods cause meanders to migrate. This contrasts with the cuspate meandering reach of the South Alligator, where channel widths are locally highly variable but the average channel width is relatively constant through a reach of about 30 km. This seems anomalous because Q_t/Q_f is larger in the South Alligator than in the Daly, and further, detailed analysis of flow regimes is needed in order to explain this.

Inherited channels

Cuspate and active estuarine meanders are dynamic forms within macrotidal estuaries which lie within Holocene sediment bodies, and contrast with inherited fluvial channels which persist through Holocene estuarine development, although they are modified by tidal flows and originate in prior riverine plains which may outcrop at the surface or even may be buried by estuarine sediments. The Adelaide River estuary is a fine example, which includes a 60 km reach of inherited sinuous meanders (Figs. 5.3b, 5.6a and 5.10c). The prior riverine plain lies beneath about 8 m of Holocene estuarine and mangrove sediment while the tidal river channel is 10–12 m deep and, except for minor progradation under mangroves on insides of bends, it generally conforms to the prior fluvial river course. Inherited meanders in some cases appear to have become cutoff progenitors of cuspate estuarine meanders, after postglacial sea level stabilised and estuarine plains became fully developed.

Anabranches and distributaries

Not all tidal river systems are single, funnelling or meandering channels, but can be compounds of multiple channels formed of anabranches and distributaries. When present, distributary channels occur in lower reaches of tidal rivers and extend to the coast, whereas anabranches, which are alternative channels which diverge only to rejoin further downstream, occur in upper reaches of tidal rivers (examples in the South Alligator are described by Woodroffe *et al.*, 1986). Distributary channels are funnelled and typically have estuarine meanders or dog-leg bends, while anabranches are of rather constant width and apparently form through avulsion during fluvial floods entering the upper tidal river.

Extinguished systems

Although some macrotidal rivers historically have been rather stable and changed only by channel migration, sinuous-cuspate meander transition or by anabranching in late Holocene times, such as the Alligator group in northern Australia, others changed more dramatically. The Adelaide River estuary (Fig. 5.3*b*) switched from its prior, highly sinuous meandering course to an alternative distributary funnel, and back again to the sinuous channel in late Holocene times (Woodroffe *et al.*, 1993). During the same period, the nearby Mary River estuary developed two major, meandering funnels (Fig. 5.7). These later became filled with sediment and died as tidal systems, so that the fluvial discharge of the Mary River largely dissipated into extensive swamplands and evaporated with relatively minor discharge to the sea through lesser, avulsive distributaries which diverted the fluvial flow. The former estuarine system of the Mary is thought to have infilled with sediment driven tidally upstream, after fluvial discharge was diverted (Woodroffe *et al.*, 1993;

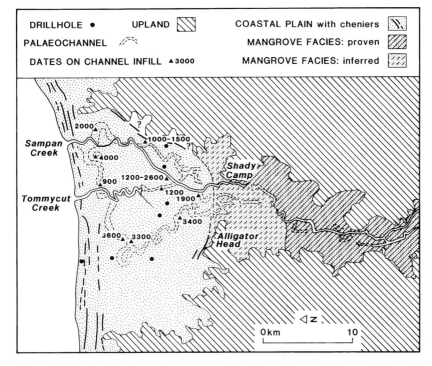

Figure 5.7. Extinguished estuarine system of the Mary River, with times of channel extinction based on radiocarbon dates (years BP) of sediment infills (after Woodroffe, *et al.*, 1993).

Woodroffe & Mulrennan, 1993). Radiocarbon dates support the chronology of successive phases of channel activity and infilling in the Mary system, shown in Fig. 5.7.

Estuarine floodplains

Macrotidal estuarine floodplains in north Australian regions of higher rainfall typically have a low convex profile, descending at a very low gradient away from the tidal river with backwater swamps where the floodplain meets the adjacent hills. Sedges and grasses with low herbs cover the plains and swamp forests occur near the margins. Tree-lined levées occur on the banks in upper reaches of the tidal rivers and mangrove creeks, often draining infilled palaeochannels, join the main channel.

Today, there is slow tidal invasion of some floodplain areas, shown by headward-extending tidal creeks colonised by juvenile mangroves (Figs. 5.8 and 5.10*d*). Owing to convexity of the floodplain, invading tidal waters flow readily into low-lying areas within reach of these creeks and the freshwater sedgelands and swamp forests are overtaken by mangrove or by bare saline mudflats. Attrition of freshwater vegetation and extending saline flats is occurring on floodplains adjoining parts of the lower reaches of several estuaries. This is particularly conspicuous in the lower Mary River, where salinisation of the plains is activated by tidal creek extension, cutting into the extensive freshwater floodplains (Knighton, Mills & Woodroffe, 1991; Knighton, Woodroffe & Mills, 1992), and by erosion of sedgeland where scarp retreat at the interface with saline mudflat leaves small remnants of sedgeland vegetation within the extending salt flats (Fig. 5.8).

Chenier and beach-ridge plains

Prograded coastal plains with chenier and/or beach ridges adjoin most of those north Australian estuaries which are associated with substantial bodies of Holocene sediment (Rhodes, 1982; Chappell & Thom, 1986). The transition from estuarine plain to coastal plain often is indicated by a chenier or beach ridge with a ^{14}C age around 5000–6000 years BP, marking the coastline of or soon after the peak postglacial sea level. Widths of coastal plains adjoining the larger north Australian estuaries range from about 1 to 20 km and this variation depends more on depth to the buried prior landsurface than the fluvial discharge from nearby rivers.

Progradation of the coastal plain causes seaward translation of the entrance and funnel of an adjacent estuary, which enlarges the tidal prism. In flood-tide

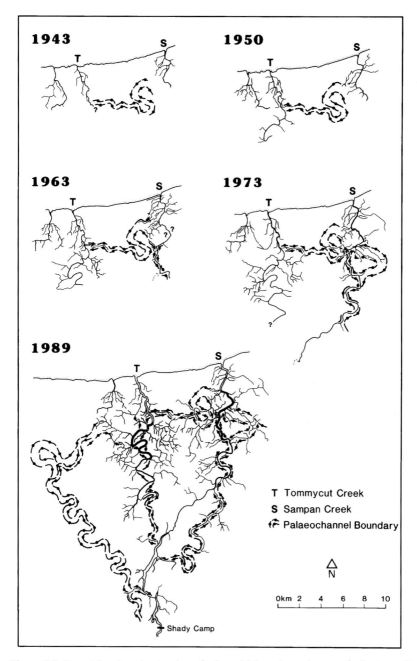

Figure 5.8. Recent headward extension of minor tidal creeks and areas of tributary salt flats which have formed from prior freshwater floodplain, in lower Mary River area (after Knighton *et al.*, 1991).

dominated systems, the consequent increase of tidal dominance (increased Q_t/Q_f) is likely to enhance the tendency for sediment to move landward up the funnel. This process, associated with coastal progradation, may have contributed to extinction of the Mary estuary in the past few thousand years (Fig. 5.7). Radiocarbon chronologies of the subsurface sediments of chenier plains of most north Australian macrotidal systems indicate most rapid progradation around 5000–3000 years BP, with negligible net change in the last 2000 years (Woodroffe *et al.*, 1989).

Net sediment movement and estuary development

Sediment movements upstream by tidal transport, downstream by fluvial flooding and overbank onto the estuarine floodplain (Fig. 5.9) are implicated in morphodynamics of most of the elements of macrotidal estuaries reviewed in the preceding paragraphs. The balance between opposed tidal and fluvial transport, which influences morphodynamics of the middle reaches of macrotidal estuaries (Anhert, 1960; Dalrymple, *et al.*, 1992), is affected by the Q_t/Q_f ratio. Morphodynamics of north Australian macrotidal systems with different Q_t/Q_f values differ, accordingly. Funnelled estuaries in this region include one case with very active sinuous meanders and scroll plains (the Daly), several with long reaches of stable, cuspate estuarine meanders (e.g., South and East Alligator Rivers), and the extinguished estuarine system of the Mary River. The smaller the fluvial influence, the weaker the seaward transport of sediment during wet-season floods and the stronger the tendency for upstream-moving sediment to remain within the estuary. Seaward-migrating sinuous meanders in the Daly estuary, which develop rapidly from cusps after meander cutoff, reflect net downstream sediment transport forced by high wet-season discharges from the relatively large catchment of this river. The South Alligator, with its smaller catchment, has stable cuspate meanders and probably maintains a neutral sediment budget. The yet smaller Mary estuaries have become extinguished through sedimentary infill from seaward. The Adelaide system, with a catchment similar to that of the Mary, formerly had a funnelled estuary which probably infilled in the same manner, redirecting the tidal river back to its former, fluvially defined channel (Fig. 5.3*b*)

Sedimentary infill of the lower Mary and the former Adelaide estuary may partly reflect an increase of Q_t/Q_f either through coastal progradation, as suggested earlier, or from diminution of rainfall since mid Holocene times. Alternatively, the former estuarine funnels may have developed under an ebb-dominated tidal regime, within extensive mangrove swamp; in this scenario,

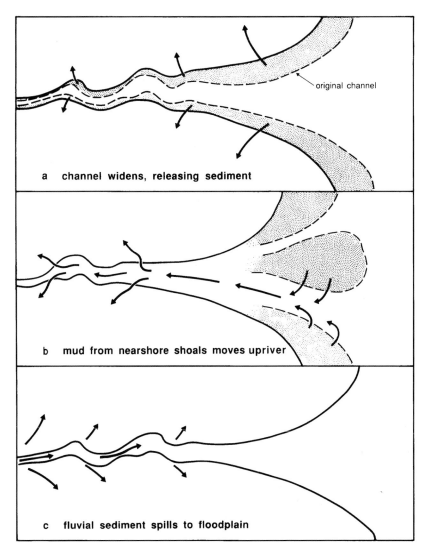

original channel

a   **channel widens, releasing sediment**

b   **mud from nearshore shoals moves upriver**

c   **fluvial sediment spills to floodplain**

Figure 5.9. Schematic diagram of macrotidal funnel showing (*a*) overbank flooding and sediment deposition associated with estuary widening, (*b*) upstream sediment transport in flood-tide dominated regime, and (*c*) downstream transport and overbank deposition due to fluvial flooding. Intertidal shoals are shaded.

subsequent transition from mangrove to floodplain, through continued sedimentation, would induce a flood-tide dominated regime, which eventually led to extinction of the former estuaries. These and other aspects of Holocene evolution of macrotidal systems are explored in the next section.

Postglacial sea-level changes and estuary evolution

Rising sea level invaded the rather shallow, broad valleys of far northern Australia towards the end of the postglacial transgression (Chappell & Polach, 1991). The following summary of ensuing sedimentation and estuary development is based on extensive drillhole studies reported by Woodroffe *et al.* (1986, 1989, 1993), Woodroffe & Mulrennan (1993), and Chappell (1990, 1993). The prior valley floor is indicated by either quartzose alluvial sands, or lateritic gravel, or highly oxidised sandy clay. This prior surface is shown by thermoluminescence dating to be of Late Pleistocene age (unpublished data: D. Price, pers. commun., 1992), and lies about 14 m below the estuarine plains of the South Alligator and Daly systems and at 8–10 m depth in the Adelaide and Mary systems. Mangrove facies composed of organic muds containing mangrove pollen and wood, with local mangrove peat, commonly overlies the prior valley floors, recording the inundation of the valley by the rising sea level around 7000–8000 years BP. Transgressive basal sediments pass upwards into mangrove facies showing that these macrotidal estuaries became areas of vertical accretion under widespread mangrove forest, which has been termed the 'big swamp', throughout the last 8 m or more of sea-level rise (Woodroffe, Chappell & Thom, 1985; Woodroffe *et al.,* 1987; Chappell, 1993). The presence of tidal channels within these swamps is shown by buried sediment bodies of sands and muds containing foraminifera with estuarine affinites. Most of these lie close to the present tidal rivers but some are up to 2–3 km away (Wang & Chappell, in press). Facies relationships are illustrated in Fig. 5.5*b*, using the example of the South Alligator system.

These vertically accreting sedimentary facies underlie a thin cover (0.5–2 m) of freshwater floodplain sediments which accumulated after sea level stabilised. The transition from mangrove-dominated 'big swamp' sedimentation to freshwater sedge and grassland plains is identified by pollen analysis of drillcores, which typically show succession, at shallow depth, from Rhizophoraceous communities to mixed mangroves, especially *Avicennia*, followed by a sharp passage to freshwater sedges and grasses (Woodroffe *et al.,* 1985; Russell-Smith, 1985; Chappell & Grindrod, 1985; Grindrod, 1988; Clark & Guppy, 1988). Transition from mangrove 'big swamps' to freshwater estuarine plains occurred soon after sea level stabilised, around 5000–6000 years BP, in the Daly and South Alligator systems, and somewhat later in the Mary and other smaller systems (see Fig. 1.5). Sediments of the rising-sea-level phase are interrupted near the axis of each valley by deposits of the meandering tidal rivers; the extent of the meander tract reflects the balance between fluvial input and tidal energy, being most extensive at the Daly River, narrower and less active at the South Alligator, and least on the Adelaide plains. Changes of positions and geometries of the estuarine channels during the last 6000 years are shown by surface traces of palaeochannels and subsurface facies of channel sands and muds which often show lamination and point-bar structures. Progradation of the coastal plains seawards from the innermost fossil shoreline occurred in the last 6000 years and decreased in rate in the last 2000–3000 years in most systems other than the Daly.

Extinguished estuarine funnels of the Mary and Adelaide rivers imply that these formed under fluvio-tidal regimes different from those of today, as noted earlier. Radiocarbon-dated estuarine sediments within lower reaches of the Daly and South Alligator indicate that estuarine funnels existed during the 'big swamp' phase of rising sea level (Woodroffe *et al.*, 1986; Chappell, 1993). These are likely to have been ebb-dominated due to tidal flooding within the extensive mangroves; the transition from mangrove 'big swamp' to freshwater plains would have been accompanied by a change to flood-dominated regimes and enhancement of upstream sediment transport leading to infilling of smaller systems.

Comparison of Australian macrotidal rivers with other systems

Estuaries are a sink for sediment, and gain sediment both from upstream by direct riverine input and from seaward by tidal and salt-wedge circulations (Guilcher, 1967; Nichols and Biggs, 1985; Dronkers, 1986). North Australian macrotidal rivers lie within Holocene sediment bodies dominantly derived from seaward; in terms of the classification proposed above (Fig. 5.4), systems such as the South Alligator and Daly lie in the region defined by high Q_t/Q_f and low V_f/V_e. Other, comparable north Australian macrotidal systems have been described, including the Ord in Western Australia (Wright *et al.*, 1973; Wright, Coleman & Thom, 1975; Thom, Wright & Coleman, 1975; Coleman and Wright, 1978) and the Fitzroy River estuary (Jennings, 1975; Semeniuk, 1980, 1982). Stratigraphic relationships and chronology suggest a similar pattern of development to that of the South Alligator and neighbouring rivers in the Northern Territory (Woodroffe, 1988). It is useful to compare the north Australian systems with others which either are macrotidal with relatively higher fluvial sediment yields from their catchments (higher V_f/V_e), or are mesotidal with similar fluvial sediment yields (lower Q_t/Q_f). There are systematic morphodynamic differences in both cases which relate to V_f/V_e and Q_t/Q_f values, rather than to other boundary conditions.

Rivers entering the Gulf of Carpentaria drain catchments with geomorphic and climatic regimes which are similar to those of the Daly–Alligators macrotidal suite but their estuaries have substantially smaller values of Q_t/Q_f as tides are dominantly diurnal and have maximum tidal range of about 3 m. Wave climate varies with exposure and affects estuarine morphology; the deltaic McArthur estuary, for example, is protected by an offshore group of islands and has two major and several minor distributary channels (Woodroffe & Chappell, 1993), while the more exposed Norman estuary has a single channel traversing a very broad chenier plain (Rhodes, 1982). The Gilbert River, on the eastern margin of the Gulf of Carpentaria, similarly is dominated by riverine processes and morphology throughout much of its tidal reach (Jones, Martin & Senapati, 1993). All have relatively low funnelling coefficients (k), reflecting lesser tidal influence than the macrotidal cases. These systems lack the extensive mid Holocene 'big swamp' mangrove sediments that characterise the north Australian macrotidal systems, described above, and prograded late Holocene coastal sediments overlie mid Holocene nearshore and estuarine sands and muds.

The Bay of Fundy, with perhaps the world's largest tidal range, includes the Cobequid Bay macrotidal estuarine system which shares many morphological similarities with north Australian systems (Dalrymple *et al.*, 1990). Dalrymple *et al.* (1992) identify three estuarine segments with different channel morphology. A relatively straight seaward funnel, dominated by tidal and marine processes, has sediment moving in a net landward direction; a central reach is a zone of sediment convergence and has a sinuous, highly meandering channel; and the upstream reach is relatively straight, least influenced by marine processes and has net seaward sediment movement. Similarly, an estuarine funnel and upper estuarine meandering channels are recognised in the Gironde estuary in southwestern France but, in this case, coastal wave energy is high and the outer region is being transgressed by a sandy barrier (G. Allen, 1991; Allen & Posamentier, 1993). In this system all the fluvially derived sand, and about 75% of the mud, is deposited in the estuary with only a minor output of mud to seaward (Allen *et al.*, 1980; Castaing & Allen, 1981). The macrotidal Severn estuary in southwestern Great Britain has similar features. Studies, which commenced last century, suggest landward movement of sand (Sollas, 1883; Murray & Hawkins, 1977; Wang & Murray, 1983). The exact pathways of landward sand transport are still not clear (Harris & Collins, 1985, 1988, 1991; Stride & Belderson, 1990), whereas it appears that the dominant source of mud in the Severn at present is from rivers which feed the estuary (J. Allen, 1991), there may be a net export of mud to the continental shelf through cliff erosion of channel margins related to relative sea-level rise (Allen, 1987, 1990).

Despite similarities between these systems and those of northern Australia, there are differences. Longitudinal sand bars and shoals are more prominent in the outer funnels and estuary-embayments of the Severn and Bay of Fundy than in most north Australian macrotidal systems (with the partial exception of the Daly: Fig. 5.3*a*). Perhaps the most significant difference is represented by mid-Holocene development, in north Australian systems, of mangrove 'big swamps' in which sedimentation kept pace with rising sea level. We note that extensive Holocene swamp facies occur in cool temperate systems such as the English Fenlands (Godwin, 1978), but these appear to be time-trangressive horizons associated with sea-level changes. Our tropical, vertical sequences may have no exact counterpart in cool temperate regions, not only by virtue of their mangrove vegatation. Earlier, we suggested that vertical sedimentation of mangrove-facies muds in north Australian palaeo-estuaries was sustained by landward 'pumping' of muds in this highly seasonal, highly evaporative environment, even in ebb-dominated regimes. We note that extensive mid Holocene mangrove sediments overlain by freshwater sediments occur in other tropical, western Pacific estuaries such as the macrotidal Klang River (Coleman, Gagliano & Smith, 1970) and other sites in southeast Asia (Anderson & Müller, 1975; Haseldonckx, 1977; Bosch, 1988; Kamaludin, 1993; Woodroffe, 1993). Not all of these are macrotidal and, in the light of tropical, mesotidal cases which lack a mid Holocene mangrove 'big swamp' phase, such as those outlined above in the Gulf of Carpentaria, it appears that further work is required before the boundary conditions for vertically accreting 'big swamps' are identified.

Several recent facies models of macrotidal estuaries have been linked to the procedures of sequence stratigraphy (Cobequid Bay, Dalrymple *et al.,* 1992; Daly River, Chappell, 1993; Gironde estuary, Allen & Posamentier, 1993). These adopt similar approaches, identifying a lowstand system tract (equivalent to the pre-Holocene alluvial valley floor in the north Australian examples), a transgressive tract, and various highstand system tracts. Differentiation of transgressive and highstand system tracts is not always easy; in the case of the Daly, Chappell (1993) considered that the maximum flooding surface separates the transgressive tract from a vertical sedimentation tract, within which there is an estuarine–freshwater transition which appears synchronous across the estuarine plains. Within the highstand tract, different rivers in north Australia demonstrate different degrees of reworking of their floodplains with different extents of meander tract. A regressive tract is represented by progradation of the coastal plain.

Likely effects of future sea-level rise

Freshwater ecosystems of the estuarine plains and backwater swamps of northern Australia are endangered if sea level rises in future, which is a widely anticipated consequence of global warming. Tops of channel banks are close to

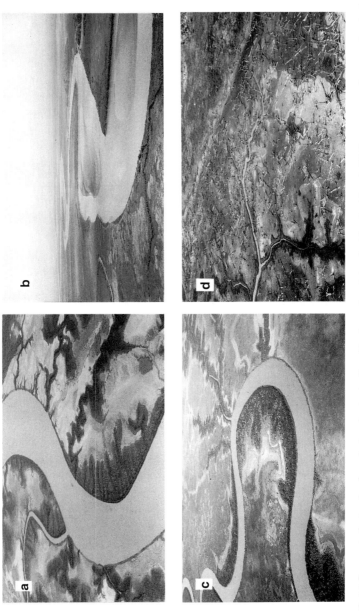

Figure 5.10. (*a*) Segment of estuarine funnel, showing typically wide channel and simple, 'dog-leg' bend. (*b*) Cuspate estuarine meanders in the East Alligator tidal river. Note large, mid-channel shoals either side of nearest cuspate point. (*c*) Sinuous estuarine meander, Adelaide River, inherited from prior fluvial channel which was drowned by rising sea level in early Holocene times. Note mangrove forests on the inner bend of river. (*d*) Tidal creeks invading freshwater floodplains, Mary River plains, with consequent salinisation and death of swamp forests (cf. Fig. 5.8).

the level of high spring tide throughout lower reaches of most macrotidal rivers of the region and, away from main tidal-river channels, floodplain surfaces and backwater swamps lie at or below the elevation of highest spring tide levels at the coast. Saline waters today intrude into these areas where small, high-tide creeks breach the low, broad levées. Networks of small creeks which have extended rapidly in the lower Mary system, causing extensive salinisation of former freshwater floodplain over the last 40 years, are shown in Fig. 5.8; this example dramatically illustrates a process which has occurred in parts of many estuarine plains in northern Australia (see Fig. 5.10d). Although this historical process arguably is attributed to compaction of sediments and the effects of feral buffalo rather than sea-level rise, as outlined earlier (see Woodroffe & Mulrennan, 1993), it indicates the most likely first stage of floodplain inundation which would occur if sea level were to rise in future.

Substantial overbank flooding would occur at high tide if highest tide level were increased by only 30–50 cm, overtopping the levées, whether as the result of global eustatic sea-level rise or a local increase of tidal amplitude. The extra volume of water, flooding overbank, would increase the total tidal prism and flow velocities in the channel would increase if all the additional flow was carried by the channel and its dimensions remained constant. Evidence cited earlier indicates that channel cross-section throughout estuarine funnels of the north Australian macrotidal type is adjusted to the maximum flow velocity. When the tidal prism is contained within the tidal river, maximum velocities occur at low- to mid-flood tide. Overbank flooding with rising sea level will enlarge the tidal prism but channel response is not necessarily immediate, because a modest degree of overbank flooding close to high water will not affect maximum flow velocities. Ebb velocities will be enhanced by return flow from the floodplain, however, and rising sea level will cause a gradual shift from flood- to ebb-dominance. A threshold is crossed when enhanced ebb velocities exceed the peak velocity which prevailed before sea-level rise commenced, and it is likely that channel adjustment to rising sea level will be slight until this occurs. Once the threshold is exceeded, channel widening is likely to proceed more rapidly. Alteration of the floodplain ecology will commence before the threshold is reached, however, by salt-water invasion similar to the historical processes in the lower Mary system (Fig. 5.10d).

Once initiated by rising sea level, channel widening may contribute sediment to the estuarine plains. Sediment tends to remain in the estuarine system of long macrotidal, tropical rivers even when ebb dominated, and sediment released by channel enlargement should find its way overbank onto the plains. This will offset the effect of flooding and may lead to steady vertical sedimentary accretion. In this respect, the past is a partial key to the future:

patterns of development of macrotidal estuaries under rising sea level in earlier Holocene times may be revisited in future. This review has outlined some of the morphodynamic factors which need more precise study before the consequences of any future scenario can be forecast accurately.

Acknowledgements

We particularly wish to thank Eric Wolanski and Rob Vertessy for valuable discussions in the course of this work, and all those who participated with us in research in northern Australia in the past decade, especially Monica Mulrennan, Eugene Wallensky and Colin Campbell.

References

Ahnert, F. (1960). Estuarine meanders in the Chesapeake Bay area. *Geographical Review*, **50**, 390–401.

Allen, G.P. (1991). Sedimentary processes and facies in the Gironde estuary: a recent model for macrotidal estuarine systems. In *Clastic tidal sedimentology*, ed. D.G. Smith, G.E. Reinson, B.A. Zaitlin & R.A. Rahmani, pp. 29–40. Canadian Society of Petroleum Geologists Memoir, 16.

Allen, G.P. & Posamentier, H.W. (1993). Sequence stratigraphy and facies model of an incised valley fill: the Gironde estuary, France. *Journal of Sedimentary Petrology*, **63**, 378–91.

Allen, G.P., Salomon, J.C., Bassoullet, P., Du Penhoat, Y. & De Grandpre, C. (1980). Effects of tides on mixing and suspended transport in macrotidal estuaries. *Sedimentary Geology*, **26**, 69–90.

Allen, J.R.L. (1987). Reworking of muddy intertidal sediments in the Severn Estuary, southwestern U.K.: a preliminary survey. *Sedimentary Geology*, **50**, 1–23.

Allen, J.R.L. (1990). Salt-marsh growth and stratification: a numerical model with special reference to the Severn Estuary, southwest Britain. *Marine Geology*, **95**, 77–96.

Allen, J.R.L. (1991). Fine sediment and its sources, Severn Estuary and inner Bristol Channel, southwest Britain. *Sedimentary Geology*, **75**, 57–65.

Anderson, J.A.R. & Müller, J. (1975). Palynological study on a Holocene peat and a Miocene coal deposit from NW Borneo. *Review of Palaeobotany and Palynology*, **19**, 291–351.

Bagnold, R.A. (1966). An approach to the sediment transport problem from general physics. *US Geological Survey Professional Paper*, **422**(1).

Bosch, J.H.A. (1988). The Quaternary deposits in the coastal plains of Peninsular Malaysia. *Geological Survey of Malaysia, Quaternary Geology Section, Report* QG/1 of 1988.

Boyd, R., Dalrymple, R. & Zaitlin, B.A. (1992). Classification of clastic coastal depositional environments. *Sedimentary Geology*, **80**, 139–50.

Byrne, R.J., Gammisch, R.A. & Thomas, G.R. (1981). Tidal prism – inlet areas relationships for small tidal inlets. In *17th International Conference on Coastal Engineering, Sydney*, pp. 244–5. New York: American Society of Civil Engineers.

Castaing, P & Allen, G.P. (1981). Mechanisms controlling seaward escape of suspended sediment from the Gironde: a macrotidal estuary in France. *Marine Geology*, **40**, 101–18.

Chappell, J. (1982). Evidence for smoothly falling sea levels relative to north Queensland, Australia, during the past 6000 years. *Nature*, **302**, 406–8.

Chappell, J. (1987). Late Quaternary sea–level changes in the Australian region. In *Sea–level changes*, ed. M.J. Tooley & I. Shennan, pp. 296–331. New York: Blackwells.

Chappell, J. (1990). Some effects of sea level rise on riverine and coastal lowlands. In *Lessons for human survival: nature's record from the Quaternary*. pp. 37–49. ed. P. Bishop Geological Society of Australia Symposium Proceedings, 1.

Chappell, J. (1993). Contrasting Holocene sedimentary geologies of lower Daly River, northern Australia, and lower Sepik–Ramu, Papua New Guinea. *Sedimentary Geology*, **83**, 339–58.

Chappell, J., Chivas, A., Wallensky, E., Polach, H. A. & Aharon, P. (1983). Holocene palaeo-environment changes, central to north Great Barrier Reef inner zone. *BMR Journal of Australian Geology and Geophysics*, **8**, 223–35.

Chappell, J. & Grindrod, J. (1985). Pollen anlysis: key to past mangrove communities and successional changes in North Australian coastal environments. In *Coastal and tidal wetlands of the Australian monsoon region*, ed. K.N. Bardsley, J.D.S. Davie & C.D. Woodroffe, pp. 225–36, North Australia Research Unit (Monograph), Darwin: ANU press.

Chappell, J & Polach, H. (1991). Postglacial sea–level rise from a coral record at Huon Peninsula, Papua New Guinea. *Nature*, **349**, 147–9.

Chappell J., Rhodes E.G., Thom B.G. & Wallensky E. (1982). Hydro–isostasy and the sea–level isobase of 5500 B.P. in north Queensland, Australia. *Marine Geology*, **49**, 81–90.

Chappell, J. & Thom, B.G. (1986). Coastal morphodynamics in north Australia: review and prospect. *Australian Geographical Studies*, **24**, 110–27.

Chappell, J. & Woodroffe, C.D. (1985). Morphodynamics of Northern Territory tidal rivers and floodplains. In *Coastal and tidal wetlands of the Australian monsoon region*, ed. K.N. Bardsley, J.D.S. Davie & C.D. Woodroffe, pp. 85–96, North Australia Research Unit (Monograph), Darwin: ANU press.

Clark, R.L. & Guppy, J.C. (1988). A transition from mangrove forest to freshwater wetland in the monsoon tropics of Australia. *Journal of Biogeography*, **15**, 665–84.

Coleman, J.M., Gagliano, S.M. & Smith, W.G. (1970). Sedimentation in a Malaysian high tide tropical delta. In *Deltaic sedimentation: modern and ancient*, ed. J.P. Morgan, pp. 185–97. Society of Economic Palaeontologists and Mineralogists, Special Publication, 15.

Coleman, J.M. & Wright, L.D. (1978). Sedimentation in an arid macro-tidal alluvial river system: Ord River, Western Australia. *Journal of Geology*, **86**, 621–42.

Dalrymple, R.W., Knight, R.J., Zaitlin, B.A. & Middleton, G.V. (1990). Dynamics and facies model of a macrotidal sand–bar complex, Cobequid Bay–Salmon River estuary (Bay of Fundy). *Sedimentology*, **37**, 577–612.

Dalrymple, R.W., Zaitlin, B.A. & Boyd, R. (1992). Estuarine facies models: conceptual basis and stratigraphic implications. *Journal of Sedimentary Petrology*, **62**, 1030–43.

Dronkers, J. (1986). Tide–induced residual transport of fine sediment. In *Physics of shallow estuaries and bays, Lecture notes on coastal and estuarine studies*, vol. 16, ed. J. van de Kreeke, pp. 228–44, Berlin: Springer-Verlag.

Godwin, H. (1978). *Fenland; its ancient past and uncertain future*. Cambridge University Press.

Grindrod, J. (1988). The palynology of Holocene mangrove and saltmarsh sediments, particularly in northern Australia. *Review of Palaeobotany and Palynology*, **55**, 229–45.

Guilcher, A. (1967). Origins of sediments in estuaries. In *Estuaries*, ed. G.H. Lauff, pp. 149–57. American Association for the Advancement of Science Publication, 83.

Hacker, J.L.F. (1988). Rapid accumulation of fluvially derived sands and gravels in a tropical macrotidal estuary: the Pioneer River at Mackay, North Queensland, Australia. *Sedimentary Geology*, **57**, 299–315.

Harris, P.T. & Collins, M.B. (1985). Bedform distributions and sediment transport paths in the Bristol Channel and Severn Estuary, U.K. *Marine Geology*, **62**, 153–66.

Harris, P.T. & Collins, M.B. (1988). Estimation of annual bedload flux in a macrotidal estuary: Bristol Channel, U.K. *Marine Geology*, **83**, 237–52.

Harris, P.T. & Collins, M.B. (1991). Sand transport in the Bristol Channel: bedload parting zone or mutually evasive transport pathways. *Marine Geology*, **101**, 209–16.

Haseldonckx, P. (1977). The palynology of a Holocene marginal peat swamp environment in Johore, Malaysia. *Review of Palaeobotany and Palynology*, **24**, 227–38.

Jennings, J.N. (1975). Desert dunes and estuarine fill in the Fitzroy estuary, north–western Australia. *Catena*, **2**, 215–62.

Jones, B.G., Martin, G.R. & Senapati, N. (1993). Riverine–tidal interactions in the monsoonal Gilbert River fandelta, northern Australia. *Sedimentary Geology*, **83**, 319–37.

Kamaludin, b.H. (1993). The changing mangrove shorelines in Kuala Kurau, Peninsular Malaysia. *Sedimentary Geology*, **83**, 187–97.

Knighton, A.D., Mills, K. & Woodroffe, C.D. (1991). Tidal creek extension and saline intrusion in northern Australia. *Geology*, **19**, 831–4.

Knighton, A.D., Woodroffe, C.D. & Mills, K. (1992). The evolution of tidal creek networks, Mary River, northern Australia. *Earth Surface Processes and Landforms*, **17**, 167–90.

Lambeck, K. & Nakada, M. (1990). Late Pleistocene and Holocene sea–level change along the Australian coast. *Palaeogeography Palaeoclimatology and Palaeoecology (Global and Planetary Change Section)*, **89**, 143–76.

Murray, J.W. & Hawkins, A.B. (1976). Sediment transport in the Severn Estuary during the past 8000–9000 years. *Journal of the Geological Society of London*, **132**, 385–98.

Nakada, M. & Lambeck, K. (1989). Late Pleistocene and Holocene sea-level change in the Australian region and mantle rheology. *Geophysical Journal*, **96**, 497–517.

Nichols, M.M. & Biggs, R.B. (1985). Estuaries In *Coastal sedimentary environments*, 2nd edn, ed. R.A. Davis, Jr. pp. 77–186, New York: Springer-Verlag.

Nielsen, A.F. & Gordon, A.D. (1981). Tidel inlet behavioural analysis. In *Proceedings 17th International Coastal Engineering Conference, Sydney*, pp. 2517–33. New York: American Society of Civil Engineers.

Pirazzoli, P.A. (1991). *World atlas of Holocene sea–level changes*. Amsterdam: Elsevier.

Rhodes, E.G. (1982). Depositional model for a chenier plain, Gulf of Carpentaria, Australia. *Sedimentology*, **29**, 201–21.

Roy, P. S. (1984). New South Wales estuaries: their origin and evolution. In *Coastal geomorphology in Australia*, ed. B.G. Thom, pp. 99–122, Sydney: Academic Press.

Russell–Smith, J.J. (1985). A record of change: studies of Holocene vegetation history in the South Alligator River region, Northern Territory. *Proceedings of the Ecological Society of Australia*, **13**, 191–202.

Semeniuk, V. (1980). Quaternary stratigraphy of the tidal flats, King Sound, Western Australia. *Journal of the Royal Society of Western Australia*, **63**, 65–78.

Semeniuk, V. (1982). Geomorphology and Holocene history of the tidal flats, King Sound, north–western Australia. *Journal of the Royal Society of Western Australia*, **65**, 47–68.

Semeniuk, V. (1985a). Development of mangrove habitats along ria coasts in north and northwestern Australia. *Vegetatio*, **60**, 3–23.

Semeniuk, V. (1985b). Mangrove environments of Port Darwin, Northern Territory: the physical framework and habitats. *Journal of the Royal Society of Western Australia*, **67**, 81–97.

Sollas, W.J. (1883). The estuaries of the Severn and its tributaries: an enquiry into the nature and origin of their tidal sediment and alluvial flats. *Quarterly Journal of the Geological Society of London*, **38**, 611–28.

Stride, A.H. & Belderson, R.H. (1990). A reassessment of sand transport paths in the Bristol Channel and their regional significance. *Marine Geology*, **92**, 227–35.

TAPSM (Task Committee for Preparation of Sediment Manual) (1971). Sediment transportation mechanics. H. Sediment discharge formulae. *Journal of the Hydraulics Division, American Society of Civil Engineers*, **97**, 523–67.

Thom, B.G. & Roy, P.S. (1985). Relative sea levels and coastal sedimentation in southeast Australia in the Holocene. *Journal of Sedimentary Petrology*, **55**, 257–64.

Thom, B.G., Wright, L.D. & Coleman, J.M. (1975). Mangrove ecology and deltaic–estuarine geomorphology, Cambridge Gulf–Ord River, Western Australia. *Journal of Ecology*, **63**, 203–22.

Vertessy, R. (1990). Morphodynamics of macrotidal rivers in far northern Australia. Unpublished PhD thesis, Australian National University.

Wang, P. & Chappell, J. (in press). Foraminifera as Holocene environmental indicators in the South Alligator River, Australia. *Marine Geology*.

Wang, P. & Murray, J.W. (1983). The use of foraminifera as indicators of tidal effects in estuarine deposits. *Marine Geology*, **51**, 239–50.

Wolanski, E. (1986). An evaporation-driven salinity maximum zone in Australian tropical estuaries. *Estuarine Coastal and Shelf Science*, **22**, 415–24.

Wolanski, E., Chappell, J., Ridd, P. & Vertessy, R. (1988). Fluidisation of mud in estuaries. *Journal of Geophysical Research*, **93**, 2351–61.

Woodroffe, C.D. (1988). Changing mangrove and wetland habitats over the last 8000 years, northern Australia and Southeast Asia. In *Northern Australia: progress and prospects*, vol. 2, *Floodplains research*, ed. D. Wade-Marshall & P. Loveday, pp. 1–33, North Australia Research Unit, ANU Press.

Woodroffe, C.D. (1993). Late Quaternary evolution of coastal and lowland riverine plains of Southeast Asia and northern Australia: an overview. *Sedimentary Geology*, **83**, 163–75.

Woodroffe, C.D. & Chappell, J. (1993). Holocene emergence and evolution of the McArthur River Delta, southwestern Gulf of Carpentaria, Australia. *Sedimentary Geology*, **83**, 303–17.

Woodroffe, C.D., Chappell, J., Thom, B.G. & Wallensky, E. (1986). *Geomorphological dynamics and evolution of the South Alligator River and plains, N.T.* North Australia Research Unit Monograph, ANU Press.

Woodroffe, C.D., Chappell, J., Thom, B.G. & Wallensky, E. (1989). Depositional model of a macrotidal estuary and floodplain, South Alligator River, Northern Australia. *Sedimentology*, **36**, 737–56.

Woodroffe, C.D. & Mulrennan, M.E. (1993). *Geomorphology of the Lower Mary River Plains, Northern Territory*. Darwin: North Australia Research Unit. 152 pp.

Woodroffe, C.D., Mulrennan, M.E. & Chappell, J. (1993). Estuarine infill and coastal progradation, southern van Diemen Gulf, northern Australia. *Sedimentary Geology*, **83**, 257–75.

Woodroffe, C.D., Thom, B.G. & Chappell, J. (1985). Development of widespread mangrove swamps in mid–Holocene times in northern Australia. *Nature*, **317**, 711–13.

Woodroffe, C.D., Thom, B.G., Chappell, J., Wallensky, E., Grindrod, J. & Head, J. (1987). Relative sea level in the South Alligator River region, North Australia, during the Holocene. *Search*, **18**, 198–200.

Wright, L.D. (1985). River deltas. In *Coastal sedimentary environments*, 2nd edn, ed. R.A. Davis, pp. 1–76. New York: Springer–Verlag.

Wright, L.D. & Coleman, J.M. (1973). Variations in morphology of major river deltas as functions of ocean wave and river discharge regimes. *American Association of Petroleum Geologists Bulletin*, **57**, 370–98.

Wright, L.D., Coleman, J.M. & Thom, B.G. (1973). Processes of channel development of a high–tide range environment: Cambridge Gulf–Ord River Delta, Western Australia. *Journal of Geology*, **81**, 15–41.

Wright, L.D., Coleman, J.M. & Thom, B.G. (1975). Sediment transport and deposition in a macrotidal river channel: Ord River, Western Australia. In *Estuarine research*, vol. II, *Geology and engineering*, ed. L.E. Cronin, pp. 309–21. London: Academic Press.

6

Lagoons and microtidal coasts

J.A.G. COOPER

Introduction

This chapter deals with the evolution of lagoonal coasts. Despite being widely distributed around the globe, coastal lagoons have been studied in a fragmented manner and few co-ordinating studies have been undertaken. Lagoons exhibit many different morphologies and form under a variety of environmental conditions which determine not only the principal morphodynamic processes but also the evolutionary path which a lagoon follows. This chapter outlines various types of lagoon. It then reviews several documented examples of lagoonal evolution and assesses the main evolutionary processes. Variations between documented studies of lagoon evolution are assessed and the main controls on the evolutionary path are discussed. Existing models of lagoonal evolution are reviewed and current research deficiencies are highlighted.

Coastal lagoon distribution

Lagoons are present on many coasts of the world in a variety of environmental settings; they are most common in microtidal environments, although examples do occur in mesotidal and even macrotidal environments (Hayes, 1975). A broad range of physical and chemical characteristics consequently exists. Current estimates (Table 6.1) suggest that lagoons border between 17% (Cromwell, 1971) and 56% (Berryhill, Kendall & Holmes, 1969) of North America, and 13% of the world's coastline (Berryhill et al., 1969; Cromwell, 1971; Barnes, 1980) from the polar to tropical latitudes. The wide discrepancy in the estimates for North America highlights the difficulty in definition of a barrier lagoon. Lagoons may be even more widespread as many occur on bedrock coasts in discrete embayments, rather than linear barrier island coasts.

Table 6.1. *Distribution of barrier/lagoon coasts by continent*

Continent	Barrier/lagoon coast (km)	% of continent's coastline	% of world's lagoonal coast
N. America	10765	17.6	33.6
Asia	7126	13.8	22.2
Africa	5984	17.9	18.7
S. America	3302	12.2	10.3
Europe	2693	5.3	8.4
Australia	2168	11.4	6.8
Total	32038		

From Barnes (1980), originally after Cromwell (1971).

Coastal lagoons are generally considered ephemeral features on a geological time scale as they form, evolve and infill within a short timespan. This chapter is concerned with the processes by which they form and evolve. Evolutionary processes vary according to environmental setting, and as lagoons occur in most latitudes and occupy a transitional location between land and sea, their morphodynamic evolution varies accordingly. Most studies of lagoon morphodynamics have centered on one or a few examples (Emery & Uchupi, 1972) and relatively few studies (Lankford, 1977; Nichols & Allen, 1981; Kjerfve, 1986; Carter *et al.*, 1989; Nichols, 1989) have synthesised information on a variety of lagoons.

Precise definition of a coastal lagoon is problematic and many definitions have been proposed. Considerable overlap between lagoons and estuaries has been identified. As morphodynamic systems, lagoons have been defined as 'coastal water bodies which are physically separated, to a greater or lesser extent, from the ocean by a strip of land' (Ward & Ashley, 1989). The imprecise definition of coastal lagoons is perhaps the main problem in the lack of co-ordinated research, as many features variously termed estuaries, blind estuaries, embayments, coastal ponds, coastal lakes, bays and sounds may alternatively be regarded as lagoons. Carter (1988) identified three types of lagoon according to the flux of water within them. In one group the inflow of seawater equals the outflow over a tidal period. In the second, evaporation of lagoon water means that the inflow of water exceeds the outflow. This produces a sabkha. In the third group there is a net seaward discharge due to the addition of freshwater discharge. Nichols & Allen (1981) identified four main lagoon types based on dominant processes: estuarine lagoon; open lagoon; partly closed lagoon; closed lagoon. Many other classifications of lagoons exist, tailored to the needs of the proposing author. Several morphodynamic realms exist within lagoons and the typical features are depicted in a generalised model in Fig. 6.1.

Processes of lagoon formation

Lagoons form where coastal embayments or depressions are separated from the adjacent sea by a barrier. Barriers comprise either clastic material (sand or gravel) or are created by vegetation, coral growth or tectonics (Lankford, 1977). Lagoons are best formed on transgressive coasts, particularly where the continental margin has a low gradient and sea-level rise is slow (Emery, 1967); however, they are also present on embayed rocky coasts (Roy, 1984; Cooper, 1991a).

Clastic barriers originate in different ways. They may arise through spit growth across an embayment through longshore drift (Hoyt 1967; Martin, Gadel & Barusseau, 1981), behind cuspate barriers produced by wave-reworked deltaic sediment, or through flooding of coastal lowlands behind former dune ridges or beaches (Hoyt, 1967). In the latter scenario the original width and depth of a lagoon depends on the antecedent slope of the land surface and extent of sea-level rise. In other cases lagoon enclosure results

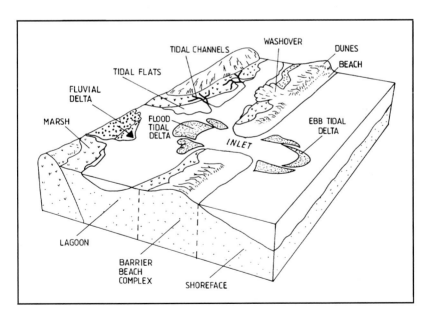

Figure 6.1. Generalised depiction of morphosedimentary features of coastal lagoons (modified after Reinson, 1980). Rarely will all features or salinity zones be present in a single lagoon and the relative importance of each may vary with time.

from landward transport of a barrier across the shoreface during transgression until it becomes stabilised. If excess transgressive sediment is available the barrier may then prograde seaward, typically by beach-ridge formation (Phleger & Ewing, 1962; Thom, 1983; see also Fig. 1.6). If additional sediment is unavailable the lagoon may be stressed such that it breaches, or the barrier may degenerate as barrier sediment is eroded (Phleger, 1981). In some cases clastic barriers completely enclose a lagoon, leaving no connection with the sea. This typically arises where the sediment supply and transport capacity of littoral drift or wind exceeds tidal or fluvial currents which act to maintain a connecting channel.

Early concepts of lagoon evolution are typified by the work of Lucke (1934) whose conceptual model was termed the 'Lucke Model' by Oertel *et al.* (1989). According to this model (Fig. 6.2) coastal lagoons evolve from a marine embayment to a relatively deep, partially enclosed, back-barrier lagoon. Sedimentation in the lagoon occurs through development of flood-tidal deltas into marsh islands, a process recently documented in detail by Cleary, Hosser & Wells (1979), and fluvial sedimentation in the form of deltas. Ultimately, the originally deep lagoon is transformed into a marsh or deltaic plain through which rivers discharge to the sea. This is the most widely held concept of lagoon evolution at present (Barnes, 1980) but, as will be shown below, several alternative views have recently been expounded. In some cases, however, unquestioning acceptance of this model has led to misconceptions regarding lagoonal processes and consequent attribution of apparent deviations to human impacts. Begg (1978, 1984), for example, attributed the shallow nature of Natal lagoons to increased sediment yield from recently cultivated catchments, without recognising that naturally high-sediment yields from the humid, subtropical hinterland throughout the Holocene had rendered most lagoons in a mature stage of evolution several thousand years before widespread cultivation (Cooper, 1991a).

Examples of lagoonal evolution

In this section a variety of examples is used to illustrate the variation in processes of evolution in different settings and under different conditions. In the subsequent section they are compared.

South-east African lagoons

The microtidal coastline of Natal and Zululand represents an excellent natural laboratory for the study of coastal lagoons in a humid (annual rainfall 1000 mm) subtropical setting. Mean tidal range at Durban is 1.72 m on spring

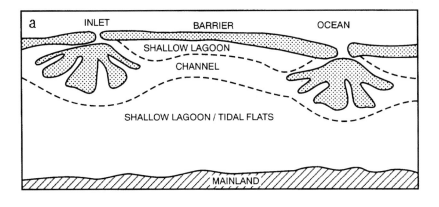

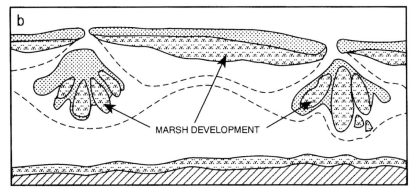

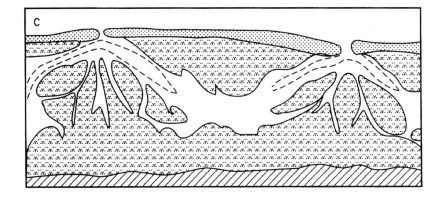

Figure 6.2. A summary diagram of the 'Lucke model' of lagoon evolution (after Oertel *et al.*, 1989). This represents one of the earliest, and perhaps still most widely held, concepts of lagoon evolution. An initially deep lagoon (*a*) becomes progressively infilled by sediment accumulation (*b*) (in this case from marine sources) and is ultimately transformed into a swamp with fluvial channels (*c*).

and 0.5 m at neap tide. The Natal coast, whose northern boundary is the Tugela
River mouth, is a linear, rock-based coastline with shallowly indented
embayments and sandy beaches overlying rock outcrops. The coastal
hinterland rises steeply to over 3000 m and the continental shelf ranges in
width from 10 km to 40 km (Fig. 6.3). The 100 m isobath marks the edge of the
shelf. The Zululand coast, north of the Tugela River, has a low-lying sandy

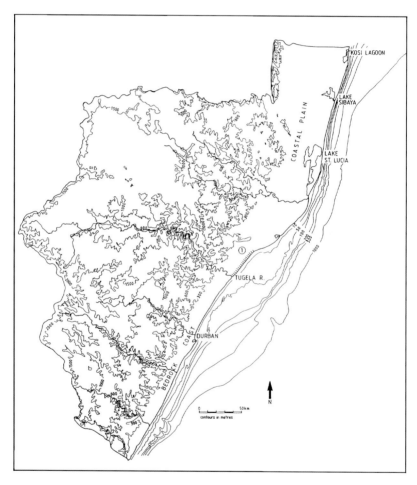

Figure 6.3. The coastal hinterland of Natal and Zululand, showing continental shelf
width and hinterland topography. Only the major rivers are shown. Note the steep
gradients in Natal (south of Tugela River), compared with those in Zululand (north of
Tugela River), where an extensive coastal plain occurs. The Natal coast is mainly rocky
and owes its linear nature to its tectonic orgin. The Zululand coastal plain comprises
unconsolidated sands and muds. Aeolianite and beachrock form the main solid
outcrops. ① indicates beach-ridge plain.

coastal plain which widens northward to 80 km and the continental shelf is as narrow as 3 km. There is a prograding beach-ridge plain for the 45 km north of the Tugela River. On this strongly wave-dominated section of the coast, river courses are diverted northward in the direction of net littoral drift to produce a number of shore-parallel coastal lagoons.

Siyai Lagoon

The Siyai Lagoon in southern Zululand (Fig. 6.4) is located on a coastal sector where historical progradation rates average up to 5 m per year, but are strongly episodic (Cooper, 1991b). The lagoon drains an 18 km^2 catchment and is generally isolated from the sea. Barrier breaches caused by heavy rainfall are generally sealed within a week (Begg, 1978). Between 1937 and 1977 the barrier of the Siyai Lagoon prograded by about 125 m through beach-ridge accretion and subsequent stabilisation by dune vegetation. This coastal progradation through northward longshore transport of sediment from the Tugela River caused the northward elongation of the Siyai Lagoon by 800 m (Fig. 6.4), a 35% increase in total length. Over the same period rapid growth of reeds in the back-barrier caused loss of open water area as the channel narrowed and apparently shallowed. Reduction in species diversity of fish and aquatic birds was also noted (Begg, 1978; Benfield, 1985) and attributed to decreased frequency of outlet formation and incursion of seawater. The changes could be attributed to reduced stream gradient and loss of stream power, and to the use of fertilisers and destruction of riparian vegetation in the catchment.

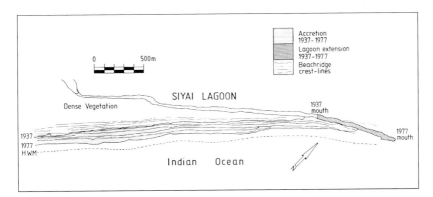

Figure 6.4. The Siyai Lagoon on the prograding coast of southern Zululand, showing the seaward edge of stabilised beach ridges in 1937 and 1977. Progradation of the coast by 125 m (light shading) was accompanied by elongation of the lagoon through 800 m (dark shading), causing reduced frequency of outlet formation, reed encroachment, shallowing of the lagoon and decreased species diversity.

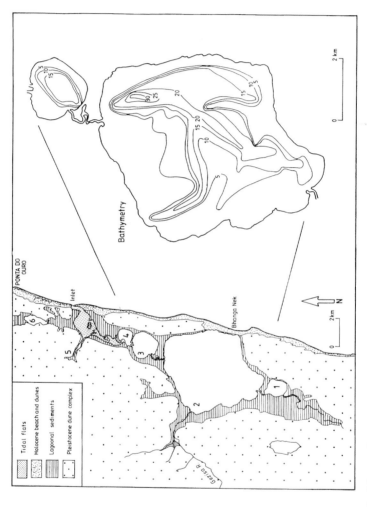

Figure 6.5. The Kosi lagoon in northern Zululand. Segmentation has given rise to a unique set of aquatic environments and habitats in each segment. The fetch and depth of each segment determines morphodynamic processes. Key: 1. Amanzimnyama. 2. Nhlange. 3. Mpungwini. 4. Makawulani. 5. Khalu Inlet. 6. Zilonde. The modern bathymetry (right, in metres) preserves the outline of an incised Pleistocene river valley in which the lagoon formed.

Kosi Lagoon

The Kosi lagoon in northern Zululand is an excellent example of the effects of segmentation (Zenkovich, 1959) on lagoon evolution (Fig. 6.5). The lagoon is situated on the sandy Zululand coastal plain (see Fig. 6.3) and is fed by several small streams with negligible sediment input, other than suspended vegetal matter. The evolution of the lagoon has not been studied in detail but preliminary results of seismic profiling and vibracoring suggest that Holocene sediments are very thin (<1 m), and comprise mainly gyttja. Underlying and adjacent sediments of probable Late Pleistocene age (Cooper, Kilburn & Kyle, 1989) show that the modern lagoon reoccupied the site of a Pleistocene lagoon in an inter-dune depression. Present water depths up to 35 m in the lagoon reflect incision during the last glacial maximum (Hill, 1969). Closure of a former outlet at Bhanga Nek by aeolian deposition isolated the southern part of the lagoon from the sea. Contact with the sea is maintained through a narrow northern outlet.

Closure of the southern outlet may have been associated with regression from a Holocene highstand for which there is ample geomorphological evidence in the area. Low sediment supply from the narrow shelf and coastal streams prevented coastal progradation during regression; instead, the coastline has evolved into a series of log spiral bays in apparent equilibrium with the ambient wave field, backed by high aeolian dunes.

Segmentation of the Kosi Lagoon was enabled by its uncohesive, sandy margins from which sediment was reworked by wind-generated lagoon waves to form cuspate spits which ultimately coalesced, dividing the lagoon into a series of segments. Preliminary results from seismic profiling suggest that progradation of the spits into deeper water occurred through avalanching and turbidite deposition. The course of the original incised channel is still evident from the deep sections of adjoining segments (Fig. 6.5). Segmentation of the lagoon was aided by fluctuations in late Holocene sea levels which elevated formerly subtidal parts of the cuspate spits. Narrow channels connect the segments. The origin of these channels is uncertain: they may have arisen by incision associated with falling water levels after floods or sea-level highstands, or as paths followed by hippopotami.

Segmentation of the lagoon has produced a series of distinctive environments and habitats. Lake Amanzimnyama (diameter 1000 m) is a 1.5 m deep marginal basin whose water is entirely fresh due to its distance from the tidal inlet. Lake Nhlange, (diameter 5000 m) is dominated by wind waves which cause mixing throughout its 35 m depth. Long-term (decade-scale) salinity changes, between 5 and 0‰, occur (Kyle, 1986). The lake margins are dominated by wave-rippled sand while fine organic matter settles in the deeper

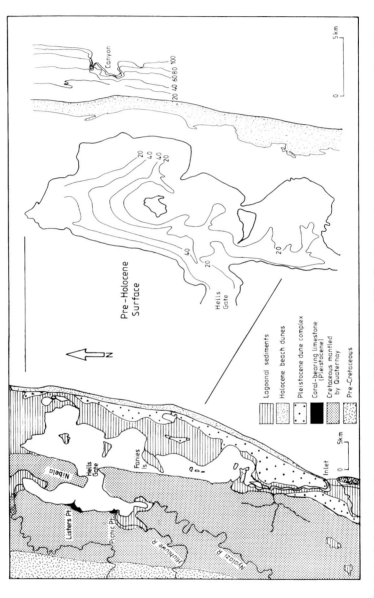

Figure 6.6. Lake St Lucia is the largest lagoon on the Zululand coast. Like Kosi Lagoon it occupies the same site as a Late Pleistocene lagoon into which a river channel was incised. The channel is now infilled and the lagoon averages 1.5 m deep. Holocene sedimentation has infilled a formerly 40 m deep incised river channel (right, bathymetry, in metres).

basins. Lake Mpungwini occupies a central position in the lake system and receives seawater on each tidal cycle. Its great depth (maximum 21 m) and small fetch (1800 m) has produced marked salinity stratification ranging from over 20‰ at depth to less than 5‰ at the surface. Anoxic and sulphidic conditions occur at depth (Ramm, 1992). During equinoctial high tides accompanied by strong winds the stability of the water column is broken down and fish kills occur when sulphidic water reaches the surface. Lake Makawulani is 8 m deep and displays marked stratification. Salinities range from 25 ‰ at the bottom to 10‰ at the surface. Between this lake and the tidal inlet is a 5 km stretch of low-gradient sandy intertidal flats with a narrow channel. Comparison of historical aerial photographs indicates that intertidal deposition is minimal at present but former landward progradation of the tidal flats is indicated by small deltas in Lake Makawulani and the Khalu inlet. The current lack of tidal sedimentation may be attributed to lack of longshore supply of sand on the equilibrium planform coastline.

Kosi lagoon has preserved its incised channel almost intact through low depositional rates from the catchment. Its stability is further enhanced by the log-spiral coastline, which minimises littoral drift and prevents inlet closure and flood-tidal delta buildup. Longshore sand supply is further reduced by the large distance from the closest sediment supplying river (175 km to the south) and interception of shelf-sediment pathways by submarine canyons (Ramsay, 1990).

Lake St Lucia

Lake St Lucia is the largest lagoon in Zululand and covers an area of 380 km^2 (Fig. 6.6). Several large rivers form deltas in the lagoon which has an average depth of 1.5 m (Hobday, 1976). The modern lagoon occupies the same site as a Late Pleistocene lagoon, deposits of which (Port Durnford Formation) are preserved locally (Hobday & Orme, 1974). During the last glaciation these deposits were partly eroded and incised to depths up to 40 m (Sydow, 1987); seismic profiling reveals the presence of a former river channel under the lagoon (Fig. 6.6). For much of the Holocene an outlet was maintained through a northern channel. This was ultimately sealed by aeolian deposition, perhaps associated with slight regression during the late Holocene (Sydow, 1987). Subsequently, an outlet formed in the present position in the southern end of the lagoon.

Cores through the lagoonal fill show various mixtures of clay, silt and fine sand containing molluscan and foraminiferal fauna typical of brackish lagoonal conditions (Hobday, 1976). Lagoonal confinement was evidently established at an early stage of the transgression behind an earlier Pleistocene barrier dune. Sedimentation probably kept pace with transgression as there is no evidence of particularly deep water sedimentation. In marginal areas of the lagoon, brackish lagoonal deposits overlie pre-Holocene freshwater swamp deposits, reflecting progressive drowning of surrounding swamps during transgression.

Due to its shallow nature and large water area Lake St Lucia has been reported to become hypersaline when fluvial discharge is low (Orme, 1973; Begg, 1978). The presence of gypsum in cores from 20 m below the present sediment surface (Hobday, 1976) provides evidence of similarly shallow conditions early in the evolution of the lagoon. Segmentation of the lake is only weakly developed (Orme, 1973) a fact which may be linked to rocky outcrops on the landward lagoon margin and to the presence of cohesive muddy sediment whose cohesion limits wave-transport and shoreline realignment.

The entire lake is now above wave base (Hobday, 1976) and wave-generated currents and direct wave action are the principal sediment dispersal agents. The high sediment level may be in part a reflection of sedimentation during high Holocene levels, since which time fluvial discharge has been unable to erode the accumulated muddy sediment.

Around the St Lucia Lagoon, Late Pleistocene lagoon and barrier sediments of the Port Durnford Formation are preserved. Focusing of drainage channels into low-lying areas during the last glacial maximum caused differential erosion of Pleistocene deposits from the back-barrier area but preserved them under the Pleistocene barrier. Holocene transgression drowned the incised valley and deposited additional sand against the Pleistocene barrier while a new set of lagoonal deposits were laid down in the incised valley. The nature of the Holocene transgressive sequence has not been investigated but may well duplicate that of the Late Pleistocene, documented by Hobday & Orme (1974) and Hobday & Jackson (1979). This sequence comprises a lower lagoonal facies, overlain by inter-layered washover sands and lagoonal muds and a barrier with an upper aeolian component (Fig. 6.7).

Small bedrock-confined lagoons in Natal

On the mainly rocky continental margin of Natal south of the Tugela River, small lagoons occupy fluvial palaeovalleys incised in bedrock during lowered sea levels. The lateral confinement of these channels has led to complete erosion of pre-Holocene deposits (Maud, 1968). Most of these lagoons are relatively narrow (<500 m) and coast-normally orientated but some exhibit limited coast-parallel extension behind sandy barriers. In most cases river

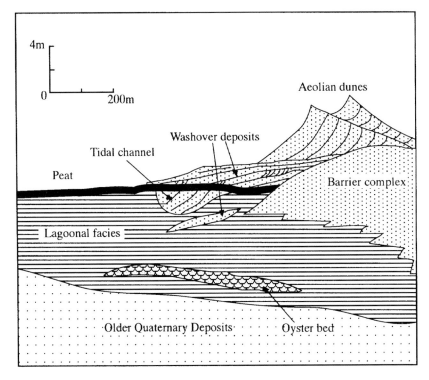

Figure 6.7. A model of the trangressive barrier lagoon deposits of the Pleistocene Port Durnford Formation preserved under the Pleistocene barrier of Lake St Lucia (after Hobday & Jackson, 1979).

discharge is insufficient to maintain an outlet. The sandy barriers are dominated by overwash lamination and have an upper aeolian component. Flood-tidal deltas are absent and ephemeral ebb-tidal deltas form opposite flood-generated breaches (Cooper, 1990b). The ephemeral outlet channels are shallow and are rapidly reworked after formation. In this respect they are similar to the outlets described from the Oregon coast (Clifton, Phillips & Hunter, 1973)

Stable conditions are characterised by high water levels and low-energy conditions in which suspension settling dominates sedimentation. The main sediment sources are the catchment, surrounding vegetation and local bank erosion. Marine sediment input is limited to overwash and deposition is restricted to the immediate barrier area. Sediment supply from the hinterland is high and can be correlated directly with catchment size (Cooper, 1991a). The volume of overwash input depends on the length and height of the barrier. In uMgababa Lagoon (Grobbler, 1987; Grobbler, Mason & Cooper, 1987) marine sediment input through overwash is low due to the short (100 m) barrier while in Mhlanga Lagoon the 900 m-long barrier promotes overwash sedimentation (Cooper 1989).

Periodic, short-lived, high-energy phases are precipitated by barrier breaching and outlet formation (Cooper, 1990b). This appears to be caused mainly by increased fluvial discharge as noted by Carter, Johnson & Orford, (1984) in Irish lagoons. In the cell-confined barriers of Natal, where long-term changes are minimal (Cooper, 1991c) alternative processes of breaching such as stretching and separation (Carter *et al.*, 1989) have not been documented. Overwash may lower the barrier and enable rising water levels to form an outlet there (Begg, 1984; Cooper, 1989). Rapid outflow of water following breaching causes erosion of accumulated fine sediment from the bed of the lagoon and promotes selective retention of sand (Cooper 1989; Grobbler 1987).

Floodplain deposits comprise vegetated wetlands, typically fringed with either *Phragmites* reeds or *Barringtonia*, the 'freshwater mangrove'. True mangroves are generally absent. Back-barrier channels which are comparatively unconfined are shallow and sandy, while those which are laterally confined by vegetated or muddy banks are deeper.

Cores from such lagoons reveal impounded brackish water conditions early in their evolution. Incursions of the sea during late Holocene highstands may have promoted breaching more frequently, as evidenced by the presence of brackish water molluscan remains in some lagoonal deposits above present sea level. At present most lagoons appear to have reached equilibrium (Cooper, 1989, 1991a); fine-grained fluvial sediment resides temporarily in the lagoon before being scoured during breaching. This preferentially preserves sand in the sedimentary record. Slow buildup of overwash sands adjacent to some barriers occurs as outflowing currents are insufficient to remove such grain sizes.

In Mdloti Lagoon (Grobbler, 1987), the preservation of a widespread lagoonal mud in the lagoon fill (Fig. 6.8*a*) reflects a period when deeper-water conditions prevailed and fluvial currents were unable to scour deeply enough to remove this sediment. Brackish-water conditions are indicated by the molluscan shell remains. This is overlain by fluvial sands, which now dominate the lagoon. In contrast, the valley fill of the Mtwalume Lagoon (Fig. 6.8*b*), which has similar dimensions to the Mdloti, is almost entirely sandy. This was interpreted by Cooper (1991a) as a result of higher sediment supply from the inflowing river which continually kept pace with drowning of the bedrock valley in a similar manner to estuaries of the southeast United States (Nichols & Biggs, 1985).

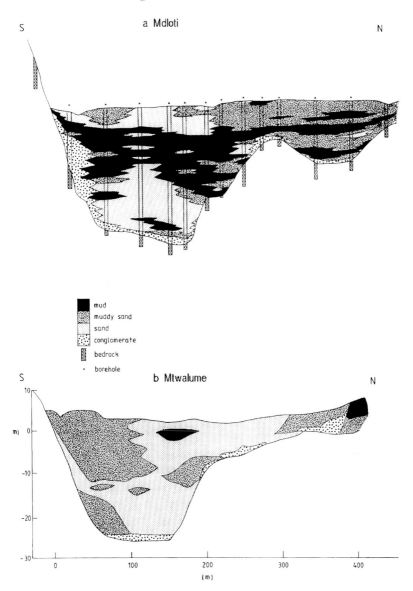

Figure 6.8. Cross-sections of sedimentary fills of the bedrock-confined (*a*) Mdloti Lagoon (after Grobbler, 1987) and (*b*) Mtwalume Lagoon (after Orme, 1974) on the mainly rocky Natal coast. The presence of a well-defined muddy horizon in the Mdloti fill indicates deep-water brackish lagoonal conditions. The upper sands are fluvial and were deposited in shallow water: mud is preferentially eroded during floods. In the Mtwalume the entire fill is dominated by sand, suggesting that deep-water conditions were never attained and coarse-grained fluvial sedimentation kept pace with drowning of the bedrock valley.

Coastal lagoons of Virginia, USA

The coastal lagoons of Virginia receive relatively low fluvial inflow and are largely filled with marsh deposits and narrow tidal channels. Inlets in the sandy barriers are maintained by tidal currents. Their evolution is characterised by barrier retreat, resulting from rising sea level and a nearshore sediment deficit (Finkelstein & Ferland, 1987), which has led to lagoon narrowing and reduction of tidal prisms. This in turn caused inlet constriction and reduction in tidal current strength. A corresponding increase in sedimentation was noted on marshes and tidal flats. Stratigraphic cross-sections show changes from a high-energy lagoon to muddy tidal flat and marsh environments, accompanied by increasingly quiescent conditions, during a sea-level rise of about 6 m in the past 5000 years. The resultant 'regressive' back-barrier stratigraphy was produced under transgressive conditions through sedimentation under progressively lower-energy conditions.

Infilling of these lagoons occurred principally through fluvial inputs, which were probably increased by deforestation since the Colonial period. Under slowly rising sea level the lagoon infilled while under more rapid sea-level rise it was drowned. In the past 40 years an increased rate of sea-level rise has been matched by increased sedimentation, which has prevented lagoon rejuvenation.

Port Royal Bay, Bermuda

Port Royal Bay is a sheltered lagoon, surrounded by high ground and sheltered from wind action. A summary of its Holocene evolution is included here as an example of lagoonal development under low-energy conditions and low sedimentation rates (Ashmore & Leatherman, 1984).

At approximately 9500 years BP, rising sea level increased the groundwater table to support a freshwater swamp on top of pre-Holocene soils in a depression in aeolianite bedrock. Continued rise of sea level outstripped swamp aggradation and by 8200 BP a freshwater pond was formed in which 2.5 m of gyttja accumulated. This pond became increasingly brackish with time as seawater percolated through the surrounding aeolianite. The lagoon was drowned about 7000 BP when the sea overtopped its margins. This enabled colonisation by marine molluscs and corals. Low sedimentation rates accompanying continued sea-level rise produced deeper water until reducing conditions and development of a seasonal thermocline led to the demise of the molluscan population and deposition of lime mud in the deeper sections. Coral growth was maintained in shallow marginal areas. The lagoon is over 25 m deep as a result of low sedimentation rates but Ashmore & Leatherman (1984) predict continued infilling of the lagoon by coral growth under present sea-level conditions.

Warnbro Sound, Australia

Warnbro Sound, a semi-enclosed lagoon in Western Australia, represents a lagoon with very low terrigenous sedimentation in which wave action affects the marginal areas (Carrigy, 1956). The lagoon is protected from ocean waves by a line of aeolianites, which are the remnant of a former barrier. A self-generating sedimentary system of calcareous sands and silts derived from pelagic foraminiferal and benthic molluscan debris has led to a sedimentary equilibrium in the lagoon. Fine sediment is restricted to the deeper sections (>14 m) while the shallow, wave-dominated marginal areas are sandy. Warnbro Sound receives negligible clastic sediment input either from marine or terrestrial sources and carbonate precipitation and dissolution balance to preserve the equilibrium. Excess marine sediment is removed by longshore drift.

Lagoons in New South Wales, Australia

On the embayed bedrock coast of New South Wales (NSW), lagoons occupy incised fluvial valleys, often protected from direct ocean influences by large embayments. Roy (1984) presented an evolutionary model of the evolution of lagoons or 'saline coastal lakes' based on widespread drilling records and observation of modern lagoons at various stages of development (Fig. 6.9). These evolved from a relatively deep back-barrier basin of marine to marginal marine salinity. Water area and volume were reduced gradually by fluvial delta progradation and barrier overwash and at the same time salinity variations increased, favouring prolonged fresh to brackish conditions. This change was accompanied by swamp encroachment and deposition of freshwater muds. Ultimately, the lagoons were infilled by swampy floodplains surrounding a narrow channel. The possibility of increased barrier breaching as a result of channelisation in the back-barrier was cited as a possible evolutionary trend.

Coastal lagoons in Rio de Janeiro State, Brazil

The lagoons of the embayed rocky coastline of Rio de Janeiro State, Brazil are completely isolated from the sea by sandy barriers. They formed during the Holocene and the earliest dates for their formation are about 7800 years BP (Ireland, 1987). Subsequently, sea level has oscillated around its present level in a series of minor transgressions and regressions. In most cases Holocene lagoonal sediments are only 4 to 5 m thick, and are underlain by similar pre-Holocene lagoonal facies. The typical sedimentary succession comprises a basal non-marine sand or peat, passing upward into clay and clayey sand of brackish to marine affinity in a low-energy environment. Diatoms indicated

236   *J.A.G. Cooper*

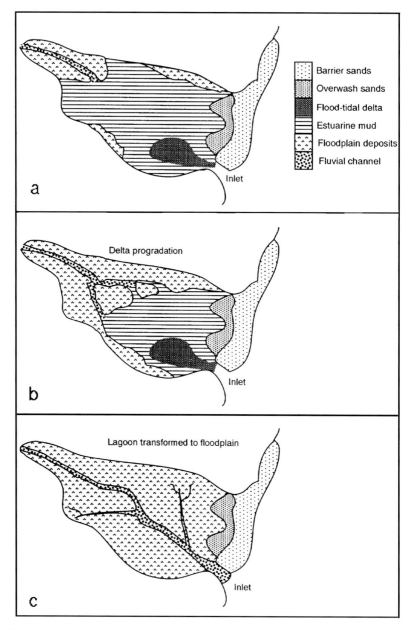

Figure 6.9. Model of evolution of saline coastal lakes in New South Wales (after Roy, 1984; see also Fig. 2.6 and Fig. 4.24). This model predicts gradual infilling of an initially deep basin (*a*) and subsequent transformation into a swamp (*b*) with confined fluvial channels (*c*). It is similar to the Lucke (1934) Model; however, the principal sediment source is fluvial rather than marine.

steadily declining marine influence with time and a concomitant increase in freshwater influence. The sequence is capped by a layer of freshwater peat and organic detritus, indicating complete closure of the barrier inlet. Salinity variation in the back-barrier deduced from diatom assemblages could be linked to periods of increased or decreased connection with the sea via barrier inlets. The general evolutionary trend was for barriers to be breached as sea level rose and to close when it stabilised. Although no explanation was offered for barrier breaching during transgression, it was probably a result of low sediment supply to the barrier. Lagoonal fills for each transgressive phase suggest that sedimentation rates were comparatively rapid.

Lagoons of Maine, USA

The coastal lagoons of Maine occur on an irregular rocky coastline and number over 200 (Duffy, Belknap & Kelley, 1989). Their coarse-grained barriers are typically <1000 m long and assume a number of forms including barrier spits, looped barrier, cuspate barriers, double tombolos and pocket barriers. Sedimentation in these barrier lagoon systems began about 9000 years BP and reached maximum development when sea level rise slowed after 5000 BP. Barrier sediments were eroded from glacial till as sea level rose. Limited sediment supply and rapid sea-level rise caused the barriers to retreat into embayments where they became sheltered from littoral drift. Holocene lagoonal facies accumulated on top of glaciomarine sediments and a thin regressive tidal flat or beach deposit. The 3 to 4 m thick Holocene facies comprises gyttja and three types of marsh with distinctive flora. Up to 0.8 m of gyttja accumulated in shallow brackish-water lagoons early in their evolution. It is typically overlain by swamp deposits whose plant remains indicate increasingly marine conditions. This transgressive sequence, which is similar to those identified in adjacent areas of New England and Boston, therefore reflects increased marine influence during relative sea-level rise and is the most common stratigraphic sequence found. Fresh- to brackish-water swamps accumulated as lagoons infilled or as water levels were lowered by outlet formation.

Some lagoons showed prolonged periods of brackish-water conditions, without increasing marine influence with time, while a few showed variations in marine influence with time. In a few lagoons a reverse sequence of progressively fresher water conditions was indicated by marsh progradation over back-barrier flats. The latter, regressive, situation is interpreted as a result of abundant sediment or a slow rate of sea-level rise, which permitted marsh expansion. Duffy *et al.* (1989) concluded that the presence of a barrier inlet was the most important control on back-barrier stratigraphy in Maine. Variations in barrier morphology, rate of sea-level rise, tidal range and bedrock morphology were not reflected in back-barrier stratigraphy.

Discussion and comparison

Comparison of the examples above shows that there is wide variation in the evolution of coastal lagoons. Lagoon evolution was summarised by Kjerfve & Magill (1989) as 'the combined action of marine and fluvial processes ... [leading to] ... trapping and infilling of semi-enclosed coastal systems, including coastal lagoons, and the re-shaping of seaward boundaries'. It is, however, clear that evolution progresses in a variety of ways and with different results in different lagoons.

The size of coastal lagoons varies greatly. In Zululand, for example, much larger coastal lagoons are present on the sandy coastal plain than on the steep rocky coast of Natal. This is a direct result of shallower gradients and soft surrounding sediment in Zululand.

Where the continental shelf is narrow, erosion during lowered sea levels incises fluvial valleys to great depths. This produces deep depressions in which to form lagoons. In contrast, areas with wide continental shelves offer only shallow depressions in which to form lagoons. In Natal and Zululand, Holocene lagoonal sequences are up to 40 m thick, while in Maine and Brazil lagoonal deposits are only 3 to 4 m thick.

The maximum age of Holocene lagoonal sediments in Virginia is 5000 years (Finkelstein & Ferland, 1987) and in Brazil, 7800 years (Ireland, 1987). The Canet–St Nazaire Lagoon in the Mediterranean formed only during the past 6000 years (Martin *et al.*, 1981). Young ages and shallow sedimentary fills are apparently more typical of areas where lagoons occupy coastal depressions rather than incised river valleys. Holocene lagoons formed in incised river valleys on steep-gradient high-latitude coasts may have originated as much as 10 000 years BP (Thom, 1983) and have required much greater sediment volumes to be infilled. This impacts on the age of sediments and the length of time for which a lagoon may persist.

The role of sediment supply is illustrated by comparison of Lake St Lucia and Kosi Bay (Cooper, 1990a), which clearly indicates the influence of sediment availability on lagoon coast evolution. The bedrock valley of St Lucia Lagoon is filled by 40 m of fluvially derived Holocene sediments which reduced its mean depth to 1.5 m, while in Kosi Lagoon the incised valley is largely preserved by the modern bathymetry as no major rivers discharge into the lagoon.

The transgressive back-barrier sequence found in most Maine lagoons contrasts with the apparently more typical regressive back-barrier fills described from Brazil and New South Wales. The differences may be attributed to different balances between sediment supply and rate of sea-level rise. In Maine, sediment supply was low and sea-level rise rapid, while in Brazil and New South Wales, sediment supply was comparatively higher and prevented complete inundation of back-barrier areas. In Natal, ubiquitous lagoon fills such as that of the the Mtwalume (Fig. 6.8b), suggest an oversupply of sediment, which maintained shallow fluvial conditions through most of the lagoon's evolution.

The coastal lagoons of New South Wales infilled from an initially deep basin through successive shallowing with time to a swamp-surrounded river channel (Roy, 1984). Lagoons can be found at various stages of development in an evolutionary sequence. This compares directly with the progressive infilling of initially deep lagoons in other areas such as Brazil (Ireland, 1987) and southeast France (Martin *et al.*, 1981) along the lines of the Lucke model.

In subtropical Natal extended periods of chemical weathering since the Miocene (Partridge & Maud, 1987) have produced large quantities of fluvial sediment. Consequently, in lagoons such as the Mtwalume, sedimentation matched the rate of infilling and so deepwater lagoonal conditions never developed. Furthermore, later Holocene fluctuations of sea level in bedrock-confined Natal lagoons caused incision of former subtidal deposits and their emergence as floodplains during minor regressions. This has had the effect of increasing evolutionary rates: consequently, such phases in infilling could not be recognised in the Natal lagoons, as all are geomorphologically mature.

The role of fluvial sediment supply on lagoonal evolution processes is particularly well illustrated by comparison of the Kosi and St Lucia lagoons on the Zululand coastal plain. In Lake St Lucia, rapid fluvial sedimentation accompanied the Holocene transgression and even promoted desiccation early in its Holocene evolution. In contrast, in Kosi Lagoon the Late Pleistocene channels are preserved as a result of low fluvial influx. The idea that segmentation reduces sedimentation rates by trapping inflowing sediment (as in Lake Amanzimnyama) can be rejected from Kosi as seismic profiling reveals no Holocene sediment accumulation, but this may be an effective process elsewhere in reducing lagoonal infilling rates.

Sediment supply is in part influenced by climate. Martin *et al.* (1981), for example, noted a sixfold increase in sedimentation rates in Mediterranean lagoons during the Subboreal Period through increased rates of soil erosion and increased 'storminess' in the hinterland. Sediment supply rates in Natal and Zululand are high in relation to New South Wales, for example, but factors such as catchment size also influence sediment supply to particular lagoons. It is probable that a relationship exists between catchment size, runoff volume and lagoon volume; if runoff to a particular lagoon were to increase a breach may form and even persist, whereas if discharge were reduced the breach may close.

Closure of tidal inlets as a result of decreased tidal prism, documented by Finkelstein & Ferland (1987) from Virginia, contrasts with the situation in Natal and New South Wales where normally closed outlets may open more frequently as a result of increased competence arising from fluvial channelisation (Roy, 1984; Cooper 1991c). This is a direct result of the importance of fluvial discharge in Natal and New South Wales in comparison with tidal currents, which maintain inlets in Virginia and elsewhere.

The response of a barrier to sea-level rise has a direct bearing on lagoonal evolution. If a transgressive lagoonal barrier stabilises against a break in slope or topographic high and accretes upward rather than landward then the volume of the lagoon may increase if sea-level rise exceeds fluvial sediment supply. If the barrier continues to move landward as sea level rises then the actual lagoon volume may remain constant. In certain cases the barrier progrades seaward and this may reduce contact with the sea, as in the case of the Siyai lagoon.

Barrier grain size apparently limits the influence of washover in the back-barrier environment (Duffy et al., 1989). The lagoons of Maine contain limited volumes of overwash-derived sediment, in contrast to lagoons with sandy barriers where washover fans may extend far into the lagoon (Andrews 1970; Cooper & Mason, 1986) and become an important component in the back-barrier stratigraphy (Hobday & Orme 1974) and a major sediment source (Bartberger, 1976). Carter & Orford (1984) noted that swash percolates rapidly into porous coarse-grained barriers and deposits entrained material. This limits washover by building up the barrier. Eventually, the accumulated material slumps into the barrier as a series of small lobes (Orford & Carter 1982).

Most of the examples in the literature (for example New South Wales, Brazil, Virginia) depict the evolution of a lagoon from an initial basin, which is progressively isolated from the sea and is ultimately infilled by a swamp or floodplain. The source of sediment in some cases (New South Wales) is fluvial and differs from the Lucke Model of marine-derived sediment accumulation in this respect. The examples cited from Maine, which show increasing marine influence, differ from this model because of low sedimentation rates in relation to sea-level rise while some of those in Natal differ in that sedimentation matched or exceeded the rate of drowning during sea-level rise.

Processes of lagoon evolution

Numerous distinct evolutionary processes contribute to the evolution of lagoons. In some cases a single process dominates, while in others a variety of processes play roles of differing importance. The relative importance of each necessarily influences the evolution of a particular lagoon. The major processes are discussed below under back-barrier and barrier headings. The division is for ease of discussion only as one may impact on the other.

Back-barrier changes

Segmentation

This is the process whereby a single lagoon is converted into a series of isolated basins. It arises through the reorientation of a lagoon shoreline by erosion and deposition to assume equilibrium with the ambient wave field. Zenkovich (1959) proposed a model (Fig. 6.10) for complete segmentation of

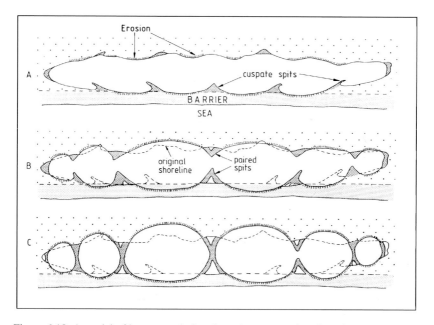

Figure 6.10. A model of lagoon evolution through segmentation (based on Zenkovich, 1959). This process operates best on uncohesive sandy shores in micro- or non-tidal environments. An open lagoon (*a*) becomes progressively segmented (*b*) and (*c*). It is well illustrated by the Kosi Lagoon (Fig. 6.5).

lagoons in which cuspate forelands develop into paired spits, which ultimately divide the lagoon into a series of roughly circular basins. Rosen (1975) subsequently noted that while cuspate spits form only in microtidal areas, complete segmentation only occurs in non-tidal areas because elsewhere tidal currents limit spit growth. Zenkovich's (1959) suggestion that segmentation occurred when the long axis of a lagoon extended parallel to the dominant winds was disputed by Rosen (1975), who argued that fetch limitation in an elongate lagoon was itself sufficient to force wave approaches to be perpendicular to the shoreline regardless of wind approach direction. Segmentation has been reported from many parts of the world from the Arctic to the tropics and occurs in lagoons ranging from a few tens of metres in diameter to the 350 km long Lagoa dos Patos in southern Brazil. One of the main impacts of segmentation is to create distinct sedimentary cells, which in the case of Kosi Lagoon exhibit distinctive sedimentary, chemical and biological characteristics. Segmentation may potentially preserve greater depths in certain segments by isolating them from fluvial sediment inputs, or create different evolutionary responses in adjacent segments.

Vertical accretion

Vertical accretion in lagoons is mediated by the rate and nature of sediment supply. Lagoon sediments may originate from the sea or land or as biogenic particles generated within the lagoon. Quiet water deposition of silt and clay occurs in lagoons in a manner analogous to pro-delta environments, where material is carried in by a river. Deposition from suspension of fluvial sediment is often enhanced by flocculation in the more saline lagoonal waters (Hobday, 1976). In certain cases increases in sedimentation rates have been attributed to human impacts (Finklestein & Ferland, 1987).

Organic or skeletal material is produced within lagoons below water level and on marginal intertidal flats by lagoonal organisms and vegetation. In some cases (Warnbro Sound and Port Royal Bay) this sediment is skeletal carbonate, while in others it is vegetal, originating from fringing vegetation such as mangroves in the tropics or salt-marsh vegetation in more temperate regions.

Fluvial delta progradation

Growth of fluvial deltas into lagoons is one mechanism by which the volume of a lagoon may be reduced. The size of a delta and its morphology depend largely on the volume of sediment carried by the inflowing river and energy levels within the lagoon. In theory, lagoonal deltas could exhibit most of the morphologies exhibited by deltas deposited directly in the sea (Galloway, 1975).

Chemical precipitation

Precipitation of minerals can occur both within and on the surface of lagoonal sediments. In the arid sabkhas the process is most important and may result in thick sequences of evaporite deposits (Friedman & Sanders, 1978). Short-lived periods of hypersalinity due to evaporation may also occur in normally brackish lagoons, as in Lake St Lucia (Hobday, 1976).

Lateral accretion

Reduction in the open water area of lagoons due to vegetation encroachment is a widespread problem in coastal zone management. It is apparently most common in climatic zones which favour rapid vegetal growth. In subtropical areas reeds and mangroves are the most common offenders. Orme (1973) reported loss of water area in Lake St Lucia through swamp encroachment during the Holocene. He found that swamp encroachment occurred when water depths were reduced by sedimentation to between 1 and 1.5 m. In the Siyai Lagoon in southern Zululand, reduction in stream power in the elongating lagoon, coupled with increased nutrients from cultivation, caused the rapid growth of reeds in marginal areas.

Tidal flat advance often precedes swamp encroachment and produces a predictable sequence (Semeniuk, 1981; Bird, 1986) which is related to tidal range (Klein, 1972). In microtidal lagoons, tidal flats are generally less extensive than in estuaries but they act to reduce the surface water area (Reineck & Singh, 1973). Rates of deposition on salt marsh and tidal flats may vary seasonally (Letzsch & Frey, 1980).

The advance of reeds may be limited by salinity and periodic hypersalinity may decimate reedbeds. This happens in Lake St Lucia during drought periods when hypersaline conditions occur (Begg, 1978). Sikora & Kjerfve (1985) ascribed reedswamp dieback and associated coastal erosion on the landward side of Lake Pontchartrain, a coastal lagoon in Louisiana, to subsidence and lack of compensatory sediment input. The erosion increased the wave fetch and this, coupled with subsidence, permitted brackish lagoon water to invade the freshwater swamps, causing dieback and erosion.

Marine sedimentation

Deposition of marine sediment in lagoons may arise through tidal delta progradation, barrier overwash or deposition from suspension. In some cases the system is self generating and may result in equilibrium (Carrigy, 1956) or eventual infilling (Ashmore & Leatherman, 1984)

Growth of tidal deltas occurs where an abundant supply of sand in the

J.A.G. Cooper

nearshore is supplied by alongshore transport on the open coast and ebb-tidal erosion does not match deposition of the flood tide. These may ultimately become emergent and vegetated (Boothroyd, Friedrich & McGinn, 1985; Cleary, Hosier & Wells, 1979).

Barrier overwash may instantaneously introduce large quantities of marine sediment directly into a lagoon. Overwash is frequently associated with storms, particularly hurricanes (Andrews, 1970). Thom (1984) noted that although washover was a dominant process during transgression in southeast Australia it ceased to be important during the past 6000 years as barriers accreted through dune development to levels above storm surge elevations.

Reworked muds from the continental shelf are the principal sediment source in the back-barrier lagoons of the Cape May Peninsula in New Jersey (Kelley, 1980). There, suspension settling of this mud dominates evolutionary processes.

Barrier changes

Barrier erosion

Examples of barrier erosion have been cited from the Robe/Beachport Lagoons in South Australia (Bird, 1970) and Warnbro sound in Western Australia (Carrigy, 1956), where formerly enclosed lagoons have been transformed into open marine embayments by erosion of the former barrier. Lithified remnants of former barriers at Warnbro Sound are preserved offshore as aeolianites, supporting the contention that early diagenesis may promote overstepping (Cooper 1991d). Complete barrier erosion appears to be limited to areas of low contemporary sediment supply and high wave energy (Barnes, 1980). Ireland (1987) cited rising sea level as a cause of barrier breaching and reinstatement of marine conditions in the lagoons of Rio de Janeiro State in Brazil. Carter *et al.* (1989) demonstrated the typical evolutionary sequence in southeast Irish lagoons to be one of increased marine influence from terrestrial through fresh- and then salt-water wetlands to intertidal flats and eventually open marine conditions as sediment supply is low and coastal facies are reworked *in situ*. The situation in Port Royal Bay (Ashmore & Leatherman, 1984), where sea level exceeded the surrounding rock level, produced similar results to those expected when a barrier is eroded or drowned.

Lagoon isolation

Sustained accretion of a barrier may eventually lead to isolation of a lagoon through closure of its outlet. Closure requires that stream discharge is unable to maintain an outlet against sediment buildup. Finkelstein & Ferland (1987) documented outlet closure in the lagoons of Virginia through reduction in tidal prism leading to reduced inlet efficiency. Similarly, Carter (1988) cited infilling of the Dutch Waddens as a case of reduced inlet efficiency and eventual closure as an example of this process.

Alternatively, lagoon isolation may arise through coastal progradation and associated lagoon extension as in the Siyai Lagoon. Sydow (1987) also proposed a reduction in stream gradients through coastal progradation and channel elongation as the mechanism by which Lake St Lucia was impounded. Aeolian deposition may also act to close a lagoon outlet and isolate it from the sea. Lake Sibaya in Zululand is now separated from the sea by a 100 m high dune and its water surface is 20 m above sea level. Evidence of its former connection to the sea is the presence of estuarine organisms, including crabs and fish, which have adapted to the isolated condition of the lake (Bruton, 1979).

Barrier translation

Migration of a barrier is one of the major influences in lagoon evolution. Landward migration may reduce the size of the tidal prism and generate a series of changes which cause inlet closure and water freshening, and ultimately obliterate the lagoon if the barrier is joined to the mainland shore. In this situation, a lagoon which is elongated perpendicular to the coast is more likely to prevail than one which is aligned parallel to the coast. The mode of barrier translation is an important factor in lagoon evolution, as discussed below (see also Chapter 4).

Controls on lagoon evolution

Lagoon evolution is the result of the balance between those processes which act to reduce the size of a lagoon and those which act to increase it. Clearly, the relative importance of a particular process in a lagoon depends upon the environmental setting in which the lagoon is located and the evolutionary path followed by a lagoon depends upon the magnitude and relative importance of each of the operative processes. This section attempts to determine the factors which control evolution. Such factors ultimately need to be incorporated in a comprehensive lagoon coast evolutionary model.

Roy (1984) hypothesised that diversity in New South Wales estuaries is a function of two classes of factor: inherited geological factors (bedrock type, tectonic setting, coastal morphology); and contemporary processes (tidal currents, waves, river discharge). To differentiate lagoonal evolutionary processes over a wider geographical area climatic controls must be added.

Variations in relative importance of each of the evolutionary processes summarised above are controlled by a variety of factors. These may be divided into macro-scale and local or micro-scale variations. Macro-scale controls include sea-level changes, climate and tectonic stability. Local controls include fluvial sediment supply, longshore drift, salinity, surrounding geology, coastal morphology, lagoon orientation, wave energy and tidal range. While many authors identify controls on the importance of various processes of lagoon evolution, there has been little effort to synthesise these and combine them into universal models. Carter *et al.* (1989) proposed a hierarchy of controls on the effect of sea-level changes on coarse-grained barrier/lagoon evolution. Such a hierarchical arrangement is often difficult and Fig. 6.11 gives an impression of the complexities and interrelationships between the various controls. The following discussion assesses the role of each variable.

Fluvial sediment supply varies around the globe but is at its highest where chemical weathering is accompanied by sufficient rainfall to transport eroded debris. In some cases this is hampered by dense vegetation (Milliman & Meade, 1983). Steep hinterlands aid in fluvial transport and enhance sediment supply. Where sediment supply is high there is increased likelihood of lagoon infilling, reduction of tidal prism, coastal progradation and outlet closure. Folger (1972) considered sediment supply from fluvial, overwash and flood-tidal deposition to be major variables in lagoon morphological development. In three out of the four climatic zones (high latitude, mid latitude, and low latitude humid) cited by Nicholls & Allen (1981) fluvial sediment supply was the primary source of sediment. Only in arid low latitudes was biological carbonate production more important. Carter *et al.* (1989) concluded 'It is apparent that the abundance or scarcity of sediment (and the situation may switch rapidly) is perhaps the strongest determinant of barrier-lagoon evolution'.

Wave energy plays an important role in barrier location and lagoon morphology. In Australia (Thom, 1984), northwest Ireland (Carter, 1988) and on the southern coast of Natal (Cooper, 1991a), barrier sediment has been confined to discrete embayments protected from high wave energy. This contrasts with the east coast of the United States where wave energy is lower and sandy barriers form elongate chains separated from the mainland.

Barrier location is, however, also controlled by antecedent topography. Halsey (1979) showed from the eastern United States that barriers formed along topographic highs on interfluves, while inlets were associated with former valleys. Boyd & Penland (1984) demonstrated the influence of basement slope on the position in which transgressive barriers stabilised. This can sometimes be traced to a break in slope or point of inflection in the

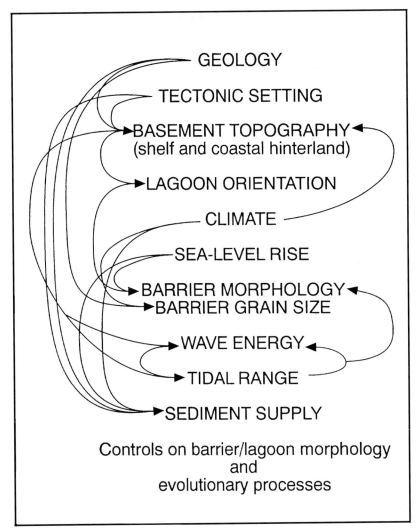

Figure 6.11. The complex interaction of processes which control evolutionary processes in a lagoon. The arrangement is broadly hierarchical but feedback between different variables renders the interaction complex.

underlying surface (Evans *et al.*, 1985). In St Lucia Lagoon, the Holocene barrier stabilised against a Late Pleistocene barrier remnant. The slope of the pre-existing basement also determines the width and depth of a lagoon. In Virginia, where basement slopes are low, lagoons are up to 20 km wide (Finklestein & Ferland, 1987). In Natal, where the basement slope is steep, and wave energy high, lagoons seldom extend more than 50 to 60 m seaward of the rocky mainland shore or simply occupy embayments. Similar embayment-constrained barriers occur in the wave-dominated coasts of New South Wales and northwestern Ireland (Carter, 1988). Boyd & Penland's (1984) comparative study of coasts with differing bedrock slopes illustrated the inability of present models to cope with the range of conditions that exist in nature.

Hayes (1979) outlined a conceptual model of variation in barrier morphology with tidal range, a factor which he considered to be of prime importance in lagoon morphology. Lagoons with greater tidal range have stronger currents and hence tidal channels are better developed. Those with smaller tidal range are dominated by waves and have less pronounced tidal channels. In addition, the thickness of a tidal flat deposit is controlled by tidal range (Klein, 1972) and depositional rates on salt marshes generally increase with tidal range (Thorbjarnarson *et al.*, 1985). Tidal range also affects the volume of flood-tidal deposition by influencing inlet spacing and barrier island length (Hayes & Kana, 1976).

Lagoon orientation is a further control on the wave fetch and ability of wave action to redistribute sediment along the lagoon shore. A lagoon in a narrow valley running perpendicular to dominant winds will experience much less effect from wind waves than one which is elongated parallel to dominant winds. The lagoon orientation depends to a large extent on the surrounding geology (Warme, Sanchez-Barreda & Biddle, 1977; Barnes, 1980) and lagoons may occupy joints or other structurally controlled depressions in the coast. In a lagoon which is aligned perpendicular to the coast, fluvial influences may be much more important, particularly if a lagoon is confined between steep rocky margins. The loss of volume through barrier translation is much reduced in a coast-normal lagoon than in one which is orientated coast-parallel. The latter type may even be completely obliterated (Barnes, 1980).

In lagoons which are confined to bedrock valleys, regressions may remove all sediment from previous depositional episodes. In cases where wave energy is low and lagoons are not constricted, deposits from several time periods may accumulate in the same basin (Oertel *et al.*, 1989). This situation is well illustrated by comparison of the bedrock-confined lagoons of Natal, whose entire valley fill is Holocene, with the wide lagoons of the Zululand coastal plain where sediments from pre-Holocene lagoons are preserved.

The material forming the lagoon margins exerts a strong control on its evolution. Bird (1970) noted that lagoons with well-developed fringing vegetation seldom exhibit segmentation. This is also true of lagoons with cohesive muddy or rocky banks, such as Lake St Lucia. In the small bedrock-confined lagoons of Natal, Cooper (1991c) found that those in softer lithologies generally had wider valleys than those in resistant lithologies and this affected the ability of the inflowing river to effect downward scour or lateral (overbank) deposition during flood events.

The importance of climate on lagoon evolution arises mainly through the relative importance of freshwater inflow and evaporation, which may promote evaporite precipitation or deposition of clastic sediments. In addition, coral growth occurs only in warm waters and the nature of fringing lagoonal vegetation (mangroves or salt marsh) and rate of growth is strongly climate dependent. Sediment supply and fluvial discharge are also controlled by climate.

Sediment source controls barrier grain size and this can commonly be correlated with climate. In regions which were ice covered or ice marginal during the last glaciation, sources of sediment are frequently carse clastic diamicts. In warmer regions where fluvial sediment supply onto the shelf is greater during glaciations, reworked fluvial sands are transported across the shelf to form lagoonal barriers during transgressive phases.

The rate of sea-level rise or fall is of particular importance in lagoon coast evolution and in the preservation potential of facies (Kraft *et al.*, 1987; Davis & Clifton, 1987). In a comparative study Carter *et al.* (1989) examined the effects of variation in rates of sea-level rise on coarse-grained barrier/lagoon evolution (see Fig. 1.4). With rapid sea-level rise barriers were destroyed either by breaching or overstepping as they failed to adjust rapidly to changes in sea level and sediment budgets. Antecedent topography was reduced to a minor role as rapid erosional fronts formed. Under slow sea-level rise barriers adjusted according to antecedent topography and a cell structure evolved as headlands protruded further into the wave field.

The conclusion of Dickinson, Berryhill & Holmes (1972) that several basic controls including climate, sediment supply, tidal range and tectonism produce different types of barrier island sand body also applies to lagoon coasts in general. Thorbjarnarson *et al.* (1985) also concluded that sedimentary characteristics of certain lagoons are affected by local factors such as tides, rivers and waves. This approach was used to differentiate lagoons into wave, tide and fluvially dominated in the same manner as deltas (Cooper, 1991a). Nichols & Allen (1981) regarded regional factors such as climate and tectonic setting as major controls on lagoon morphology, and hence evolutionary progression. Carter *et al.* (1989) identified a hierarchy of processes which control lagoon coast evolution. These included sea-level change, basement geology, sediment supply, wave and tide regimes and textural parameters. A comprehensive lagoon evolutionary model would have to include all of these variables and enable recognition of the importance of each process. The value of such a model in palaeoenvironmental reconstruction cannot be over-emphasised.

Evolutionary models

Although many studies exist of individual lagoons, they have been the subject of only a limited number of comparative sedimentological studies (e.g. Orme, 1973; Lankford, 1977; Ireland, 1987; Duffy *et al.*, 1989; Pilkey *et al.*, 1989) and synthesis of sedimentary models is still at an early stage. This may be in part due to their frequent occurrence in less-developed or sparsely populated areas (Kjerfve, 1986). Recently a number of papers (Kjerfve, 1986; Nichols, 1989; Kjerfve & Magill, 1989) have attempted to assess lagoonal variability and to place lagoons in a conceptual framework. Such an approach could ultimately lead to the formulation of general sedimentary models.

A facies model is a general summary of a specific sedimentary environment, written in terms that make the summary useable in at least four different ways: it must act as a *norm*, for purposes of comparison; as a *framework* and guide for future observations; as a *predictor* in new geological situations; and form a basis for *hydrodynamic interpretation* for the environment or system that it represents (Walker, 1980). Barrier islands and mainland-attached barriers, which can either be viewed separately or as an integrated component of the lagoon environment, have received more attention than lagoons as a whole (Schwartz, 1973; Leatherman, 1979) and several coastal barrier models have been formulated. These can be basically divided into sand and gravel barriers. Sandy barriers are best studied due to their widespread distribution around the world's coastline (Kraft 1971, 1978; Kraft, Biggs & Halsey, 1973; Kraft & John, 1979; Halsey 1979; Thom, 1984). Coarse-grained barriers, in comparison, are restricted to high latitudes on glacial and paraglacial coasts (Carter *et al.*, 1987; Duffy *et al.*, 1989). Evolutionary models for coarse-grained barriers and associated lagoons in temperate latitudes (Carter *et al.*, 1987, 1989) show that they differ from sandy barriers (Carter & Orford, 1984). For example, they are more prone to overstepping due to a slower response time to sea-level change than sand barriers (Forbes *et al.*, 1991; see also Chapters 1 and 10).

Given the variety of controls on lagoonal evolution presented above, development of a conceptual model for lagoonal evolution appears a daunting task. However, several basic models have been proposed. The main groups of models are discussed below.

Transgressive barrier models

As sea level rises a barrier may undergo one of three responses: erosional, translational (rollover), or stationary (overstepping) (Fig. 6.12). In an erosional response (the Bruun rule), material eroded from the shoreface is deposited below wave base in the nearshore zone and the same cross-sectional geometry

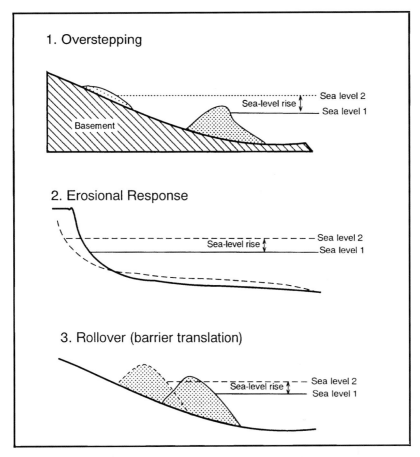

1. Overstepping

Basement

Sea-level rise

Sea level 2
Sea level 1

2. Erosional Response

Sea-level rise

Sea level 2
Sea level 1

3. Rollover (barrier translation)

Sea-level rise

Sea level 2
Sea level 1

Figure 6.12. Models of barrier response to rising sea level (after Carter, 1988).

is maintained but the whole profile moves upward by the same amount as sea level has risen. Problems with this approach have been discussed by Kraft *et al.* (1987). In a rollover-type response the entire barrier migrates across the basement slope without loss of material. The driving mechanism is barrier overwash, which transports material landward from the shoreface into the back barrier and enables its continual reworking as transgression proceeds. Landward translation could be strongly episodic. In certain cases the barrier migrates across bedrock (Evans *et al.*, 1985) or across finer-grained back-barrier deposits (Oertel *et al.*, 1989). Rollover is particularly effective in coarse clastic barriers where seaward-directed currents are unable to retard the process by transporting washover sediment back to the shoreface (Carter &

Orford, 1984) and on strongly wave-dominated coasts where transport through tidal inlets is reduced to a subordinate role (Hobday & Jackson, 1979).

Overstepping of a barrier during transgression leaves it drowned on the seabed as a relict feature. Carter (1988) listed five possible situations for barrier overstepping: increase in rate of sea-level rise; high sediment influx; interplay of landward and seaward transport processes; stranding on a topographic high; and interplay of opposing wave fields. To these can be added early diagenesis, particularly as beachrock or aeolianite in low latitudes, which prevents mobility of a barrier or a portion of a barrier (Cooper, 1991d) .

Controls on each type of response have not been identified but a knowledge of such controls could give additional information in interpretation of ancient sequences where a particular response can be recognised. Further research is necessary to identify the conditions which cause an erosional response rather than a rollover response, for example. Fisher (1980) argued from field evidence from Rhode Island that the erosional model (Bruun Rule) was ineffective as a model of transgressive shoreline development as more than 76% of eroded shoreface material moved landward rather than seaward. Kraft et al. (1987) attributed the inappropriateness of the Bruun Rule on the Delaware coast in part to the presence of fine-grained shoreface sediments but conceded that an alternative model would have to take into account many additional factors (Fig. 6.13).

Kraft & John (1979) presented a model for transgressive barrier/lagoon coasts, in which landward movement of barrier sands by washover was dominant (Fig. 6.14). Similarly, in New South Wales the models of Roy (1984) and Thom (1983) of barrier/lagoon evolution depict landward barrier transport during transgression.

Kraft et al. (1987) found that transgressive barriers along most of the USA Gulf and Atlantic coasts were generally similar in overall form and processes but varied in thickness and width. In developing a three-dimensional model of transgressive barrier/lagoon shoreline development they recognised several sources and sinks of sediment had to be incorporated. Hobday & Jackson (1979) presented sections from the Pleistocene, Port Durnford Formation of Zululand which show typical vertical sequences for transgressive barriers. Their model illustrates landward migration of the barrier by overwash into a back-barrier lagoon and swamp, followed by aeolian reworking of barrier sands to build high coastal dunes on top of the washover sands. Thom (1983) discussed differences between barrier models from Australia with those from other areas of lower wave energy. The main differences were in the thickness of the barrier deposit, the presence of embayments which limited alongshore sand transport, and the lack of mud on the high-wave-energy shelf. Otherwise, the general facies arrangement was similar to those found elsewhere.

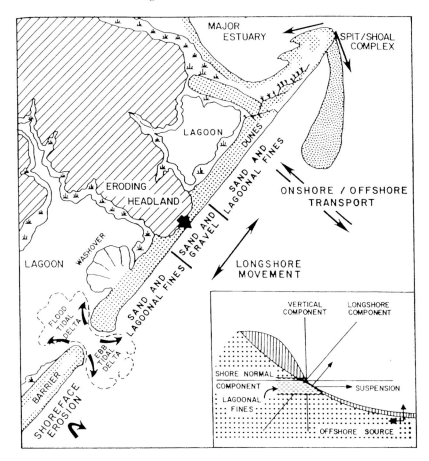

Figure 6.13. An illustration of the factors to be accounted for in an improved alternative to the Bruun rule model of transgressive coasts (after Kraft *et al.*, 1987), reproduced from *Sea-level Fluctuations and Coastal Evolution* with permission, SEPM).

Friedman & Sanders (1978) illustrated a model of transgressive barrier/ lagoon deposits showing the landward migration of deeper water deposits of both lagoon and nearshore across more landward environments

Regressive (falling sea-level) models

A number of characteristics distinguish regressive barriers, of which the presence of beach ridges is the most prominent (Thom, 1983). The classical regressive models are Galveston Island in Texas (Bernard *et al.*, 1970; Kraft & John, 1979; Morton & McGowen, 1980) and the Nayarit coast of Mexico

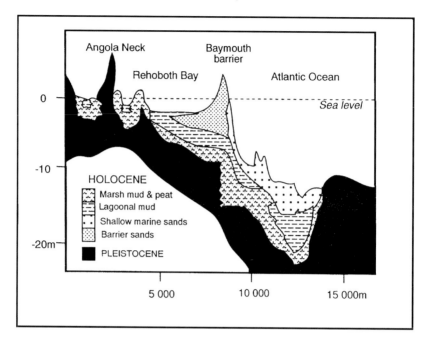

Figure 6.14. Simplified cross-section of part of the barrier/lagoon coast of Delaware (after Kraft, 1971). This figure is frequently cited as an appropriate model for transgressive, barrier/lagoon coasts.

(Curray, Emmel & Crampton, 1969). Morton (1979) demonstrated that Galveston comprised both a transgressive and regressive (accretionary) portion. On the Nayarit coast of Mexico (Curray *et al.*, 1969) a lower transgressive sequence of littoral sand overlies back-barrier lagoonal muds deposited between 7000 and 4500 years BP as the sea rose to its present level. Subsequently, the coastline prograded by up to 10 km through successive accretion of beach ridges with an upper aeolian component. Sediment for progradation originated from the rivers reaching the coast locally and was transported alongshore in littoral drift. The sequence described (Fig. 6.15) was considered by Reineck & Singh (1973) to be an appropriate model for an aggradational barrier/lagoonal coastline.

In Natal, active coastal progradation producing regressive barriers north of the Tugela River reflects influx of sediment from that source (Cooper, 1991a). Lagoons located behind prograding barriers become increasingly isolated from the sea on embayed coasts such as that of southeast Australia (Roy, 1984), while those which occur on linear clastic shorelines are extended in the direction of longshore drift and may ultimately close or be captured by other lagoons. In the embayments of southeast Australia prograded barriers consist of a series of beach ridges which display swash alignment. The lagoons confined behind these barriers are occasionally elongated along the length of the embayment (Thom, 1983). On the rocky Natal coast, narrow, shallow embayments limit coast-parallel extension (Cooper, 1991c). The Siyai Lagoon represents an example of coastal extension accompanied by lagoonal isolation.

Friedman & Sanders (1978) proposed two possibilities for lagoon evolution during coastal progradation (Fig. 6.16), one during stable sea level and the

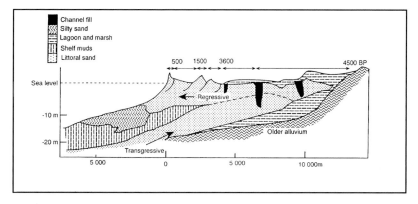

Figure 6.15. Cross-section of the barrier/lagoon coast of Nayarit, Mexico (after Curray *et al.*, 1969), considered an appropriate model for prograding barrier/lagoon coasts. The lower transgressive sequence is overlain by shoreface sediments, which are in turn overlain by a progradational sequence of beach ridges and interdune lagoon sediments.

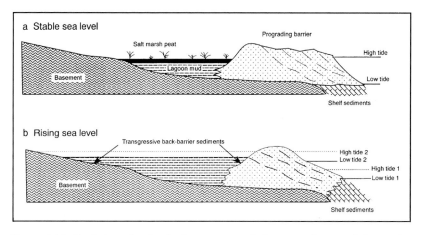

Figure 6.16. Evolution of a barrier/lagoon coast during coastal progradation under conditions of stable (*a*) and rising sea level (*b*) (after Friedman & Sanders, 1978).

other during a rise in sea level. Under a stable sea level progressive infilling transforms the lagoon into a marsh. Under a rising sea level the back-barrier sediments become transgressive over marginal areas and the landward side of the barrier.

Lucke Model

The Lucke Model of progressive infilling of an initially deep basin, outlined above, can be shown to hold for many modern lagoons; however, recently documented studies show that it is not universally applicable and that other factors may intervene to produce different stratigraphies. The model proposed by Roy (1984) is essentially similar in that a lagoon gradually infills and shallows but differs in that fluvial rather than marine sediment dominates the sedimentary fill. In areas with high sedimentation rates, deep water conditions may never be attained while in others with low sedimentation rates, lagoons deepen with time.

Nichols Model

Nichols (1989) presented a model of transgressive lagoon development, based upon the relative importance of sedimentation rates and relative sea-level rise. He identified three types: a deficit lagoon; a surplus lagoon; and an equilibrium lagoon (Fig. 6.17). A deficit lagoon is one in which the rate of sea-level rise exceeds sediment supply and the lagoon deepens, a surplus lagoon is one in which sediment supply exceeds relative sea-level rise and the lagoon is infilled and exports excess sediment to the adjacent continental shelf. In an equilibrium lagoon, morphology is maintained by a balance between sediment supply and relative sea-level rise. In this scheme an 'equilibrium lagoon' may persist for extended periods, in contrast to the Lucke model which predicts eventual infilling. Increased evidence of dynamic equilibrium in lagoons has indeed recently been found (Carson *et al.*, 1988; Ward & Ashley, 1989). In a study of lagoons and river mouths in Natal, Cooper (1991a) proposed a similar model of development but relating changes in *basin volume* to sediment supply and lagoonal morphology (Fig. 6.18). In this scheme changes in volume occur at different rates under constant sea-level rise according to cross-sectional changes in basins as a basin is drowned. It is probably most applicable to lagoons orientated coast-normal in drowned river valleys. When the cumulative sediment supply exceeds basin volume excess sediment is transported through the lagoon and into the sea. When basin volume increases more rapidly than sediment supply, a deep-water situation arises where suspension settling dominates sedimentation. It is under these circumstances

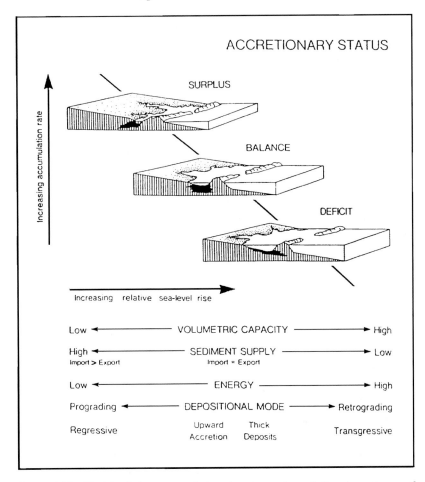

Figure 6.17. Model of lagoon evolution based on the relative importance of sedimentation and relative sea-level rise (after Nichols, 1989). Two end members are recognised, deficit and surplus lagoons, with an intermediate type of equilibrium lagoon. (Reproduced from *Marine Geology* with permission from Elsevier Science Publishers.)

that thick accumulations of muddy sediments may accumulate. In most modern Natal lagoons south of the Tugela River bedload predominates due to a series of Holocene regressions which reduced the volume of lagoon basins during the past 5000 years. However, in the sedimentary record of many lagoons a muddy sequence is present in the middle of the column, indicating a period of rapid volume increase as water drowned increasingly large cross-sectional areas.

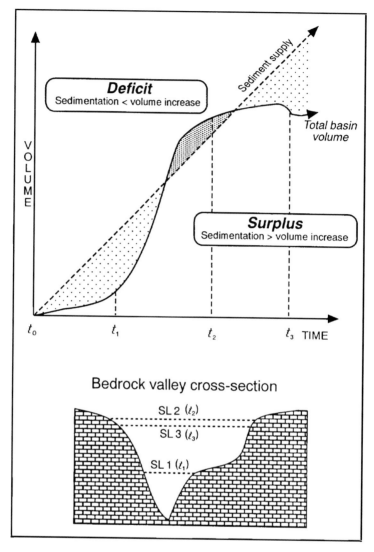

Figure 6.18. Lagoon evolution viewed as a balance between basin volume and sediment supply (after Cooper 1991a). In the example shown, sediment supply is regarded as constant with time. Total bedrock basin volume (including the portions already filled) varies according to the rate of sea-level rise *and* the cross-sectional area of the valley. Up to time t_1 the rate of volume increase is slow. From t_1 to t_2, the rate increases markedly as the valley widens. By time t_3 a minor regression occurred, reducing the total volume, but excess fluvial sediments continue to pass through the lagoon into the sea. The example depicted is representative of conditions in the Mdloti Lagoon (Fig. 6.8a), where the muddy sediments, in the middle of the valley-fill succession, accumulated during a deficit period when sedimentation could not keep pace with the rapidly increasing basin volume.

Current problems/ gaps in knowledge

The case studies of individual lagoon evolution and resultant sedimentary models documented above illustrate such variation that no single model can yet be applied to all circumstances. The major problem in this regard is the general lack of integrating studies on lagoons from different environmental settings. This hampers the formulation of generally applicable evolutionary models. The few existing conceptual models deal mainly with vertical changes in morphology and sedimentation. Few attempts have been made to link physical evolution with chemical and more especially faunal changes which reflects a lack of integration of disciplinary research in lagoons. Indeed, many biological comparisons of lagoons have been made from standpoints of ignorance, not recognising inherent differences in functioning and morphology from one lagoon to another. One of the few attempts to link these processes, albeit at a low level, is that of Roy (1984) whose conceptual model links species faunal diversity and abundance, and vegetation structure to stage of lagoonal infilling. Such an approach could provide valuable clues as to processes of lagoonal evolution. Duffy *et al.* (1989) and Ireland (1987), for example, used floral remains in back-barrier swamp deposits to interpret stratigraphy and former water levels.

In many cases the barrier and back-barrier are treated separately but studies to link back-barrier sedimentation changes to changes in the barrier would enhance the overall understanding of the lagoon ecosystem. Landward translation of the barrier would reduce the area of a lagoon and thus cause an apparent increase in vertical sedimentation rates through the same volume of sediment being deposited over a smaller area. Boyd & Penland (1984) compared descriptive models of coastal sedimentation from several areas and concluded that these may in fact represent only well-documented case studies rather than true, generalised models of coastal sequences. Such is also the case as far as lagoons are concerned and the problems arise from a lack of integrating studies from different environmental settings.

Carter *et al.* (1989) noted that 'a key goal in recent investigations of barrier evolution has been to understand the spatial and temporal dynamics, ultimately through evolutionary models that emphasise the likelihood of multiple outcomes (stochastic sequences) rather than assuming rigid divisions into cause and effect'. The illustrations (Fig. 6.13; Kraft *et al.*, 1987) of the morphodynamic elements which must be addressed when creating appropriate models and the complexities of relationships between controlling variables (Fig. 6.11) show the difficulties involved in formulation of universally applicable models. Further comparative studies, particularly of an inter-disciplinary nature, are clearly necessary if appropriate models of lagoon coast evolution are to be achieved. In view of the increasing human disturbance of lagoonal ecosystems, ongoing sea-level rise, and the concomitant need for sound environmental management guidelines, the need for such models is pressing.

References

Andrews, P.B. (1970). *Facies and genesis of a hurricane washover fan, St Joseph Island, central Texas coast*. Bureau of Economic Geology Report, 67, University of Texas at Austin. 147 pp.

Ashmore, S. & Leatherman, S.P. (1984). Holocene sedimentation in Port Royal Bay, Bermuda. *Marine Geology*, **56**, 289–98.

Barnes, R.S.K. (1980). *Coastal lagoons*. Cambridge University Press. 106 pp.

Bartberger, C.E. (1976). Sediment sources and sedimentation rates, Chincoteague Bay, Maryland and Virginia. *Journal of Sedimentary Petrology*, **46**, 326–36.

Begg, G.W. (1978). *The estuaries of Natal*. Pietermaritzburg: Natal Town and Regional Planning Commission Report, 41. 657 pp.

Begg, G.W. (1984). *The estuaries of Natal, part 2*. Pietermaritzburg: Natal Town and Regional Planning Commission Report, 55. 631 pp.

Benfield, M.C. (1985). *Management of* Phragmites *in the Siyaya Lagoon*. Siyaya Workshop, Oceanographic Research Institute, Durban, Working Document, 13B. 16 pp.

Bernard, H.A., Major, C.F., Parrott, B.S. & LeBlanc, R.J. (1970). *Recent sediments of Southeast Texas*. Guidebook 11, Bureau of Economic Geology, University of Texas at Austin. 132 pp.

Berryhill, H.L., Kendall, A.D. & Holmes, C.W. (1969). Criteria for recognizing ancient barrier coastlines. *American Association of Petroleum Geologists Bulletin*, **53**, 706–7.

Bird, E.C.F. (1970). *Coasts*. Australian National University Press. 246 pp.

Bird, E.C.F. (1986). Mangroves and intertidal morphology in Westernport Bay, Victoria, Australia. *Marine Geology*, **69**, 251–71.

Boothroyd, J.C., Friedrich, N.E. & McGinn, S.R. (1985). Geology of microtidal coastal lagoons, Rhode island. *Marine Geology*, **63**, 35–76.

Boyd, R. & Penland, S. (1984). Shoreface translation and the Holocene stratigraphic record: examples from Nova Scotia, the Mississippi Delta and eastern Australia. *Marine Geology*, **60**, 391–412.

Bruton, M.N. (1979). The fishes of Lake Sibaya. In *Lake Sibaya, Monographiae Biologicae*, vol. 36, ed. B.R. Allanson, pp. 162–245. The Netherlands: Dr W. Junk.

Carrigy, M.A. (1956). Organic sedimentation in Warnbro Sound, Western Australia. *Journal of Sedimentary Petrology*, **26**, 228–39.

Carson, B., Ashley, G.M., Lennon, G.P., Weisman, R.N., Nadeau, J.E., Hall, M.J., Faas, R.W., Zeff, M.L., Grizzle, R.E., Schuepfer, F.E., Young, C.L., Meglis, A.J., Carney, K.F. & Gabriel, R. (1988). Hydrodynamics and sedimentation in a back–barrier lagoon–salt marsh system, Great Sound, New Jersey: a summary. *Marine Geology*, **82**, 123–32.

Carter, R.W.G. (1988). *Coastal environments*. London: Academic Press. 609 pp.

Carter, R.W.G., Forbes, D.L., Jennings, S.C., Orford, J.D., Shaw, J. & Taylor, R.B. (1989). Barrier and lagoon coast evolution under differing relative sea–level regimes: examples from Ireland and Nova Scotia. *Marine Geology*, **88**, 221–42.

Carter, R.W.G., Johnston, T.W. & Orford, J.D. (1984). Stream outlets through mixed sand and gravel coastal barriers: examples from southeast Ireland. *Zeitschrift für Geomorpologie, N.F.*, **28**, 427–42.

Carter, R.W.G. & Orford, J.D. (1984). Coarse clastic barrier beaches: a discussion of their distinctive dynamic and morphosedimentary characteristics. *Marine Geology*, **60**, 377–89.

Carter, R.W.G., Orford, J.D., Forbes, D.L. & Taylor, R.B. (1987). Gravel barriers, headlands and lagoons: an evolutionary model. In *Coastal Sediments '87*, pp. 1776–92. New Orleans: American Society of Civil Engineers.

Cleary, W.J., Hosier, P.E. & Wells, G.R. (1979). Genesis and significance of marsh islands within southeastern North Carolina lagoons. *Journal of Sedimentary Petrology*, **49**, 703–10.

Clifton, H.E., Phillips, R.L. & Hunter, R.E. (1973). Depositional structures and processes in the mouths of small coastal streams, southwestern Oregon. In *Coastal geomorphology*, ed. D.R. Coates, pp. 115–40. New York: New York State University.

Cooper, J.A.G. (1989). Fairweather versus flood sedimentation in Mhlanga Lagoon, Natal: implications for environmental management. *South African Journal of Geology*, **92**, 279–294.

Cooper, J.A.G. (1990a). Late Pleistocene marine embayments of Zululand, South Africa. *Palaeoecology of Africa*, **21**, 49–59.

Cooper, J.A.G. (1990b). Ephemeral stream–mouth bars at flood–breach river mouths on a wave–dominated coast: comparison with ebb–tidal deltas at barrier inlets. *Marine Geology*, **95**, 57–70.

Cooper, J.A.G. (1991a). Sedimentary models and geomorphological classification of river-mouths on a subtropical, wave-dominated coast, Natal, South Africa. Unpublished PhD thesis, University of Natal. Durban, 401 pp.

Cooper, J.A.G. (1991b). *Shoreline changes on the Natal coast: Tugela River Mouth to Cape St Lucia.* Natal Town & Regional Planning Report, 76. Pietermaritzburg: Natal Town & Regional Planning Commission. 57 pp.

Cooper, J.A.G. (1991c). *Shoreline changes on the Natal coast: Mkomazi River Mouth to Tugela River Mouth.* Natal Town & Regional Planning Report, 77. Pietermaritzburg: Natal Town & Regional Planning Commission. 57 pp.

Cooper, J.A.G. (1991d). Beachrock formation in low latitudes: implications for coastal evolutionary models. *Marine Geology*, **98**, 145–54.

Cooper, J.A.G., Kilburn, R.N. & Kyle, R. (1989). A Late Pleistocene molluscan assemblage from Lake Nhlange, Zululand, and its palaeoenvironmental implications. *South African Journal of Geology*, **92**, 73–83.

Cooper, J.A.G. & Mason, T.R. (1986). Barrier washover fans in the Beachwood mangrove area, Durban, South Africa: cause morphology and environmental effects. *Journal of Shoreline Management*, **2**, 285–303.

Cromwell, J.E. (1971). Barrier coast distribution: a worldwide survey. In *Abstracts 2nd National Coastal and Shallow Water Research Cofnerence*, p. 50.

Curray, J.R., Emmel, F.J. & Crampton, P.J.S. (1969). Holocene history of a strandplain, lagoonal coast, Nayarit, Mexico. In *Coastal lagoons: a symposium*, ed. A.A. Castanares & F.B. Phleger, pp. 63–100. Universidad Nacional Autónoma de Mexico.

Davis, R.A. & Clifton, H.E. (1987). Sea–level change and the preservation potential of wave-dominated and tide-dominated coasts. In *Sea-level fluctuation and coastal evolution*, ed. D. Nummedal, O.H. Pilkey, & J.D. Howard, pp. 167–78. Society of Economic Paleontologists and Mineralogists Special Publication, 41.

Dickinson, K.A., Berryhill, H.L. & Holmes, C.W. (1972). Criteria for recognizing ancient barrier islands. In *Recognition of ancient sedimentary environments*, ed. J.K. Rigby & W.K. Hamblin, pp. 192–214. Society of Economic Paleontologists and Mineralogists Special Publication, 16.

Duffy, W., Belknap, D.F. & Kelley, J.T. (1989). Morphology and stratigraphy of small barrier–lagoon systems in Maine. *Marine Geology*, **88**, 243–62.

Emery, K.O. (1967). Estuaries and lagoons in relation to continental shelves. In *Estuaries*, ed. G.H. Lauff, pp. 9–11. Washington DC: American Association for the Advancement of Science.

Emery, K.O. & Uchupi, E. (1972). *Western north Atlantic Ocean; topography, rocks, structure, water, life and sediments.* American Association of Petroleum Geologists Memoir, 17. 532 pp.

Evans, M.W., Hine, A.C., Belknap, D.F. & Davis, R.A. (1985). Bedrock controls on barrier island development: west–central Florida coast. *Marine Geology*, **63**, 263–83.

Finkelstein, K. & Ferland, M.A. (1987). Back–barrier response to sea–level rise, eastern shore of Virginia. In *Sea-level fluctuation and coastal evolution*, ed. D. Nummedal, O.H. Pilkey, & J.D. Howard, pp. 145–55. Society of Economic Paleontologists and Mineralogists Special Publication, 41.

Fisher, J.J. (1980). Shoreline erosion, Rhode Island and North Carolina coasts: tests of Bruun rule. In *Proceedings of the Per Bruun Symposium, Newport, Rhode Island, November 1979*, ed. M.L.Schwartz & J.J. Fisher, pp. 32–54. Bellingham: Western Washington University.

Folger, D.W. (1972). Texture and organic carbon content of bottom sediments in some estuaries of the United States. *Geological Society of America Memoir*, **133**, 391–408.

Forbes, D.L., Taylor, R.B., Orford, J.D., Carter, R.W.G. & Shaw, J. (1991). Gravel barrier migration and overstepping. *Marine Geology*, **97**, 305–13.

Friedman, G.M. & Sanders, J.E. (1978). *Principles of sedimentology.* New York: John Wiley & Sons. 792 pp.

Galloway, W.E. (1975). Process framework for describing the morphological and stratigraphic evolution of deltaic depositional systems. In *Deltas, models for exploration*, ed. M.L. Broussard, pp. 87–98. Houston Geological Society.

Grobbler, N.G. (1987). Sedimentary environments of Mdloti, uMgababa and Lovu lagoons, Natal, South Africa. Unpublished M.Sc. Thesis, University of Natal, Durban. 171 pp.

Grobbler, N.G., Mason, T.R. & Cooper, J.A.G. (1987). *Sedimentology of Mdloti Lagoon.* Sedimentation in Estuaries and Lagoons (S.E.A.L. Report no. 3), Dept. Geology & Applied Geology, University of Natal, Durban. 43 pp.

Halsey, S.D. (1979). Nexus: new model of barrier island development. In *Barrier islands*, ed. S.P. Leatherman, pp. 185–209. New York: Academic Press.

Hayes, M.O. (1975). Morphology of sand accumulations in estuaries. In *Estuarine research*, vol. 2, ed. L.E. Cronin, pp. 2–22. New York: Academic Press.

Hayes, M.O. (1979). Barrier island morphology as a function of tidal and wave regime. In *Barrier islands*: from the Gulf of St. Lawrence to the Gulf of Mexico, ed. S.P. Leatherman, pp. 1–27. New York: Academic Press.

Hayes, M.O. & Kana, T.W. (1976). *Terrigenous clastic depositional environments.* University of South Carolina, Coastal Research Division, Technical Report 11–CRD. 184 pp.

Hill, B.J. (1969). The bathymetry and possible origin of Lakes Sibaya, Nhlange and Sifungwe in Zululand (Natal). *Transactions of the Royal Society of South Africa*, **38**, 205–16.

Hobday, D.K. (1976). Quaternary sedimentation and development of the lagoonal complex, Lake St Lucia, Zululand. *Annals of the South African Museum*, **71**, 93–113.

Hobday, D.K. & Jackson, M.P.A. (1979). Transgressive shore zone sedimentation and syndepositional deformation in the Pleistocene of Zululand, South Africa. *Journal of Sedimentary Petrology*, **49**, 145–58.

Hobday, D.K. & Orme, A.R. (1974). The Port Durnford Formation: a major Pleistocene barrier–lagoon complex along the Zululand coast. *Transactions of the Geological Society of South Africa*, **77**, 141–9.

Hoyt, J.H. (1967). Barrier island formation. *Geological Society of America Bulletin*, **78**, 1125–36.

Ireland, S. (1987). The Holocene sedimentary history of the coastal lagoons of Rio de Janeiro State, Brazil. In *Sea level changes*, ed. M.J. Tooley & I. Shennan, pp. 25–66. London: Blackwell.

Kelley, J.T. (1980). Sediment introduction and deposition in a coastal lagoon, Cape May, New Jersey. In *Estuarine perspectives*, ed. V.S. Kennedy, pp. 379–88. New York: Academic Press.

Kjerfve, B. (1986). Comparative oceanography of coastal lagoons. In *Estuarine variability*, ed. D.A. Wolfe, pp. 63–81. New York: Academic Press.

Kjerfve, B. & Magill, K.E. (1989). Geographic and hydrodynamic characteristics of shallow coastal lagoons. *Marine Geology*, **88**, 187–99.

Klein, G. de V. (1972). A model for establishing paleotidal range. *Geological Society of America Bulletin*, **82**, 2585–92.

Kraft, J.C. (1971). Sedimentary facies patterns and geologic history of a Holocene marine transgression. *Geological Society of America Bulletin*, **82**, 2131–58.

Kraft, J.C. (1978). Coastal stratigraphic sequences. In *Coastal sedimentary environments*, 1st edn, ed. R.A. Davis, pp. 361–83. New York: Springer–Verlag.

Kraft, J.C., Biggs, R.B. & Halsey, S.D. (1973). Morphology and vertical sedimentary sequence models in Holocene transgressive barrier systems. In *Coastal geomorphology*, ed. D.R. Coates, pp. 321–54. New York State University.

Kraft, J.C., Chrzastowski, M.J., Belknap, D.F., Toscano, M.A. & Fletcher, C.H. (1987). The trangressive barrier–lagoon coast of Delaware: morphostratigraphy, sedimentary sequences and responses to relative rise in sea-level. In *Sea-level fluctuation and coastal evolution*, ed. D. Nummedal, O.H. Pilkey, & J.D. Howard, pp.129–43. Society of Economic Paleontologists and Mineralogists Special Publication, 41.

Kraft, J.C. & John, C.J. (1979). Lateral and vertical facies relations of transgressive barriers. *American Association of Petroleum Geologists Bulletin*, **63**, 2145–63.

Kyle, R. (1986). Aspects of the ecology and exploitation of the fishes of the Kosi Bay system, KwaZulu, South Africa. Unpublished PhD thesis, University of Natal, Pietermaritzburg. 245 pp.

Lankford, R. (1977). Coastal lagoons of Mexico: their origin and classification. In *Estuarine processes*, vol. II, ed. M. Wiley, pp. 182–215. New York: Academic Press.

Leatherman, S.P., ed. (1979). *Barrier islands*. New York: Academic Press. 319 pp.

Letzsch, W.S. & Frey, R.W. (1980). Deposition and erosion in a Holocene salt marsh, Sapelo Island, Georgia. *Journal of Sedimentary Petrology*, **50**, 529–42.

Lucke, J.B. (1934). A theory of evolution of lagoon deposits on shorelines of emergence. *Journal of Geology*, **42**, 561–84.

Martin, R.T., Gadel, F.Y. & Barusseau, J.P. (1981). Holocene evolution of the Canet–St Nazaire lagoon (Golfe du Lion, France) as determined from a study of sediment properties. *Sedimentology*, **28**, 823–36.

Maud, R.R. (1968). Quaternary geomorphology and soil–formation in coastal Natal. *Zeitschrift für Geomorpologie, N.S., Suppl. Bd.,* **7**, 155–99.

Milliman, J.D. & Meade, R.H. (1983). World–wide delivery of river sediments to the oceans. *Journal of Geology,* **91**, 1–21.

Morton, R.A. (1979). Temporal and spatial variations in shoreline changes and their implications, examples from the Texas coast. *Journal of Sedimentary Petrology,* **49**, 1101–11.

Morton, R.A. & McGowen, J.H. (1980). *Modern depositional environments of the Texas coast.* Guidebook 20, Bureau of Economic Geology, University of Texas at Austin. 167 pp.

Nichols, M.M. (1989).Sediment accumulation rates and relative sea–level rise in lagoons. *Marine Geology,* **88**, 201–19.

Nichols, M.M. & Allen, G. (1981). Sedimentary processes in coastal lagoons. In *Coastal lagoon research, present and future.* pp. 27–80. Paris: UNESCO Technical Papers in Marine Science, 33.

Nichols, M.M. & Biggs, R.B. (1985). Estuaries. In *Coastal sedimentary environments,* 2nd edn, ed. R.A. Davis, pp. 77–186. New York: Springer-Verlag.

Oertel, G.F., Kearney, M.S., Leatherman, S.P. & Woo, H–J. (1989). Anatomy of a barrier platform: outer barrier lagoon, southern Delmarva Peninsula, Virginia. *Marine Geology,* **88**, 303–18.

Orford, J.D. & Carter, R.W.G. (1982). Crestal overtop and washover sedimentation on a fringing sandy gravel barrier coast, Carnsore Point, southeast Ireland. *Journal of Sedimentary Petrology,* **52**, 265–78.

Orme, A.R. (1973). Barrier and lagoon systems along the Zululand coast, South Africa. In *Coastal Geomorphology,* ed. D.R. Coates, pp. 181–217. State University of New York.

Orme, A.R. (1974). *Estuarine sedimentation along the Natal coast, South Africa.* Technical Report, 5, Office of Naval Research, U.S.A. 53 pp.

Partridge, T.C. & Maud, R.R. (1987). Geomorphic evolution of southern Africa since the Mesozoic. *South African Journal of Geology,* **90**, 179–208.

Phleger, F.B. (1981). A review of some general features of coastal lagoons. In *Coastal lagoon research, present and future.* pp. 7–14. Paris: UNESCO Technical Papers in Marine Science, 33.

Phleger, F.B. & Ewing, G.C. (1962). Sedimentology and oceanography of coastal lagoons in Baja California, Mexico. *Geological Society of America Bulletin,* **73**, 145–82.

Pilkey, O., Heron, D., Harbridge, W., Keer, F. & Thronton, S. (1989). The sedimentology of three Tunisian lagoons. *Marine Geology,* **88**, 285–301.

Ramm, A.E.L. (1992). Aspects of the biogeochemistry of sulphur in Lake Mpungwini, southern Africa. *Estuarine, Coastal and Shelf Science,* **34**, 253–61.

Ramsay, P.J. (1990). A new method for reconstructing late Pleistocene coastal environments, Sodwana bay, Zululand. *Geological Survey of South Africa Report,* 1990–0227. 44 pp.

Reineck, H.E. & Singh, I.B. (1973). *Depositional sedimentary environments.* Berlin: Springer–Verlag. 439 pp.

Reinson, G.E. (1980). Barrier island systems. In *Facies models,* ed. R.G. Walker, pp. 57–74. Geoscience Canada, Reprint series 1, Third printing. Geological Association of Canada.

Rosen, P.S. (1975). Origin and processes of cuspate spit shorelines. In *Estuarine research,* vol II, ed. L.E. Cronin, pp. 76–92. New York: Academic Press.

Roy, P.S. (1984). New South Wales estuaries: their origin and evolution. In *Coastal geomorphology in Australia,* ed. B.G. Thom, pp. 99–121. Australia: Academic Press.

Schwartz, M.L., ed. (1973). *Barrier islands.* Benchmark Papers in Geology. Stroudsburg, PA: Dowden, Hutchinson & Ross. 451 pp.

Semeniuk, V. (1981). Sedimentology and the stratigraphic sequence of a tropical tidal flat, northwestern Australia. *Sedimentary Geology*, **29**, 195–221.

Sikora, W.B. & Kjerfve, B. (1985). Factors influencing the salinity regime of Lake Pontchartrain, Louisiana, a shallow coastal lagoon: analysis of a long term data set. *Estuaries*, **8**, 170–80.

Sydow, C.J. (1987). Stratigraphic control of slumping and canyon development on the Zululand continental margin, east coast, South Africa. Unpublished BSc thesis, University of Cape Town. 58 pp.

Thom, B.G. (1983). Transgressive and regressive stratigraphies of coastal sand barriers in southeast Australia. *Marine Geology*, **56**, 137–58.

Thom, B.G. (1984). Sand barriers of eastern Australia: Gippsland: a case study. In *Coastal geomorphology in Australia*, ed. B.G. Thom, pp. 233–61. Australia: Academic Press.

Thorbjarnarson, K.W., Nittrouer, C.A., DeMaster, D.J. & McKinney, R.B. (1985). Sediment accumulation in a back–barrier lagoon, Great Sound, New Jersey. *Journal of Sedimentary Petrology*, **55**, 856–63.

Walker, R.G. (1980). Facies and facies models: general introduction. In *Facies models*, ed. R.G. Walker, pp. 1–7, Geoscience Canada, Reprint Series 1, Third printing. Geological Association of Canada.

Ward, L.G. & Ashley, G.M. (1989). Introduction: coastal lagoonal systems. *Marine Geology*, **88**, 181–5.

Warme, J.E., Sanchez-Barreda, L.A. & Biddle, K.T. (1977). Sedimentary processes in west coast lagoons. In *Estuarine processes*, vol. II, ed. M. Wiley, pp. 167–81. New York: Academic Press.

Zenkovich, V.P. (1959). On the genesis of cuspate spits along lagoon shores. *Journal of Geology*, **67**, 269–77.

7

Coral atolls

R.F. McLEAN AND C.D. WOODROFFE

Introduction

Coral atolls consist of an annular reef rim surrounding a central lagoon. On the atoll rim there may be reef islands; either sandy *cays* or shingle *motu* (Stoddart & Steers, 1977). They appear particularly fragile constructions, exposed to a range of oceanographic, atmospheric and anthropogenic processes, and yet the prototypes of present day sea-level atolls have endured in a variety of forms for millions of years. Of the 425 atolls in the world (Stoddart, 1965), most are in the Indian and Pacific Oceans, though isolated atolls do occur in the Caribbean (e.g. Stoddart, 1962a)

The evolution of atolls is of particular relevance in the context of coastal evolution studies, because one of the most central theories of atoll evolution dates back to Charles Darwin and the voyage of *HMS Beagle*. This voyage gave rise to the concept of natural selection and evolution of species, with the publication of *On the Origin of Species* in 1859, but Darwin's view on evolution of atolls, was published much earlier with his observations on the geology of reefs (Darwin, 1842).

Darwin had formulated his theory of reef development shortly after leaving South America, and before seeing the reefs of the Pacific. He refined his ideas as a consequence of visiting reefs in the Society Islands, and had completed a draft of his theory before reaching New Zealand (see Stoddart, 1962b). The only atoll that Darwin landed on was the Cocos (Keeling) Islands in the Indian Ocean, and at that stage he was keen to verify his intuition. Darwin's theory was later tested first by deep-drilling that was undertaken on the atoll of Funafuti in Tuvalu (previously the Ellice Islands), and later as a consequence of atomic testing on other Pacific atolls.

The Cocos (Keeling) Islands in the Indian Ocean and Funafuti in the Pacific Ocean were the best-known atolls at the turn of the century. In this chapter the history of theories of reef development is examined, with particular emphasis on the Cocos (Keeling) Islands and Funafuti. The long-term development of atoll structure is considered, based upon the considerations which Darwin outlined, and atoll evolution is accommodated into the concepts of plate tectonics. Late Quaternary development of atolls, during the oscillations of sea level, is examined, and the Holocene morphodynamics of reef growth, lagoonal infill and reef island formation are considered.

At the time of Darwin's voyage, the most widely held view of atolls was that they were submerged volcanic crater rims veneered with a thin cover of coral (Lyell, 1832). Darwin recognised the improbability that these volcanic rims would everywhere reach into the shallow depths required for reef growth, and proposed that there:

is but one alternative; namely the prolonged subsidence of the foundations on which the atolls were primarily based, together with the upward growth of the reef-constructing corals. On this view every difficulty vanishes; fringing reefs are thus converted into barrier-reefs; and barrier-reefs, when encircling islands, are thus converted into atolls, the instant the last pinnacle of land sinks beneath the surface of the ocean.

(Darwin, 1842: p. 109.)

This view, illustrated by Darwin's own sketch (see Fig. 7.1), envisaged fringing reefs, barrier reefs, and atolls as stages in an evolutionary sequence, and that the process driving the sequence was gradual subsidence of the volcanic basement on which the reef initially established, in combination with vertical reef growth.

On Cocos, Darwin's attention was drawn to erosion of the shoreline, which had resulted, as it still does, in toppling of coconut palms. He considered this 'tolerably conclusive evidence' of subsidence. We now know, as John Clunies Ross knew at the time, that localised shoreline erosion on Cocos, as on other atolls, is easily explicable by other means. Indeed, concerned at Darwin's suggestion that the islands of which he was in possession were about to subside beneath the sea, Ross (1855: p. 9) claimed that a

moderately attentive investigation of the Cocos islets affords ample reasons for believing that they have stood up to the present time above the level of the ocean during hundreds if not thousands of years.

Darwin's theory has proved robust, and is widely supported by drilling results from other atolls, and by evidence that is described below from both Cocos and Funafuti. Surface geomorphological evidence from Cocos, however, indicates that there has been a relative fall of sea level (emergence) over the last few thousand years, quite the converse of what Darwin had

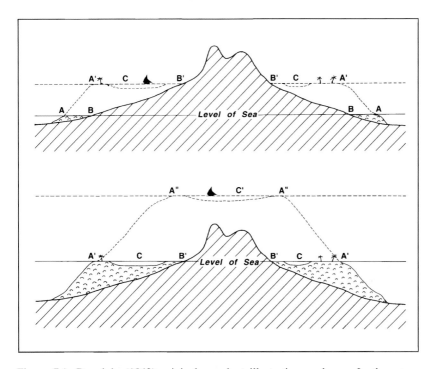

Figure 7.1. Darwin's (1842) original woodcut illustrations redrawn. In the upper diagram the original volcanic island is shown with a *fringing-reef* A–B. A section of the reef and island after subsidence is given by the dotted lines which show an encircling *barrier-reef* with A' the outer edge of the reef, B' the shore of the encircled island, and C the lagoon-channel between the reef and island's shore. In the lower diagram the newly formed barrier reef is represented by unbroken lines. As the island continues subsiding the coral reef will continue growing up on its own foundation and after the sea covers the highest volcanic pinnacle a *'perfect atoll'* is formed with A" representing the outer edge and C' the lagoon of the newly formed atoll.

concluded. To understand this it is necessary to discriminate between two contrasting aspects of the evolution of coral atolls: the *structure* of atoll foundations, including the relationship between reefal limestone and volcanic basement, on the one hand; and *surface morphology* of atolls, including modern reefs, surficial deposits, lagoonal infill and the origin of reef islands, on the other (Stoddart, 1973). It is important to realise that 'the movement of subsidence, if coral growth was to keep pace with it, would have to be too slow to be observable' (Stoddart, 1983: p. 525). The distinction between structure and surface morphology is of significance because the rates of operation of the formative processes and the timescales involved are so different.

The Cocos (Keeling) Islands and Funafuti Atoll

The Cocos (Keeling) Islands comprise the South Keeling Islands, a horseshoe shaped atoll (12° 12' S, 96° 54' E), and North Keeling, an isolated atollon (almost entirely closed reef-top island) (11° 50' S, 96° 49' E), 27 km to the north. These rise as a single feature from the ocean floor, which is about 5000 m deep, and 60–90 Ma old (Jongsma, 1976). They are dominated by the southeast tradewinds which blow strongly during most of the year, and maintain a large swell particularly on the southern and eastern margins of the atoll.

The South Keeling Islands, henceforth referred to as Cocos, consist of a series of reef islands on a horseshoe-shaped atoll rim, with two deep passages southwest and southeast of Horsburgh Island, the northernmost island (Fig. 7.2). Shallow interisland passages separate the other reef islands, with a broad reef flat between the southern elongate islands of West Island and South Island. The reef front is relatively barren of living scleractinian corals; the reef crest is algal veneered, with surge channels at intervals of 50–250 m. The reef flat is covered by 1–2 m of water at high tide and much of it dries at low tide. The atoll is microtidal with a spring tidal range of 1.2 m. The northern part of

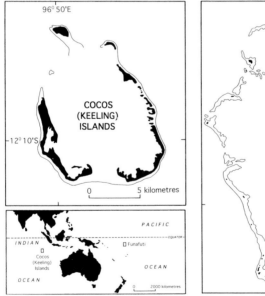
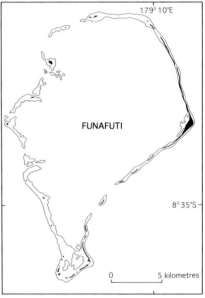

Figure 7.2. The Cocos (Keeling) Islands and Funafuti Atoll, Tuvalu, showing location and reef and island outlines.

the lagoon averages around 15 m deep and is covered with sand, or in places dead coral. The southern part of the lagoon is shallow, but contains a network of 'blue holes', individual deep holes with coral rims, and a muddy fill, up to 15 m deep in their centres. Extensive sand flats and sand aprons occur around the margin of the lagoon, especially towards the southern end of the lagoon (Smithers *et al.*, 1994).

Funafuti Atoll (8° 30' S, 179° 10' E) is substantially larger than Cocos (Fig. 7.2). Its long axis, aligned approximately north–south, is about 25 km long, and the atoll is about 18 km west to east at its widest. The lagoon comprises a deep central basin of average depth 40–50 m, which although generally of low relief is in places punctuated with deeper depressions and pinnacles some of which reach to the surface. The basin slopes gently upwards initially and then much more steeply (slopes of 10–20°) towards the atoll rim. Depths of 30 m are reached within 500 m of the shore in the north and east and within 2 km of the shore in the west. Much of the narrow southern lagoon is less than 10 m in depth and in the extreme south the lagoon shoals at low tide.

Funafuti's rim is cut by two narrow and deep (40–50 m) passages in the west and several shallower ones in the north and southeast. Compared with Cocos, the atoll rim is narrow and solid reef flat is exposed at low tide on both the ocean and lagoon margins of islands in the east and south. In the west the wider more discontinuous reef flat is occupied by a number of small more-compact islands. Maximum tidal range is 1.7 m and like Cocos, Funafuti is subject to tradewind-generated seas from the east and southeast.

Atoll structure

Although not the first attempts to drill reefs (Stoddart, 1992), deep drilling was undertaken on Funafuti in 1896–1898 in order to examine Darwin's interpretation of atolls that they developed as a result of a gradually subsiding volcanic basement around which vertical reef growth had been established. After several attempts, 333 m of coral rock was drilled, with the upper 194 m in coral limestone, and the rest in dolomite, which Judd (1904) and others regarded as being of shallow-water origin throughout. While volcanic basement was not encountered, the Funafuti borehole provided some substantiation of Darwin's ideas. Volcanic basement was further indicated by seismic refraction experiments carried out in the Funafuti lagoon during the *H.M.S. Challenger* expedition (Gaskell & Swallow, 1953; Locke, 1991). Gaskell & Swallow concluded that at Funafuti there is about 550 m of coral limestone overlying the volcanics.

Davis (1928), in an extensive review of the development of coral reefs, considered that the Darwinian subsidence theory was the most credible of the competing theories. He discounted the solutional theory of atoll development put forward by Murray (1889), and the sedimentation view, propounded by Wood-Jones (1912), who spent over a year on the Cocos Islands as the medical doctor.

Although the Funafuti boring did not confirm that basalt underlaid the atoll, it did show greater thicknesses of shallow-water carbonates than could occur in the water depths that were drilled, implying subsidence. Similar results were obtained from the drilling at Kita Daito Zima, southeast of Kyushu, Japan (Hanzawa, 1938), which penetrated 431 m without encountering basalt.

Post-war deep-drilling of atolls, such as that on Bikini and Enewetak in the Marshall Islands, which revealed more than 1000 m of shallow-water carbonates overlying a basalt core (Ladd, Tracey & Lill, 1948; Ladd *et al.*, 1953; Emery, Tracey & Ladd, 1954), has generally substantiated the subsidence theory of coral atoll development proposed by Darwin (1842).

On Midway Atoll in the Hawaiian Islands, volcanic basement was encountered at less than 200 m in the lagoon and around 400 m on the atoll rim (Ladd, Tracey & Gross, 1970), and in the Tuamtous on Mururoa a series of drillholes encountered basalt at just below 400 m (Lalou, Labeyrie & Delibrias, 1966; Labeyrie, Lalou & Delibrias, 1969) and on Fangataufa Atoll boreholes reached the volcanic basement at 360 m under the atoll rim and 230 m under the lagoon (Guillou *et al.*, 1993). The combined results demonstrate 'with a certainty rarely obtained in geomorphology, that these ... atolls were built by corals and associated organisms during a long-continued subsidence' (Guilcher, 1988).

That the Cocos Islands are underlain by a volcanic basement can be inferred from magnetic surveys showing an anomaly (Chamberlain, 1960; Finlayson, 1970), and a basalt and tuff pebble dredged from the western end of the Cocos Rise (Bezrukov, 1973).

The Darwinian sequence of coral atoll development, in which the driving forces are gradual subsidence of the volcanic basement and vertical reef growth (Braithwaite, 1982), has been incorporated into plate tectonic theory, and much of the subsidence can be explained in association with the aging and contraction of the ocean floor as the oceanic plate moves (Parsons & Sclater, 1977; Scott & Rotondo, 1983). Atolls in the Pacific are often found in linear chains, some of which (i.e. the Hawaiian and Society Islands) demonstrate stages in the Darwinian sequence – fringing reefs, barrier reefs, and coral atolls – along their length (see Chapter 12). Reef growth, however, has not been continuous, but has been interrupted by a series of solutional unconformities (Schlanger, 1963), indicating periods of subaerial erosion resulting from fluctuations of sea level during the Tertiary and Quaternary. Atolls have been exposed above the sea and subjected to weathering and erosion during low stands of sea level, and submerged with reefs re-establishing over their previously exposed surface during high stands of sea level (Lincoln & Schlanger, 1991).

Late Quaternary sea-level fluctuations

Superimposed upon the gradual subsidence of the atoll basement there have been pronounced fluctuations of sea level of around 120–150 m amplitude over the Quaternary. It is these alternate periods of subaerial exposure during glacial lowstands and drowning and reef re-establishment over the upper surface of the platform during interglacial highstands which have given rise to the solutional unconformities in atoll stratigraphy.

The importance of sea-level fluctuations was realised by several reef geologists, particularly Reginald Daly (Daly, 1915, 1925, 1934). Daly's glacial control theory of coral reef development maintained not only that there was contraction of reefs from marginal seas during glacial periods, but that reef platforms were planed off at a lower sea level, explaining the consistency of lagoon depths, with postglacial reef growth constructing the atoll rim. These views were perpetuated in the interpretations of atoll development propounded by Wiens (1959, 1962). In fact the rate of planation of reefs is generally too slow for such dramatic truncation to occur (Stoddart, 1969). Furthermore, outcrops of Pleistocene reef limestone which have clearly not been planated indicate that wave abrasion is not as effective as envisaged by Daly. More effective than wave abrasion are karst erosion processes, particularly at lower sea levels, and their importance has been stressed by Purdy in what has become known as the antecedent karst hypothesis (Purdy, 1974).

In the Cocos Islands shallow boreholes (up to 30 m deep), undertaken as part of water resource investigations and geomorphological studies of the development of the atoll, have encountered an unconformity at around 10–12 m below sea level. The lower limestone is well lithified, porous, and contains micritic cements indicating subaerial diagenesis. The pattern of boreholes and the depth of this unconformity are shown in Fig. 7.3*a*; the shallowest at which it is encountered is at 6–7 m at the southern and far western sides of the atoll. Uranium-series disequilibrium dating of coral from below the unconformity has yielded dates of around 120 000 years BP (Woodroffe *et al.*, 1991*a*), indicating that it was deposited as part of a Last Interglacial reef. The level of the sea at that time was roughly comparable to that at present or even slightly above present sea level. Solution of this thickness of

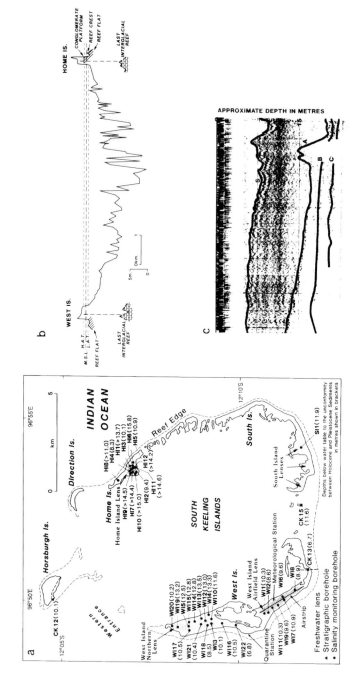

Figure 7.3. (a) The location of boreholes on islands on the rim of Cocos, showing the depth to the Thurber unconformity (updated from Woodroffe et al., 1991a). (b) schematic cross-section of lagoon, showing Last Interglacial reef beneath West and Home Islands (after Woodroffe et al., 1991a). HAT = Highest Astronomical Tide; LAT = Lowest Astronomical Tide. (c) seismic reflection profile of western margin of Cocos lagoon, representing about 250 m of trace running southeast from WI4. S = Lagoon floor; A, B and C are reflectors; A is taken to represent the Last Interglacial surface (after Searle, 1994, Fig. 2b).

limestone over this time appears unlikely, and the depth of this Thurber discontinuity (cf. Thurber *et al.*, 1965) appears to support subsidence since its deposition.

Much of the lagoon at the northern end of Cocos is 12–15 m deep (Fig. 7.3*b*). However, seismic reflection profiling across the lagoon (Fig. 7.3*c*) indicates a distinct reflector which occurs at a similar depth as the unconformity in the vicinity of the islands, but which slopes relatively steeply into the lagoon, and which occurs at depths of 20–22 m throughout much of the lagoon (Searle, 1994).

On Funafuti, unlike Cocos, the Last Interglacial reef has not been positively identified through radiometric dating. Nevertheless, extensive seismic reflection profiling in the Funafuti lagoon has indicated the presence of several unconformities, the uppermost being interpreted as a 'coral surface', the base of which is coincident with a widespread unconformity, U1 (Gibb Australia, 1985). Overlain by 20–25 m of unconsolidated sediment in the deep lagoon, the morphology of the 'coral surface' is highly variable with depressions and pinnacles; some of the latter extend to the lagoon floor. We suggest that this karst surface represents the Last Interglaical solutional unconformity.

From the deep lagoon basin the surface rises towards the atoll rim along with a decline in thickness of the overlying unconsolidated sediments. In the east we have attempted to correlate the seismic reflection results with previous interpretations of data from the 1896–98 drilling program on Fongafale Island and the adjacent lagoon (Fig. 7.4). Ladd (1948) identified a fossil land snail *Ptychodon* sp. A in partially leached coralliferous limestone from a depth of 50.6–51.8 m in the main bore, which he suggested was Pleistocene to Recent in age and represented a time during a glacial epoch when sea level was lower and when 'the top of the atoll stood above the sea'. Ginsburg *et al.* (1963) identified a 'probable surface of subaerial exposure' at about 42 m in the main borehole and 55 m in the lagoon borehole. The latter coincides with the depth of the main unconformity (U1) distinguished in the seismic record. The relationships between this unconformity and the 'coral surface' (Gibb Australia, 1985), the 'coral reefs and blocks' facies between depths of 10–26 m in Borehole No 1 (Sollas, 1904) and the 'first calcite cemented skeletal debris' at 40 m in the main borehole (Cullis, 1904; Ginsburg *et al.*, 1963) remain unclear and need to be resolved through further analysis of the cores and absolute dating techniques.

During the 125 000 years or so that it has taken for the sea to return to its present level, Cocos and Funafuti appear to have subsided, perhaps partly through cooling and contraction of the ocean floor, but also as a result of other factors, so that the former (Last Interglacial) surface is everywhere below

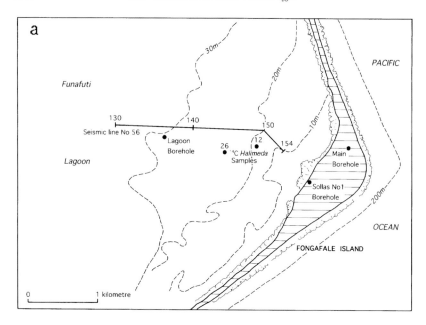

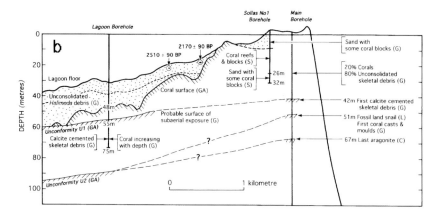

Figure 7.4. Shallow subsurface structure of Fongafale Island and adjacent lagoon, Funafuti Atoll. (*a*) Location of the Main, Sollas No 1 and lagoon boreholes drilled during the 1896–98 Royal Society Expedition, and the seismic reflection line number 56, stations 130–154 and the radiocarbon dated sediment (*Halimeda*) samples taken from 3 m below the surface at depths of 18 m (location 12) and 22 m (location 26) by Gibb Australia (1985). (*b*) Sedimentary facies and subsurface unconformities based on several sources: (C) Cullis, 1904; (G) Ginsburg *et al*, 1963; (L) Ladd, 1948; (S) Sollas, 1904; (GA) Gibb Australia, 1985. See text for details.

present sea level. To what extent this is true of other atolls is discussed further below. The sea rose rapidly during the postglacial marine transgression (Fairbanks, 1989; Chappell & Polach, 1991; Eisenhauser *et al.*, 1993), in each case drowning the former atoll surface. The upper 10–15 m of the present rim on these atolls has formed as a result of Holocene reef growth over the last few thousand years. These late Quaternary sea-level fluctuations 'were not envisaged by Darwin though they fit perfectly into later versions of the subsidence theory, and are easily explainable by almost any version of the glacial control theory' (Ladd, 1948: p. 197).

Holocene reef development

Reef response to sea-level change

Reefs are not necessarily able to track sea level exactly. At least three patterns of reef response to sea level can be envisaged. These have been termed keep-up, catch-up and give-up (Neumann & Macintyre, 1985). Keep-up reefs are reefs which do grow close to sea level and which are able to accrete vertically at rates fast enough to remain at sea level. They are characteristic of gradually or rapidly uplifting coasts (i.e. Barbados, Bard *et al.*, 1990 and Huon Peninsula, Chappell & Polach, 1991), but can also be seen on some stable coasts, even close to the latitudinal limit to reef growth (Collins *et al.*, 1993). Catch-up reefs are reefs where vertical growth initally lags behind sea level and then catches up when sea level stabilises or the rate of rise declines. Most reefs on Australia's Great Barrier Reef have grown by catch-up mode (see Chapter 8). Give-up reefs are reefs that founder as accretionary growth falls behind sea level until growth declines or ceases.

The pattern of sea-level change over the late Quaternary has been driven primarily by changes in the volume of water in the oceans. However, the earth's crust and upper mantle have adjusted to variations in the load of both polar ice and ocean water, and these glacioisostatic and hydroisostatic adjustments, together with local tectonic movements, mean that land–sea changes have varied around the world. These variations have become especially apparent over the mid and late Holocene, because ocean water volume has changed little during that period, and therefore subtle isostatic adjustments are not masked by rapid eustatic change (Pirazzoli, 1991).

Geophysical modelling of the response of a viscoelastic earth to changes in ice and water loads, calibrated with reconstructed ice-melt histories and key sea-level data, allows the extrapolation of relative-sea-level curves to regions from which data are presently scant (Clark, Farrell & Peltier, 1978; Peltier, 1988; Nakada & Lambeck, 1988; Tushingham & Peltier, 1991). This modelling supports the recognition of a 'Caribbean sea-level curve', which has been rising, but at a decelerating rate over the last 6000 years, and a 'Indo-Pacific sea-level curve', which had reached a level close to present 6000 years ago, rose above that level, and which has been falling for the last few thousand years (Fig. 7.5, see Adey, 1978; Davies & Montaggioni, 1985), though some variability also occurs at a regional scale. Coral atolls, together with other

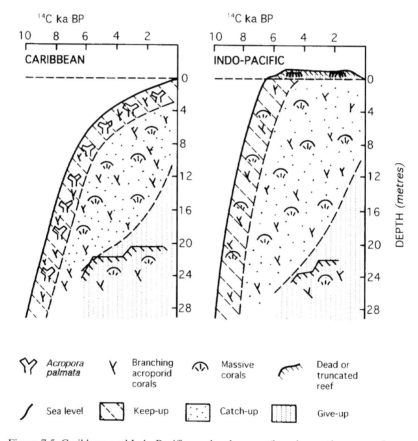

Figure 7.5. Caribbean and Indo-Pacific sea-level curves (based upon the curves shown in Davies & Montaggioni, 1985: Caribbean curve after Lighty *et al.*, 1982; Indo-Pacific curve is a combination of curves for Australia and the Polynesian curve of Pirazzoli *et al.*, 1988b). Keep-up, catch-up and give-up reef growth strategies in relation to sea level are shown. Give-up reefs are represented by drowned reefs, or in the case of Indo Pacific reefs, growth may be truncated by a fall of sea level.

islands less than 25 km in diameter, are likely to provide one of the least-complicated records of relative sea-level change, because they should be relatively immune to local hydroisostatic effects (Nakada, 1986).

The keep-up, catch-up and give-up reef growth strategies are most often expressed in terms of Caribbean reefs, where they often form distinct facies. However, similar growth patterns can be identified in the Indo-Pacific reef province. In the Indo-Pacific, keep-up reefs are generally characterised by branching framework corals, whereas there are greater proportions of massive coral heads in catch-up reef, though without any as yet recognised patterning of genera or species. Give-up reefs can be found as dead reefs in 20–30 m of water depth, upon which there may be sparse modern coral growth but which appear to have been drowned and to have accreted slowly if at all. The drowned banks and atolls, such as the Chagos Bank (Stoddart, 1971a), and Saya de Malha (Guilcher, 1988), are examples of probable give-up reefs, though less work has been done on these than on equivalent structures in the Caribbean. A second type of give-up reef found in the Indo-Pacific region is a truncated reef which has given up because it has been exposed by a relative fall in sea level (emergence). Areas of such give-up reef, often with scattered flat-topped intertidal corals, termed microatolls, which emerge at lowest tide, characterise many reef flats in the region and are described in more detail below.

Holocene reef development in the Cocos (Keeling) Islands

Fig. 7.6 shows details of stratigraphy and radiocarbon chronology of three transects from a series of drillhole transects across reef flat, and reef islands on Cocos (from Woodroffe *et al.*, 1994). Recovery in drillholes is generally poor, but is dominated by shingle clasts of branching acroporid corals apparently in a sandy matrix. Except immediately beneath the reef flat, or cemented conglomerate platform, this coral shingle is rarely lithified. The oldest dates obtained are around 7000 years BP at 14 m below sea level in boreholes on Home Island, recording reef establishment over the Pleistocene surface after flooding by the rapidly rising postglacial sea level. Similar dates (around 6800 years BP) have been obtained at shallower depths (around 9 m) in a drillhole on an island in the southern passage. At the northern end of the atoll, however, dates at similar depths in drillholes on Horsburgh Island yielded ages of 4600–5500 years BP, suggesting later establishment of reefs at this end of the atoll (profile X, Fig. 7.6).

Vertical reef accretion appears to have been rapid. Ages do not necessarily record *in situ* reef growth, as some of the reef fabric could have accumulated

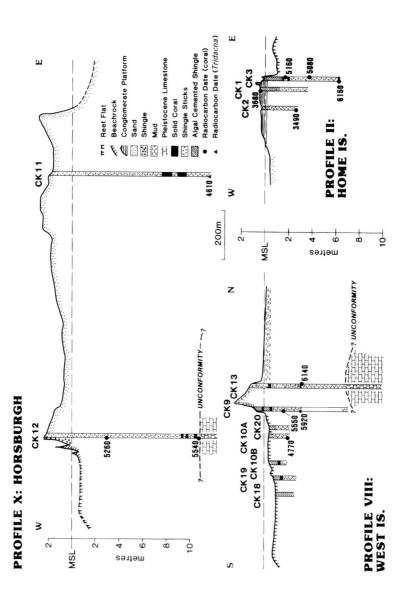

Figure 7.6. Transects of boreholes on Cocos, showing stratigraphy and radiocarbon chronology. Location of transects can be seen in Fig. 7.3a.

as detrital material. Nevertheless, dates in CK3 and CK12 suggest average rates of accretion of 4–40 mm a^{-1}. In Fig. 7.7 radiocarbon ages from a series of drillholes and from surface exposures of conglomerate platform are shown on an age–depth plot (updated from Woodroffe, McLean & Wallensky, 1990b). Several dates indicate that reefs were 2–3 m below present sea level around 6000 years ago, though the sea appears to have reached present level by this time elsewhere in the Indo-Pacific region. This disparity might either result from a different hydro-isostatic response for mid-ocean sites to those in continental settings (Chappell, 1974), or more probably might reflect reef growth lagging behind the rapidly rising sea (Neumann & Macintyre, 1985) and only later catching up when sea level had stabilised.

Data on shallow subsurface stratigraphy and radiometric chronology are now available from several Pacific atolls. Fig. 7.7 also summarises age–depth

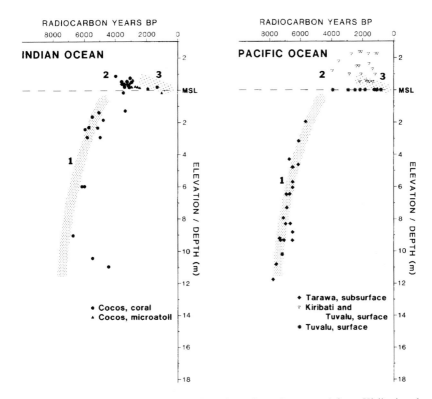

Figure 7.7. Age-depth plot of radiocarbon dates from Cocos, and from Kiribati and Tuvalu. Dates are on coral (or in some instances on *Tridacna*). Only those labelled as microatolls are known to be in growth position. Three phases of Holocene development, numbered 1–3, can be identified; see text for details.

relationships of radiocarbon dates on coral and *Tridacna* samples from Tarawa, Kiribati, (Marshall & Jacobson, 1985), together with surface samples from Tuvalu and Kiribati (Schofield, 1977a; McLean & Hosking, 1991) which are discussed later. In Tarawa, drilling undertaken on reef islands on the atoll rim several hundred metres from the reef crest indicates that Holocene reefs established over the former atoll surface between 7000 and 8000 years ago, and there was then a period of rapid vertical growth of reefs, catching up with the rapidly rising sea.

Surface features of Indian and Pacific Ocean atolls do not appear to record the attainment of modern sea level at around 6000 years BP, as is generally observed for sea-level curves from the region (Pirazzoli, 1991) and at several sites close to the atolls under consideration (i.e. Fiji and Tonga, Miyata et al., 1990; Nunn, 1991, in the Pacific; Sri Lanka, Katupotha, 1988, in the Indian Ocean). Instead, throughout Pacific atolls there is evidence to indicate that reefs on many atolls were at or above present level 4500–4000 years BP (Hopley, 1987; Pirazzoli & Montaggioni, 1988a, 1988b), with dates from Makatea, Mataiva and Takapoto indicating that in French Polynesia they may have grown to above present sea level by 5500 years BP. We believe that the first phase of Holocene reef development on atolls was characterised by catch-up reef growth.

Three stage model of development of the Cocos Islands

Based on the data in Fig. 7.7, a three-stage model can be proposed for the Holocene evolution of the Cocos Islands (Woodroffe et al., 1990a). The first phase from about 8000 to 4500 years BP was a phase of rapid vertical reef growth as the reefs strived to 'catch-up' with a rapidly rising sea level (Neumann & Macintyre, 1985). The radiocarbon-dated coral samples portrayed in Fig. 7.7 do not directly indicate the position of sea level, except for the microatoll samples shown from Cocos, because corals can grow in a range of water depths (Davies & Montaggioni, 1985). Nevertheless, there is some consistency of dates from around the atoll rim at Cocos. Dates from a couple of cores do lag significantly behind the general pattern, but these were from cores on the lagoonward side of a leeward reef (i.e. those shown in transect X, Fig. 7.6).

The second phase from about 4500 to 3000 years BP was a phase of reef flat formation as reefs caught up with sea level and consolidated. This former reef flat is indicated by the presence of fossil corals, particularly specimens of massive and branching *Porites* in growth position, which underlie a conglomerate platform of cemented coral boulders, which is widespread around the island (Fig. 7.8*a*). *In situ Porites* microatolls (Fig. 7.8*b*), radiocarbon-dated at about 3000 years BP, growing up to 80 cm above their modern living counterparts, indicate that the sea level was higher and that there has been late Holocene emergence (Woodroffe *et al.*, 1990b). This represents a give-up reef. The boulder conglomerate appears in many places on Cocos to have developed as a reef flat deposit under higher sea level, though on other atolls the origin of similar boulder conglomerates is more controversial and is discussed below.

The third phase, perhaps starting around 3500 years ago and continuing to the present, was a phase of reef island formation. It is clear that islands are geologically very young. Radiocarbon dating of shingle from within the sands of West Island on Cocos indicates that the core of the island was formed around 3500–3000 years BP overlying the conglomerate platform, and that there has been gradual accumulation of island sediments since then, with the extension of a series of recurved spits flanking the larger interisland passages. Guppy (1889) speculated on how the reef islands might have formed, putting particular island shapes into an evolutionary sequence. There are insufficient dates on other reef islands around the atoll rim, however, to substantiate his theories. It has been suggested by several researchers that reef islands on other atolls may have formed as a result of a net fall in sea level in the last few thousand years (Gardiner, 1903; Kuenen, 1933; Schofield, 1977a). It remains unclear how important changes in sea level are to patterns of island accumulation, but reef islands are now recognised as being very dynamic, and are examined below.

Lagoonal infill occurred throughout the three stages of Holocene reef development, but it changed in nature. During the first phase, as reefs were catching up with sea level, we infer that the lagoon was a very open environment characterised by uninhibited water exchange, and probably covered with flourishing coral cover. During the second, phase the catch-up of the atoll rim and the development of reef flats decreased the exchange of water and appears to have permitted extensive branching coral growth in the relatively sheltered lagoon. Coral from vibrocores in the southern lagoon indicate abundant acroporid corals around 3000 years BP. During the third phase, reef island formation has meant that the input of sediment into the lagoon from ocean reef and reef flats is restricted to interisland passages, and quiet water environments and enclosed lagoonlets (termed teloks on Cocos, and barachois elsewhere) have developed in which finer grained sediments accumulate (Smithers *et al.*, 1994).

The foregoing three-stage model relating sea-level change to reef growth, reef flat formation and island formation appears applicable to many atolls. In Fig. 7.7 the same interpretation may be applied to atolls in Kiribati; the catch-

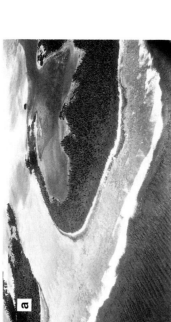

Figure 7.8. (a) An oblique aerial view of Pulu Pandan on the eastern rim of Cocos. The island sits upon a conglomerate platform which can be seen outcropping along the eastern shore. (b) Microatolls of massive *Porites*, above the modern upper limit to growth of coral and dating around 3000 years BP. These corals indicate that at that time the sea was around 80 cm above present relative to Cocos. (c) Emergent reef of the blue coral, *Heliopora*, along the southern shore of Bairiki, Tarawa, Kiribati. This fossil coral underlies the conglomerate platform (background), and is evidence of a Holocene highstand of sea level.

up phase of reef growth is clear from dating boreholes on Tarawa (Marshall & Jacobson, 1985), reef-flat development may be indicated by the extensive areas of fossil *Heliopora* reef exposed on modern reef flats of southern Tarawa (see Fig. 7.8*c*), and reef islands would seem to have developed after the fossil reef and conglomerates had formed.

However, the rates of upward reef growth can also vary between (and within) atolls, and this variation is likely to have ramifications for the absolute timing of stages 2 and 3. In fact, lags in all three stages can be expected. For instance, on Funafuti there is no indisputable evidence of emergence in the form of *in situ* emergent corals. Though fossil corals were reported to imply a higher sea level in the past by David & Sweet (1904), and several Holocene oscillations in sea level were inferred by Schofield (1977b), it has not been proven that any of these are beyond the reach of modern processes, (McLean & Hosking, 1991). Radiometric dates from the modern reef flats both on the ocean and lagoon sides of the atoll and from the fossil reef flat (at the same elevation) beneath reef islands indicate that the major phase of reef flat development took place 2000–3000 years BP on this atoll (McLean & Hosking, 1991). Late catch-up is implied, the reef reaching sea level at a time when elsewhere sea level may have been descending from its higher position or was approaching its present level. A corollary of this is that the main phase of island building on Funafuti also lagged behind, and this is confirmed by the fact that all of the 21 samples from island deposits around the atoll so far radiocarbon dated have ages less than 2000 years BP.

Surface morphology

The reef rim of atolls, from seaward reef crest to the first significant break of slope into the lagoon, frequently comprises three fundamental units: oceanside reef flat, reef island (cay or motu), and lagoonside reef flat or sand flat. Sollas (1904) suggested that on Funafuti the 'belt of land which separates the ocean from the lagoon presents much the same character as in the case of Keeling Atoll, described by Darwin' and he presented two diagrams (Fig. 7.9) as being 'sufficient to show this'. Our surveys and observations indicate that the margins of these two atolls have some important differences which were not highlighted by Sollas but are of significance when considering the recent evolution of the reefs and particularly the islands.

First, the width of the atoll rim is generally greater on Cocos, at least where there is a wide reef flat at the southern end. Similarly, reef islands are much wider on Cocos, where islands frequently are more than 300 m wide, than on Funafuti, where they are rarely more than 100 m wide (see Fig. 7.2).

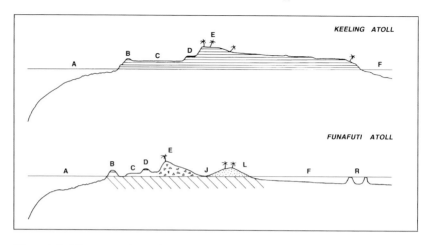

Figure 7.9. Diagrammatic sections through the rim of Keeling (Cocos) and Funafuti atolls, redrawn with double vertical exaggeration from Sollas (1904, Fig. 4). The upper diagram was originally in Darwin (1842) and reproduced by Sollas unmodified. A. Constantly submerged portions of reef. B. Nullipore (algal) rim. C. Reef flat of coral rock. D. Ledge of coral rock (conglomerate platform, Cocos; breccia sheet, Funafuti). E. Seaward or outer ridge (hurricane bank, Funafuti). F. Floor of lagoon. J. Central flat of island. L. Lagoon mound. R. Growing reefs of lagoon

Second, the 'ledge of coral rock' (D on Fig. 7.9) has quite different characteristics. The conglomerate platform on Cocos is a flat surface, with few coral clasts of more than 50 cm diameter, and with *in situ* corals representing a former reef flat formed around 3000 years BP. In the case of Funafuti, it is a narrow breccia sheet of loosely cemented coral rubble which does not appear to extend far beneath the islands. This breccia is interpreted as the base of an early island beach formed around 1500 years BP.

Third, on Funafuti reef flats there are large storm-tossed blocks and extensive boulder tracts on the eastern and western sides of the atoll respectively. Such coarse storm-generated deposits are virtually absent on Cocos. A similar difference applies to the islands also, with coral gravel and boulders the dominant island material on motu in Funafuti (excepting sandy cays to the north of the atoll), while on Cocos islands are predominantly sandy, with rubble on the modern beach but rare elsewhere.

Fourth, on Funafuti, the lagoon shore drops steeply from solid reef flat into the lagoon with a distinct break of slope 3–4 m high. On Cocos there is a broad sand flat flanking the southern islands, with sand spilling into and filling a complex of lagoonal blue holes. Beachrock outcrops are found on the steeper lagoon beaches of islands on Funafuti, but beachrock is rare in Cocos.

Finally, although the reef rims of Cocos and Funafuti display similar gross characteristics, in detail their surface morphologies are quite different and such contrasts are common on atolls elsewhere around the world.

Synthesis

A pattern of atoll development over an interglacial–glacial–interglacial cycle, the last 125 000 years, is shown in Fig. 7.10. Such a pattern is presumed to have also occurred in a similar manner over previous sea-level cycles. The sea-level curve shown for that period is based upon oxygen isotope analysis of foraminifera from deep-sea cores, and correlation with sequences of raised reef terraces on uplifted shorelines (Chappell & Shackleton, 1986). The response of an atoll is shown schematically. When sea level was high, as during the Last Interglacial (marine oxygen isotope substage 5e), an atoll similar to that presently found would have existed. The extent to which the lagoon was infilled and to which the rim contained reef islands probably varied from atoll to atoll, as it does between atolls today, though as the Last Interglacial appears to have contained a period of around 12 000 years of relatively unchanging sea level (Lambeck & Nakada, 1992), many atolls probably filled in to a far greater extent than they have been able to during the last 6000 years of relatively stable sea level. During interstadials (such as oxygen isotope substages 5c and 5a) shallow lagoonal sediments and peripheral reefs were presumably exposed. This seems to have been associated with phases of aeolianite building in marginal reef areas such as Bermuda (Hearty, Vacher & Mitterer, 1992), but has not yet been recognised in the subsurface stratigraphy of atolls.

As the sea became progressively lower, the atoll would have been exposed as an emergent limestone island, such as present day Niue. During this emergence, it underwent solution giving rise to a highly eroded karst surface, with karst and speleothem formation (Li *et al.*, 1989; Lundberg *et al.*, 1990) being possible at depths of tens of metres below present sea level, at the peak of the last glaciation (18 000 radiocarbon years BP, corresponding to around 21 000 years BP determined by uranium-series dating, Bard *et al.*, 1990). Solutional features, such as blue holes, were formed, or enlarged during this emergence, and internal drainage is likely to have concentrated solution into the lagoon (Purdy, 1974). As the sea rose so the three phases of Holocene reef development were experienced (indicated by 3, 4 and 5 in Fig. 7.10), with inundation of the last interglacial platform occurring before sea level reached its present level because the atoll had subsided.

While this model may apply to Cocos, it is not equally applicable to all atolls, and we consider below some of the factors which may lead to a slightly

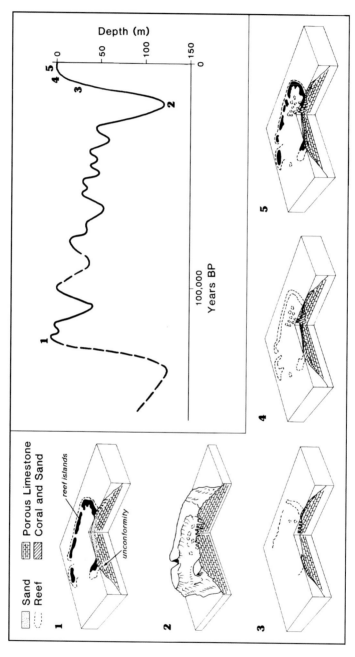

Figure 7.10. A summary of the late Quaternary evolution of an atoll such as Cocos. The sea-level curve for the last 140 000 years is from Chappell & Shackleton (1986). Stage 1 is the Last Interglacial with an atoll like that at present. Stage 2 is the peak of the glacial. the previous unconsolidated lagoonal and reefal sediments are lithified or eroded, and the island is highly karstified. Stages 3, 4 and 5 correspond to 1, 2 and 3 of Fig. 7.7, and are described in the text.

different atoll history. On some atolls the Last Interglacial surface is exposed at the surface on the modern atoll, such as in the drier Line (i.e. Christmas Island, unpublished results) and Phoenix Islands of eastern Kiribati (Tracey, 1972, and pers. commun., 1992), Anaa atoll in the Tuamotu Archipelago (Veeh, 1966; Pirazzoli *et al.*, 1988a), and Aldabra Atoll in the western Indian Ocean (Braithwaite, Taylor & Kennedy, 1973). On these atolls there might be either no subsidence, or subsidence might be countered by lithospheric flexure (see Chapter 12). Last Interglacial reefs occur at 7–14 m beneath present sea level on Enewetak (Tracey & Ladd, 1974; Szabo, Tracey & Goter, 1985), at 6–11 m beneath Mururoa Atoll in the Tuamotus (Labeyrie, *et al.*, 1969; Trichet, Repellin & Oustrière, 1984; Perrin, 1990), at 8–17 m on Tarawa in Kiribati (Marshall & Jacobson, 1985), and at 7–22 m beneath the lagoon on Aitutaki, Pukapuka and Rakahanga in the Cook Islands (Gray *et al.*, 1992). Some lowering of the previous atoll surface can undoubtedly be attributed to solution, which is likely to have been least on the drier atolls (Menard, 1982). However, if Last Interglacial reefs were all at sea level (and some may have lagged behind as catch-up reefs), the occurrence of the Thurber discontinuity at different depths indicates either flexure of the crust and upper mantle (Lambeck & Nakada, 1992) or differential subsidence (see Chapter 12). Differential vertical movements appear particularly likely for the Cook Islands, where Last Interglacial limestones are emergent on the makatea islands and range in elevation from 10–20 m above sea level (Woodroffe *et al.*, 1991b), but occur at 7–22 m below sea level on atolls (Gray *et al.*, 1992). Although there are karst features developed on the surface of the emergent makatea reefs (Stoddart, Woodroffe & Spencer, 1990), differences in the elevation of the Last Interglacial surface between islands in a similar climatic setting (i.e. Aitutaki and Atiu, which are <200 km apart), must result from vertical movements and cannot be entirely ascribed to solution.

Holocene reef growth rates are known to differ substantially from place to place (Hopley, 1982; Buddmeier & Smith, 1988). Individual reefs may adopt a keep-up, catch-up or give-up strategy, and there may be considerable variations from one part of a reef to another. As a result the time at which a reef reaches sea level also varies. Whereas in Cocos that time seems to have been around 3500 years BP, in other parts of the Pacific it may be around 4500 years BP (Hopley, 1987), or, as in parts of French Polynesia, around 5500 years BP (Pirazzoli *et al.*, 1988a, 1988b). On Funafuti, we have suggested that it may have been as recently as 2000 years BP. In the Maldives there is also some evidence for reef flat development around modern level 3000 years BP, but evidence of higher sea level is ambiguous (Woodroffe, in press). It is tempting to speculate that this late time of catch-up is linked to a deep Last Interglacial surface from which vertical Holocene reef growth commenced; perhaps this might be a consequence of more rapid subsidence. However, there would need to be far more drilling and dating results from Funafuti before this could be substantiated.

Variations in the timing of reef catch-up to sea level may also explain the variety of surface rubble conglomerate deposits on atolls. Boulder deposits, cemented into a conglomerate platform, on tropical islands have been interpreted either as evidence of a sea level higher than present (Daly, 1934 ; Newell, 1961), or alternatively as a result of storm-wave action (Shepard *et al.*, 1967; Newell & Bloom, 1970). In addition, observations of fossil coral reefs at higher elevations than their modern, living counterparts appeared to indicate that the sea stood 1–2 m above its present level with respect to many of the coral atolls of the Pacific and Indian Oceans (Stearns, 1945; Cloud, 1952), and that it has subsequently fallen relative to those islands. Cemented coral conglomerates on the reef flats and islands of atolls and above the present limit to coral growth have been radiometrically dated on many islands to about 4000–3000 years BP, e.g. from the Marshall Islands (Tracey & Ladd, 1974 ; Buddemeier, Smith & Kinzie, 1975), Kiribati and Tuvalu (Tracey, 1972; Schofield, 1977a, 1977b; Valencia, 1977; Guinther, 1978; McLean & Hosking, 1991), Fiji (Ash, 1987; Miyata *et al.*, 1990; Shepherd, 1990; Nunn, 1991), and the Cook Islands (Scoffin *et al.*, 1985; Yonekura *et al.*, 1988; Woodroffe *et al.*, 1990c). The evidence has been examined in great detail through French Polynesia, where the pattern of higher sea level from atolls in the Tuamotu Archipelago can be demonstrated at around +0.8 ± 0.2 m from prior to 4000 years BP to at least 1500 years BP (Montaggioni & Pirazzoli, 1984; Pirazzoli & Montaggioni, 1986, 1988a, 1988b; Pirazzoli *et al.*, 1987a, 1987b, 1988a). Similar boulder conglomerates have been described from atolls in the Indian Ocean, and again have been inferred to indicate a sea level higher than present in the Maldives (Gardiner, 1903; Sewell, 1935, 1936a, 1936b; Stoddart, Spencer Davies & Keith, 1966), and Diego Garcia (Stoddart, 1971b).

Where extensive conglomerate platform occurs as a result of fossil reef flat becoming emergent, then it forms an anchor upon which reef islands may develop. That combined with the relatively long period of potential accumulation may explain the larger islands in Cocos than in atolls such as Funafuti which have had reef islands forming for a shorter time. In addition, there are different processes operative.

There have been few studies of when reef islands began to form, or of their chronology of development (Stoddart, 1969). From the northern Great Barrier Reef, radiocarbon dates on shingle clasts indicate that similar reef islands formed a core at least 4000 years ago, with a major phase of island building forming an upper terrace on shingle islands, sand cays and composite islands 3500–3000 years BP, and some further building 1500 years BP (McLean & Stoddart, 1978; McLean *et al.*, 1978; Stoddart, McLean & Hopley, 1978). On the other hand, a study based upon extensive radiocarbon dating on an island in the southern Great Barrier Reef, has demonstrated that shingle islands appear to have accreted uniformly over the last few thousand years (Chivas *et al.*, 1986).

However, it is clear that not all atolls have experienced similar storm and sea-level histories to islands on the Great Barrier Reef, and that given the generally larger tidal range, sediment movement processes may be different on the Great Barrier Reef to those experienced on atolls. Little is known about the pattern of sediment accretion on atoll reef islands over the ~3500 years over which there has been a substrate upon which islands could form. At least three different possibilties exist: first, islands may have accumulated uniformly over that time, and thus be continuing to build up at a similar gradual rate (as demonstrated in the case of the study of Chivas *et al.*, 1986); second, islands may have formed initially around 3500 years ago as conditions were first favourable for supratidal accumulation, and then have received progressively less sediment with time, or have been subject to less storm activity as a result of reef crest growth dissipating the energy from the open sea (analogous to the Holocene high-energy window suggested for the Great Barrier Reef by Hopley, 1982, 1984; see also Chapter 8); third, islands may have accreted during one or more episodes within that period (as suggested for the northern Great Barrier Reef islands, McLean *et al.*, 1978). It would be extremely useful to know which, if any, of these simplified models of island growth is appropriate, as it would give an insight into modern and future island dynamics.

It is important to stress that reef islands are naturally dynamic, and that sediment production occurs around reef islands, and that erosion, deposition and cementation can occur concurrently on atolls today (Wiens, 1962). Some islands may be in a stable equilibrium with neither addition nor loss of sediment. However, on most islands, sediment is added and lost over time, and there is more likely to be a dynamic equilibrium between inputs and outputs. Islands adjust over a range of timescales. They may adjust to seasonal changes; thus in the Maldives the seasonal reversal of the prevailing monsoon can lead to sediment accumulation at one end of an island for a part of the year, and its redistribution to the other end for the rest of the year. In Kiribati, during the 1983 El Niño, mean monthly sea level rose 40 cm. Shoreline erosion was detected in places on Tarawa, but was not especially rapid or devastating (Howorth & Radke, 1991).

By contrast, one severe tropical cyclone can totally devastate an entire atoll, destroying vegetation and reducing (or increasing) island area. Catastrophic storms play an important role in both construction and destruction in those areas which experience hurricanes over a longer timescale (Stoddart, 1971c; Bourrouilh-Le Jan & Talandier, 1985), though the effect of these episodic but high-magnitude events is also likely to differ between islands. Rubble is usually an important constituent of islands in the storm belt, a legacy of past storms. Sand on the other hand tends to be stripped off islands by storms, but to be moved back onto islands by the more regular processes that operate between storms (see Fig. 1.1). Sand and shingle motus, typical of high-energy settings of atolls which experience storms, are built up in part by rubble-sized material moved by storms onto the reef flat. The impact of Hurricane Bebe on the atoll of Funafuti, Tuvalu has been studied in particular detail (Maragos, Baines & Beveridge, 1973). A rubble rampart, composed of corals ripped from off the reef front, was thrown onto the reef flat or onto the elongate reef islands, adding about 10% to the total land area of the atoll. Regular less severe storms have broken down and redistributed the storm rubble (Baines, Beveridge & Maragos, 1974; Baines & McLean, 1976a, 1976b).

Not all islands are affected equally by storms and the morphology of islands differs accordingly (McLean, 1980). Fig. 7.11 illustrates schematically some of the features of the morphology of islands in areas of different storm frequency and intensity. Strong tradewinds outside hurricane belts can also serve to modify islands, as in Kiribati or the Maldives where storms of hurricane force are not frequent. Where storms are frequent, reef flats generally contain rubble ramparts or degraded rubble deposits on the motus on the more-exposed side of the atoll. There is often a well-developed algal ridge (though this may not be related to storm frequency), and conglomerate platforms are prominent and may extend across much of the reef flat, underlying entire islands (i.e. northern Cook Islands); sand cays are found in the less-exposed areas. Where storms are not as frequent or as severe, there are less-extensive rubble deposits; the algal ridge is less prominent and the conglomerate platform is not as extensive across the reef flat (i.e. Cocos Islands; some islands in Kiribati). In storm-free, low-energy areas, rubble is not a major component in island construction; instead sand cays are found even on the outer atoll rim (i.e. many islands in Kiribati, Maldives).

Fig. 7.11 also depicts a series of schematic responses of island form to processes (modified from Bayliss-Smith, 1988). Storms result in loss of sand from cays and motus, but an input of rubble to the motus in the form of ramparts. The motus which receive an input of rubble adjust with the medium-term

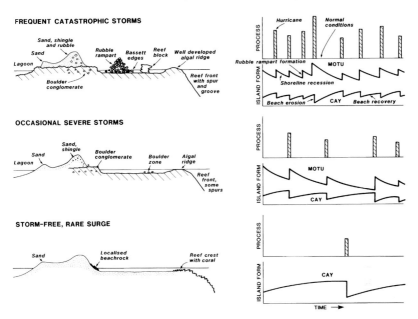

Figure 7.11. The morphology of reefs and reef islands in areas of different storm occurrence, and the response of form to process. Where there are frequent storms there is usually abundant coral shingle and boulders, and there may be rubble ramparts and reef blocks. Boulders are often cemented into a boulder conglomerate. Where there are only occasional storms of less severity, there are generally less-extensive boulder deposits on the reef flat, and where there are only rare storms, islands are predominantly composed of sand (after Woodroffe, 1994). Storm events are indicated by bars in the process diagram, and storms generally add material to shingle motu, but erode sand cays (following Bayliss-Smith, 1988). Rapid change during storms is followed by redistribution of coarse material on motus, and slow recovery of beaches on cays between storms.

breakdown and redistribution of that material as they return to equilibrium. Cays lose sand during storms, but are rebuilt towards equilibrium by beach recovery through normal processes. Relaxation time (time taken to readjust between high energy events) and recurrence interval (frequency of such events) are important in controlling reef island morphology. When storms are very frequent (or very severe) motus and cays may be in disequilibrium for most of the time. When storms are occasional, complete recovery is possible between storms and islands may be in dynamic equilibrium. Where there are no storms cays should reach a stable equilibrium; a rare disturbing event such as a storm surge can cause devastation on these islands, and such catastrophes will require a long time for recovery (Fig. 7.11).

Conclusion

In this chapter the response of atolls to large-scale movements of sea level during the Quaternary has been discussed first, followed by an examination of how atoll reefs adjusted to sea-level rise during the Holocene and how reef islands were subsequently formed. Atoll structure appears relatively similar between atolls and can be explained as a function of long-term geological processes, upon which large-scale sea-level changes (100–150 m amplitude) have been superimposed throughout the Quaternary. Surface morphology of atoll reefs, reef islands, and lagoons, on the other hand, varies considerably within and between atolls, and reflects finer-scale sea level (0–5 m amplitude) and other environmental changes as well as contemporary processes.

There is an intimate relationship between coral atolls, reef islands and sea level, which has been the focus of studies of two types. First, there have been studies on the *determination* of sea-level histories based upon data from atolls which serve as 'dipsticks' in mid-oceanic locations (i.e. Bloom, 1967). Such data contribute to the *record* of sea-level change itself and have also been used to test and constrain geophysical models of the response of a viscoelastic earth to changes in ice and water loads (i.e. Clark *et al.*, 1978; Peltier, 1988). Second, there have been studies of how atoll environments have *responded* to sea-level changes. For coral atolls adjustments (responses) have been associated with major shifts in sea level operating over long periods of geological time (10^5 years), while for reef islands adjustments have been associated with much more recent and subtle movements of sea level (10^3 years). Although other physical and biological processes are also involved, it is clear that the development of both coral atolls and reef islands cannot be understood without an assessment of the role of sea-level change.

Recognition of the sensitivity of atoll environments and reef islands to variations in sea level has resulted in widespread concern that future sea-level rise, as a consequence of global warming (Houghton, Jenkins & Ephraums, 1990; Houghton, Callander & Varney, 1992), will have devastating effects. Coral atolls, including several entire nations (e.g. Kiribati, Tuvalu, Maldives), appear particularly vulnerable (Commonwealth Secretariat, 1989; Lewis, 1990); indeed, the total disappearance of atoll reef islands has been foreshadowed under the more extreme predictions (Roy & Connell, 1989, 1990, 1991), along with the imminent displacement of the populations of atoll nations (Connell & Roy, 1989). The challenge for atoll geomorphology is to develop robust models of the natural morphodynamics of reefs, reef flats and particularly reef islands, which can then be used to assess the impact of changes in boundary conditions, whether those be associated with sea-level rise, storm incidence, or anthropogenic modification (usually degradation) of reef ecosystems.

References

Adey, W.H. (1978). Coral reef morphogenesis: a multidimensional model. *Science,* **202**, 831–7.

Ash, J. (1987). Holocene sea levels in northern Viti Levu, Fiji. *New Zealand Journal of Geology and Geophysics,* **30**, 431–5.

Baines, G.B.K., Beveridge, P.J. & Maragos J.E. (1974). Storms and island building at Funafuti Atoll, Ellice Islands. In *Proceedings Second International Coral Reef Symposium, Brisbane,* vol. 2, pp. 485–96.

Baines, G.B.K. & McLean, R.F. (1976a). Resurveys of 1972 hurricane rampart on Funafuti atoll. *Search,* **7**, 36–7.

Baines, G.B.K. & McLean, R.F. (1976b) Sequential studies of hurricane bank evolution at Funafuti atoll. *Marine Geology,* **21**, M1–M8.

Bard, E., Hamelin, B., Fairbanks, R.G. & Zindler, A. (1990). Calibration of the ^{14}C timescale over the past 30 000 years using mass spectrometric U–Th ages from Barbados corals. *Nature,* **345**, 405–10.

Bayliss–Smith, T.P. (1988). The role of hurricanes in the development of reef islands, Ontong Java Atoll, Solomon Islands. *Geographical Journal,* **154**, 377–91.

Bezrukov, P.L. (1973). Principal scientific results of the 54th cruise of the R.V. Vitian in the Indian and Pacific Oceans (Feb–May 1973). *Oceanology,* **13**, 761–6.

Bloom, A.L. (1967). Pleistocene shorelines: a new test of isostasy. *Geological Society of America Bulletin,* **78**, 1477–94.

Bourrouilh–Le Jan, F. & Talandier, J. (1985). Sédimentation et fracturation de haute énergie en milieu récifal: tsunamis, ouragans et cyclones, leurs effets sur la sédimentation et la géometrie d'un atoll. *Marine Geology,* **67**, 263–333.

Braithwaite, C.J.R. (1982). Progress in understanding reef structure. *Progress in Physical Geography,* **6**, 505–23.

Braithwaite, C.J.R., Taylor, J.D. & Kennedy, W.J. (1973). The evolution of an atoll: the depositional and erosional history of Aldabra. *Philosophical Transactions of the Royal Society of London, Ser. B,* **266**, 307–40.

Buddemeier, R.W. & Smith, S.V. (1988). Coral reef growth in an era of rapidly rising sea level: predictions and suggestions for long-term research. *Coral Reefs,* **7**, 51–6.

Buddemeier, R.W., Smith, S.V. & Kinzie, R.A. (1975). Holocene windward reef–flat history, Enewetak Atoll. *Geological Society of America Bulletin,* **86**, 1581–4.

Chamberlain, N.G. (1960). *Cocos Island magnetic survey, 1946.* Bureau of Mineral Resources, Record, 1960/124.

Chappell, J. (1974). Late Quaternary glacio– and hydro–isostasy on a layered earth. *Quaternary Research,* **4**, 429–40.

Chappell, J. & Polach, H. (1991). Post glacial sea level rise from a coral record at Huon Peninsula, Papua New Guinea. *Nature,* **349**, 147–9.

Chappell, J. & Shackleton, N.J. (1986). Oxygen isotopes and sea level. *Nature,* **324**, 137–40.

Chivas, A., Chappell, J., Polach, H., Pillans, B. & Flood, P. (1986). Radiocarbon evidence for the timing and rate of island development, beach–rock formation and phosphatization at Lady Elliot Island, Queensland, Australia. *Marine Geology,* **69**, 273–87.

Clark, J.A., Farrell, W.E. & Peltier, W.R. (1978). Global change in post glacial sea level: a numerical calculation. *Quaternary Research,* **9**, 265–87.

Cloud, P.E. (1952). Preliminary report on geology and marine environments of Onotoa Atoll, Gilbert Islands. *Atoll Research Bulletin,* **12**, 1–73.

Collins, L., Zhu, Z.R., Wyrwoll, K.H., Hatcher, B.G., Playford, P., Chen, J. H., Eisenhauser, A. & Wasserburg, G.J. (1993). Late Quaternary facies characteristics and growth history of a high latitude reef complex: the Abrolhos carbonate platforms, eastern Indian Ocean. *Marine Geology*, **111**, 203–12.

Commonwealth Secretariat (1989). *Climate change: meeting the challenge*. Report of a Commonwealth group of experts, London, September 1989.

Connell, J. & Roy, P. (1989). The greenhouse effect: the impact of sea level rise on low coral islands in the South Pacific. In *Studies and reviews of greenhouse related climatic change impacts on the Pacific Islands*, ed. J.C. Pernetta & P.J. Hughes, pp. 106–33. Majuro: Association of South Pacific Environmental Institutions.

Cullis, C.G. (1904). The mineralogical changes observed in the cores of the Funafuti borings. In *The Atoll of Funafuti: borings into a coral reef and the results*, pp. 392–420. Royal Society, London.

Daly, R.A. (1915). The glacial–control theory of coral reefs, *Proceedings of the American Academy of the Arts and Science*, **51**, 155–251.

Daly, R.A. (1925). Pleistocene changes of level. *American Journal of Science, Series 5*, **10**, 281–313.

Daly, R.A. (1934). *The changing world of the Ice Age*. New Haven: Yale University Press. 271 pp.

Darwin, C. (1842). *The structure and distribution of coral reefs*. London: Smith, Elder & Co.

Darwin, C. (1859). *On the Origin of Species*. London: Smith, Elder & Co.

David, T.W.E. & Sweet, G. (1904). The geology of Funafuti. In *The Atoll of Funafuti: borings into a coral reef and the results*, pp. 61–124. Royal Society, London.

Davies, P.J. & Montaggioni, L.F. (1985). Reef growth and sea-level change: the environmental signature. In *Proceedings Fifth International Coral Reef Congress, Tahiti*, vol. 3, pp. 477–515.

Davis, W.M. (1928). *The coral reef problem*. American Geographical Society, Special Publication, 9. 596 pp.

Eisenhauser, A., Wasserburg, G.J., Chen, J.H., Bonani, G., Collins, L.B., Zhu, Z.R. & Wyrwoll, K.H. (1993). Holocene sea-level determination relative to the Australian continent: U/Th (TIMS) and ^{14}C (AMS) dating of coral cores from the Abrolhos Islands. *Earth and Planetary Science Letters*, **114**, 529–47.

Emery, K.O., Tracey, J.I., Jr. & Ladd, H.S. (1954). *Geology of Bikini and nearby atolls*. United States Geological Survey, Professional Paper, 260 A. 265 pp.

Fairbanks, R.G. (1989). A 17,000-year glacio-eustatic sea level record: influence of glacial melting rates on the Younger Dryas event and deep-ocean circulation. *Nature*, **342**, 637–42.

Finlayson, D.M. (1970). *First-order regional magnetic survey at Cocos Island, Southern Cross and Augusta*. Bureau of Mineral Resources, Record, 1970/101.

Gardiner, J.S., ed. (1903). *The fauna and geography of the Maldive and Laccadive Archipelagoes, being an account of the work carried on and of collections made by an expedition during years 1899 and 1900*. Cambridge University Press (2 volumes).

Gaskell, T.F. & Swallow, J.C. (1953). Seismic experiments on two Pacific atolls. *Occasional Papers, Challenger Society*, **3**, 1–8.

Gibb Australia (1985). *Tuvalu Lagoon Bed Resources Survey*. Report to Australian Development Assistance Bureau, Canberra. 261 pp.

Ginsburg, R.N., Lloyd, R.M., Stockman, K.W. & McCallum, J.S. (1963). Shallow-water carbonate sediment. In *The sea: ideas and observations on progress in the study of the seas*, vol. 3, ed. M.N. Hill, pp. 554–82. New York: Wiley Interscience.

Gray, S.C., Hein, J.R., Hausmann, R. & Radtke, U. (1992). Geochronology and subsurface stratigraphy of Pukapuka and Rakahanga atolls, Cook Islands: late Quaternary reef growth and sea level history. *Palaeogeography, Palaeoclimatology, Palaeoecology*, **91**, 377–94.

Guilcher, A. (1988). *Coral reef geomorphology*. Chichester: Wiley.

Guillou, H., Brousse, R., Gillot, P.Y. & Guille, G. (1993). Geological reconstruction of Fangataufa Atoll, South Pacific. *Marine Geology*, **110**, 377–91.

Guinther, E.B. (1978). Observations on terrestrial surface and subsurface water as related to island morphology at Canton atoll. *Atoll Research Bulletin*, **221**, 171–83.

Guppy, H.B. (1889). The Cocos–Keeling Islands. *Scottish Geographical Magazine*, **5**, 281–97, 457–74, 569–88.

Hanzawa, S. (1938). Studies on the foraminifera fauna found in the bore cores from the deep well in Kita-Daito-Zima (North Borodino Island). *Proceedings of the Imperial Academy, Tokyo*, **14**, 384–90.

Hearty, P.J., Vacher, H.L. & Mitterer, R. (1992). Aminostratigraphy and ages of Pleistocene limestones of Bermuda. *Geological Society of America Bulletin*, **104**, 471–80.

Hopley, D. (1984). The Holocene 'high energy window' on the central Great Barrier Reef. In *Coastal geomorphology in Australia*, ed. B.G. Thom, pp.135–50. Sydney: Academic Press.

Hopley, D. (1987). Holocene sea-level changes in Australasia and the Southern Pacific. In *Sea surface studies: a global review*, ed. R.J.N. Devoy, pp. 375–408. London: Croom Helm.

Houghton, J.T., Callander, B.A. & Varney, S.K., eds. (1992) *Climate change 1992: the supplementary report to the IPCC scientific assessment*. Cambridge University Press.

Houghton, J.T., Jenkins, G.J. & Ephraums, J.J., eds. (1990). *Climate change: the IPCC scientific assessment*. Cambridge University Press.

Howorth, R. & Radke, B. (1991). Investigation of historical evidence of shoreline changes: Betio, Tarawa Atoll, Kiribati, and Fongafale, Funafuti Atoll, Tuvalu. Proceedings workshop on coastal processes in the South Pacific nations, Lae, 1987, *CCOP/SOPAC Technical Bulletin*, **7**, 91–8.

Jongsma, D. (1976). *A review of the geology and geophysics of the Cocos Islands and Cocos Rise*. Bureau of Mineral Resources, Record, 1976/38.

Judd, J.W. (1904). General report on the materials sent from Funafuti and the methods of dealing with them. In *The Atoll of Funafuti: boring into a coral reef and the results*, pp. 167–85. Royal Society, London.

Katupotha, J . (1988). Evidence of high sea level during the mid–Holocene on the southwest coast of Sri Lanka. *Boreas*, **17**, 209–13.

Kuenen, P.H. (1933). Geology of coral reefs. In *Snellius Expedition in the eastern part of the Netherlands East Indies, 1929–30*. Vol. 5. *Geological Results*, No.2. Utrecht: Kemink. 125 pp.

Labeyrie, J., Lalou, C. & Delibrias, G. (1969). Etude des transgressions marines sur l'atoll de Mururoa par la datation des différents niveaux de corail. *Cahiers du Pacifique*, **13**, 59–68.

Ladd, H.S. (1948). Fossil land snails from western Pacific atolls. *Journal of Paleontology*, **32**, 183–98.

Ladd, H.S., Ingerson, E., Tonsend, R.C., Russell, M. & Stephenson, H.K. (1953). Drilling on Eniwetok Atoll, Marshall Islands. *American Association of Petroleum Geologists Bulletin*, **37**, 2257–80.

Ladd, H.S., Tracey, J.I., Jr. & Gross, M.G. (1970). Deep drilling on Midway Atoll. *U.S. Geological Survey Professional Paper,* 680A, A1–22.
Ladd, H.S., Tracey, J.I., Jr. & Lill, G.G. (1948). Drilling on Bikini Atoll, Marshall Islands. *Science,* **107**, 51–5.
Lalou, C., Labeyrie, J. & Delibrias, G. (1966). Datation des calcaires coralliens de l'atoll de Mururoa (Archipel des Tuamotu) de l'epoque actuelle jusqu'a-500,000 ans. *Comptes Rendus Academie des Sciences, Paris,* **263**, 1946–9.
Lambeck, K. & Nakada, M. (1992). Constraints on the age and duration of the last interglacial period and on sea-level variations. *Nature,* **357**, 125–8.
Lewis, J. (1990). The vulnerability of small island states to sea level rise: the need for holistic strategies. *Disasters,* **14**, 241–8.
Li, W–X., Lundberg, J., Dickin, A.P., Ford, D.C., Schwarz, H.P., McNutt, R. & Williams, D. (1989). High precision mass spectrometric uranium–series dating of cave deposits and implications for paleoclimate studies. *Nature,* **339**, 534–6.
Lighty, R.G., Macintyre, I.G. & Stuckenrath, R. (1982). *Acropora palmata* reef framework: a reliable indicator of sea level in the Western Atlantic for the last 10,000 years, *Coral Reefs,* **1**, 125–30.
Lincoln, J.M. & Schlanger, S.O. (1991). Atoll stratigraphy as a record of sea level change: problems and prospects. *Journal of Geophysical Research,* **96**, 6727–52.
Lloyd, J.W., Miles, J.C., Chessman, G.R. & Bugg, S.F. (1980). A ground–water resources study of a Pacific Ocean atoll – Tarawa, Glbert Islands. *Water Research Bulletin,* **16**, 646–53.
Locke, C.A. (1991). Geophysical constraints on the structure of Funafuti. *South Pacific Journal of Natural Sciences,* **11**, 129–40.
Lundberg, J., Ford, D.C., Schwarc, H.P., Dickin, A.P. & Li, W–X. (1990). Dating sea level in caves. *Nature,* **343**, 217–18.
Lyell, C. (1832). *Principles of geology.* London: Murray.
McLean, R.F. (1980). Spatial and temporal variability of external physical controls on small island ecosystems. In *Population–environment relations in tropical islands: the case of eastern Fiji,* ed. H.C. Brookfield, pp. 149–175, Paris: UNESCO.
McLean, R.F. & Hosking, P.L. (1991). Geomorphology of reef islands and atoll motu in Tuvalu. *South Pacific Journal of Natural Science,* **11**, 167–89.
McLean, R.F. & Stoddart, D.R. (1978). Reef island sediments of the northern Great Barrier Reef. *Philosophical Transactions of the Royal Society of London, Ser. A,* **291**, 101–17.
McLean, R.F., Stoddart, D.R., Hopley, D. & Polach, H.A. (1978). Sea level change in the Holocene on the northern Great Barrier Reef. *Philosophical Transactions of the Royal Society of London, Ser. A,* **291**, 167–86.
Maragos, J.E., Baines, G.B.K. & Beveridge, P.J. (1973). Tropical cyclone creates a new land formation on Funafuti atoll. *Science,* **181**, 1161–4.
Marshall, J.F. & Jacobson, G. (1985). Holocene growth of a mid–plate atoll: Tarawa, Kiribati. *Coral Reefs,* **4**, 11–17.
Menard, H.W. (1982). Influence of rainfall upon the morphology and distribution of atolls. In *The Ocean floor,* ed. R.A. Scrutton & M. Talwanii, pp. 305–11. New York: Wiley.
Miyata, T., Maeda, Y., Matsumoto, E., Matsushima, Y., Rodda, P., Sugimura, A & Kayanne, H. (1990). Evidence for a Holocene high sea-level stand, Vanua Levu, Fiji. *Quaternary Research,* **33**, 352–9.
Montaggioni, L.F. & Pirazzoli, P.A. (1984). The significance of exposed coral conglomerates from French Polynesia (Pacific Ocean) as indicator of recent relative sea-level changes. *Coral Reefs,* **3**, 29–42.

Murray, J. (1889). Structure, origin, and distribution of coral reefs and islands. *Nature*, **39**, 424–8.

Nakada, M. (1986). Holocene sea levels in oceanic islands: implications for the rheological structure of the Earth's mantle. *Tectonophysics*, **121**, 263–76.

Nakada, M. & Lambeck, K. (1988). The melting history of the late Pleistocene Antarctic ice sheet. *Nature*, **333**, 36–40.

Neumann, A.C. & Macintyre, I.G. (1985). Reef response to sea level rise: keep–up, catch–up or give–up. In *Proceedings Fifth International Coral Reef Congress, Tahiti*, vol. 3, pp. 105–10.

Newell, N.D. (1961). Recent terraces of tropical limestone shores. *Zeitschrift für Geomorphologie, N.F., Suppl. Bd.*, **3**, 87–106.

Newell, N.D. & Bloom, A.L. (1970). The reef flat and 'two–meter eustatic terrace' of some Pacific atolls. *Geological Society of America Bulletin*, **81**, 1881–94.

Nunn, P. (1991). Sea-level changes during the last 6000 years in Fiji, Tonga and Western Samoa: implications for future coastline development. *SOPAC Technical Bulletin*, **7**, 79–90.

Parsons, B & Sclater, J.G. (1977). An analysis of the variation of ocean floor bathymetry and heat flow with age. *Journal of Geophysical Research*, **82**, 803–27.

Peltier, W.R. (1988). Lithospheric thickness, Antarctic deglaciation history, and ocean basin discretization effects in a global model of postglacial sea level change: a summary of some sources of nonuniqueness. *Quaternary Research*, **29**, 93–112.

Perrin, C. (1990). Genèse de la morphologie des atolls: le cas de Mururoa (Polynésie Française). *Comptes Rendus des Academie de Sciences, Paris*, **311**, II, 671–8.

Pirazzoli, P.A. (1991). *World atlas of Holocene sea-level curves*. Amsterdam: Elsevier.

Pirazzoli, P.A., Delibrias, G., Montaggionin, L.F., Saliège, J.F. & Vergnaud-Grazzini, C. (1987a). Vitesse de croissance latérale des platiers et évolution morphologique récente de l'atoll de Reao, îles Tuamotu, Polynésie française. *Annales de l'institut oceanographique, ns*, **63**, 57–68.

Pirazzoli, P.A., Koba, M., Montaggioni, L.F. & Person, A. (1988a). Anaa (Tuamotu Islands, central Pacific): an incipient rising atoll? *Marine Geology*, **82**, 261–9.

Pirazzoli, P.A. & Montaggioni, L.F. (1986). Late Holocene sea-level changes in the northwest Tuamotu Islands, French Polynesia. *Quaternary Research*, **25**, 350–68.

Pirazzoli, P.A. & Montaggioni, L. (1988a). The 7000 yr sea-level curve in French Polynesia: geodynamic implications for mid–plate volcanic islands. In *Proceedings Sixth International Coral Reef Symposium, Townsville*, vol. 3, pp. 467–72.

Pirazzoli, P.A. & Montaggioni, L.F. (1988b). Holocene sea-level changes in French Polynesia. *Palaeogeography, Palaeoclimatology, Palaeoecology*, **68**, 153–75.

Pirazzoli, P.A., Montaggioni, L.F., Salvat, B. & Faure, G. (1988b). Late Holocene sea level indicators from twelve atolls in the central and eastern Tuamotus (Pacific Ocean). *Coral Reefs*, **7**, 57–68.

Pirazzoli, P.A., Montaggioni, L.F., Vergnaud–Grazzini, C. & Saliège, J.F. (1987b). Late Holocene sea levels and coral reef development in Vahitahi Atoll, eastern Tuamotu Islands, Pacific Ocean. *Marine Geology*, **76**, 105–16.

Purdy, E.G. (1974). Reef configurations, cause and effect. In *Reefs in time and space*, ed. L.F. Laporte, pp. 9–76. Society of Economic Palaeontologists and Mineralogists, Special Publication, 18.

Ross, J.C. (1855). Review of the theory of coral formations set forth by Ch. Darwin in his book entitled: Researches in Geology and Natural History. *Natuurkundig Tijdschrift voor Nederlandsch Indië*, **8**, 1–43.

Roy, P. & Connell, J. (1989). *Greenhouse: impact of sea level rise on low coral islands in the South Pacific.* Research Institute for Asia and the Pacific Occasional Paper 6. 55 pp.

Roy, P. & Connell, J. (1990). Greenhouse effects on atoll islands in the South Pacific. *Geological Society of Australia Symposium Proceedings*, **1**, 57–70.

Roy, P. & Connell, J. (1991). Climatic change and the future of atoll states. *Journal of Coastal Research*, **7**, 1057–75.

Schlanger, S.O. (1963). Subsurface geology of Eniwetok Atoll. *United States Geological Survey, Professional Paper*, **260-BB**, 991–1066.

Schofield, J.C. (1977a). Effect of Late Holocene sea-level fall on atoll development. *New Zealand Journal of Geology and Geophysics*, **20**, 531–6.

Schofield, J.C. (1977b). Late Holocene sea-level, Gilbert and Ellice Islands, west central Pacific Ocean. *New Zealand Journal of Geology and Geophysics*, **20**, 503–29.

Scoffin, T.P., Stoddart, D.R., Tudhope, A.W. & Woodroffe, C.D. (1985). Exposed limestone of Suwarrow Atoll. In *Proceedings Fifth International Coral Reef Congress, Tahiti*, vol. 3, pp. 137–40.

Scott, G.A.J. & Rotondo, G.M. (1983). A model to explain the differences between Pacific plate island atoll types. *Coral Reefs*, **1**, 139–50.

Searle, D.E. (1994) Late Quaternary morphology of the Cocos (Keeling) Islands. *Atoll Research Bulletin*, **401**, 1–13.

Sewell, R.B.S. (1935). Studies on coral and coral formations in Indian waters. *Memoirs of the Royal Asiatic Society of Bengal*, **9**, 461–540.

Sewell, R.B.S. (1936a). An account of Addu Atoll. *John Murray Expedition 1933–1934, Science Reports*, **1**, 63–93.

Sewell, R.B.S. (1936b). An account of Horsburgh or Goifurfehendu Atoll. *John Murray Expedition 1933–1934, Science Reports*, **1**, 109–25.

Shepard, F.P., Curray, J.R., Newman, W.A., Bloom, A.L., Newell, N.D., Tracey, J.I. & Veeh, H.H. (1967). Holocene changes in sea level: evidence in Micronesia. *Science*, **157**, 542–4.

Shepherd, M.J. (1990). The evolution of a moderate energy coast in Holocene time, Pacific Habour, Viti Levu, Fiji. *New Zealand Journal of Geology and Geophysics*, **33**, 547–56.

Smithers, S.G., Woodroffe, C.D., McLean, R.F. & Wallensky, E. (1994). Lagoonal sedimentation in the Cocos (Keeling) Islands. In *Proceedings Seventh International Coral Reef Symposium*, Guam.

Sollas, W.J. (1904). Narrative of the expedition in 1896. In *The Atoll of Funafuti: borings into a coral reef and the results*, pp 1–28. Royal Society, London.

Stearns, H.T. (1945). Decadent coral reef on Eniwetok Island, Marshall Group. *Bulletin of the Geological Society of America*, **56**, 283–8.

Stoddart, D.R. (1962a). Three Caribbean atolls: Turneffe Islands, Lighthouse Reef and Glover's Reef, British Honduras. *Atoll Research Bulletin*, **87**, 1–151.

Stoddart, D.R. (1962b). Coral islands by Charles Darwin. *Atoll Research Bulletin*, **88**, 1–20.

Stoddart, D.R. (1965). The shape of atolls. *Marine Geology*, **3**, 369–83.

Stoddart, D.R. (1969). Sea-level change and the origin of sand cays: radiometric evidence. *Journal of the Marine Biological Association of India*, **11**, 44–58.

Stoddart, D.R. (1971a). Environment and history in Indian Ocean reef morphology. *Symposium of the Zoological Society of London*, **28**, 3–38.

Stoddart, D.R. (1971b). Geomorphology of Diego Garcia Atoll. *Atoll Research Bulletin,* **149**, 7–26.

Stoddart, D.R. (1971c). Coral reefs and islands and catastrophic storms. In *Applied coastal geomorphology,* Ed. J.A. Steers, pp. 155–97, London: Macmillan.

Stoddart, D.R. (1973). Coral reefs: the last two million years. *Geography,* **58**, 313–23.

Stoddart, D.R. (1983). Grandeur in this view of life: Darwin and the ocean world. *Bulletin of Marine Science,* **33**, 521–7.

Stoddart, D.R. (1992). The foundations of atolls: first explorations. In *Geology and offshore mineral resources of the central Pacific basin.* ed. B.H. Keating & B.R. Bolton, pp. 11–19. Circum-Pacific Council for Energy and Mineral Resources Earth Science Series, vol. 14. New York: Springer-Verlag.

Stoddart, D.R., McLean, R.F. & Hopley, D. (1978). Geomorphology of reef islands, northern Great Barrier Reef. *Philosophical Transactions of the Royal Society of London, Ser. B,* **284**, 39–61.

Stoddart, D.R., Spencer Davies, P. & Keith, A.C. (1966). Geomorphology of Addu Atoll. *Atoll Research Bulletin,* **116**, 13–41.

Stoddart, D.R. & Steers, J.A. (1977). The nature and origin of coral reef islands. In *Biology and geology of coral reefs,* vol. IV, ed. O.A. Jones & R. Endean, pp. 59–105. Geology II. New York & London: Academic Press.

Stoddart, D.R., Woodroffe, C.D. & Spencer, T. (1990). Mauke, Mitiaro and Atiu: geomorphology of Makatea islands in the Southern Cooks, *Atoll Research Bulletin,* **341**, 1–61.

Szabo, B.J., Tracey, J.I. & Goter, E.R. (1985). Ages of subsurface stratigraphic intervals in the Quaternary of Eniwetak Atoll, Marshall Islands. *Quaternary Research,* **23**, 54–61.

Thurber, D.L., Broecker, W.S., Blanchard, R.L. & Potratz, H.A. (1965). Uranium-series ages of Pacific atoll coral. *Science,* **149**, 55–8.

Tracey, J.I. (1972). Holocene emergent reefs in the central Pacific. In *Second National Conference, American Quaternary Association, Abstracts,* pp. 51–2. Miami: University of Miami.

Tracey, J.I. & Ladd, H.S. (1974). Quaternary history of Eniwetok and Bikini atolls, Marshall Islands. In *Proceedings Second International Coral Reef Symosium, Brisbane,* vol. 2, pp. 537–50.

Trichet, J., Repellin, P. & Oustrière, P. (1984) Stratigraphy and subsidence of the Mururoa Atoll (French Polynesia) *Marine Geology,* **36**, 241–57.

Tushingham, A.M. & Peltier, W.R. (1991). Ice–3G: a new global model of late Pleistocene deglaciation based upon geophysical predictions of post–glacial relative sea level change. *Journal of Geophysical Research,* **96**, 4497–523.

Valencia, M.J. (1977). Christmas Island (Pacific Ocean): reconnaissance geologic observations. *Atoll Research Bulletin,* **197**, 1–14.

Veeh, H.H. (1966). Th^{230}/U^{238} and U^{234}/U^{238} ages of Pleistocene high sea level stand. *Journal of Geophysical Research,* **71**, 3379–86.

Wiens, H. (1959). Atoll development and morphology. *Annals of the Association of American Geographers,* **49**, 31–54.

Wiens, H. (1962). *Atoll environment and ecology.* New Haven: Yale University Press.

Wood-Jones, F. (1912). *Coral and atolls: a history and description of the Keeling–Cocos Islands, with an account of their fauna and flora, and a discussion of the method of development and transformation of coral structures in general.* London: Reeve and Co.

Woodroffe, C.D. (1994). Morphology and evolution of reef islands in the Maldives. In *Proceedings Seventh International Coral Reef Symposium, Guam.* pp. 1229–38.

Woodroffe, C.D., McLean, R.F., Polach, H. & Wallensky, E. (1990c) Sea level and coral atolls: Late Holocene emergence in the Indian Ocean, *Geology,* **18**, 62–6.

Woodroffe, C.D., McLean, R.F. & Wallensky, E. (1990b) Darwin's coral atoll: geomorphology and recent development of the Cocos (Keeling) Islands, Indian Ocean. *National Geographic Research,* **6**, 262–75.

Woodroffe, C.D., McLean, R.F. & Wallensky, E. (1994). Geomorphology of the Cocos (Keeling) Islands. *Atoll Research Bulletin,* **402**, 1–33.

Woodroffe, C.D., Short, S.A., Stoddart, D.R., Spencer, T. & Harmon, R.S. (1991b). Stratigraphy and chronology of late Pleistocene reefs in the southern Cook Islands, South Pacific. *Quaternary Research,* **35**, 246–63.

Woodroffe, C.D., Stoddart, D.R., Spencer, T., Scoffin, T.P. & Tudhope, A.W. (1990a). Holocene emergence in the Cook Islands, South Pacific. *Coral Reefs,* **9**, 31–9.

Woodroffe, C.D., Veeh, H. H., Falkland, A., McLean, R.F. & Wallensky, E. (1991a) Last interglacial reef and subsidence of the Cocos (Keeling) Islands, Indian Ocean. *Marine Geology,* **96**, 137–43.

Yonekura, N., Ishii, T., Saito, Y., Maeda, Y., Matsushima, Y., Matsumoto, E. & Kayanne, H. (1988) Holocene fringing reefs and sea-level change in Mangaia Island, southern Cook Islands. *Palaeogeography, Palaeoclimatology, Palaeoecology,* **68**, 177–88.

8

Continental shelf reef systems

D. HOPLEY

Introduction

Major advances in the understanding of landforms occurred when it was recognised that environment could control, largely through climate, the suite of processes operating on the landscape and in extreme instances produce distinctive landforms. Climatic geomorphology was further advanced when it was realised that climates have not remained stable during the period in which modern landforms had evolved, and that modern processes may be operating on relict landscapes (Stoddart, 1968). For example, in the areas peripheral to the great ice sheets of the major glaciations, present fluvial processes are superimposed over landscapes which may have been largely determined by periglacial processes. In the tropics, etchplanation and deep weathering may have alternated with periods of regolith stripping in response to wetter and drier climatic phases to produce the characteristic savanna landforms of the present day (Thomas, 1974).

In the coastal zone, landforms are produced by combination of subaerial processes (in which climate may play a part) and oceanic processes which have been presumed to be much more stable through time. Only very broad morphogenetic regions have been recognised, for example, the humid tropics (Bird & Hopley, 1969). Coastal landforms as we know them today however, are young, being produced only during the relatively short period when sea level has been at, or close to, its present position. Probably because of this, inheritance within the coastal zone through the re-occupation of early interglacial coastline positions and the reworking of landforms produced 125 000 years ago or earlier has been recognised as being of great importance (e.g. Hopley, 1985).

Coral reefs at their broadest level of distribution are the ultimate example of climatically controlled landforms being limited to largely tropical or immediately subtropical waters. However, within the tropics zonality is

303

difficult to recognise and it has been advocated that there are greater contrasts across continental shelves and into the open oceans, than there are in a latitudinal direction (Hopley, 1989a). Environmental control on the biological organisms which produce coral reefs is therefore strong, and not surprisingly the fringing reefs and shelf reefs found along continental margins are in many ways very different when compared with those of volcanic islands and atolls in open oceans (see Chapter 7).

This chapter examines the distinctive features of shelf reefs, with particular reference to the Great Barrier Reef of northeastern Australia. It also suggests that coral reefs, like landforms, inherit many of their characteristics from earlier events and that just as changes to climate can pass through important landscape-forming thresholds, changes within the ocean may have similar effects on coral reefs.

The special features of fringing and shelf reefs

The proximity of the continental landmass is of far greater importance to coral reefs than any kind of zonal gradient which can be found within tropical waters. Continentality produces amplified temperature ranges in nearshore waters which can extend across a continental shelf, though much reduced towards the outer margins. Nearshore areas are also influenced by run-off from the landmass and reefs have a restricted tolerance to reduced salinities, turbidity, sediment loading and increases in nutrients which may emanate from the land (for review see Hopley, 1989a).

Sea-level changes also have much greater impact on the shelf reef systems. At the maximum of each glaciation, with a fall of 100 metres or more in sea level, entire reef systems are left high and dry. Along the Great Barrier Reef such a fall in sea level would put the coastline on the top of the continental slope, most of which is steeply sloping producing a cliffed coastline at glacial maxima, along which reef formation would have been difficult and very limited (Hopley, 1982). Recolonisation during the postglacial transgression may well have been from more distant refuges such as the Queensland Plateau reefs and have required the establishment of suitable oceanic circulation patterns for larval transport (Davies, Marshall & Hopley, 1985). Whilst low sea levels may also have limited coral development to steeply sloping sides of atolls and open ocean volcanic islands, these were immediately adjacent to the reef systems which were subsequently colonised during the transgression. There, the effect of changing sea level has been one of moving a vertical zonation up and down a slope with little horizontal migration. In contrast, on continental shelves, horizontal migration is the major effect of changing sea

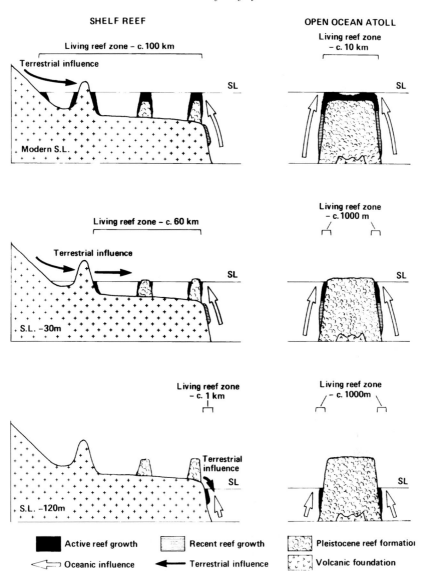

Figure 8.1. A schematic comparison of the reef development on a continental shelf and an atoll during Quaternary sea-level changes.

level and, as it has been established that the most influential gradient for coral reefs is one which is normal to the continental shoreline, the critical zone of freshwater influence, sediments and nutrients will have migrated to and fro across the shelf with each transgression/regression cycle.

Oceanic circulation is important in many ways. It provides the mechanism for larval dispersal, may produce upwelling and nutrient enhancement, and determines the residence time of water bodies which may not be ideal for reefs. Circulation around open ocean reefs will be little different at low sea-level stages as the obstruction formed by a volcanic island or atoll will have changed little (presuming that major oceanic circulation has not changed). In contrast, shelf reefs show major changes, and some circulation may be completely restricted. For example, at any sea level lower than −20 m Torres Strait between Australia and New Guinea is closed and circulation across northern Australia between Indian and Pacific Oceans is halted (Jennings, 1972). Major ocean currents are also brought up hard against the steeply sloping shelf margin, rather than being able to flood across the outer shelf region as during an interglacial high. Changes to upwelling patterns may be expected.

Shelf reefs therefore have a steeper environmental gradient to withstand at the present time and also experience a far greater contrast in ambient conditions during sea-level change than do oceanic reefs. The history of evolution of such reefs would be expected to show responses to changes in the environment. The possibility that these changes pass across important thresholds is discussed below.

The major ecological controls on coral reef growth

Within the general intertropical area defined by temperature, hermatypic or reef building corals are limited by a number of ambient environmental parameters. In many instances these controls are applied indirectly via the zooxanthellae, the minute algal symbionts which inhabit the endodermal tissue of the coral polyp. These algae are important in removing waste, providing the coral with nutrients and most importantly, through controlling a number of biochemical processes, influence the rate of calcification, or skeletal deposition, as the coral colony grows. It is important that many of these ecological controls are applied via plants not animals. Plants require light for photosynthesis and respond directly to nutrients such as phosphate and nitrate. Thus, whilst salinity as a control may act directly on the coral polyps (a tolerance range between 30‰ and 40‰ is required), the normal response to prolonged reduced salinity is expulsion by the coral colonies of their symbiotic zooxanthellae, resulting in coral bleaching and ultimately death. Similarly, temperature extremes may also affect the polyps, but the response is also zooxanthellae expulsion and bleaching. (Veron, 1986)

Other controls may be applied to the corals via the zooxanthellae. Although heavy sedimentation may smother coral polyps and cause mortality, Great

Barrier Reef corals have been shown to be able to withstand sedimentation rates and levels of turbidity an order of magnitude higher than those published for Caribbean and Pacific reefs (e.g. Pastorok & Bilyard, 1985), (Table 8.1). At lower levels of turbidity, the important factor is the more rapid attenuation of light through the water column and whereas coral may be found at depths of more than 100 metres in clear open ocean waters with a widely spaced and intricate vertical zonation, in inshore waters the lower level of coral growth may be as little as 8 metres with a greatly compressed vertical zonation. The important factor is the amount of light required for photosynthesis by the zooxanthellae. On the Great Barrier Reef, corals have been found on Myrmidon Reef, located on the outermost shelf, at a depth of 100 metres, and

Table 8.1. *Measured rates of sedimentation on reefs of the Great Barrier Reef Province* and elsewhere*

Location	Mean or range of sedimentation ($mg\,cm^{-2}\,day^{-1}$)	Source
Great Barrier Reef		
1. Cape Tribulation, mainland adjacent to disturbed rainforest catchments	145.8	Hoyal (1986)
2. Cape Tribulation, mainland adjacent to partially disturbed rainforest catchment	88.4	Hoyal (1986)
3. Cape Tribulation, mainland undisturbed rainforest catchment	26.1	Hoyal (1986)
4. Magnetic Island, high granitic inshore island	20–114	Mapstone *et al.* (1989)
5. Low Isles, inner-shelf reef	69.7 (0.6–899.9)	Marshall & Orr (1931)
6. John Brewer Reef, mid-shelf reef lagoon	0.17–2.87	Hoyal (1986)
Other		
Kaneohe Bay, Hawaii, oceanic volcanic island	35–41	Maragos (1972)
Puerto Rico, large continental island	12.8–1179.9	Cortes & Risk (1985)
Canton Atoll, open ocean atoll	0.13 (based on calcification and metabolism only)	Smith & Jokiel (1975)

*Rates quoted here are an order of magnitude greater than those quoted in Pastorok & Bilyard (1985). This frequently quoted reference based on data from Guam does not appear applicable to the Great Barrier Reef.

almost 100% coral cover of *Leptoseris, Pachyseris* and *Endophyllia* spp. was found between 60–80 metres depth (Fig. 8.2). On the inner shelf of the Cumberland and Northumberland Groups of islands, where tidal ranges in excess of 6 metres are found and high turbidity as a result of re-suspension of sediments is constant, the lower limit of coral growth may be as little as 4 metres (van Woesik, 1992).

Nutrients may also be limiting. Whilst the traditional view of oceanic reefs living in a 'nutrient desert' has now been challenged and reefs can grow in a wide range of naturally occurring nutrient environments (Kinsey, 1991), inshore reefs may be growing close to their tolerance limits with respect to phosphate and nitrogen and are therefore susceptible to anthropogenically enhanced nutrient levels or spikes in the natural system caused by disturbances such as tropical cyclones (Furnas & Mitchell, 1986).

Thus, there may be a relatively wide envelope of oceanographic conditions within which reefs will grow and within which the metabolic performance is predictable. Reef metabolic performance may be expressed as gross primary production, but more significantly for geological purposes as net calcification (Kinsey, 1985). Over shallow areas with 100% coral cover this may reach $10 \, kg \, m^{-2} a^{-1}$ which, taking into account the density of the aragonite being laid down and the porosity of the reef fabric, converts to a vertical growth potential of approximately 7 to 8 mm per year (7 to 8 metres per thousand years). This correlates well with the geological record as determined by dated coral cores (Davies & Hopley, 1983). In other areas of the reef, net calcification rates are proportionately slower: for example, $4 \, kg \, m^2 a^{-1}$ on algal pavements and as low as $0.5 \, kg \, m^2 a^{-1}$ on sandy lagoon floors.

However, reefs under stress may show different metabolic performances. These stress situations may be produced directly by environmental perturbations such as tropical cyclones. Shelf reefs, particularly those close inshore, may be living continuously in conditions which are at the lower end of their tolerance limits and thus may be under a continuous chronic stress situation. These reefs are therefore much more susceptible to acute stresses whether natural or human induced (Kinsey, 1988). Such stresses may be expressed through expulsion of zooxanthellae and bleaching with the resultant decreases in net calcification rates.

The response to nutrient increases is particularly interesting and relevant to the geological history of shelf reef systems. For example, small increases in phosphate have been shown to result in an increase in coral growth rates. However, the skeleton laid down may be more porous and the actual rate of calcification may be reduced. (Rasmussen, Ness & Cuff, 1994a) The coral

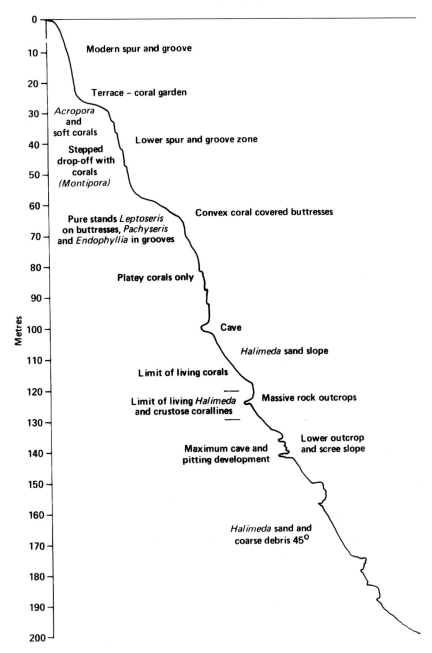

Figure 8.2. Shelf-edge transect to 200 metres off Myrmidon Reef. Determined from 1986 *Platypus* submersible transect.

skeleton becomes more susceptible to mechanical breakage by waves and tidal surge. Simultaneously, algae, particularly macroalgae, which are competing with corals for substrate space, may gain significantly from the fertilisation effect. Within the normally accepted envelope of ambient nutrient levels algae may be kept under competitive control by grazing fish. Once beyond a particular threshold, however, they may grow to the extent that they smother the corals and in particular limit the amount of light reaching the corals living in their shade. Competition from macroalgae is as important as water temperature in limiting the poleward limits of coral reefs (Wiebe *et al.*, 1982).

The concept of thresholds is of particular importance to coral reefs. Beyond a particular limit, a reef may change from coral dominated to algal dominated. The concept of catastrophe theory has been applied to such situations in coral reef areas with the cusp model, in which the point of change from one condition to another may be different dependent on the direction of change, being particularly appropriate (Buddemeier & Hopley, 1988). Thus, the establishment of a coral reef community may require far higher levels of water quality with respect to nutrients, salinity, turbidity and light than is required to maintain that reef community, once established. The demise of the coral reef community will take place in water quality conditions far inferior to those in which the reef became established.

It is suggested that during a transgressive/regressive sea-level cycle shelf reef systems may pass through important thresholds several times and that this can be identified in the present constituent components of shelf communities and in the vertical column of the Holocene reefal veneer. In recognising such environmental flips it is important that one of the algal response communities may also result in the formation of bioherms. The calcium carbonate producing alga *Halimeda* under certain circumstances can be extremely prolific with cover of up to 90%. Growth can be extremely rapid with dense stands having productivity rates of $4\,\mathrm{g\,C\,m^{-2}\,d^{-1}}$ gross (Hillis-Collinvaux, 1974). On the Great Barrier Reef dense *Halimeda* meadows have been shown to have densities as great as $4637\,\mathrm{g\,m^{-2}}$ (Drew & Abel, 1988), with the potential for vertical sedimentation rates of 1 metre in about 500 years. Such a figure has been confirmed by radiocarbon dating of sediments from *Halimeda* bioherms with a rate of 1.75 m per 1000 years quoted (Marshall & Davies, 1988). Littler, Littler & Lapointe, (1988) showed that shallow species of *Halimeda* are adapted to take advantage of episodic nutrient pulses. Although they believed that light was not a major controlling factor, the limited distribution of *Halimeda* on fringing and nearshore reefs of the Great Barrier Reef (Drew & Abel, 1988) suggests that nutrient enhancement when combined with low salinity and sediment yield will not produce luxurious *Halimeda* meadows.

Origin and evolution of the Great Barrier Reef

The Great Barrier Reef (Fig. 8.3) is the world's largest coral reef system, extending off northeastern Australia for a distance of 2300 km over 14° of latitude (Hopley, 1982). Even excluding the Torres Strait region, the shelf waters exceed 230 000 km² and of this almost 9% is occupied by reefs or submerged shoals (Hopley, Parnell & Isdale, 1989).

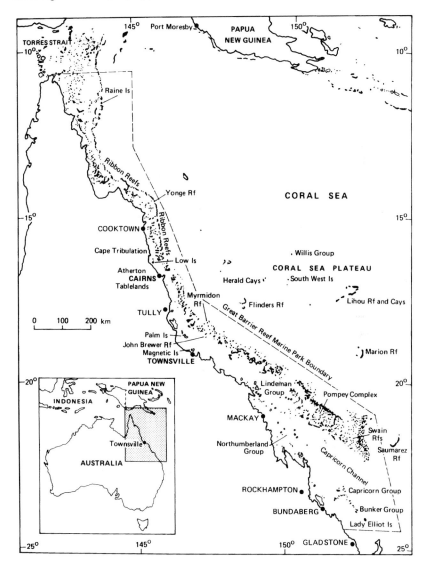

Figure 8.3. The Great Barrier Reef province.

The northern region of the reef has developed on a narrow shelf no more than 50 km wide. Its most distinctive features are the linear or ribbon reefs running parallel to the edge of the continental shelf almost as far south as Cairns. They form a near continuous barrier of individual reefs up to 25 km long. Inside many of the ribbon reefs are large banks at depths of 20–40 metres. These are formed almost entirely of *Halimeda*, which appears to have built structures equal in size to many of the coral reefs (Fig. 8.4). The middle shelf is occupied by large platform reefs with extensive reef flats, some up to 25 km in length (Fig. 8.5*a*). Closer to shore is a more open area of inner shelf, where a number of small reefs capped by distinctive low wooded (mangrove) islands are found. Most high islands of this region, which are not large in number, have fringing reefs and in some areas fringing reefs also extend along the mainland, most distinctly in the Cape Tribulation area some 100 km north of Cairns (Fig. 8.5*b*).

As the continental shelf widens south of Cairns, the Great Barrier Reef occupies only the outer one-third of the shelf. Reefs are more widely spaced and generally have less-well-developed reef flats. Most reefs are irregular reef patches, or crescentic features, aligned across the dominant southeasterly trade winds. No ribbon reefs are found on the outer shelf but in some areas at least, notably just south of Cairns and outside the Pompey Complex in the south, there are linear outer shoals rising from depths of about 70 metres. South of approximately 20° S the continental shelf widens even further to about 300 km. With an increase of tidal range (to more than 4 metres on even the outermost reefs in the Pompey complex) narrow, well-defined tidal channels up to 70 metres deep intersect the reefs. Even the innermost reefs are 100 km from the mainland. The outermost reefs of the Pompey Complex stretch for 200 km as a solid mass of reef and lagoons 15 km wide with narrow, intricate channel systems (Fig. 8.5*c*). Individual reefs may be more than 100 km^2 in area. The southern extent of the Pompey Complex is a distinctively T-line junction of

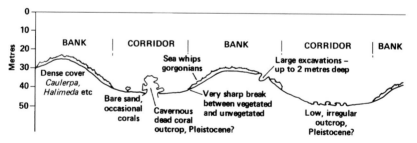

Figure 8.4. Idealised morphology of the *Halimeda* banks behind Ribbon 5, northern Great Barrier Reef. Determined from 1986 *Platypus* submersible transect.

reefs, to the south of which are the contrasting Swain Reefs, smaller flat-top reefs closely spaced and with numerous sand cays. *Halimeda* shoals have been reported in this area (Searle & Flood, 1988).

Although no fringing reefs are found along the mainland coast in the central and southern section of the Great Barrier Reef, they are exceptionally well developed around all continental islands north of about 21° S. At this point, the tidal range increases to such an extent (more than 5 metres) that resuspension of sediments appears to have depressed major reef development (Kleypas, 1991; van Woesik, 1992).

South of the Capricorn Channel the shelf narrows again to less than 100 km. The Bunker Capricorn Group of reefs, the southernmost of the Great Barrier Reef, are a series of 22 reefs and 11 shoals of only moderate size but with numerous vegetated sand cays.

In spite of its size, the Great Barrier Reef is one of the youngest reef systems in the world, most of it being less than 1 million years old (Davies, 1992). Until about 75 million years ago, Australia and Antarctica were joined. Most of Australia lay south of 40° S far from waters warm enough for coral growth. About 65 million years ago Australia began to split from Antarctica and moved northwards. Subsequently, northeastern Australia was formed by rifting between the Australian and Pacific plates and by the time the continental shelf had formed, northern Australia lay close to 30° S latitude. Uplift, rifting and vulcanism produced a complex rift basin system that was to control the location and form of the continental shelf. As Australia continued to move north, shelf evolution was dominated by fluvial sediment yield. Current annual sediment input from north Queensland rivers alone is estimated at 28 million tonnes (Pringle, 1986). Deltaic progradation produced extension of the shelf margin (Symonds, Davies & Parist, 1983).

Surprisingly, even though Australia was within sufficiently warm waters by 26 million years ago, seismic records show little reef development apart from on the northernmost area of the Great Barrier Reef. Reef building began less than 500000 years ago over most of the Great Barrier Reef (Davies, 1992). The reef sequence is therefore thin, less than 300 metres in thickness. Its entire development has taken place during periods of rapidly fluctuating sea levels produced by high-latitude glaciation. Reef growth has occurred during short periods of high sea level. During the intervening periods of low sea level, the reefs were subaerially eroded by karstic processes and continual recolonisation of older foundations has produced reefs that are composite features made up of a series of remnant reefs separated by unconformities, a process aided by general shelf subsidence. The Holocene has been the final period of high sea level, during which the final layer has been added to the reefal column. On

Figure 8.5. (*a*) Mid-shelf reefs, although growing in shallower water than their outer-shelf counterparts, probably started to grow at around about the same time, i.e. 8000 years BP. However, on the central Great Barrier Reef in particular, their foundations tended to be deeper (15 to 30 metres). They therefore trailed sea level by 10 metres or more as they grew upwards and only reached modern sea level subsequent to 6000 years BP. Nonetheless, extensive reef flats have developed though deep lagoons still remain. (Darley Reef, central Great Barrier Reef.) (*b*) Mainland fringing reefs are rare and occur only north of Cairns. These reefs near Myall Creek south of Cape Tribulation commenced to grow shortly before 7000 years BP on foundations provided by gravel fans laid down by steep coastal streams. By 5400 years BP most reef growth had taken place. Subsequently, as coastal water conditions deteriorated, these reefs have been in a

Caption for fig. 8.5 (*cont.*).
delicate state of balance typical of many mainland and nearshore island reefs in north Queensland. (Cape Tribulation reefs.) (*c*) Even the largest reefs of the Great Barrier Reef accreted principally in the short period between 8000 years BP, and the maximum of the transgression about 6500 years BP. These reefs are part of the Pompey Complex in the south central Great Barrier Reef with much of the growth and carbonate production since 6500 years BP being directed towards lagoon infilling. (*d*) Flood plume of the Murray River near Tully, north Queensland, shortly after the passage of Cyclone Winifred in 1986. Such plumes carrying massive amounts of sediment have increased during the period of European occupation due to extensive clearance of natural rainforest catchments for agriculture.

average this is approximately 20 metres in thickness but can vary from zero to more than 30 metres. The evolution of the reef during the Holocene throws much light on the overall development of this reef system.

Recolonisation, growth and maturation of a shelf reef system in response to changing water quality

Hopley (1982) has documented the stages in reef growth following the inundation of older Pleistocene foundations during the Holocene transgression. Following inundation, growth on elevated portions of the foundations and in particular on the windward margins enhances the original relief. With stabilisation of sea level and vertical reef growth catching up, reef flat development commences and, after extending around the margins of the platform, enclosing single or multiple lagoons, infilling of the lagoons results in a final reef form which is planar.

Whilst drilling has largely confirmed this model of reef development, it has also been recognised that reef growth did not commence immediately the antecedent platform was inundated. Initially (Davies & Marshall, 1979), this was attributed to poor water-quality conditions over reef tops as regolith, developed during the low sea-level phase, was reworked. However, during the 1980s extensive work on the Great Barrier Reef highlighted three important factors:

- Holocene reef growth commenced on the Great Barrier Reef almost exclusively between 8320 and 7500 years BP regardless of the time of inundation. A timelag of between 1200 and 2000 years separates transgression from initial colonisation (Davies *et al.*, 1985).
- extensive areas of the Great Barrier Reef, formerly thought to be submerged coral reefs are now known to be *Halimeda* bioherms (Orme, Flood & Sargent, 1978; Orme, 1985; Orme & Salama, 1988; Davies & Marshall, 1985; Drew & Abel, 1985, 1988; Marshall & Davies, 1988; Phipps, Davies & Hopley 1985; Searle & Flood, 1988). These algal bioherms commenced to grow more than 10 000 years ago, i.e. more than 2000 years before reef growth was initiated.
- further, recent studies (e.g. Kleypas & Hopley, in press) have also indicated that the rate of reef growth has not been constant over the last 8000 years.

Various explanations have been put forward to explain these phenomena, many of them related to water quality. However, no holistic attempt has been made to explain the relationship between the changes in water quality and reef growth during the Holocene. Geographical changes produced by rising sea level were also accompanied by changes in climate, water circulation and water quality. Changes are still taking place at the present time, exacerbated by anthropogenic influences which ultimately have the potential for greatest impact.

The major environmental changes which may have affected reef growth include:

1. Flooding of the continental shelf

Up to about 11 000 years BP the postglacial transgression had risen against the steeply sloping shoulder of the continental shelf and for much of the reef, particularly in the north, a cliff shoreline would have predominated. Subsequently, however, as the transgression flooded onto the shelf, environmental conditions changed rapidly. There is evidence that at least locally extensive fluvial sediments have been laid down over the shelf and these were available for reworking. Soils which had formed over these deposits and over older Pleistocene coastal plain and reefal features would have stored nutrients which were also released into the nearshore waters. Initially at least, the strong upwelling which may have taken place against the steep continental slope would have remained, but especially north of Cairns the outer shelf would have consisted of a linear range of limestone hills, the precursors of the present ribbon reefs with only narrow passes through which the sea flooded. Strong nutrient jetting through these is suggested during the very early Holocene. Marshall (1988) has suggested that this upwelling of cold deep water also played a major role in modulating surface temperature along the margins of the shelf.

2. Flooding above the level of the older reefal foundations

This took place about 8500 years ago as the average depth of reefal foundations is approximately 20 metres. Although some of the foundations in the northern Great Barrier Reef, southernmost Great Barrier Reef and some fringing reefs are shallower, once sea level had reached approximately -20 metres freer movement of water from the Coral Sea onto the shelf was possible. As the older Pleistocene foundations no longer formed a barrier, this exchange of water would have almost certainly increased the water quality of the middle shelf. It also produced the mid-Holocene high-energy window (Hopley 1984), which produced more turbulent mixing of waters than had previously been possible and existed until the modern reefs grew up to sea level, generally 5000–4000 years ago.

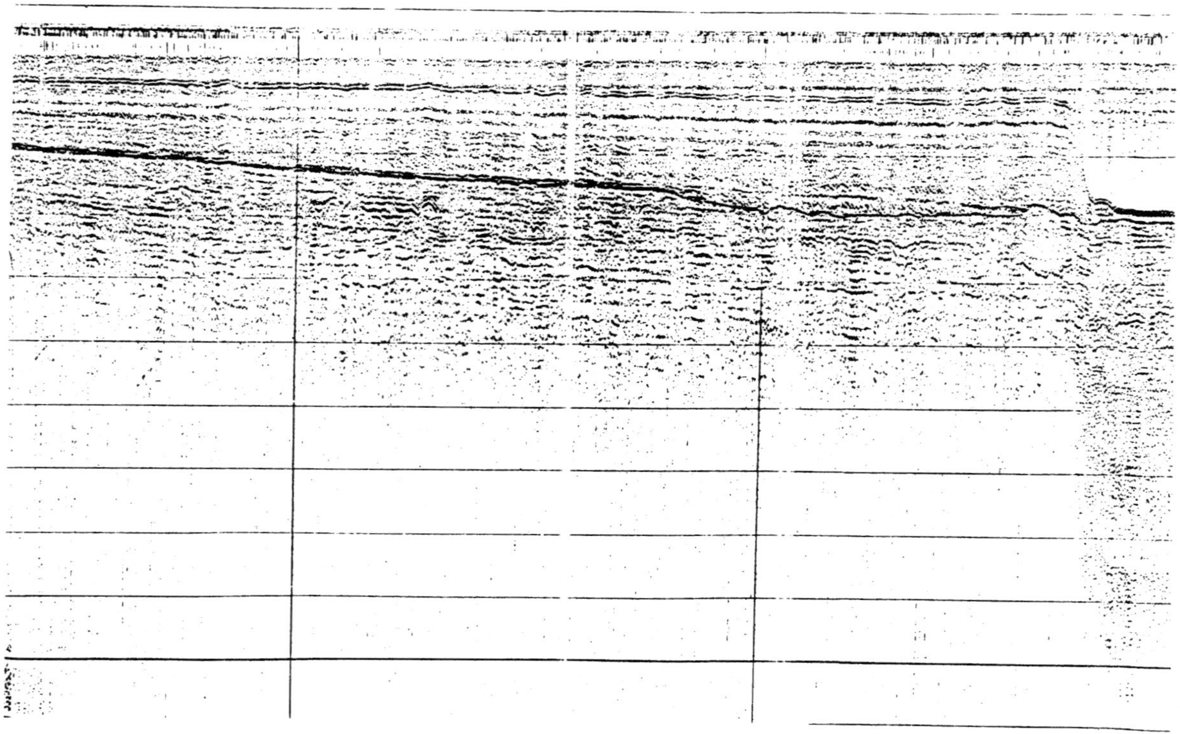

Figure 8.6. Seismic reflection profile inshore area northern Great Barrier Reef showing a mud silt wedge of up to 8 metres thickness overlying Pleistocene surface.

3. The commencement of the postglacial stillstand

Modern sea level was achieved between 6000–6500 years BP along most of the Great Barrier Reef. For reef growth this was important as it allowed reefs which had formerly lagged behind the rise in sea level to catch up and commence the lateral development of reef flats. Strong energy gradients developed from windward to leeward margins of reefs and lagoons became sinks for not only sediments but also nutrients. On the mainland, the stabilisation of sea level had a further important effect. In contrast to the previous 20 000 years or more the coastline remained in almost exactly the same position, apart from progradation in areas of rapid sediment yield. Initially at least the area of sea floor immediately in front of the coastline had little sediment. Pleistocene sediments had largely been driven onshore and seismic surveys have shown that over much of the sea floor, the Pleistocene surface outcropped (Searle *et al.*, 1982; Johnson & Searle, 1984). However, in the subsequent 6000 years the coastal streams have been delivering enormous amounts of sediment to the coastline (Pringle, 1986) and these have built up a mud and silt wedge (Fig. 8.6) adjacent to the coast up to 20 metres in thickness and extending as much as 20 km from the mainland, i.e. out to at least the innermost of the high islands with fringing reefs. This sediment, often in less than 10 metres of water depth, is resuspended even in moderate southeasterly conditions of 20–25 knots as can be found for up to 50% of the time during the winter months. Nearshore waters have therefore changed from relative clarity to highly turbid over the period of stillstand. Stability of the coastline has also allowed the accumulation of nutrients not only in lagoons, but also in coastal estuaries, particularly in mangrove areas.

4. Climatic change

Details of climatic change during the Holocene are available for at least the Atherton Tableland behind Cairns (Kershaw, 1978) and may be indicative of much of the central and northern Great Barrier Reef. Prior to 7800 years BP sclerophyll vegetation existed on the tableland and rainfall totals are estimated as being less than 1800 mm. Between 7800 and 6500 years BP the vegetation changed to simple notophyll vine forest and rainfall totals rose to about 2500 mm. Between 6500–3000 years BP a complex mesophyll vine forest existed and rainfall totals were as high as 3500 mm. Subsequently, rainfall totals have declined to about 2500 mm annually and a complex notophyll vine forest has become established. These results suggest that the major growth phase for coral reefs took place during a period of increasing rainfall when totals have been less than present. The period of stillstand was one of maximum rainfall totals during which reefs reached modern sea level, whilst the period of reef flat development has been one of declining rainfall. These changes are important for natural sediment yield to the coast.

However, the relationships are not simple. Maung Maung Aye (1976) investigated drainage basin parameters related to erosion in north Queensland catchments. Minimum denudation was related to mean annual rainfall totals of between 2500–3000 mm. An increase to totals above those of present (>3000 mm) would automatically have resulted in delivery of greater sediment loads and fresh water by small coastal catchments. The reason is that the rainforest canopy gives no greater protection than it does between 2500–3000 mm annual rainfall, and the throughput of water is greater. Similarly, a decrease in rainfall totals below 2500 mm is also demonstrated to be related to an increase in sediment yield. Such rainfall totals are usually related to seasonal rainfall and open eucalypt woodland vegetation, which gives far less protection to the ground surface than rainforest. The implications for changing natural sediment yield from coastal catchments for reef development is discussed further below.

5. Human impacts

Although Aborigines occupied Australia more than 40 000 years ago and almost certainly wandered the coastal plains which were the exposed continental shelf at low sea-level stages, their impact on water quality, even allowing for practices which included firing of native vegetation, would have been minimal. In contrast, 200 years of white settlement has seen major changes to the environment, resulting in increased sediment yield, more rapid run-off and higher flood peaks and delivery of nutrients from agricultural catchments through run-off (Fig. 8.5*d*). Other changes may be more visible (Hopley, 1988, 1989b), but these are the ones which have caused a decline in water quality, particularly since the middle of the twentieth century.

Phases in reef development in relation to water quality changes

Although not absolutely synchronised, many of these changes have taken place close enough to the same time to produce important responses in coral reef and associated biohermal growth, largely because most changes have been reinforcing, i.e. have been in the same direction in terms of water quality, or have been large enough on their own to pass through a reefal threshold. The main phases recognised are:

1. Commencement of the Holocene to c.8000 years BP

Water quality was low during this period as seas were shallow and probably highly turbid, as previously developed soils and regolith were reworked. Mainland river mouths were sufficiently close to what is now the main reef tract to allow their plumes to reach out to the older Pleistocene foundations of the modern-day reefs. Nutrients released from the soils were reinforced by nutrient upwelling on the shelf edge and jets extending through the passes on the shelf edge. Coral growth during this period appears to have been negligible. Problems with water quality may have been compounded by incursions of cold upwelled water on the shelf edge and by larger temperature extremes on the shelf because of its shallower nature compared with the present.

It has also been suggested that the absence of larval replenishment centres locally may also have delayed recolonisation processes with major reef refuges occurring on the Queensland Plateau. Davies *et al.* (1985) have suggested that the likely low sea-level water circulation pattern would have been dominated by north to south flows across the Queensland Plateau, which would have inhibited the westward dispersal of planulae and would have persisted until sea-level rise was advanced sufficient to flood the Torres Strait between Australia and New Guinea. At such time, a new oceanographic pattern would have been established comparable to that which operates during the winter months today, i.e. westward flows from the Coral Sea onto the Queensland continental shelf. Such a change in circulation pattern would have occurred around 8000 years BP when the shelf had been largely flooded.

Thus, coral reef growth to about 8000 years BP was negligible. Instead, carbonate productivity appears to have been concentrated in what are now recognised as major *Halimeda* banks. First recognised on the northern Great Barrier Reef (Orme *et al.*, 1978), extensive *Halimeda* deposits probably occur along much of the outer third of the shelf in water depths exceeding 30 metres and have been recently recognised as far south as the Swain Reef Complex (Searle & Flood, 1988). The banks are up to 20 metres in thickness, overlying pre-Holocene substrate, and began to grow about 10 000 years BP. Living algae still cover these banks (not always predominantly *Halimeda*) and although accumulation rates could be as high as 2 metres per thousand years, i.e. sufficient to produce the present thickness of the banks, this is probably at the upper end of the scale at the present time. The indications are that the bulk of these banks formed in the first half of the Holocene very rapidly and in response to water conditions which favoured their growth, rather than that of coral. Growth may have commenced soon after the immediate terrestrial influences (which currently appear to restrict *Halimeda*) had migrated away from the outer shelf.

Not surprisingly, several metres of loosely cemented *Halimeda* gravels are also found at the base of some Holocene reefal sequences, mainly on horizontal surfaces, e.g. Yonge and Raine Island Reefs (Fig. 8.7). Although there are no ages for the Holocene *Halimeda* gravels, the oldest dates obtained for overlying corals are 6580 ± 120 years BP (Beta 40822) on Yonge Reef and 7040 ± 140 years BP (ANU 6621) on Raine Island. Notably amongst all cores from the Great Barrier Reef to which this author has access, although *Halimeda* may occur as individual fragments within the core, *Halimeda* gravels are found nowhere except at the base.

On both Yonge Reef and Raine Island, the top of the Pleistocene is also composed of *Halimeda* grainstones. Radiocarbon ages indicate a Pleistocene age, but are not absolute and the *Halimeda* may have accumulated either during an interstadial high sea level, or during the fall from the Last Interglacial high sea-level stand. In both instances, the palaeogeography would have been similar to that during the early Holocene with similar low water quality and apparently the same ecological response.

Coral growth on fringing reefs of the present mainland and high continental islands did not take place pre 8000 years BP for more obvious reasons, i.e. these shallower depths had yet to be inundated (e.g. see Partain & Hopley, 1989)

2. 8000 years to 6500 years BP: The final period of transgression

Sea level was close to its modern position by 6500 years BP along much of the Great Barrier Reef. In the final 1500 years of the transgression the rate of rise may have been little more than 10 mm per year. Nonetheless, the shoreline retreated rapidly over the low-gradient inner-coastal shelf, removing terrestrial influence away from the outer third of the shelf where most reef growth takes place. With the removal of terrestrial run-off influences, conditions for coral reef growth became ideal. All regolith by this time had been removed from the older reefal foundations which, when colonised, may have been as much as 10 metres below the rising sea level. As the Torres Strait became flooded the present oceanic circulation within the Coral Sea became established and larval recruitment from the Coral Sea Plateau low sea-level refuges was more easily achieved. The Atherton Tableland palaeoclimatic record (Kershaw, 1978) shows increasing rainfall during this period, with the vegetation changing from sclerophyll woodland to a simple notophyll vine forest. However, this is likely to have restricted the terrestrial influence as the greater vegetational cover may

RAINE ISLAND REEF   YONGE REEF

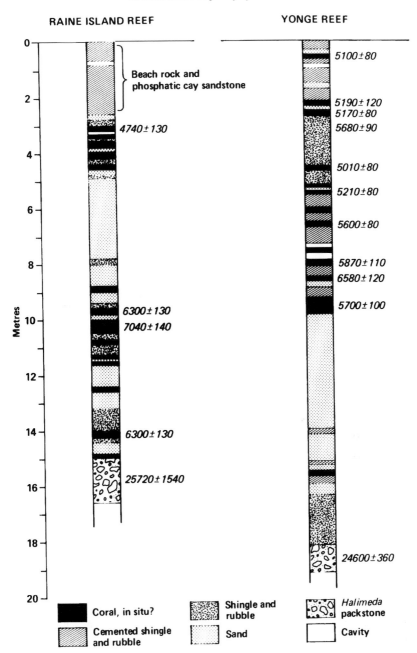

Figure 8.7. Dated cores from Yonge and Raine Island Reefs, in the far northern section of the Great Barrier Reef.

have reduced sediment yield and nutrient recycling by the rainforest system could well have reduced the natural flow of nutrients to the ocean.

There is a strong possibility that these conditions saw a decline in the growth of the previously flourishing *Halimeda* banks. The early Holocene bathymetry of the shelf edge, formed by linear barriers of Pleistocene reefs and narrow passes, was now replaced by a far more open developing shelf-edge reef system. Although there is some evidence that parts of the outer ribbon reefs may have kept up with sea-level rise, it is highly unlikely that the outer shelf retained the full barrier morphology. The tidal jet vortex systems which had previously delivered nutrient-rich water from below the thermocline to the outermost shelf, as modelled by Thomson & Wolanski (1984) and Wolanski *et al.* (1988), may have been substantially reduced. Until the outer reefs grew up to sea level and re-established a continuous outer barrier, *Halimeda* growth was probably retarded.

In contrast, coral reef growth was prolific and analysis of drilling results suggests that up to 80% of framework construction in the Holocene reefal veneer took place during this period and immediately afterwards. The reason for this is clear. Kinsey (1985) has indicated sharp zonality in the production of calcium carbonate over reefs, with maximum amounts of about $10\,\mathrm{kg}$ per m^2 per year occurring in areas of 100% coral growth. Today, such areas are limited to the outermost perimeters of the reefs, particularly on windward sides. Leeward margins may be made up of large amounts of sediment and talus. In contrast, during this rapid growth period, reefs were still largely below the rising sea level and 100% coral cover across the entire reef platform may have been the norm. Much of the more fragile coral growth may have been removed during cyclonic events but regrowth in such environments has been shown to be extremely rapid.

The pattern for the inshore fringing reefs shows many similarities (e.g. Hopley *et al.*, 1983; Hopley & Barnes, 1985; Johnson & Risk, 1987; Partain & Hopley, 1989; Kleypas, 1991) although growing in shallow water and therefore inundated much later than the outer shelf foundations. Removal of regolith does not seem to have retarded reef growth in many areas. Indications are of almost immediate colonisation once sea level had risen sufficiently. By definition, these reefs would still have experienced terrestrial influences, but in contrast to the outer reef, which had been in similar situations 2000 years or more earlier, larval recruitment for the fringing reef would not have been a problem. However, common occurrence of alcyonarian spiculite (Konishi, 1982) most probably formed from the soft coral *Sinularia* in the basal section of the Holocene in many reefs of the Cumberland and Northumberland Group and also in the Palm Islands may well have been the result of shallower and more turbid inshore conditions between 8000 and 6500 years BP (e.g. Kleypas, 1991).

Another important environmental factor during this period was the much lower amount of protection given to the mainland coastline by reefs, many of which were several metres below sea level. Today, most oceanic swells are excluded by the reefal barrier along much of the north and central Queensland coastline and fetch during both normal weather conditions and tropical cyclones is limited by the reefal barrier. These influences would have been far less during the final part of the transgression and even during the next phase the mainland coastline may have experienced far higher energy conditions than at the present time. This has been referred to as the Holocene 'high energy window' (Hopley, 1984) and may have also been one of the factors promoting reef growth. Far better circulation of inshore waters may have resulted and, in particular, widespread settlement of extensive mud areas close inshore may have been restricted.

3. 6500 years to 3000 years BP: stillstand and lateral growth

Between 6500 years and 3000 years BP the majority of reef reached modern sea level. A slowing down of reef growth within the zone affected by stormwave activity has long been recognised (Davies & Marshall, 1979) and much of the carbonate productivity of this period has gone into detrital accumulation rather than main framework construction. Not surprisingly, as Kleypas (1991) has noted, most ^{14}C date reversals in coral cores are found within this period.

After stillstand the gross morphology of reefs was changing rapidly. As reefs reached sea level reef flat construction commenced and maximum carbonate productivity of $10\,kg\,m^{-2}a^{-1}$ became limited to the reef perimeters. Outer reef flats now produced at about $4\,kg\,m^{-2}a^{-1}$ and much of the inner sandy reef flats and lagoons may have been reduced to $0.5\,kg\,m^{-2}a^{-1}$. As a result, reefs became much more complex ecological systems, with zones and habitats determined by energy gradients from windward to leeward coming into existence. For the first time in the Holocene, extensive reef flats were available for the accumulation of sediments and the coral cays of the Great Barrier Reef came into existence during this period.

Although reef growth was predominantly on the leeward sides as calcium carbonate produced on windward margins was transported to the leeside, leeside framework construction became much more limited. There is considerable seismic evidence which shows that flourishing reef patches to the leeward side of reefs became rapidly buried by sediment once the adjacent reef structure reached sea level and this horizontal transport of carbonate commenced, a process which Goreau & Land (1974) have termed the endogenous control of reef growth.

A further factor which aided in the emergence of reef flats and the production of reef flat sediments, at least on reefs of the inner shelf, was a fall in sea level of 1.0 to 1.5 metres (Chappell *et al.*, 1983; Hopley, 1983a,b). Attributed to hydroisostatic processes, this relative emergence has affected the inshore fringing reefs in particular leaving inner reef flats still within the modern tidal range, but with numerous dead microatolls. Subsequent reef growth strategy has relied on framework construction on the reef front only aided at times by detrital accumulation (e.g. Hopley *et al.*, 1983). Retarded growth becomes even more prominent close to the mainland and has been particularly noted on the Cape Tribulation reefs (Fig. 8.8) along the mainland north of Cairns (Partain & Hopley, 1989). The slight fall in relative sea level is partly responsible but there are other factors. By 3000 years BP the stability of sea level had meant that the position of the mainland coastline had been more or less the same for 3500 years and the closing of the Holocene high energy window had allowed the accumulation of a significant part of the nearshore mud–silt wedge. Serious decline in water-quality conditions due to re-suspension of sediments now took place whenever wind speeds were sufficient to produce significant wave activity (circa 15–20 knots). In the very high tidal range area of the Cumberland and Northumberland Islands, tidal motion alone became sufficient to maintain highly turbid waters almost continuously (Kleypas, 1991).

The climatic record, at least from north Queensland, also suggests that this was the period of maximum rainfall during the Holocene, with rainfall totals on the Atherton Tableland being approximately 3500 mm, or more than double what they had been some 4000 years previously. This is recorded in the change in the vegetation to a complex mesophyll vine forest, but from a geomorphological viewpoint this would have given no more protection than the earlier notophyll vine forest and it is likely therefore that there was greater run-off and possibly even higher sediment yield during this period.

4. 3000 years BP to 200 years BP: a reef system under increasing natural stress

All the trends of declining water quality noted in the previous 3500 years were intensified during the millennia leading up to the present time. Declining energy conditions with increasing reef barrier development accompanied the expansion of nearshore mud wedges and increases in turbidity. Reef flat continued to expand and reef framework development became even more restricted to reef margins.

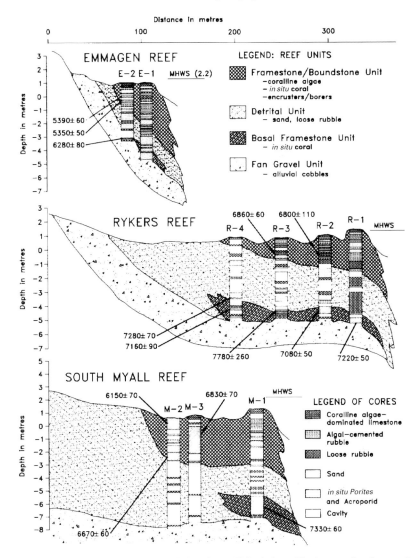

Figure 8.8. Age and lithology of the Cape Tribulation fringing reefs, far north Queensland. MHWS = Mean High Water Springs.

Accompanying the accumulation of fine sediments in the nearshore zone has been the increasing storage of nutrients in sinks in the natural system. These include the mangroves of expanding estuarine and deltaic systems on the mainland and lagoonal deposits within the reefs themselves. Release of these nutrients is possible during occasional high-energy events. This was illustrated during cyclone Winifred in 1986 when disturbance of lagoonal deposits resulted in a short-lived period of algal blooms (Furnas & Mitchell, 1986). Short periods of acute nutrient stress may therefore be superimposed on a regime in which water quality under natural conditions, though not reaching chronic stress proportions, is certainly declining.

These conditions are most pertinent to the inshore reefs. However, similar though less-intensive trends may also have taken place on the outer shelf. The outer barrier became fully established and morphological conditions returned to those which initiated the early growth of *Halimeda* banks during the first part of the Holocene. Certainly, living *Halimeda* is found over most of the present banks and productivity levels appear to be high (Drew & Abel, 1988). Research has clearly shown that cold nutrient-enriched water enters passages from the Coral Sea and is transported to the areas in which the banks are currently found. Although the water is only enriched from the normal depleted level of 0.07 μM to about 0.25 μM nitrate with very little ammonia present, it is suggested that even this input on three days of spring tides every two weeks could still satisfy the entire annual nitrogen requirements of this very productive algal vegetation. (Thomson & Wolanski, 1984; Wolanski, Drew & Abel, 1987; Wolanski *et al.*, 1988). Moreover, geochemical and geomechanical signals which have been correlated with nutrient increases in coral skeletons clearly show a variability which has been linked to the El Niño phenomenon, which is also considered to be a driving force in this nutrient upwelling (Rasmussen *et al.*, 1994a). Whether or not this nutrient enhancement is sufficient to affect reef communities is debatable. Reef communities of the middle shelf are the most prolific and most diverse and although this has been attributed to other factors such as energy levels, it is the middle-shelf area which is best buffered from the terrigenous influences to the west and the upwelling phenomena to the east.

5. 200 years BP to the present: an anthropogenic overlay

Although there was an Aboriginal presence in the Great Barrier Reef region throughout the Holocene and impact on the landscape may have been significant through seasonal firing of the vegetation, the impact on the Great Barrier Reef was almost certainly minimal. (However, a recent media report on an Ocean Drilling Programme core taken just off Cairns suggested that an increase in charcoal in the sediments, immediately off the Great Barrier Reef, during the last 100 000 years was related to Aboriginal settlement.)

The early exploitation of Great Barrier Reef resources and the settlement by Europeans over the last 200 years has had a widespread impact, as reviewed by Hopley (1988, 1989b). These impacts range from direct, such as the mining of guano from reef islands and damage to reefs by anchor chains, to indirect, such as the enhancement of nutrients close to sewage outfalls. It has been generally regarded that because of its size, most of these impacts are highly local and that the general health of the reef, particularly under the protection given since 1975 by the Great Barrier Reef Marine Park Act has left the Great Barrier Reef in a better state than most other reef systems in the world.

Whilst this is probably so, anthropogenic activities have still tended to aggravate the natural chronic stresses which have built up over the reef over the last 3000 years. Land clearance and development of unsealed roads close to fringing reefs, such as those of Cape Tribulation, have clearly increased sediment yield. In the example of Cape Tribulation reefs north of Cairns, the sediment yield increase in the adjacent streams may be as much as 14 fold due to roadworks (Hopley *et al.*, 1993). However, the increasing stress levels of the late Holocene may have acclimatised the reefs to naturally occurring high turbidity. Sedimentation levels which can be tolerated by Great Barrier Reef corals (Table 8.1) appear far higher than those reported elsewhere in the world (Pastorok & Bilyard, 1985).

There are increasing signs of an impact on widespread areas of the Great Barrier Reef. Damage to the reef as the result of at least three major outbreaks of the coral-eating crown-of-thorns starfish over the last 30 years has been attributed by some to anthropogenic causes. Whilst this is yet to be proven, it is still a possibility and research into the water-quality record retained by the coral skeleton itself in the annual growth rings laid down in the colonies suggests a drastic decline in water quality over the last 30 years on the inshore reefs within the reach of river plumes in the Cairns area, an area of major agricultural activity and fertiliser application. (Rasmussen, Cuff & Hopley, 1994b). Such results appear to confirm anecdotal accounts of reef deterioration during the twentieth century, as for example, suggested by Sir Maurice Yonge for Low Isles between his visits in 1928/29 as leader of the Royal Society Expedition and his return to the area during the 1970s. The effects of the decline of water quality on the Great Barrier Reef from sources such as agricultural fertilisers and sewage have been the focus of great debate in Australia in recent years. One result have been a major $1 million research programme initiated by the Great Barrier Reef Marine Park Authority in 1991 into water-quality problems.

The Great Barrier Reef over the last 3000 years has clearly been growing in what may be considered deteriorating water-quality conditions. Just as clearly, there is a record of degradation of these water-quality conditions over the last 30 years or more as the result of human activities. Thresholds at which reef systems can change direction very quickly (Buddemeier & Hopley, 1988) may not yet have been reached, but there are clear signals that they may be approaching.

Over the next century or so effects of global environmental change as the result of the Greenhouse Effect may be superimposed over these recent trends. The effects for coral reefs may not be as drastic as those for other ecosystems (Hopley & Kinsey, 1988; Kinsey & Hopley, 1991; Buddemeier & Smith, 1988). A slight rise in sea level and increase in storminess as predicted for Great Barrier Reef waters may lead to renewed coral growth and greater flushing of lagoon and other nutrient sink areas, producing conditions which are analogous to those of the latter part of the postglacial marine transgression 7000 years ago. Such changes ultimately may do much to counteract the negative human impacts of the last century.

Worldwide comparisons

There is no single major shelf reef system which has available as comprehensive a history as that for the Great Barrier Reef. However, there is sufficient evidence from the Caribbean to suggest that changing water-quality conditions during the Holocene transgression have similarly been responsible for many fossilised features of the continental shelves (e.g. Macintyre, 1967, 1972, 1988; Macintyre & Milliman, 1970; Adey *et al.*, 1977; Lighty 1977; Lighty, Macintyre & Stuckenrath, 1978; Neumann & Macintyre, 1985; Fairbanks, 1989, and Macintyre *et al.*, 1991).

Macintyre (1988) summarised four stages of development of eastern Caribbean reefs as follows:

a) Pleistocene shelves were exposed to subaerial weathering and a soil zone developed on their surfaces.
b) With rising seas, fringing reefs became established along the upper slopes or shelf edges. In some areas, fringing reefs developed into barrier reefs with associated lagoons.
c) The continuing rising seas that flooded broad areas of the shelves caused widespread erosion of soil and lagoon deposits. The resulting high turbidity and high nutrient levels terminated reef growth on the outer edges of these shelves. In northerly areas, such as the east coast of Florida, the intermittent cooling of shallow-shelf waters could also have led to the demise of shelf-edge reefs.
d) By the time water conditions returned to normal, the advancing seas were too deep to allow shallow-water communities of *Acropora palmata* to establish. The rapid reef growth associated with these communities was thus transferred to shallow-water substrates adjacent to coastlines. As a result, 'modern barrier reefs are notably absent in this part of the world' (Macintyre, 1972, p.737). Instead, this area is characterised by relict 'give-up reefs' (Neumann & Macintyre, 1985 [see also Fig. 7.5]) along the upper slopes and shelf edges and by relatively young (less than 7000 years BP) late Holocene reefs fringing most coastlines. The age limit of inner shelf reef development is related to the time of flooding of these shallow areas and has nothing to do with regional cooling related to continental glaciation.

Probably the most distinctive feature of this sequence is the demise of Late Pleistocene and early Holocene shelf-edge reefs spanning depths of up to 80 metres and dated up to 12000 years BP. The death of these reefs has variously been referred to 'being shot in the back by your own lagoon' (Neumann & Macintyre, 1985) or 'killed off by inimical bank waters' (e.g., Schlager, 1981). Hallock & Schlager (1986) attribute the demise of these reefs almost entirely to eutrophication due to release of nutrients from a submerging shelf at a time when sea level was still rising rapidly.

The eastern Caribbean and Great Barrier Reef appear to be different in two major ways.

1. the pattern outlined above for the Great Barrier Reef did not incorporate 'give-up' shelf edge reefs; and
2. the eastern Caribbean appears to have no major equivalent of the *Halimeda* banks of the Great Barrier Reef.

However, deep shelf-edge reefs have been described along the entire length of the Great Barrier Reef for over 65 years (Paradice, 1925; Hopley, 1982 pp. 271–3; Harris & Davies, 1989). Depth ranges are between 40 and 150 metres (Fig. 8.9). Very little is known about these reefs, not even to the extent of whether or not they support living deepwater coral communities and nothing

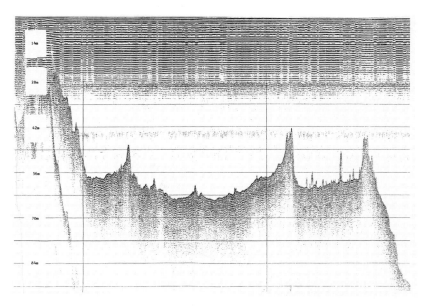

Figure 8.9. Seismic profiles of shelf-edge reefs off the Pompey Reefs, south central section of the Great Barrier Reef.

is known about their age or sea-level relationship. Nonetheless, their ubiquitous presence and similar shelf-edge location to features in the eastern Caribbean suggests that they could also be Late Pleistocene or early Holocene in age and could have been largely retarded in their growth at the time when the shelf was first flooded and *Halimeda* banks began to accumulate.

   The lack of *Halimeda* banks or equivalent structures in the Caribbean story may also be more apparent than real. *Halimeda* is present in the Caribbean and some *Halimeda* structures have been described, but the depth of the shelf in relation to the different history of sea-level rise probably meant that conditions of eutrophication were present for a relatively short period, perhaps less than 2000 years. This was sufficient to kill off shelf reefs, but insufficient for major bioherm construction. Hine *et al.* (1988), however, have described *Halimeda* bioherms which form a nearly continuous band bordering the margins of the Miskito Channel in the southwest Caribbean. Banks with a relief of 20–30 metres and depths of 40–50 metres appear to be built of largely dead *Halimeda* material. Their location and depth suggests that they could have been formed at a time when bank flooding was first taking place and upwelling was also present from the adjacent deep water. Similar deep bank edge *Halimeda* banks have also been described from the Java Sea (Phipps & Roberts, 1988; Roberts, Aharon & Phipps, 1988). Banks have depths of 20–100 metres and upwelling and nutrient overloading are suggested as explanations for the remarkable algal growth at the expense of reef-building corals during early platform drowning.

A model for tropical carbonate accumulations during a transgression–stillstand cycle (Fig. 8.10)

The data from the Great Barrier Reef and worldwide are sufficiently consistent to suggest that there are predictable changes to water quality during a transgression and stillstand period. Although the pattern could apply to any transgression and stillstand period, the Late Pleistocene/Holocene stages are as follows:

1. Late Pleistocene/Early Holocene
   Oceanic waters suitable for coral growth rising against a steeply sloping continental shelf leading to extensive narrow fringing reef development.
2. Early Holocene
   Shelf flooding with rapid decline in water-quality conditions, an increase in nutrients and turbidity in particular leading to a decline in reef growth whilst sea level is still rising rapidly and, in suitable locations, the development of algal bioherms at the expense of reef growth.

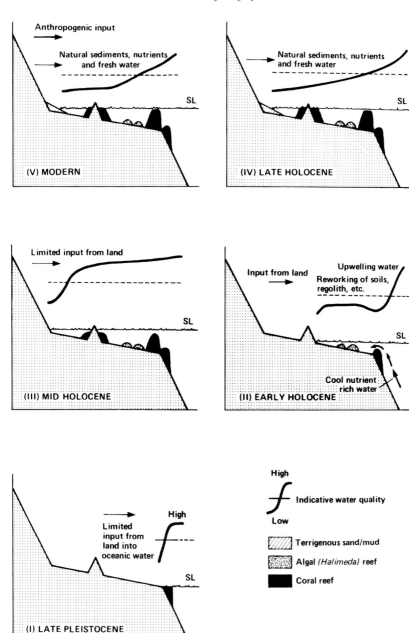

Figure 8.10. A model for tropical carbonate accumulations during a transgression–stillstand cycle. Water-quality conditions are closely linked to the sea-level stage. Note that pre-existing reefal platforms are not illustrated in this schematic representation.

3. Mid Holocene

A period of sea-level stabilisation, but at a level too high for regeneration of the former shelf edge reefs. Water quality has improved and the major phase of coral reef growth commences.

4. Late Holocene

A late Holocene phase in which reef growth on the innermost shelves becomes restricted as terrigenous sediments build outwards from the coastline (for a Caribbean example the Galeta Point reef, Panama shows many features similar to the Cape Tribulation reefs of the Great Barrier Reef (Macintyre & Glynn, 1976; Macintyre, 1988).

5. Modern

At a global scale, the anthropogenic overprint which has been recognised on the Great Barrier Reef is even more distinctive. The decline of Caribbean reefs in particular as a result of several centuries of human impact is well documented (e.g. Brown, 1987; Lapointe, 1989; Shinn, 1989; Glynn 1990). Many of these reefs are showing the classic, chronic stress symptoms as described by Kinsey (1988), resulting in widespread bleaching and death at times of acute stress such as produced by abnormally high temperatures during recent El Niño events.

Conclusion

It is suggested that this morphogenetic approach to shallow-water marine ecosystems will have the same advantages as a climatic or morphogenetic approach in terrestrial geomorphology. Explanation is given to what may seem anomalous features on continental shelves, whilst simultaneously these features may provide insight into changing environmental conditions. Applications are possible to the longer-term geological record. Changing ecological patterns have been described at geological timescales (e.g., Copper, 1988). Whilst many large-scale changes are related to plate tectonics (for example the movement of coral reef platforms on the Pacific plate northwards out of the zone of modern reef growth to produce the drowned guyots of the Emperor Seamount chain (Grigg & Hey, 1992)), other changes have been related to changing water-quality conditions (Copper, 1989). These may be cyclic, as for example the 100 000 year ocean nutrient cycle occurring during glacial and interglacial cycles as suggested by Wilson (1989) resulting in rapid changes in low-latitude nutrient levels, or individual and catastrophic as the Miocene phosphate spike (Riggs, 1984). Nonetheless, the sequence of events described in this chapter appear to be logical and distinctive enough to warrant research for similar patterns covering transgressive–stillstand periods in the long-term geological record.

Acknowledgements

This paper draws on the work of many of my Honours and Postgraduate students. In particular, I would like to acknowledge Bruce Partain, Cecily Rasmussen, Joanie Kleypas and Rob van Woesik.

References

Adey, W.H., Macintyre, I.G., Stuckenrath, R. & Dill, R.F. (1977). Relict barrier reef system of St. Croix: its implications with respect to late Cenozoic local reef development in the western Atlantic. In *Proceedings Third International Coral Reef Symposium, Miami*, vol. 2, pp. 15–21.

Bird, E.C.F. & Hopley, D. (1969). Geomorphological features on a humid tropical sector of the Australian coast. *Australian Geographical Studies*, **7**, 87–108.

Brown, B.E. (1987). Worldwide death of corals: natural cyclical events or man-made pollution? *Marine Pollution Bulletin* **18**, 9–13.

Buddemeier, R.W. & Hopley, D. (1989). Turn-ons and turn-offs: causes and mechanisms of the initiation and termination of coral reef growth. In *Proceedings Sixth International Coral Reef Symposium, Townsville*, vol. 1, pp. 253–62.

Buddemeier, R.W. & Smith, S.V. (1988). Coral reef research in an era of rapidly rising sea level: predictions and suggestions for long-term research. *Coral Reefs,* **7**, 51–6.

Chappell, J., Chivas, A., Wolanski, E., Polach, H.A. & Aharon, P. (1983). Holocene palaeoenvironmental changes, central to north Great Barrier Reef, inner zone. *Bureau of Mineral Resources Journal of Australian Geology and Geophysics,* **8**, 223–36.

Copper, P. (1988). Ecological succession in Phanerozoic reef ecosystems: is it real? *Palaios*, **3**, 136–52.

Copper, P. (1989.) Enigmas in Phanerozoic reef development. *Memoirs of the Association of American Palaeontologists,* **8**, 371–85.

Cortes, J.N. & Risk, M.J. (1985). A reef under siltation stress, Cahuita, Costa Rica. *Bulletin of Marine Science,* **36**, 339–56.

Davies, P.J. (1992). Origins of the Great Barrier Reef. *Search*, **23**, 193–6.

Davies, P.J. & Hopley, D. (1983). Growth facies and growth rates of Holocene reefs in the Great Barrier Reef. *Bureau of Mineral Resources Journal of Australian Geology and Geophysics,* **8**, 237–51.

Davies, P.J. & Marshall, J.F. (1979). Aspects of Holocene reef growth – substrate age and accretion rate. *Search*, **10**, 276–9.

Davies, P.J. & Marshall J.F. (1985). *Halimeda* bioherms, low energy reefs, northern Great Barrier Reef. In *Proceedings Fifth International Coral Reef Symposium, Tahiti*, vol. 5, pp. 1–8.

Davies, P.J., Marshall, J.F. & Hopley, D. (1985). Relationships between reef growth and sea level in the Great Barrier Reef. In *Proceedings Fifth International Coral Reef Symposium, Tahiti*, vol. 3, pp. 95–103.

Drew, E.A. & Abel, K.M. (1985). Biology, sedimentology and geography of the vast inter reefal *Halimeda* meadows within the Great Barrier Reef province. In *Proceedings Fifth International Coral Reef Symposium, Tahiti*, vol. 5, pp. 15–20.

Drew, E.A. & Abel, K.M. (1988). Studies of *Halimeda*. I. The distribution and species composition of *Halimeda* meadows throughout the Great Barrier Reef province. *Coral Reefs*, **6**, 195–205.

Fairbanks, R.G. (1989). A 17,000 year glacio-eustatic sea level record: influence of glacial melting rates on the Younger Dryas event and deep-ocean circulation. *Nature*, **342**, 637–42.

Furnas, M. & Mitchell, A. (1986). Oceanographic aspects of cyclone Winifred. *Great Barrier Reef Marine Park Authority workshop on offshore effects of cyclone Winifred*, pp. 41–2.

Glynn, P.W. (1990). Coral mortality and disturbances to coral reefs in the tropical eastern Pacific. In *Global ecological consequences of the 1982–83 El Niño–Southern Oscillation*, ed. P.W. Glynn, pp. 52, 55–126, Amsterdam: Elseiver Oceanography Series.

Goreau, T.F. & Land, L.S. (1974). Fore-reef morphology and depositional processes, North Jamaica. In *Reefs in time and space*, ed. L.F. Laporte, pp. 77–89. Society of Economic Palaeontologists and Mineralogists Special Publication **18**.

Grigg, R.W. & Hey, R. (1992). Palaeoceanography of the tropical eastern Pacific Ocean. *Science*, **255**, 172–8.

Hallock, P. & Schlager, W. (1986). Nutrient excess and the demise of coral reefs and carbonate platforms. *Palaios*, **1**, 389–98.

Harris, P.T. & Davies, P.J. (1989). Submerged reefs and terraces in the shelf edge of the Great Barrier Reef, Australia: morphology, occurrence and implications for reef evolution. *Coral Reefs,* **8**, 87–98.

Hillis-Collinvaux, L. (1974). Productivity of the coral reef alga *Halimeda* (Order Siphonales). In *Proceedings Second International Coral Reef Symposium, Brisbane*, vol. 1, pp. 35–42.

Hine, A.C., Hallock, P., Harris, M.W., Mullins, H.T., Belknap, D.F. & Jaap, W.C. (1988). *Halimeda* bioherms along an open seaway: Miskito Channel, Nicaraguan Rise, Southwest Caribbean Sea. *Coral Reefs,* **6**, 173–8.

Hopley, D. (1982). *Geomorphology of the Great Barrier Reef: Quaternary development of coral reefs*. New York: John Wiley-Interscience. 453 pp.

Hopley, D., ed. (1983a). *Australian sea levels in the last 15 000 years*. James Cook University of North Queensland, Department of Geography Monograph Series Occasional Paper, **3**. 104 pp.

Hopley, D. (1983b). Deformation of the north Queensland continental shelf in the late Quaternary. In *Shorelines and isostasy*, ed. D.E. Smith & A.G. Dawson, pp. 347–66. London: Academic Press.

Hopley, D. (1984). The Holocene 'high energy window' on the central Great Barrier Reef. In *Coastal geomorphology in Australia*, ed. B.G. Thom, pp. 135–50. Sydney: Academic Press.

Hopley, D. (1985). Geomorphological development of modern coastlines: a review. In *Themes in geomorphology*, ed. A.F. Pitty, pp. 56–71. London: Croom Helm.

Hopley, D. (1988). Anthropogenic influences on Australia's Great Barrier Reef. *Australian Geographer*, **19**, 26–45.

Hopley, D. (1989a). Coral reefs: zonation, zonality and gradients. *Essener Geographische Arbeiten*, **18**, 79–123.

Hopley, D. (1989b). *The formation, use and management of the Great Barrier Reef*. Melbourne: Longman Cheshire. 54 pp.

Hopley, D. & Barnes, R.G. (1985). Structure and development of a windward fringing reef, Orpheus Island, Palm Group, Great Barrier Reef. In *Proceedings Fifth International Coral Reef Symposium, Tahiti.*, vol. 3, pp. 141–6.

Hopley, D. & Kinsey, D.W. (1988). The effects of rapid short term sea level rise on the Great Barrier Reef. In *Greenhouse: planning for climatic change*, ed. G.E. Pearman, pp189–201. Melbourne: CSIRO Division of Atmospheric Research.

Hopley, D., Parnell, K.E. & Isdale, P.J. (1989). The Great Barrier Reef Marine Park: dimensions and regional patterns. *Australian Geographical Studies,* **27**, 47–66.

Hopley, D., Slocombe, A.M., Muir, F. & Grant, C. (1983). Nearshore fringing reefs in north Queensland. *Coral Reefs,* **1**, 151–60.

Hopley, D., van Woesik, R., Hoyal, D.C.J.D., Rasmussen, C.E. & Steven A.D.L. (1993). Sedimentation resulting from road development Cape Tribulation area. *Great Barrier Reef Marine Park Authority Technical Memorandum.*

Hoyal, D.C.J.D. (1986). The effect of disturbed rainforest catchments on sedimentation in an area of nearshore fringing reef: Cape Tribulation, North Queensland. Unpublished honours thesis, Department of Geography, James Cook University of North Queensland.

Jennings, J.N. (1972). Some attributes of Torres Strait. In *Bridge and barrier: the natural and cultural history of Torres Strait,* ed. D. Walker, pp. 29–38. ANU Canberra: Research School of Pacific Studies Publication BG3.

Johnson, D.P. & Risk, M.J. (1987). Fringing reef growth on a terrigenous mud foundation: Fantome Island, central Great Barrier Reef, Australia. *Sedimentology,* **34**, 275–87.

Johnson, D.P. & Searle, D.E. (1984). Postglacial seismic stratigraphy, central Great Barrier Reef province. *Sedimentology,* **31**, 335–52.

Kershaw, A.P. (1978). Record of last interglacial–glacial cycle from northeastern Queensland. *Nature,* **272**, 159–61.

Kinsey, D.W. (1985.) Metabolism, calcification and carbon production. I. Systems level studies. In *Proceedings Fifth International Coral Reef Symposium, Tahiti,* vol. 4, pp. 505–26.

Kinsey, D.W. (1988). Coral reef system response to some natural and anthropogenic stresses. *Galaxea,* **7**, 113–28.

Kinsey, D.W. (1991). The coral reef: an owner-built, high-density, fully serviced, self sufficient housing estate – or is it? *Symbiosis,* **10**, 1–22.

Kinsey, D.W. & Hopley, D. (1991). The significance of coral reefs as global carbon sinks – response to Greenhouse. *Palaeogeography Palaeoclimatology Palaeoecology,* **89**, 1–15.

Kleypas, J.A. (1991). Geological development of fringing reefs in the southern Great Barrier Reef, Australia. Unpublished PhD thesis, James Cook University of North Queensland. 199 pp. + appendices.

Kleypas, J.A. & Hopley, D. (1994). Reef development across a broad continental shelf, southern Great Barrier Reef, Australia. In *Proceedings Seventh International Coral Reef Symposium, Guam.*

Konishi, K. (1982). Alcyonarian spiculite: limestone of soft corals. In *Proceedings Fourth International Coral Reef Symposium, Manilla,* vol. 1, pp. 643–9.

Lapointe, B.E. (1989). Caribbean coral reefs: are they becoming algal reefs? *Sea Frontiers,* Mar–Apr, 1989, pp. 83–91.

Lighty, R.G. (1977). Relict shelf-edge Holocene coral reef: southeast coast of Florida. In *Proceedings Third International Coral Reef Symposium, Miami,* vol. 2, pp. 215–21.

Lighty, R.G., Macintyre, I.G. & Stuckenrath, R. (1978). Submerged early Holocene barrier reef southeast Florida shelf. *Nature,* **276**, 59–60.

Littler, M.M., Littler, D.S. & Lapointe, B.E. (1988). A comparison of nutrient and light-limited photosynthesis in psammophytic versus epilothic forms of *Halimeda* (Caulerpales, Halimedaceae) from the Bahamas. *Coral Reefs,* **6**, 219–25.

Macintyre, I.G. (1967). Submerged coral reefs, west coast of Barbados, W.I. *Canadian Journal Earth Sciences,* **4**, 461–74.

Macintyre, I.G. (1972). Submerged reefs of the eastern Caribbean. *American Association of Petroleum Geologists Bulletin,* **56**, 720–38.

Macintyre, I.G. (1988). Modern coral reefs of the western Atlantic: new geological perspective. *American Association of Petroleum Geologists Bulletin,* **72**, 1360–9.

Macintyre, E.G. & Glynn, P. (1976). Evolution of modern Caribbean fringing reef, Galeta Point, Panama. *American Association of Petroleum Geologists Bulletin,* **60**, 1054–72.

Macintyre, I.G. & Milliman, J.D. (1970). Physiographic features on the outer shelf and upper slope, Atlantic continental margin, southeastern United States. *Geological Society of America Bulletin,* **81**, 2577–2598.

Macintyre, I.G., Rutzler, K., Norris, J.N., Smith, K.P., Cairns, S.D., Bucher, K.E. & Stenick, R.S. (1991). An early Holocene reef in the western Atlantic submersible investigations of a deep relict reef off the west coast of Barbados, W.I. *Coral Reefs,* **10**, 167–74.

Mapstone, B.D., Choat, J.H., Cumming, R.L. & Oxley, W.G. (1989). The fringing reefs of Magnetic Island: benthic biota and sedimentation: a baseline survey. Unpublished report to Great Barrier Reef Marine Park Authority, 88 pp.

Maragos, J.E. (1972). A study of the ecology of Hawaiian reef corals. Unpublished PhD thesis, University of Hawaii, Honolulu.

Marshall, J.F. (1988). Potential effects of oceanic deep waters on the initiation and demise of coral reefs. In *Proceedings Sixth International Coral Reef Symposium, Townsville,* vol. 3, pp. 509–12.

Marshall, J.F. & Davies, P.J. (1988). *Halimeda* bioherms of the northern Great Barrier Reef. *Coral Reefs,* **3/4**, 139–48.

Marshall, S.M. & Orr, A.P. (1931). Sedimentation on Low Isles reef and its relation to coral growth. *Scientific Reports Great Barrier Reef Expedition,* **1**, 94–133.

Maung Maung Aye (1976). Variation in drainage basin morphometry in north Queensland. Unpublished MA thesis, James Cook University of North Queensland, Townsville. 282 pp.

Neumann, A.C. & Macintyre, I.G. (1985). Reef response to sea level rise: keep-up, catch-up or give-up. In *Proceedings Fifth International Coral Reef Symposium, Tahiti,* vol. 3, 105–10.

Orme, G.R. (1985). The sedimentological importance of *Halimeda* in the development of back reef lithofacies, northern Great Barrier Reef, Australia. In *Proceedings Fifth International Coral Reef Symposium, Tahiti,* vol. 5, pp. 31–8.

Orme, G.R., Flood, P.G. & Sargent, G.E.G. (1978). Sedimentation trends in the lee of outer (ribbon) reefs, northern region of the Great Barrier Reef. *Philosophical Transactions Royal Society London, Ser. A.,* **291**, 85–9.

Orme, G.R. & Salama, M.S. (1988). Form and seismic stratigraphy of *Halimeda* banks in part of the northern Great Barrier Reef province. *Coral Reefs,* **6**, 131–7.

Paradice, W.E.J. (1925). The pinnacle or mushroom-shaped coral growths in connection with the reefs of the outer barrier. *Reports of the Great Barrier Reef Committee,* **1**, 52–9.

Partain, B.R. & Hopley, D. (1989). Morphology and development of the Cape Tribulation fringing reefs, Great Barrier Reef, Australia. *Great Barrier Reef Marine Park Authority Technical Memorandum* **21**, 45 pp.

Pastorok, R.A. & Bilyard, G.R. (1985). Effects of sewage pollution on coral reef communities. *Marine Ecolology Progress Series,* **21**, 175–89.

Phipps, C.V.G., Davies, P.J. & Hopley, D. (1985). The morphology of *Halimeda* banks behind the Great Barrier Reef east of Cooktown, Queensland. In *Proceedings Fifth International Coral Reef Symposium, Tahiti*, vol. 5, pp. 27–30.

Phipps, C.V.G. & Roberts, H.H. (1988). Seismic characteristics and accretion history of *Halimeda* bioherms on Kalukalukuang Bank, eastern Java Sea (Indonesia). *Coral Reefs*, **6**, 149–59.

Pringle, A.W. (1986). *Causes and effects of changes in fluvial sediment yield to the northeast Queensland coast, Australia.* Department of Geography, James Cook University, Monograph Series Occasional Paper, 4.

Rasmussen, C., Cuff, C. & Hopley, D. (1994b). Evidence of anthropogenic disturbance retained in the skeleton of massive corals from Australia's Great Barrier Reef. In *Proceedings Seventh International Coral Reef Symposium, Guam.*

Rasmussen, C., Ness, S. & Cuff, C. (1994a). A correlation between the El Niño–Southern Oscillation and crystallographic variability in corals of the Great Barrier Reef. In *Proceedings Seventh International Coral Reef Symposium, Guam.*

Riggs, S.R. (1984). Palaeoceanographic model of Neogene phosphorite deposition, US Atlantic continental margin. *Science*, **223**, 123–31.

Roberts, H.H., Aharon, P. & Phipps, C.V.G. (1988). Morphology and sedimentology of *Halimeda* bioherms from the eastern Java Sea (Indonesia). *Coral Reefs*, **6**, 161–72.

Schlager, W. (1981). The paradox of drowned reefs and carbonate platforms. *Geological Society American Bulletin*, **92**, 197–211.

Searle, D.E. & Flood, P.G. (1988). *Halimeda* bioherms of the Swain Reefs, southern Great Barrier Reef. In *Proceedings Sixth International Coral Reef Symposium, Townsville*, vol. 3, pp. 139–44.

Searle, D.E., Harvey, N., Hopley, D. & Johnson, D.P. (1982). Significance of results of shallow seismic research in the Great Barrier Reef province between 16°10' S and 20°05' S. In *Proceedings Fourth International Coral Reef Symposium, Manilla*, vol. 1, pp 532–9.

Shinn, E.A. (1989). What is really killing the corals? *Sea Frontiers*, Mar–Apr, **1989**, 72–81.

Smith, S.V. & Jokiel, P.L. (1975). Water composition and biogeochemical gradients in the Canton Atoll lagoon: budgets of nitrogen, phosphorus, carbon dioxide and particulate materials. *Marine Science Communications*, **1**, 165–207.

Stoddart, D.R. (1968). Climatic geomorphology: review and reassessment. *Progress in Geography*, **1**, 161–222.

Symonds, P.A., Davies, P.J. & Parisi, A. (1983). Structure and stratigraphy of the central Great Barrier Reef. *Bureau of Mineral Resources Journal of Australian Geology and Geophysics*, **8**, 277–91.

Thomas, M.F. (1974). *Tropical geomorphology: a study of weathering and landform development in warm climates.* London and Basingstoke: MacMillan, 332 pp.

Thomson, R.E. & Wolanski, E. (1984). Tidal period upwelling within Raine Island entrance, Great Barrier Reef. *Journal Marine Research*, **42**, 787–808.

Veron, J.E.N. (1986). *Corals of Australia and the Indo-Pacific*. Sydney: Angus and Robertson. 644 pp.

Wiebe, W.J., Crossland, C.J., Johannes, R.E., Rimmer, D.W. & Smith, S.V. (1982). High latitude (Abrolhos Islands) reef community metabolism: what sets latitudinal limits on coral reef development? In *Proceedings Fourth International Coral Reef Symposium, Manilla*, vol. 1, pp. 721.

Wilson, T.R.S. (1989). Climate change: possible influence of ocean upwelling and
     nutrient concentration. *Terra Nova*, **1**, 172–6.
van Woesik, R. (1992). Ecology of coral assemblages on continental islands in the
     southern section of the Great Barrier Reef, Australia. Unpublished PhD thesis,
     James Cook University of North Queensland. 226 pp.
Wolanski, E., Drew, E.A. & Abel, K.M. (1987). The Barrier Reef *Halimeda* banks –
     why? *Australian Marine Science Bulletin,* **98**, 10–11.
Wolanski, E., Drew, E.A. & Abel, K.M. and O'Brien, J. (1989). Tidal jets, nutrient
     upwelling and their influence on the productivity of the alga *Halimeda* in the
     Ribbon Reefs, Great Barrier Reef. *Estmarine Coastal and Shelf Science*, **26**,
     169–201.

9

Arctic coastal plain shorelines

P.R. HILL, P.W. BARNES, A. HÉQUETTE AND M-H RUZ

Introduction

In this chapter we discuss shorelines which are loosely grouped under the name 'Arctic coastal plain shorelines'. These shorelines are formed in polar regions where unconsolidated and permafrost-affected sediments form low bluffs and where marine processes are affected by the seasonal presence of sea ice. The emphasis will be exclusively on coastlines where relative sea level is rising. Such coastlines are found along the perimeter of the Arctic Ocean – along the Alaskan and Canadian Beaufort Sea coast and northeastern Siberia (Fig. 9.1). Much of the remaining Arctic coastline is characterized by either

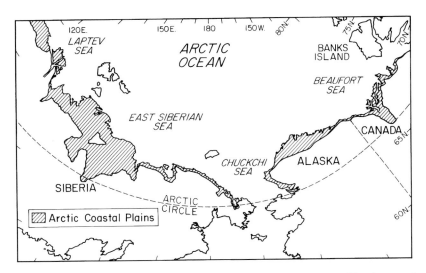

Figure 9.1. The Arctic Ocean margin showing the approximate extent of Arctic coastal plains in northwestern North America and eastern Siberia. These regions are characterized by low tundra bluffs and rising relative sea level.

bedrock cliffs or falling sea level (Taylor & McCann, 1983), and is excluded from this chapter.

The broad aims of this chapter are to review the present understanding of coastal evolution in Arctic coastal plain settings and to evaluate the importance of truly Arctic (zonal) processes to this evolution. A synthesis of the important zonal processes, using several examples from the Alaskan and Canadian Beaufort Sea coasts, will be presented. We emphasise that this does not represent all polar coasts and that, whereas the factors we have identified here may be important in other polar regions, the range of possible climatic, geologic and oceanographic conditions dictates that other factors may be important elsewhere.

One of the most important considerations when evaluating the information presented here is the problem of data acquisition in these remote regions. Although adequate aerial photography is available over most of the Beaufort Sea coastline, the distribution of ground surveys is relatively patchy, surveys being conducted from either remote field camps, small boats or aircraft. Measurement of marine processes, such as waves and currents, is extremely limited in both duration and geographic distribution, the sea ice making equipment moorings particularly hazardous. Furthermore, there is a strong bias towards data collected during the summer season. Although a few winter studies have been carried out, studies during autumn and spring have been extremely limited due to logistic difficulties during the freeze-up and break-up of the sea-ice cover. Unfortunately, these periods may have considerable significance in terms of potentially important coastal processes and the lack of information is a major limitation to the understanding of Arctic tundra coasts.

Important zonal factors

The concept of 'zonality' as it relates to latitudinal or climatic variation of coastal geomorphology has been discussed by numerous authors (including Guilcher, 1953; Davies, 1972; Kelletat, 1989). This concept recognizes that whereas some coastal processes, such as tides and currents, do not show a marked variation with latitude or climatic zones, others are strongly related to distinct latitudinal zones. Coastal processes associated with extremely cold conditions provide probably the clearest example of zonality, being restricted to polar latitudes (Kelletat, 1989). Several zonal factors characterize Arctic coastal plain shorelines and have a potentially important influence on coastal evolution. These can be grouped under the two major headings: permafrost and sea-ice phenomena (Fig. 9.2).

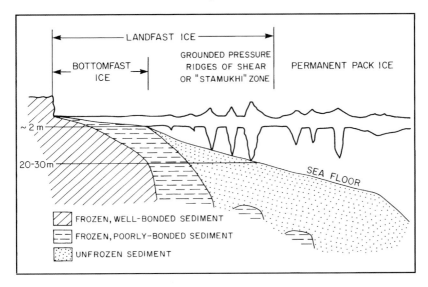

Figure 9.2. Schematic illustration of sea-ice zonation and permafrost distribution. First year sea ice in the landfast ice zone reaches up to 2 m in thickness, except in regions where pressure ridging occurs. Multi-year ice of the permanent pack is thicker, averaging 3 m. The representation of permafrost is based on studies from the Canadian Beaufort Sea (Dallimore *et al.*, 1988; Blasco *et al.*, 1990). The coastal zone represents a transition zone from continuous permafrost to patchy sub-sea permafrost.

Permafrost phenomena

Permafrost is ubiquitous throughout northern Siberia, Canada and Alaska. On land, permafrost is aggradational, the depth of frozen ground corresponding to the depth of the 0 °C isotherm, which may reach 700 m depth in Wisconsinan or older sediments (Burgess *et al.* 1982; Judge, 1986; Lachenbruch *et al.*, 1987). In terrestrial Holocene sediments such as those of the Mackenzie Delta, the thickness of permafrost may be considerably less than 100 m (Smith, 1975) due to the relatively short time period of permafrost aggradation. As summer temperatures rise above 0 °C, a seasonally active layer develops in the top 1 to 2 m of unconsolidated sediments.

Permafrost is also present in continental shelf sediments offshore (Hunter & Hobson, 1974; Neave & Sellman, 1984), but is presently degrading seaward of the bottomfast ice zone, due to the warming effect of the sea. The 0 °C isotherm occurs at up to 600 m depth below the seabed on the Canadian Beaufort Shelf, but sediments containing saline pore water may not be frozen to this level (Judge, 1986).

Most of the perennially frozen ground of the Arctic coastal plains of Alaska and Canada contains both interstitial pore ice and larger ground-ice bodies. The quantity of both depends on the particular lithology and local processes of groundwater migration (Mackay, 1971; Rampton, 1988a). Ground ice is present in several forms, including veins, lenses and massive ice bodies (Fig. 9.3*a*). The melting of pore ice and ground ice in bluff faces results in mass failure as 'thaw flow-slides' (Fig. 9.3*b*; Harry, 1985) and as thermoerosional falls along ice wedges (Barnes, Rawlinson & Reimnitz, 1988). These processes result in very rapid local and short-term bluff retreat. Certain sites monitored along the Canadian Beaufort Sea coast have retreated 23 m in two years (Hill *et al.*, 1990). Average retreat rates range from $1-2\,\mathrm{m\,a^{-1}}$ to over $15\,\mathrm{m\,a^{-1}}$ (Reimnitz, Graves & Barnes, 1988; Harper, 1990). The products of these cliff failures are deposited temporarily on the beach and shoreface (Fig. 9.3). Depending on the efficiency of shoreface erosion and sediment transport, the bluff may stabilize for several years until erosion again causes the cliff to become thermally and gravitationally unstable.

Thermally triggered slumping also contributes to the development of thermokarst topography, a morphology typical of Arctic coastal plains (Czudek & Demek, 1970; Rampton, 1988b). Caused initially by the presence of standing water, this morphology is formed by continued ground ice melting and development of an unfrozen 'talik' beneath the standing water. The pond gradually expands in size due to preferential consolidation of the unfrozen sediments, assisted by thaw flow-slide activity along the margins. Eventually, the terrain becomes dominated by numerous lakes, commonly several kilometres in diameter and a few metres deep. Rampton (1988b) has suggested that the development of thermokarst topography in northern Canada may have originated around 9000 to 10000 years BP during climatic optimum conditions. Thermokarst topography has a considerable influence on the development of shoreline morphology in Arctic coastal plain settings, through the development of narrow headlands and broad embayments (Wiseman *et al.*, 1973; Ruz, Héquette & Hill, 1992). This provides a particular form of antecedent morphology to a shoreline retreating under rising relative sea level, although its importance to coastal evolution has been questioned (Reimnitz, *et al.*, 1988).

A less-understood phenomenon related to permafrost is the subsidence associated with degrading permafrost at the coastline. In a general sense, the shoreline and nearshore zone represent the boundary between regions of aggrading and degrading permafrost. Where sediments remain frozen, the rate of consolidation, and therefore subsidence, may be reduced because the pore

Figure 9.3. (*a*) Typical fresh exposure of ground ice in low tundra bluff, North Point, NWT, Canada. Stake on the left is approximately 1 m high. (*b*) Extensive thermal erosion of ice-rich cliffs, Hooper Island, NWT, Canada. Thaw flow-slides transport cliff material onto the beach, often forming an extensive bench on the upper foreshore. Beach width is approximately 20 m.

space would be filled or partially filled with ice (Sellman, Brown & Lewellen, 1975; Reimnitz *et al.*, 1988). As sediments thaw, consolidation may accelerate as the pore spaces collapse and water is expelled. In detail, this phenomenon is complex, depending on the thermal regime, the sediment properties (thermal conductivity, porosity and permeability), the presence of massive ice, sedimentation rates and the rate of coastal retreat (Dyke, 1991).

The thermal profile of an Arctic shoreface depends principally on the overlying air or water temperature. Where sea ice is frozen to the bottom, the seabed experiences winter temperatures below 0 °C because heat is conducted to the atmosphere by the sea ice. In the Beaufort Sea, the winter thickness of sea ice reaches between 1.5 and 2.0 m. Thus, much of the shoreface and nearshore zone may be seasonally frozen. Dallimore, Kurfurst & Hunter (1988) have shown that the thermal profiles and permafrost conditions of two sections of coastline within 5 km of each other can be considerably different (Fig. 9.4). Although both sections were characterized by a gently sloping shoreface underlain by ice-bonded sediment, the permafrost table dipped seaward more steeply at the site with the steeper nearshore profile than at the site with a gently sloping nearshore profile. This suggests that the depth of permafrost in the nearshore depends on the length of time the seabed is exposed to low winter temperatures through freezing of the sea ice to the seabed and the temperature of coastal water in the summer (Dyke, 1991). In the Canadian Beaufort Sea, summer near-bottom water temperatures range from 0 to 11 °C due to the influence of Mackenzie River water and transgressed permafrost will tend to thaw. The thaw will be intermittent within the bottomfast ice zone and continuous seaward of it.

Whereas the thaw of frozen sediment may result in delayed but predictable subsidence, the thaw of massive ground ice in transgressed sediments may cause rapid subsidence. Dallimore *et al.* (1988) calculated that the melting of a massive ice body discovered in the drilling program described above could result in the seabed settling locally by as much as 10 m. Depressions in the transgressive unconformity surface may be explained by such a process (Fig. 9.4).

Sea ice phenomena

The Arctic Ocean is characterized by a permanent ice pack, consisting of aggregates of multi-year ice and rare fragments of glacier ice (ice islands). The modal thickness of the ice pack is approximately 3 m, with pressure ridge keels and ice islands reaching 45 m (Untersteiner, 1990; Hobson *et al.*, 1989). The margins of the ice pack consist of thinner 'first year' ice formed during freeze-up of a single year. This ice is generally less than 2 m thick, but may be thicker

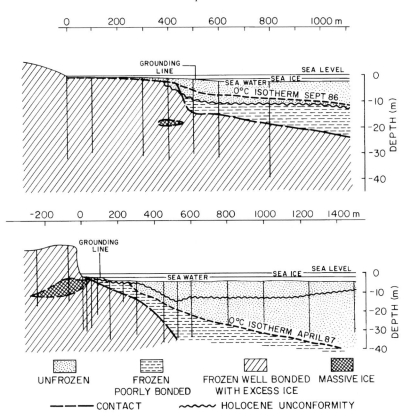

Figure 9.4. Thermal conditions of gently sloping (top) and more steeply sloping (bottom) nearshore zones, off western Richards Island, NWT, Canada. Vertical lines represent borehole penetration. Note the more steeply dipping permafrost table in the lower diagram, associated with the narrower bottomfast ice zone. (Modified from Dallimore *et al.*, 1988.)

locally where ridging occurs. Along most sections of the Arctic Ocean coast, open water is present during the summer. The extent of open water depends on both atmospheric forcing and the resultant oceanic circulation (Mysak & Manak, 1989), as well as on local conditions such as wind stress or river discharge. As a consequence, it is highly variable from year to year (Fissel, Birch & Melling, 1990).

The presence of sea ice is an important control on the fetch and consequent wave regime experienced by Arctic coastal plain shorelines (Hill *et al.*, 1991). The development of waves in ice-laden waters is poorly understood. The presence of ice inhibits the generation of waves, but it is not clear under what conditions a wave train can be formed. Pinchin, Nairn & Philpott, (1985) achieved reasonable calibration of their wave hindcast model assuming that stable wave trains form in water with less than 3/10 ice cover. In the Beaufort Sea, the fetch distance changes throughout the summer, generally reaching a maximum in late August. The most pronounced variability in fetch is alongshore. Along the Alaskan and Yukon (western Canadian Beaufort Sea) shorelines, the fetch distance rarely exceeds 30–40 km due to the presence of grounded ice along the stamukhi zone (Reimnitz *et al.*, 1988). However, in the eastern Beaufort Sea, close to the influence of the Mackenzie River, the fetch can reach up to several hundred kilometres. Where the fetch direction coincides with predominant storm winds, relatively high wave conditions can be generated. For most of the Beaufort Sea, long fetches to the northwest and strong winds from the same quadrant result in the most extreme wave conditions. Estimation of return periods for extreme waves, however, requires evaluation of the joint probability of long fetches and extreme storms. Also, the shallowness and flat profile of Arctic shelves means that substantial depth limitation of waves would occur. Hodgins (1985), based on an evaluation of five different hindcast models, estimated a deep-water significant wave height of 7 to 9 m for a return period of 100 years in the Canadian Beaufort Sea.

The presence of sea ice in the water column may modify considerably the processes of erosion, transport and deposition of sediment in the Arctic nearshore. Ice may be present as large blocks, as small newly formed crystals (frazil ice) or as a surface layer of slush ice, consisting of both frazil ice and fragments of older sea ice (Reimnitz & Kempema, 1987; Reimnitz, Kempema & Barnes, 1987).

Large blocks of sea ice may interact with the nearshore seabed through the process of ice wallow. Wallowing is the term used to describe the motion of grounded ice in waves (Reimnitz & Kempema, 1982). The wallowing of grounded ice floes in 2 to 7 m water depth, within 1 km of the coast, forms local, crater-like erosion surfaces with relief of 2 to 3 m and spacing of 50 to 100 m (Reimnitz & Kempema, 1982). These features form preferentially in granular sediments and repetitive surveys show that they are unstable over a period of three years. These observations suggest that coastal processes rapidly cause infill of wallow depressions. Ice wallow may nevertheless exert a considerable influence on beach and nearshore dynamics.

Frazil ice forms during the autumn when turbulent coastal waters become supercooled. The fine frazil crystals are characteristically sticky and are known to adhere together, to the seabed, and to other objects (Martin, 1981). Typically, frazil ice is generated during autumn storms, which mix and cool the coastal water column (Reimnitz *et al.*, 1987). The downward transport of frazil ice by turbulence causes the formation of anchor ice, where the seabed becomes coated with a mass of frazil particles (Reimnitz *et al.*, 1987). The presence of anchor ice can have an armouring effect mitigating against entrainment of sediment, but only during the storm which produces it. Heat conducted from the seabed leads to the eventual release of the anchor ice which floats up to form part of the developing ice canopy (Reimnitz *et al.*, 1987). This process can result in considerable erosion of sediment from the bottom and its incorporation into the winter ice cover.

Before complete freezing of the ice canopy, there is a period of several weeks where a layer of slush ice floats at the sea surface. Reimnitz & Kempema (1987) have documented the characteristics of this layer on coastal transects off Alaska. The slush ice effectively filters out high-frequency wave oscillations and inhibits wave breaking. Nevertheless, in the free water below the slush ice layer, Reimnitz & Kempema (1987) observed a rippled seabed, with a wavelength similar to that of the observed waves in the slush, and transport of cobble-sized clasts at 0.95 m depth, indicating that sediment transport occurs on the shoreface under these conditions. The long-term effects of freeze-up processes on beach and nearshore equilibrium profiles have not been evaluated.

Apart from modifying hydrodynamic processes, the ice cover can have a direct impact on the seabed. The phenomenon of ice scouring (or ice gouging – terminology varies from country to country) has been well documented in the Beaufort Sea (Fig. 9.5; Lewis, 1978; Barnes, Rearic & Reimnitz, 1984; Rearic, Barnes & Reimnitz, 1990; Shapiro & Barnes, 1991). Ice scouring is caused by

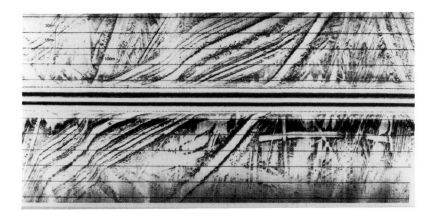

Figure 9.5. Sidescan sonar image showing extensive ice-scour erosion of lower shoreface/inner shelf, off the Tuktoyaktuk Peninsula, NWT, Canada.

P.R. Hill et al.

the impact of ice ridge keels on the seabed. The highest intensity of seabed scouring occurs on the seabed below the shear zone between the moving Arctic pack ice and the more stable landfast ice (Barnes *et al.*, 1987). This region occurs in 15–40 m water depth over most of the Beaufort Sea (Barnes *et al.*, 1984). Smaller pressure ridges and ice blocks cause scouring inside the main shear zone, within the landfast ice, in shallow water (Shapiro & Barnes, 1991). Frequent scouring also occurs in shallow water by the grounding of isolated ice floes during the summer. Héquette & Barnes (1990) have shown that, at least locally, ice-scour erosion of the nearshore seabed may change the equilibrium profile and lead to coastal retreat (Fig. 9.6). Ice scouring may bulldoze sediments in a net onshore direction. Reimnitz, Barnes & Harper (1990) calculated that a single scouring event moved 1–2 $m^3 m^{-1}$ at 8 m water depth.

Two other sea-ice processes which have a potential impact on the coastal zone are the phenomena of ice pile-up (Fig. 9.7*a*) and ice ride-up (Barnes, 1982; Kovacs, 1983, 1984; Reimnitz *et al.*, 1990). These result from the thrusting of ice onto the shoreline in winter, ice pile-up occurring where the ice buckles and fails, forming a pile of broken ice on the beach, and ice ride-up

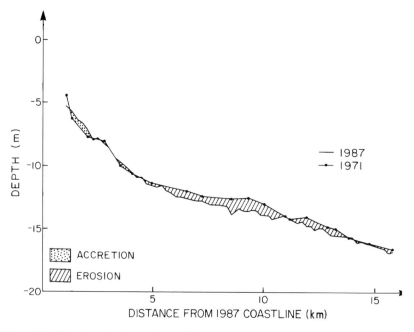

Figure 9.6. Nearshore bathymetric profile showing erosion attributed to ice scour in water depths of 12–15 m, Tuktoyaktuk Peninsula, NWT, Canada (from Héquette & Barnes, 1990).

Figure 9.7. (*a*) Ice pile-up ridge formed along the spit at Atkinson Point, Tuktoyaktuk Peninsula, NWT, Canada. Note 2 m figure on ridge showing scale (photograph by D. F. Dickins); (*b*) mounds of ice-pushed sediment resulting from an ice pile-up event, Camden Bay, Alaska, USA.

where a large sheet of ice slides smoothly up on to the shore, continuing for hundreds of metres (Reimnitz *et al.*, 1990). In some cases, ice has been documented to override bluffs of several metres high (Kovacs, 1983, 1984) and on a regional basis may occur every year. Ice pile-up and ride-up may occur very quickly, the whole process taking 15–30 minutes (Reimnitz *et al.*, 1990). Subsequent melt-out of this ice often leaves a substantial deposit of sediment on the beach (Fig. 9.7*b*) and/or bluff (Barnes, 1982; Reimnitz *et al.*, 1990).

Evolution of Arctic coastal plain shorelines

When considering the morphodynamic evolution of Arctic coastal plain shorelines, it is useful to consider two end-member cases: 1. those shorelines whose evolution is strongly influenced by zonal (i.e. Arctic) processes; and 2. those shorelines where zonal processes are subordinate to the normal (azonal) processes of wave action. Obviously, the importance of both zonal and azonal processes along any one section of shoreline varies with time, both seasonally and at longer time scales. It is impossible to define any one region to be exclusively controlled by one or other type of process. Nevertheless, the two examples presented below are thought to represent as far as possible these two end-member cases.

Shorelines strongly influenced by zonal processes: Alaskan Beaufort Sea

Background

The Alaskan Beaufort Sea (Fig. 9.8) is almost completely ice-covered for nine months of the year (October to June). Even during the summer months, when some open water exists along the coast, the maximum fetch is rarely less than 30 to 40 km and drifting ice commonly reduces the effective fetch (Barnes *et al.*, 1988). A mobile ice belt is also commonly stranded on the shoreface for extended periods. As a result, wave heights are generally less than 0.5 m with periods less than 3 s throughout the open-water season (Short, Coleman & Wright, 1974) and the impact of waves on the shoreface is intermittent and modified by the presence of ice. Higher waves are observed in coastal lagoons than in the open sea (Reimnitz *et al.*, 1988). Storm waves reach 1 to 3 m in height, with periods of 5 to 10 s, the strongest winds and longest fetches coming from the northwest (Short *et al.*, 1974; Reimnitz *et al.*, 1988). Extreme wave heights estimated at 3 m were observed during a one-in-25-year storm in September 1970, associated with a storm surge which left driftwood at elevations as high as 3.4 m above mean sea level (Reimnitz & Maurer, 1979).

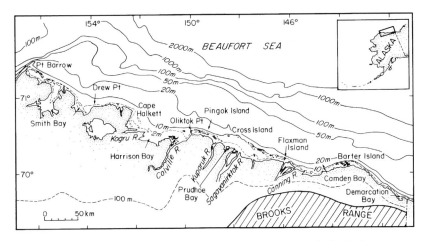

Figure 9.8. The Alaskan Beaufort coastline, showing the principal place names mentioned in the text. Note the presence of offshore barrier islands along much of the coast. (Modified after Barnes *et al.*, 1988.)

Because the maximum tidal range is less than 0.15 m, water-level fluctuations at the coast are influenced strongly by meteorological events, with positive and negative surges causing as much as 1 m of additional amplitude (Short *et al.*, 1974).

Sea level has risen in the Alaskan Beaufort shelf region from a late-Wisconsinan minimum of 96 to 116 m (Dinter, Carter & Brigham-Grette, 1990). In the absence of reliable tide gauge data, the present rate of relative sea-level rise is thought to be close to the eustatic rate (1.0 to 1.5 mm a^{-1}; Gornitz, Lebedeff & Hansen, 1982; Reimnitz *et al.*, 1988). Holocene tectonic movements are not thought to be significant contributors to relative sea-level changes because older (c. 120 000 years BP) shoreline deposits are not deformed and have a consistent elevation across the region (Reimnitz *et al.*, 1988).

Geomorphology

Most of the Alaskan coastal plain is made up of unconsolidated and ice-bearing Quaternary deposits of the Gubik Formation (Carter, Brigham-Grette & Hopkins, 1986), overlain by late Quaternary aeolian deposits and Holocene lacustrine sediments (Barnes *et al.*, 1988). The coastal plain is characterized by thermokarst topography in the form of patterned ground, thaw lakes and pingoes. Several major rivers, with sources in the Brooks Range, form alluvial fans that cross the coastal plain creating large deltas (Fig. 9.8). At the coast, all these deposits outcrop as low bluffs, generally 1 to 3 m high, but locally up to

10 m, fronted by narrow sand and gravel beaches. Along approximately half of the Alaskan coast, barrier islands of coarse sand and gravel are present several kilometres offshore, having subaerial relief of 1 to 2 m and separated by inlets up to 5 m deep. These island chains are in places anchored by low tundra islands and are separated from the coast by shallow (1 to 3 m) lagoons (Fig. 9.8).

Nearshore morphology varies considerably along the coast. Seaward of the barrier islands well-developed bar systems with crests 3 to 4 m below the sea surface are developed in water depths up to 7 m (Fig. 9.9*a*; Short *et al.*, 1974; Short, 1975). Three types of bar were identified by Short (1975): multiple parallel bars, long parallel bars attached to the shore and short en echelon (transverse) bars, also attached to the shore. The grain size characteristics of these bars are not described by Short (1975). However, the shore-attached bars join with sand and gravel beaches and Barnes *et al.* (1988) describe some transverse bars as 'clean sands'.

Bars are also present off certain sections of tundra bluffs (Short, 1975), but in other regions where bars are not present the seabed slopes gradually to the 2 m isobath, forming a shallow-water bench (Fig. 9.9*b*). Reimnitz, Graves & Barnes (1988) interpret this to be related to the presence of bottomfast ice

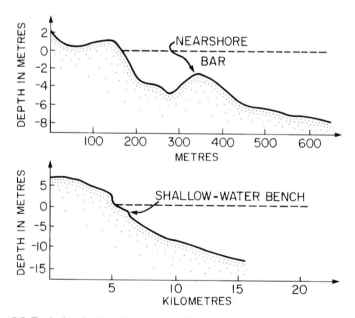

Figure 9.9. Typical seabed profiles from the Alaskan Beaufort Sea: (*a*) barred nearshore off Pingok Island (from Short, 1975); (*b*) non-barred coast showing the presence of the 2 m bench (from Reimnitz *et al.*, 1988).

during the winter. Superimposed on this bench is an irregular series of closed depressions and mounds related to ice-wallow processes. These features have relief of 2 to 3 m and are known to be rapidly formed and obliterated (Reimnitz & Kempema, 1982). Off both barred and non-barred coasts, the inner shelf shows a broadly concave profile to the 20 m isobath (Fig. 9.9). Just inside the 20 m isobath, a series of linear stamukhi shoals is present (Barnes *et al.*, 1987). Although the origin of these shoals is unclear, they are commonly the site of sea-ice grounding, which forms the main constraint on fetch for wave generation (Reimnitz & Kempema, 1984).

Coastal evolution

A geomorphologic model of coastal evolution for the Alaskan region was proposed by Wiseman *et al.* (1973). In this model, the present distribution of broad embayments, offshore islands and barrier islands develops from the thermal collapse, breaching and drowning of thaw lakes (Fig. 9.10). Continued transgression results in coalescing of breached lakes into lagoons, while erosion transforms the intermediate headlands and islands into spit and barrier island systems that eventually become isolated many kilometres from shore. Reimnitz *et al.* (1988) have raised some objections to the Wiseman *et al.* (1973) model, namely that coastal thaw lake depths are insufficient, many lake bottoms are perched several metres above sea level and lake and valley orientations are not always parallel to the shore as implied by Wiseman *et al.* (1973). Ruz *et al.* (1992) have proposed a model similar to that of Wiseman *et al.* (1973) for the geomorphologic evolution of the Canadian Beaufort Sea coast, where lake and valley orientation is, if anything, predominantly perpendicular to the shoreline. The orientation of coastal lagoons would be controlled not only by the original size of the lake, but also by the extent of erosion of the headlands formed after lake breaching. Thus, it seems that the Wiseman *et al.* (1973) model provides a reasonable starting point for understanding coastal evolution in the area. However, the model is only applicable at a relatively large scale and says little about the dynamic processes involved.

The bars present along much of the Alaskan coast attest to the importance of wave action on the shoreline, despite the limited fetch and relatively low-energy conditions (Short, 1975). The bar systems do not fit easily into classical beach and nearshore categories (Greenwood & Davidson-Arnott, 1979; Wright & Short, 1983) and there have been few measurements of wave parameters to help classification. Wiseman *et al.* (1974) measured waves at Pingok Island (Fig. 9.8) over a single summer season, including a storm with wave heights of 1.5 to 2.5 m and periods 9 to 10 s. Spectral analysis showed

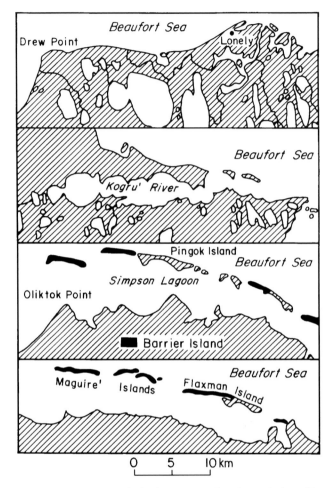

Figure 9.10. Selected coastal morphologies, representing the evolution of lagoons and barrier islands, Alaskan Beaufort Sea coast (after Wiseman *et al.* 1973). The basis of this model is the development of lagoons and barrier islands from thermokarst lakes and headlands. Reimnitz *et al.* (1988) have challenged this hypothesis.

significant infragravity wave energy and spectral peaks compatible with standing wave frequencies predicted from nearshore bar spacing.

These shorelines appear to have the dual characteristics of an apparently dissipative nearshore fronting a relatively coarse-grained, more reflective beach. This differentiation, perhaps most often associated with macrotidal coasts, may be here related to 'meteorological' tides, with the gravel beach being active only during storm surges. The importance of wave motions during freeze-up conditions and of sub-ice current flow during winter is not known. The coarse sediment may be supplied by zonal processes such as ice push. This process has been observed to transport intermittently large volumes of sediment onto barrier islands (Fig. 9.7; Reimnitz *et al.*, 1988).

Although the antecedent topography in the Wiseman *et al.* (1973) model is zonal in nature, being formed by degradation of permafrost, neither this model nor the formation of nearshore bars described by Short (1975) and Wiseman *et al.* (1974) require the short-term dynamic processes acting on this topography to be uniquely Arctic. By contrast, Reimnitz *et al.* (1988), using a sediment budget and equilibrium profile approach, have suggested that the long-term evolution of these shorelines is strongly zonal in nature. These authors calculated the sediment yield from bluff erosion, based on observed retreat rates, bluff composition and ice content. They then assumed that the present nearshore and inner-shelf profile represents an equilibrium profile and calculated the sediment yield from erosion of the inner shelf. Comparison of bathymetric profiles from 1952 to 1980 suggested erosion to approximately 5 km offshore (approximately 8 m water depth; Fig. 9.9).

The result of this analysis was that the sediment yield appears to be many times larger than the available sediment sinks in either coastal lagoons or on the shelf. Lagoons cannot act as permanent sediment sinks because they are shallow and being transgressed. Seismic profiles suggest that the shelf is predominantly erosional, except for a thin veneer, known as the 'ice-keel turbate' layer, formed by the scouring action of ice (Barnes *et al.*, 1984). Reimnitz *et al.* (1988) concluded that this turbate layer may act as a temporary storage but that, in order to maintain the equilibrium profile, sediment must be transported off the shelf and eventually into the Arctic Ocean basin.

Reimnitz *et al.* (1988) also point out that the rates of coastal erosion in the Alaskan Beaufort Sea are high compared with those of other coastlines of North America; comparable, for example, with those along the Gulf of Mexico coast, despite the very limited wave energy available. This requires a very efficient removal of sediment from the coastal zone and these authors conclude that zonal processes must be important because wave energy is insufficient. There is evidence that much of the erosion and transport of sediment occurs during autumn storms, when frazil and anchor ice is forming. When these events are intense, ice scours are completely obliterated on the inner shelf to a distance of 15 km (13 m water depth; Barnes & Reimnitz, 1979), implying extensive bottom transport of sediment. Much sediment is also incorporated into the developing ice canopy as it becomes attached to frazil particles or is lifted to the surface in anchor ice (Reimnitz *et al.*, 1987). This provides a potential mechanism for transport of sediment across the shelf and into the Arctic Ocean basin. Ice scouring may also contribute to sediment transport on the inner shelf, although the predominant transport directions appear to be alongshore or onshore (Rearic *et al.*, 1990).

Shorelines strongly influenced by azonal processes: Canadian Beaufort Sea

Background

The period of complete sea-ice cover in the Canadian Beaufort Sea is similar to that of the Alaskan Beaufort Sea, lasting from October to June. However, the extent of open water is commonly much greater in summer, the edge of the polar ice pack ranging from a few kilometres to more than 300 km offshore. Break-up of the ice cover is controlled largely by the outflow of the Mackenzie River (Fig. 9.11), although the distribution of ice during the summer months is controlled by more complex variables, including air temperature and wind patterns. As a result, wave energy at the coast is determined largely by the available fetch across open water and the wind direction. The wave regime is characterized by wave heights and periods of mostly less than 4 m and 8 s respectively (Pinchin *et al.*, 1985). The strongest winds generally blow from the northwest and fetches are commonly also longer in this direction, so that wave energy is concentrated from this orientation. Easterly winds tend to be lighter, with the result that the Tuktoyaktuk Peninsula (Fig. 9.11) and northwest-facing coastlines experience the most severe wave attack.

Water-level elevations resulting from positive storm surges are extremely important and have been studied using numerical models (Henry, 1975, 1984) and by field observations (Harper, Henry & Stewart, 1988). Mapping of drift log shoreline deposits in the Kugmallit Bay region (Fig. 9.11) suggests a maximum storm surge level of 2.4 m above mean sea level.

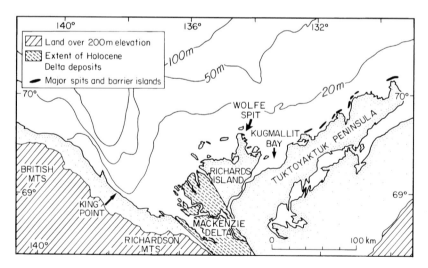

Figure 9.11. The Canadian Beaufort Sea coast, showing principal place names mentioned in text. Barrier islands are absent along the coast west of the Mackenzie Delta, but present elsewhere, particularly along the Tuktoyaktuk Peninsula.

Under open water conditions, current circulation is primarily wind-driven (Hill *et al.*, 1991), currents correlating strongly with winds in coastal waters (Fissel & Birch, 1984; Davidson, de Margerie & Lank, 1988). During periods of northwesterly winds, there is a net easterly drift of surface water (MacNeil & Garrett, 1975; Fissel & Birch, 1984). This pattern is reversed during periods of easterly winds. However, the flow is complex and, particularly in coastal waters, influenced by topography and water levels in coastal embayments (Hill *et al.*, 1990; Héquette & Hill, pers. commun.).

Relative sea level in the Canadian Beaufort Sea rose during the Holocene from a late-Wisconsinan minimum of -70 m (Forbes, 1980; Hill *et al.*, 1985). The present rate of relative rise is thought to be greater than the eustatic rate of 1.0 to 1.5 mm a^{-1}, but recent work suggests that the rate does not exceed a maximum of 2.5 mm a^{-1} (Hill, Héquette & Ruz, 1993).

Geomorphology

The coastal plain of the Canadian Beaufort Sea is also made up of Pleistocene deposits, within which the Holocene Mackenzie Delta occupies a broad erosional trough (Fig. 9.11; Rampton, 1982; 1988b). As on the Alaskan coastal plain, thermokarst topography predominates, resulting in an irregular shoreline of embayments and narrow headlands.

To the west of the Mackenzie Delta, the coastal plain is less than 50 km wide. The deposits are very similar to those of the Alaskan coastal plain (Carter *et al.*, 1986), but the coastal morphology is very different. Bluffs are relatively high (up to 50 m) and often near-vertical, reflecting erosion by a combination of thermal and wave processes. They are fringed with narrow sand and gravel beaches and small embayments may be partially or completely enclosed by gravel spits and barriers. Offshore barrier islands are absent. The barrier beach at King Point appears to have developed within the last 200 years, beginning as a spit and now accreting seaward at its distal end (Hill, 1990; Fig. 9.12). The nearshore profile off King Point is relatively steep and reflective, compared with profiles to either the west or the east (Hill *et al.*, 1986).

To the east of the Mackenzie Delta, the coastal plain widens and sandy lithologies dominate (Rampton, 1988b). Most of the bluffs east of the delta are low (<5 m) and consist of well-sorted fine to medium sand, although units of gravel-rich sediments do occur locally. Cliff-top coastal dunes are common along much of this shoreline. The abundance of thermokarst lakes in this region produces a highly indented coastline with narrow headlands separated by broad embayments (Fig. 9.11). Several small tundra islands lie offshore of Richards Island (Fig. 9.11). Headlands and islands are typically fringed by sand spits, with crestal elevations less than 1.5 m above mean sea level and

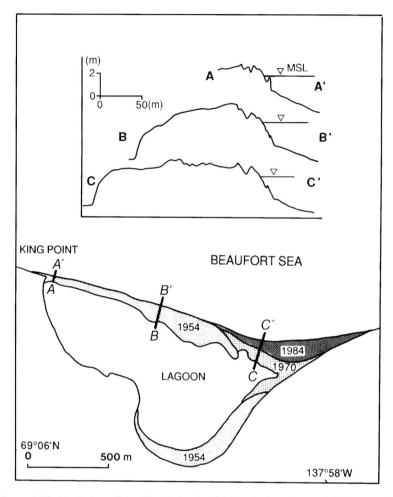

Figure 9.12. Evolution of the King Point barrier beach, based on aerial photographs from 1954 to 1984. Inset shows typical beach profiles across the barrier beach. The barrier-beach developed from a southeastward prograding spit between 1954 and 1970. Since the closing of the lagoon, accretion of the beach has occurred at the eastern end. No inlets have been observed along this gravel-rich barrier in recent years. MSL = mean sea level. (After Hill, 1990; Ruz *et al.*, 1992.)

composed predominantly of fine to medium sand (Fig. 9.13). At Wolfe Spit, where the adjacent headland supplies a proportion of gravel, a storm ridge with a somewhat higher elevation of 2 m has developed (Fig. 9.13). Along the Tuktoyaktuk Peninsula coast (Fig. 9.11), several long barrier island systems, with crestal elevations less than 1 m are present and are migrating landward at rates of 1 to 3 m a^{-1} (Hill *et al.*, 1990; Héquette & Ruz, 1991).

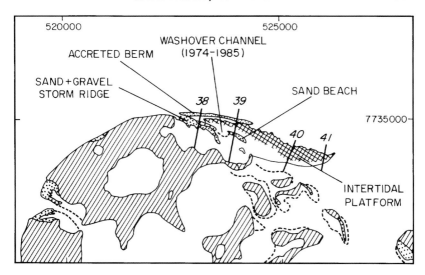

Figure 9.13. Wolfe Spit, Canadian Beaufort Sea, based on aerial photograph analysis. Grid line spacing is 5 km. Profiles 38 to 41 are shown in Fig. 9.14. Note that the spit is differentiated into a sand and gravel storm beach, a sand beach and an intertidal barred platform, suggesting partition of sediment transport between storm surge and fairweather conditions. (After Hill & Frobel 1991.)

Most of the sandy shorelines east of the Mackenzie Delta are characterized by nearshore bars (Fig. 9.14). Parallel bars are the most common form visible from aerial surveys (Harper 1990; Héquette & Ruz, 1991). Transverse bars are found commonly on the foreshore, particularly along relatively sheltered low-energy shorelines. Some low-lying spits and barrier islands, where overwash is common, are made up of accreted transverse foreshore bars (Fig. 9.13).

Coastal evolution

At the geomorphic level, Ruz *et al.* (1992) have developed a conceptual model (Fig. 9.15) for evolution of the Canadian Beaufort coast. The model is similar to that of Wiseman *et al.* (1973) in that the antecedent morphology created by thermokarst processes favours the development of headland and spit morphology. Continued erosion leaves isolated islands with flanking spits, which are transformed eventually into true barrier islands when the tundra island remnants are totally eroded. Deprived of a continued sediment supply from alongshore, the barrier islands migrate landward as a result of continued washover. Evidence from seismic records on the inner shelf suggests that at least some barrier islands are eventually inundated and partially preserved on the inner shelf within thermokarst depressions (Ruz *et al.*, 1992; Héquette &

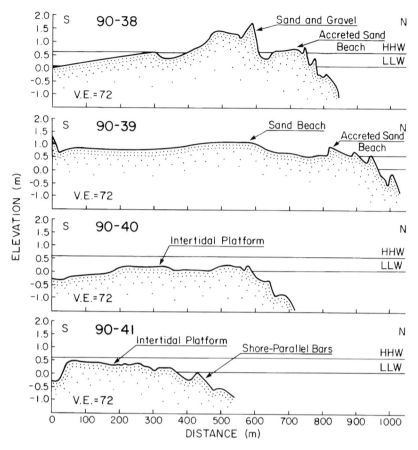

Figure 9.14. Beach profiles from Wolfe Spit, Canadian Beaufort Sea. Note the presence of several bars on the lower foreshore and the broad intertidal platform. Location of profiles shown in Fig. 9.13. V.E. = vertical exaggeration; HHW = high high water; LLW = low low water. (After Hill & Frobel 1991.)

Hill, 1989). The basal parts of some thermokarst lake basins appear to be preserved on both gravel-rich, steep-profile shorelines, such as King Point (Hill, 1990), and on the sandy, low-angle shorelines of Richards Island and the Tuktoyaktuk Peninsula (Héquette & Hill, 1989; Hill & Frobel, 1991). Thus, in contrast to the Alaskan Beaufort Sea, a relatively clear ravinement surface (Nummedal & Swift, 1987) is preserved in the shelf stratigraphic record.

The geomorphologic evolution described above is clearly zonal in nature, being strongly controlled by the presence of permafrost. However, within the above framework, the development of individual beach or spit morphology reflects local conditions of sediment supply, grain size and exposure to wave

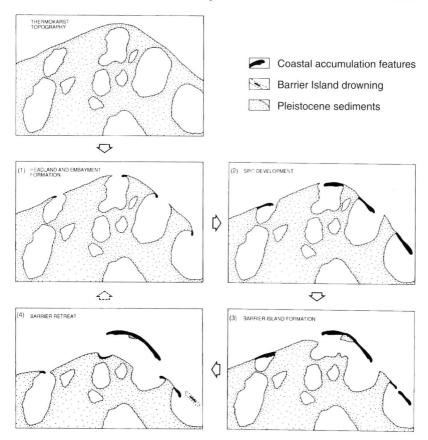

Figure 9.15. Model for coastal evolution of thermokarst terrain, Canadian Beaufort Sea coast. Note similarities to the model shown in Fig. 9.10. (From Ruz *et al.*, 1992.)

energy rather than zonal sea-ice-related processes. The barrier beach at King Point developed through southeasterly longshore transport of sand and gravel eroded from the northwest (Fig. 9.12). Sediment budget calculations suggest that the source of sediment for construction of the initial spit was the former headland, now mainly eroded, but that rapid closure of the barrier and later beach accretion at the southeastern end were related to straightening of the coast and the resultant increased sediment supply (Hill, 1990). Calculations of potential sediment transport based on a hindcast model of wave conditions at King Point have been used to suggest that this morphological evolution is limited principally by the sediment transport potential (Hill, 1990). Thermal erosion of bluffs by thaw-flow slide provides a potentially large local sediment supply, but the bluffs remain stabilized until wave energy has removed all the flow-slide material deposited on the beach. Thus, thermal erosion, although maximizing bluff retreat, does not necessarily accelerate coastal evolution in regions of low wave energy.

Mixed sand and gravel beaches east of the Mackenzie Delta show a differentiation between gravel-rich storm ridges and barred sandy foreshores. Wolfe Spit, for example, is characterized by a recurved 2 m high sand and gravel ridge that extends over 1 km eastward of the headland (Fig. 9.13). A sandy beach has accreted in front of this ridge and extends eastward almost 1 km beyond it, increasing the overall length of the spit. This part of the spit has a maximum elevation of 1.2 m and was breached by a broad washover channel between 1974 and 1985. In front of these two elevated sections of spit, a series of oblique, northeastward-facing bars form a broad platform, with a maximum elevation of 0.6 m, that presently extends over 4 km eastward. Between 1974 and 1985, a narrow sandy berm, with a maximum elevation of 1.0 m accreted seaward of the proximal end of the spit and now extends more than 2 km from the headland. Aerial photographs indicate the the presence of at least two shore-parallel nearshore bars seaward of Wolfe Spit and the headland to the west.

Analysis of this morphology suggests a complex evolution characterized by a partition of sediment transport between storm surge conditions and 'fairweather' conditions when water levels are lower. During northwesterly storms, most of the spit, with the exception of the storm ridge, is submerged. During moderate storms, waves break directly on the nearshore bars and, under extreme surge conditions, it is likely that reformed waves would break on the gravel spit, reactivating sediment transport. Sediment transport during such storms would be predominantly eastward along the nearshore bars, the sandy berm and the upper beach foreshore. The oblique foreshore bars appear to be in equilibrium with easterly wind ('fairweather') conditions and suggest that accretion of these bars reflects a return westward transport of sand.

The importance of sea-ice processes to coastal evolution in the Canadian Beaufort Sea appears to be less crucial than in the Alaskan region, but it is important to note that there have been few observations of coastal processes during the autumn and winter periods. Ice pile-up and ride-up are relatively rare at most locations, although pile-up events have been observed along the Tuktoyaktuk Peninsula (Dickins, 1987; Fig. 9.7). Ice-scouring is common in water depths less than 10 m, and undoubtedly occurs inshore, but sidescan sonar studies of the outer shoreface during the summer shows that reworking of sediments by wave and current processes obliterates ice scours (Héquette & Hill, pers. commun.; M. Desrosiers, pers. commun.). There have been no observations of ice-wallow relief on beaches, but detailed surveys of the foreshore have been by no means exhaustive.

Discussion

Both Canadian and Alaskan sectors of the Beaufort Sea coastline have clearly zonal influences, particularly at the morphologic level (Wiseman *et al.*, 1973; Ruz *et al.*, 1992). The principal difference between the two regions lies in the relative importance of zonal and azonal processes on the shoreface. These differences are observable in two characteristics: 1. the equilibrium between the rate of shoreface retreat and the littoral sediment budget; and 2. the preservation of the ravinement surface on the inner shelf. On the Alaskan coast, the fact that sediment supply from bluff retreat greatly exceeds the sediment that is stored in the coastal zone (Reimnitz *et al.*, 1988) suggests strongly that sediment transport on the shoreface is not in equilibrium with energy supplied by the relatively modest waves and currents of the region. Zonal processes, such as entrainment of sediment in the ice canopy during autumn storms (Reimnitz & Kempema, 1987) and transport of sediment by ice scouring (Rearic *et al.*, 1990) or ice push (Barnes & Reimnitz, 1988), are important over the longer time scale of coastal evolution. This is reflected in the absence of coastal deposits or ravinement surfaces on the shelf (Barnes & Reimnitz, 1974). The sedimentary record of transgression is reduced to a highly deformed ice-keel turbate layer.

In contrast, the Canadian Beaufort coast appears to be more in equilibrium over the long term with the energy supplied by wave and current processes. The sediment budget at King Point over a 30 year period, for example, indicates that the sediment supplied by bluff retreat can be accounted for in the sediments stored within the barrier beach (Hill, 1990). Although ice scouring causes considerable deformation of post-transgression sediments, a clear ravinement surface can be recognized across the shelf (Blasco *et al.*, 1990). In addition, a coastal sand sheet is present on the inner shelf and possible drowned barrier islands and thermokarst lake basins are preserved below the ravinement surface (Héquette & Hill, 1989). Zonal processes may be locally important. Héquette & Barnes (1990) have suggested, for example, that ice-scour erosion on the lower shoreface may accelerate shoreline retreat by deepening the submarine profile and requiring further erosion on the upper shoreface to maintain dynamic equilibrium. It is not clear, however, that this process is as efficient as wave and current transport, which could rapidly negate the effect of ice-scour erosion.

At shorter time scales, both regions show beach and nearshore bar morphodynamics that are not particularly zonal in nature, but rather show a morphological response to the relatively moderate wave regime, dominated by storms and associated surges. This is in the form of highly differentiated beaches and nearshore bars. Although the moderate wave regime can be considered a zonal response to fetch limitation by sea ice, the cause of this differentiation is more probably the response of a microtidal shoreline to storm surge conditions.

The conclusion from the above comparison would appear to be that the relative importance of zonal and azonal processes depends principally on the degree of fetch limitation imposed on the shoreline by the seasonal ice cover. More precisely, it is the joint probability of long fetches and extreme storms that controls the long-term evolution of Arctic shorelines. Because of the unconsolidated nature of the coastal plain deposits being eroded and the importance of thermal degradation of the exposed bluffs, sediment supply is generally high. When the fetch is very restricted, ice-related processes are certainly dominant and appear to be efficient enough agents of sediment transport to remove the sediment supplied by a rapidly eroding coast. Where the fetch is less limited, wave energy is higher and transgression proceeds as on any other wave-dominated coastline by progressive erosion of the shoreface, with the added influence of ice-related processes. The importance of these zonal processes is not clear, due largely to the lack of observations at critical times of year.

Many zonal processes, such as ice scour and ice wallow, are very intermittent in their action and therefore difficult to evaluate in terms of their efficiency. Observations, such as the obliteration of ice scours on the inner shelf during large storms (Barnes & Reimnitz, 1979) suggest that over the short term, their action is considerably less efficient than waves and currents. However, over the longer term the cumulative effect appears to be significant (Reimnitz et al., 1988; Rearic et al., 1990).

Other zonal processes, particularly those related to freeze-up conditions, are very poorly understood in terms of their efficiency as agents of sediment transport. The work of Reimnitz and others (Reimnitz & Kempema, 1987; Reimnitz et al., 1987; Reimnitz et al., 1992) has recognized the potential importance of autumn storms for removing sediment from the shoreline and that of sea ice as a transport agent. Hydrodynamic conditions may be greatly modified by the presence of frazil ice in the water column and anchor ice at the seabed. It is therefore probable that the morphology of beaches and nearshore bars along the Alaskan coast is in dynamic equilibrium with these autumn storm events. It is not known how important this would be on the more wave-influenced Canadian Beaufort Sea coast. Hydrodynamic studies have been carried out in summer, but not during this potentially critical time of the year, nor during the winter.

For a more complete understanding of beach and nearshore

morphodynamics and of the relative importance of zonal versus azonal processes, field research needs to be carried out in the autumn and the winter. Standard equipment, moored through the ice, could be used potentially to investigate winter circulation on the lower shoreface and inner shelf during the winter, after a solid ice cover has been established. However, measurements conducted in the critical autumn period are problematic due to instrument icing, hazards from drifting ice and simply access to the foreshore under potentially dangerous conditions. Nevertheless, research into the dynamic processes of this period should be a future priority.

Acknowledgements

Funding for this paper was provided by operating grants from the Natural Science and Engineering Research Council, Canada, and Research Agreements with Energy, Mines and Resources, Canada to Hill and Héquette. We would like to acknowledge the continued support and encouragement provided by colleagues from both sides of the Alaska/Yukon border, particularly Steve Blasco, Don Forbes, Erk Reimnitz, and Steve Solomon.

References

Barnes, P.W. (1982). Marine ice-pushed boulder ridge, Beaufort Sea, Alaska. *Arctic,* **35**, 312–16.

Barnes, P.W., Asbury, J.L., Rearic, D.M. & Ross, C.R. (1987). Ice erosion of a sea-floor knickpoint at the inner edge of the stamuckhi zone, Beaufort Sea, Alaska. *Marine Geology,* **76**, 207–22.

Barnes, P.W., Rawlinson, S.E. & Reimnitz, E. (1988). Coastal geomorphology of Arctic Alaska. In *Arctic coastal processes and slope protection design,* ed. A.T. Chen & C.P. Leidersdorf, pp. 3–30. New York: American Society of Civil Engineers.

Barnes, P.W., Rearic, D.M. & Reimnitz, E. (1984). Ice gouging characteristics and processes. In *The Alaskan Beaufort Sea: ecosystems and environments,* ed. P.W. Barnes, D.M. Schell & E. Reimnitz, pp. 185–212. Orlando, FL: Academic Press.

Barnes, P.W. & Reimnitz, E. (1974). Sedimentary processes on Arctic shelves off the northern coast of Alaska. In *The coast and shelf of the Beaufort Sea,* ed. J.C. Reed & J.E. Sater, pp. 439–76. Arlington, VA: Arctic Institute of North America.

Barnes, P.W. & Reimnitz, E. (1979). *Ice gouge obliteration and sediment redistribution event; 1977–1978, Beaufort Sea, Alaska.* US Geological Survey Open File Report 79–848, 22 pp.

Barnes, P.W. & Reimnitz, E. (1988). Construction of an Arctic barrier island by alternating sea-ice pileup and overwash. In *Geologic studies in Alaska by the U.S. Geological Survey during 1987,* ed. J.P. Galloway, & T.D. Hamilton, pp. 180–2. U.S. Geological Survey Circular 1016.

Blasco, S.M., Fortin, G., Hill, P.R., O'Connor, M.J. & Brigham-Grette, J.K. (1990). The Late Neogene and Quaternary stratigraphy of the Canadian Beaufort Continental Shelf. In *The Arctic Ocean region, The geology of North America* series, Vol. L, ed. A. Grantz, L. Johnson & J.F. Sweeney, pp. 491–502. Boulder, CO: Geological Society of America.

Burgess, M., Judge, A., Taylor, A. & Allen, V. (1982). Ground temperature studies of permafrost growth at a drained lake site, Mackenzie Delta. In *Proceedings of Fourth Canadian Permafrost Conference (Calgary, Alberta)*, ed. H.M. French, pp. 3–11. National Research Council of Canada, Associate Committee on Geotechnical Research.

Carter, L.D., Brigham-Grette, J. & Hopkins, D.M. (1986). Late Cenozoic marine transgressions of the Alaskan Arctic coastal plain. In *Correlations of Quaternary deposits and events around the margin of the Beaufort Sea,* ed. J.A. Heginbottom & J-S. Vincent, pp. 21–6. Geological Survey of Canada Open File Report 1237.

Czudek, T. & Demek, J. (1970). Thermokarst in Siberia and its influence on the development of lowland relief. *Quaternary Research*, **1**, 103–20.

Dallimore, S.R., Kurfurst, P.J. & Hunter, J.M. (1988). Geotechnical and geothermal conditions of nearshore sediments, southern Beaufort Sea, Northwest Territories, Canada. In *Proceedings, V International Conference on Permafrost, Trondheim, Norway*, pp. 127–31. Trondheim, Norway: Tapir Publishers.

Davidson, S., de Margerie, S. & Lank, K. (1988). *Sediment transport in the Mackenzie River plume.* Geological Survey of Canada Open File Report 2303. 123 pp.

Davies, J.L. (1972). *Geographical variation in coastal development.* London: Longman. 204 pp.

Dickins, D.F. (1987). *Aerial reconnaissance survey of ice break-up processes in the Canadian Beaufort Sea coastal zone.* Geological Survey of Canada Open File Report 1687. 16 pp.

Dinter, D.A., Carter, L.D. & Brigham-Grette, J. (1990). Late Cenozoic geologic evolution of the Alaskan North Slope and adjacent continental shelves. In *The Arctic Ocean region, The Geology of North America* series, Vol. L, ed. A. Grantz, L. Johnson & J.F. Sweeney, pp. 459–90. Boulder, CO: Geological Society of America.

Dyke, L.D. (1991). Temperature changes and thaw of permafrost adjacent to Richards Island, Mackenzie Delta, N.W.T. *Canadian Journal of Earth Sciences*, **28**, 1834–42.

Fissel, D.B. & Birch, J.R. (1984). Sediment transport in the Canadian Beaufort Sea. Unpublished report by Arctic Sciences Ltd., Victoria, B.C. for Geological Survey of Canada, Dartmouth, N.S. 165 pp.

Fissel, D.B., Birch, J.R. & Melling, H. (1990) *Interannual variability of oceanographic conditions in the southeastern Beaufort Sea.* Canadian Contractor Report on Hydrography and Ocean Sciences No. 35, Fisheries and Oceans Canada. 102 pp.

Forbes, D.L. (1980). Late-Quaternary sea levels in the southern Beaufort Sea. *Current Research, Part B.* Geological Survey of Canada, Paper 80-1B, 75–87.

Gornitz, V., Lebedeff, S. & Hansen, J. (1982). Global sea-level trends in the past century. *Science*, **215**, 1611–14.

Greenwood, B. & Davidson-Arnott, R.G.D. (1979). Sedimentation and equilibrium in wave-formed bars: a review and case study. *Canadian Journal of Earth Sciences*, **16**, 312–32.

Guilcher, A. (1953). Essai sur la zonation et distribution des formes littorales de dissolution de calcaire. *Annals de Géographie*, **62**, 161–79.

Harper, J.R. (1990). Morphology of the Canadian Beaufort Sea coast. *Marine Geology*, **91**, 75–91.

Harper, J.R., Henry, R.F. & Stewart, G.G. (1988). Maximum storm surge elevations in the Tuktoyaktuk region of the Canadian Beaufort Sea. *Arctic*, **41**, 48–52.

Harry, D.G. (1985). Ground ice slumps, Beaufort Sea coast, Yukon Territory. In *Fourteenth Arctic workshop, Arctic land-sea interactions*, Abstracts, pp. 115–17. Dartmouth, N.S., Canada: Bedford Institute of Oceanography.

Henry, R.F. (1975). *Storm surges*. Beaufort Sea Project technical report no. 19, Victoria, B.C., Canada: Department of the Environment. 41 pp.

Henry, R.F. (1984). *Flood hazard delineation at Tuktoyaktuk*. Department of Fisheries and Oceans, Canadian contractor report of hydrography and ocean sciences No. 19. Sidney, B.C., Canada: Institute of Ocean Sciences. 28 pp.

Héquette, A. & Barnes, P.W. (1990). Coastal retreat and shoreface profile variations in the Canadian Beaufort Sea. *Marine Geology*, **91**, 113–32.

Héquette, A. & Hill, P.R. (1989). Late Quaternary seismic stratigraphy of the inner shelf seaward of the Tuktoyaktuk Peninsula, Canadian Beaufort Sea. *Canadian Journal of Earth Sciences*, **26**, 1990–2002.

Héquette, A. & Ruz, M-H. (1991). Spit and barrier island migration in the southeastern Canadian Beaufort Sea. *Journal of Coastal Research*, **7**, 677–98.

Hill, P.R. (1990). Coastal geology of the King Point area, Yukon Territory, Canada. *Marine Geology*, **91**, 93–111.

Hill, P.R., Blasco, S.M., Harper, J.R. & Fissel, D.B. (1991). Sedimentation on the Canadian Beaufort Shelf. *Continental Shelf Research*, **11**, 821–42.

Hill. P.R., Forbes, D.L., Dallimore, S.R. & Morgan, P. (1986). Shoreface development in the Canadian Beaufort Sea. In *Proceedings, symposium on cohesive shores*, ed. M.G. Skafel, pp. 428–48. Ottawa: National Research Council of Canada, Publication No. 26134.

Hill, P.R. & Frobel, D. (1991). *Documentation of summer NOGAP activities, July 22– August 25, 1990*. Geological Survey of Canada Open File Report 2451. 84pp.

Hill, P.R., Héquette, A. & Ruz, M-H. (1993). Holocene sea-level history of the Canadian Beaufort shelf. *Canadian Journal of Earth Sciences*, **30**, 103–8.

Hill, P.R., Héquette, A., Ruz, M-H. & Jenner, K.A. (1990). *Geological investigations of the Canadian Beaufort Sea coast*. Geological Survey of Canada Open File Report 2387. 348 pp.

Hill, P.R., Mudie, P.J., Moran, K. & Blasco, S.M. (1985). A sea-level curve for the Canadian Beaufort shelf. *Canadian Journal of Earth Sciences*, **22**, 1383–93.

Hobson, G.D. & the Canadian Ice Island Scientific Party (1989). Ice Island field station: new features of Canadian Polar Margin. *Eos*, **70**, 833, 835, 838–9.

Hodgins, D.O. (1985). *A review of extreme wave conditions in the Beaufort Sea*. Department of Fisheries and Oceans, Canadian contractor report of hydrography and ocean sciences no. 12. Sidney, B.C., Canada: Institute of Sciences. 160 pp.

Hunter, J.A. & Hobson, G.D. (1974). Seismic refraction method of detecting subsea bottom permafrost. In *The coast and shelf of the Beaufort Sea*, ed. J.C. Reed & J.E. Sater, pp. 401–16. Arlington, VA: Arctic Institute of North America.

Judge, A. (1986). Permafrost distribution and the Quaternary history of the Mackenzie–Beaufort region: a geothermal perspective. In *Correlation of Quaternary deposits and events around the margin of the Beaufort Sea*, ed. J.A. Heginbottom & J-S. Vincent, pp. 41–5. Geological Survey of Canada Open File Report 1237.

Kelletat, D. (1989). The question of 'zonality' in coastal geomorphology – with tentative application along the east coast of the USA. *Journal of Coastal Research*, **5**, 329–44.

Kovacs, A. (1983). *Shore ice ride-up and pile-up features. Part 1. Alaska's Beaufort Sea coast.* Hanover, NH: U.S. Army Cold Regions Research and Engineering Laboratory, Report 83–9. 51 pp.

Kovacs, A. (1984). *Shore ice ride-up and pile-up features. Part 2. Alaska's Beaufort Sea coast.* Hanover, NH: U.S. Army Cold Regions Research and Engineering Laboratory, Report 84-26. 29 pp.

Lachenbruch, A.H., Sass, J.H., Lawver, L.A., Brewer, M.C., Marshall, B.V., Munroe, R.J., Kennelly, J.P., Jr. & Moses, T.H., Jr. (1987). Temperature and depth of permafrost on the Alaskan North Slope. In *Alaskan north slope geology*, ed. I.L. Tailleur & P. Weimer, pp. 544–58. Bakersfield, CA: Pacific Section, SEPM and the Alaska Geological Society.

Lewis, C.F.M. (1978). The frequency and magnitude of drift ice groundings from ice scour tracks in the Canadian Beaufort Sea. In *Proceedings 4th International Conference on Port and Ocean Engineering Under Arctic Conditions*, pp. 568–79. St. John's, Newfoundland: Memorial University of Newfoundland.

Mackay, J.R. (1971). The origin of massive icy beds in permafrost, western Arctic coast. *Canadian Journal of Earth Sciences*, **8**, 397–422.

MacNeil, M.R. & Garrett, J.R. (1975). *Open water surface currents.* Beaufort Sea Project, Technical Report No. 17, Environment Canada.

Martin, S. (1981). Frazil ice in rivers and oceans. *Annual Review of Fluid Mechanics*, **13**, 379–97.

Mysak, L.A. & Manak. D.K. (1989). Arctic sea ice extent and anomalies, 1953–1984. *Atmosphere–Ocean*, **27**, 376–405.

Neave, K.G. & Sellmann, P.V. (1984). Determining distribution patterns of ice-bonded permafrost in the U.S. Beaufort Sea from seismic data. In *The Alaskan Beaufort Sea: ecosystems and environments,* ed. P.W. Barnes, D.M. Schell & E. Reimnitz, pp. 237–58. Orlando, FL: Academic Press.

Nummedal, D. & Swift, D.J.P. (1987). Transgressive stratigraphy at sequence-bounding unconformities: some principles derived from Holocene and Cretaceous examples. In *Sea level fluctuations and coastal evolution*, ed. D. Nummedal, O.H. Pilkey & J.D. Howard, pp. 241–60. SEPM Special Publication 41.

Pinchin, B.M., Nairn, R.B. & Philpott, K.L. (1985). *Beaufort Sea coastal sediment study: numerical estimation of sediment transport and nearshore profile adjustment at coastal sites in the Canadian Beaufort Sea.* Geological Survey of Canada, Open File Report 1259. 712 pp.

Rampton, V.N. (1982). *Quaternary geology of the Yukon coastal plain.* Geological Survey of Canada Bulletin 317. 49 pp.

Rampton, V.N. (1988a). Origin of massive ground ice on the Tuktoyaktuk Peninsula, Northwest Territories, Canada: a review of stratigraphic and morphologic evidence. In *Proceedings of the V International Permafrost Conference, Trondheim, Norway*, pp. 850–5. Trondheim: Norway: Tapir Publishers.

Rampton, V.N. (1988b). *Quaternary geology of the Tuktoyaktuk Coastlands, Northwest Territories.* Geological Survey of Canada, Memoir 423. 98 pp.

Rearic, D.M., Barnes, P.W. & Reimnitz, E. (1990). Bulldozing and resuspension of shallow-shelf sediment by ice keels: implications for Arctic sediment transport trajectories. *Marine Geology*, **91**, 133–47.

Reimnitz, E., Barnes, P.W. & Harper, J.R. (1990). A review of beach nourishment from ice transport of shoreface materials, Beaufort Sea, Alaska. *Journal of Coastal Research*, **6**, 439–70.

Reimnitz, E., Graves, S.M. & Barnes, P.W. (1988). *Beaufort Sea coastal erosion, sediment flux, shoreline evolution, and the erosional shelf profile*. U.S. Geological Survey Miscellaneous Investigations Series, accompaniment to Map I-1182-G (1:82,000).

Reimnitz, E. & Kempema, E. (1982). Dynamic ice-wallow relief of northern Alaska's nearshore. *Journal of Sedimentary Petrology*, **52**, 451–61.

Reimnitz, E. & Kempema, E. (1984). Pack ice interaction with Stamukhi Shoal, Beaufort Sea, Alaska. In *The Alaskan Beaufort Sea: ecosystems and environments*, ed. P.W. Barnes, D.M. Schell & E. Reimnitz, pp. 159–83. Orlando, FL: Academic Press.

Reimnitz, E. & Kempema, E.W. (1987). Field observations of slush ice generated during freeze-up in Arctic coastal waters. *Marine Geology*, **77**, 219–31.

Reimnitz, E., Kempema, E.W. & Barnes, P.W. (1987). Anchor ice, seabed freezing and sediment dynamics in shallow Arctic seas. *Journal of Geophysical Research*, **92**, C13, 14 671–8.

Reimnitz, E., Marincovich, L., Jr., McCormick, M. & Briggs, W.M. (1992). Suspension freezing of bottom sediment and biota in the Northwest Passage and implications for Arctic Ocean sedimentation. *Canadian Journal of Earth Sciences*, **29**, 693–703.

Reimnitz, E. & Maurer, D.K. (1979). Effects of storm surges on the Beaufort Sea coast, northern Alaska. *Arctic*, **32**, 329–44.

Ruz, M-H., Héquette, A. & Hill, P.R. (1992). A model of coastal evolution in a transgressed thermokarst topography, Canadian Beaufort Sea. *Marine Geology*, **106**, 251–78.

Sellmann, P.V., Brown, J. & Lewellen, R.I. (1975) *The classification of thaw lakes on the Arctic coastal plain, Alaska*. U.S. Army Corps of Engineers, Cold Regions Research and Engineering Laboratory, Research Report 344. 21 pp.

Shapiro, L.H. & Barnes, P.W. (1991). Correlation of nearshore ice movement with seabed ice gouges near Barrow, Alaska. *Journal of Geophysical Research*, **96**, 16 979–89.

Short, A.D. (1975). Offshore bars along the Alaskan Arctic coast. *Journal of Geology*, **83**, 209–21.

Short, A.D., Coleman, J.M. & Wright, L.D. (1974). Beach dynamics and nearshore morphology of the Beaufort Sea coast, Alaska. In *The coast and shelf of the Beaufort Sea*, ed. J.C. Reed & J.E. Sater, pp. 477–88. Arlington, VA: Arctic Institute of North America.

Smith, M.W. (1975). *Permafrost in the Mackenzie Delta, N.W.T.* Geological Survey of Canada Paper 75-28. 34 pp.

Taylor, R.B. & McCann, S.B. (1983). Coastal depositional landforms in northern Canada. In *Shorelines and isostasy*, ed. D.E. Smith & A.G. Dawson, pp. 53–75. London: Academic Press.

Untersteiner, N. (1990). Structure and dynamics of the Arctic Ocean ice cover. In *The Arctic Ocean region, The geology of North America* series, Vol. L. ed. A. Grantz, L. Johnson & J.F. Sweeney, pp. 37–51. Boulder, CO: Geological Society of America.

Wiseman, W.J., Coleman, J.M., Gregory, A., Hsu, S.A., Short, A.D., Suhayda, J.N., Walters, C.D., Jr. & Wright, L.D. (1973). *Alaskan Arctic coastal processes and morphology*. Coastal Studies Institute, Technical Report 149, Baton Rouge, LA: Louisiana State University. 171 pp.

Wiseman, W.J., Jr., Suhayda, J.N., Hsu, S.A. & Walters, C.D., Jr. (1974). Characteristics of nearshore oceanographic environment of Arctic Alaska. In *The coast and shelf of the Beaufort Sea*, ed. J.C. Reed & J.E. Sater, pp. 49–64. Arlington, VA: Arctic Institute of North America.

Wright, L.D. & Short, A.D. (1983). Morphodynamics of beach and surf zones in Australia. In *Handbook of coastal processes and erosion*, ed. P.D. Komar, pp. 35–64. Boca Raton: CRC Press.

10

Paraglacial coasts

D.L. FORBES AND J.P.M. SYVITSKI

The initial shoreline, like that of the present, must have been largely dominated by drumlin outlines ... a picture the sheer beauty of which will live long in the observer's memory.

Douglas Johnson, 1925, *The New England – Acadian Shoreline*, pp. 118–19

Introduction

Marine transgression across glaciated terrain can produce striking coastal landscapes. Drumlin archipelagos such as Boston Harbour (in the northeastern USA), Mahone Bay (in southeastern Canada), or Clew Bay (in the west of Ireland) are among the most impressive such landscapes, Boston Harbour and Mahone Bay providing the inspiration for Douglas Johnson's words of reverie. Still more spectacular in many cases are the flooded glacial valleys forming fjords (Fig. 10.1), common along the margins of coastal highlands throughout the Northern and Southern Hemisphere fjord belts (Fig. 10.2). While underscoring our central assertion, that glaciation may exert a profound influence on the later evolution of a coastline, fjords and drumlin archipelagos are but two examples of the wide range of paraglacial phenomena encountered on mid- to high-latitude coasts.

Large areas of the globe have been glaciated repeatedly during late Cenozoic time and intermittently on earlier occasions in the geological record (Harland & Herod, 1975; Fulton, 1989). Major continental ice sheets have experienced cyclic growth and decay during the last two million years or more, while the record of glaciation along the Gulf of Alaska continental margin goes back to the late Miocene (Eyles, Eyles & Lagoe, 1991). Marine cores from the Norwegian shelf have yielded evidence of mountain glaciation extending back some 35 Ma (Jansen & Sjøholm, 1991) and the Antarctic has a similar history. Although ice sheets of continental scale are still present over Greenland and Antarctica, and numerous smaller ice masses persist elsewhere, most areas covered by Late Quaternary glaciation have been ice-free for many thousands of years (typically 8 to 22 ka, depending on the distance from former ice centres).

The environmental impact of glaciation in formerly ice-covered areas remains pervasive despite the passage of time (see for comparison the Arctic coastal plain shorelines described in Chapter 9). This legacy includes: distinctive

Figure 10.1. North Arm, a fjord on the northeast coast of Baffin Island in the eastern Canadian Arctic. Looking up-fjord with side-entry glacier and valley-wall talus sheets and cones. Photo courtesy R.B. Taylor.

landforms such as moraines, drumlins, fjords and other overdeepened embayments; widespread deposits of glacial or proglacial origin, including glacial diamicts, outwash sands and gravels, and glaciomarine or glaciolacustrine silts and clays; and ongoing vertical movements of land and sea that continue to influence the rates and processes of coastal change.

Natural systems thus affected by glaciation are described as *paraglacial*. This term was introduced by Ryder (1971) and Church & Ryder (1972) to describe the initial postglacial excess and subsequent relaxation of sediment supply on alluvial fans, and in river basins generally, in glaciated terrain. Sediment transport in paraglacial river systems of modest scale (Fig. 10.3) has

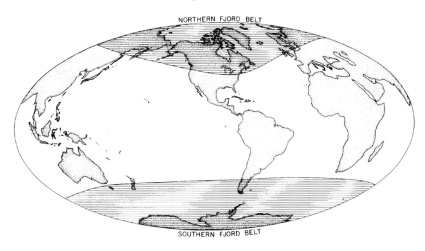

Figure 10.2. Generalized global distribution of fjords and other paraglacial coastal environments (reproduced from Syvitski *et al.*, 1987).

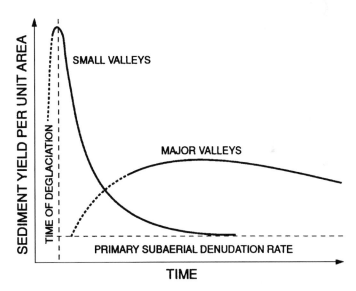

Figure 10.3. Sediment supply in paraglacial river systems (Church & Ryder, 1972; redrawn after Church & Slaymaker, 1989).

been shown to decline rapidly from late-glacial or early postglacial maxima to much lower values in the mid- to late Holocene (Church & Ryder, 1972). In larger basins (Fig. 10.3), intermediate storage in valley deposits can induce a lag in the peak sediment delivery and prolong the relaxation process (Church

& Slaymaker, 1989). This latter pattern may also occur in relatively small basins in areas of glacioisostatic rebound, where high-level deltas and terraces are later reworked as base level falls. Cold-climate effects, including winter ice cover and *periglacial* conditions such as permafrost and ground ice, may be present incidentally in paraglacial systems and play an important role in coastal development (see e.g. Taylor, 1978; Rosen, 1979; McLaren, 1982; Taylor & Forbes, 1987; Reimnitz *et al.*, 1991; Ruz, Héquette & Hill, 1992; Forbes & Taylor, 1994; see also Chapter 9), but are not essential elements.

Although glacioeustatic, glacioisostatic, and related hydroisostatic changes in relative sea level are effective on a global scale (Walcott, 1972), such that any coast in the world could be considered quasi-paraglacial in the broadest sense, we adopt a more restrictive definition in this review. Here *we define paraglacial coasts to be those on or adjacent to formerly ice-covered terrain, where glacially excavated landforms or glaciogenic sediments have a recognizable influence on the character and evolution of the coast and nearshore deposits.* In other words, the distinguishing features relate to sediment type, sediment supply, or the presence of glacial landforms. Proglacial settings in which ice is still present occupy one end of the paraglacial spectrum. Other glaciogenic effects such as changes in relative sea level are crucial to an understanding of coastal change, but are not part of the definition adopted here. Indeed, sea-level changes are known to have an almost ubiquitous influence on coastal development and related stratigraphic signatures throughout the Holocene (e.g. volume edited by Nummedal, Pilkey & Howard, 1987b) and more generally (e.g. Posamentier, Jervey & Vail, 1988; Cant, 1991).

Even under the restricted definition given above, paraglacial effects are present and frequently dominate the character and development of coasts in many parts of the world. Major areas of paraglacial influence (Fig. 10.2) include ice-free portions of Antarctica, islands in the Southern Ocean, parts of Tasmania, the South Island of New Zealand, southern South America, large areas of northern North America, northern Europe and Asia. In this review we examine a variety of paraglacial coasts, drawing our examples primarily from the North Atlantic region (eastern Canada and Iceland). We focus on cases from this part of the world because of the wide range of conditions represented here and our greater familiarity with these examples.

It is only within the past few years that paraglacial effects have been explicitly identified as such in the context of coastal development and sedimentation (Forbes, 1984; Forbes & Taylor, 1987; Syvitski, Burrell & Skei, 1987) and research activity in glaciated regions has begun to focus attention on paraglacial systems (e.g. volume edited by FitzGerald & Rosen, 1987). This is not to say that the effects of glaciation on coastal landforms had not been noted much earlier. Fjords had been recognized as coastal features of glacial origin as early as the 1820s (Esmark, 1827), although opposing theories of fluvial or tectonic origin fuelled discussion of the so-called 'fjord problem' until at least as late as 1913 (Johnson, 1915, 1919). Proglacial outwash deposits, including deltaic sinks in lakes or the sea, have received considerable attention over the years (e.g. Davis, 1890, 1896; Stone, 1899; more recently Church, 1972, 1978; papers in Jopling & McDonald, 1975; Boothroyd & Nummedal, 1978; among others), as have features associated with shoreline development in coastal drumlin fields (e.g. Johnson & Reed, 1910; Taylor *et al.*, 1986; Piper *et al.*, 1986; Wang & Piper, 1982; Boyd, Bowen & Hall, 1987; Carter & Orford, 1988; Carter *et al.*, 1990b; among others; see also Fig. 1.4). Distinctive aspects of coastal development in glaciated terrain were noted by Johnson (1919), while his 1925 monograph included chapters on 'Initial shorelines determined by glacial forms', on 'Erosional forms bordering unconsolidated [glaciogenic] deposits' (with a section on the 'Rapidity of marine erosion of glacial deposits'), and on 'Wave-built forms bordering unconsolidated [glaciogenic] deposits'.

Temporal patterns of sediment delivery or availability in paraglacial coastal systems are determined in large part by the disposition of glacial deposits relative to the coastline and by relative sea-level changes that control access to these sediments. In fjord-head and other settings where river supply plays a major role, fluvial patterns of sediment supply with relaxation scales of a few thousand years, recognized previously by Church & Ryder (1972), may prevail. Sea level plays a secondary role at the river outlet, where an isostatically induced relative drop in base level may initiate downcutting into earlier, high-level, delta deposits (e.g. Church, 1972, 1978), augmenting sediment discharge to the sea. On coasts where river inputs are unimportant, sea level may exert a major control on sediment supply, shortening the paraglacial cycle where water levels are stable and extending it where rising sea levels give access to deposits at progressively higher levels (Forbes, Taylor & Shaw, 1989). Periods of growth and decline in sediment supply play a critical role in the evolution of the coast and this aspect of paraglacial systems is considered further in a later section.

Coastal systems in the postglacial context

Coastal development in paraglacial environments occurs within a variety of distinctive physiographic frameworks (Figs. 10.2 and 10.4). These include: fjords (Fig. 10.1), fjord-head deltas, and associated shallow marine settings;

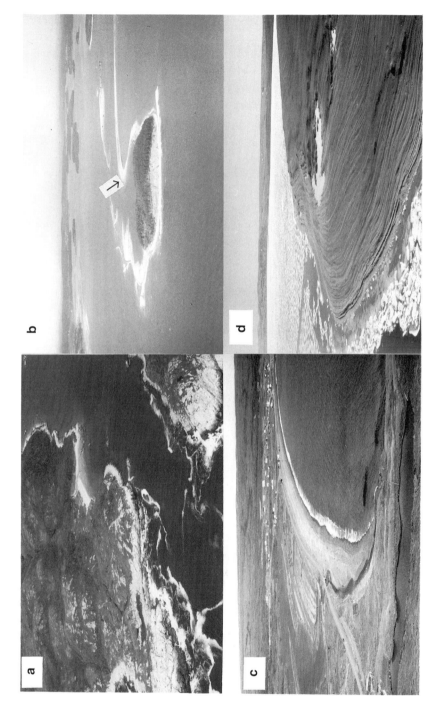

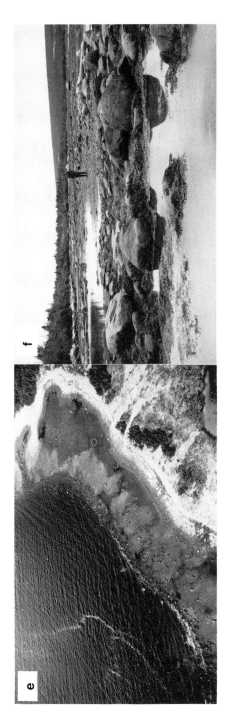

Figure 10.4. Diversity of coastal forms in paraglacial settings: (*a*) Irregular rocky coastline with pocket beach and coastal barrens, near Grand Bruit, southwest Newfoundland. (*b*) Story Head, a drumlin headland on the Eastern Shore of Nova Scotia, showing trailing drift-aligned spit on the left (Fisherman's Beach) and trailing ridge feeding sediment to swash-aligned barrier on the right. Incipient breaching of the barrier at the point of connection (arrow) may lead to substantial changes in this system over the coming few years. Note well-developed longshore cell structure on Fisherman's Beach. Chezzetcook Inlet in background. (*c*) Oblique view of bayhead barrier at Portugal Cove South, southeast Newfoundland, showing low, relict, progradational beach-ridge complex being overridden by high, transgressive, type-2, gravel storm ridge. (*d*) Raised gravel beach ridges on south shore of Inglis Bay, northwest Devon Island, eastern Canadian Arctic (photo. by S.B. McCann). (*e*) Aerial oblique view of boulder barricade in front of a thin gravel beach, west coast of Newfoundland north of Port au Choix. (*f*) Ground view of boulder-strewn tidal flat and boulder barricade at head of Freshwater Bay, Bonavista Bay, northeast Newfoundland.

outwash plains on exposed coasts; areas of extensive ice-contact or outwash deposits in eroding bluffs; areas with thin or localized sediment sources on coastlines of variable complexity (Fig. 10.4*a* and *c*); coastal drumlin fields in which glaciogenic sediments form eroding headlands (Fig. 10.4*b*); and regions in which coastal development is or has been strongly influenced by rapid uplift (Fig. 10.4*d*). Johnson (1925) distinguished erosional coastal features of glacial origin (fjords) and depositional shorelines associated with glaciogenic deposits (end moraines, ground moraine, drumlins, outwash deltas, eskers and kames).

Coastal evolution involves changes in shore-zone morphology and associated facies geometry through time. The process can be reversed or altered by changes in one or more parameters of the coastal system or when thresholds are exceeded. In other cases, subtle changes in water levels, storm frequency, sediment supply, or combinations of these factors may tip the balance between progradation and transgression in susceptible systems (Shaw & Forbes, 1987; Forbes *et al.*, 1989) or otherwise alter the response of the system to external forcing (cf. Allen, 1990). Subaerial and subaqueous parts of a coastal system are intimately related and cannot be viewed in isolation (e.g. Bruun, 1962; Shaw & Forbes, 1992).

Rates of coastal change and the form and facies characteristics of resulting deposits depend on the geological and geographical setting, including the following four major factors:

- antecedent morphology of the coastline and adjacent seabed;
- sources, physical properties, and rates of sediment supply;
- climatic and oceanographic environment; and
- rate and direction of relative sea-level change.

Changes in relative sea level (resulting from ocean-volume changes and glacioisostatic uplift or subsidence) may determine access to sediment sources and rates of sediment supply. Changes in relative sea level over the past 10 ka (Fig. 10.5) exceed 200 m at some locations (Dyke, Morris & Green, 1991). The direction and rate of relative sea-level change vary widely, from $+10\,\mathrm{mm\,a^{-1}}$ or more on the inner Scotian Shelf off eastern Canada during part of the Holocene (Forbes, Boyd & Shaw, 1991a; Shaw, Taylor & Forbes, 1993) to $-40\,\mathrm{mm\,a^{-1}}$ in Glacier Bay, Alaska, as ice ablation and retreat succeed the Neoglacial ice advance (Matthews, 1981).

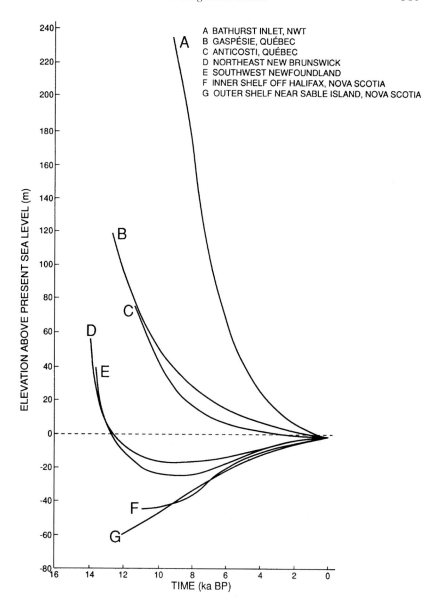

Figure 10.5. Relative sea-level curves from various sites in eastern Canada (after Grant, 1989; Scott *et al.*, 1989; Dyke *et al.*, 1991; Forbes *et al.*, 1991a), showing a range of conditions from emergence to submergence and temporal–spatial variability in rates of relative sea-level change.

Paraglacial sediments

One of the primary distinguishing features of paraglacial coasts is the predominance of glaciogenic sediment sources, often providing the only significant inputs of material to the shore-zone. This material can include a very wide range of grain sizes and sorting coefficients (Fig. 10.6). Lithologies range from varved clays and well-sorted silts to bouldery rubble. Glaciogenic source facies include tills, other ice-contact diamicts, glaciofluvial outwash, and glaciolacustrine and glaciomarine deposits, which themselves can range from diamicts through sandy turbidites to well-sorted silts and muds (e.g. Eyles, Eyles & Day, 1983; Eyles *et al.*, 1991; Miall, 1983). The grain size of ice-contact diamicts can range over at least six orders of magnitude (more than 24 ϕ units), often without distinct gaps or modes (Fig. 10.6). Flow competence is rarely an issue in glacial entrainment; subglacial erosion processes favour the production of all grain sizes from very large blocks to so-called glacial flour.

Competence does become important in proglacial and paraglacial contexts, however. As the glacial melt component of a river's discharge subsides, coarser sediment is left behind, largely immobile. Over-sized lag gravel

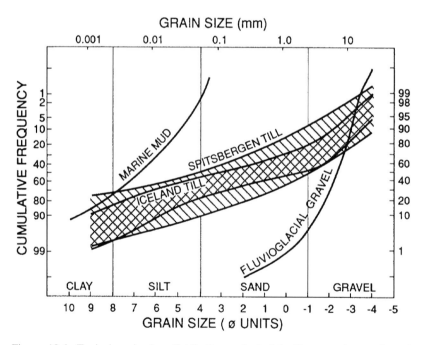

Figure 10.6. Typical grain-size distributions of glacial till, outwash gravel, and glaciomarine muds (redrawn after Boulton, 1976).

constituents can armour the transport surface, reducing and possibly eliminating the availability (or accessibility) of an underlying sediment source. Where the paraglacial sediment input to the coastal zone is below the sea surface, the same punctuation in sediment supply can develop. The winnowing action of waves or tidal currents on shallow-water deposits may remove the fines to deep or more-protected waters (e.g. Piper *et al.*, 1986), leaving behind an armouring lag of coarser particles that can ultimately reduce the availability of finer sediment. In some paraglacial environments, flow competence has increased over time. The tidal range in the Bay of Fundy has increased with rising sea level and changing basin geometry during the Holocene (Wightman, 1976; Scott & Greenberg, 1983; Amos & Zaitlin, 1985). The bay has undergone a transformation from microtidal (tidal range less than 1.5 m) in the early Holocene to the present macrotidal condition (spring tidal range up to 16 m), with concomitant high-energy currents (Amos & Long, 1980) and distinctive sedimentary environments (Dalrymple *et al.*, 1990).

Because of the wide spectrum of particle size entrained and deposited by glacial processes (Fig. 10.6), sedimentary deposits derived from glaciogenic sources often incorporate large volumes of gravel. Proglacial outwash complexes are commonly gravel-dominated, at least in their proximal reaches (e.g. Church, 1972; Church & Gilbert, 1975; Boothroyd & Nummedal, 1978; Stravers, Syvitski & Praeg, 1991). Gravel is a common constituent of beach, shoreface, and continental shelf sequences in glaciated terrain (e.g. Shepard, 1963; Stanley, 1968; Swift, Stanley & Curray, 1971; Forbes & Boyd, 1987, 1989; Forbes & Taylor, 1987; Kelley, 1987; Shipp, Staples & Ward, 1987; Orford, Carter & Jennings, 1991b; Carter & Orford, 1993). With the exception of tectonically active regions (e.g. Emery, 1955; Kirk, 1980; Massari & Parea, 1988; Moseley, Wagner & Richardson, 1992) and exceptional areas such as the south coast of England (e.g. Carr, 1969; Jennings & Smyth, 1987), coastal gravel is otherwise relatively rare and usually found only in immediate proximity to cliff sources.

On the other hand, some glacial deposits are dominated by relatively fine-grained facies and supply little gravel. Thus, sandy beaches predominate in the southern Gulf of St. Lawrence, a glaciated area of low-lying Permian and Carboniferous sandstones. The tills in the area are sandy and the rocks (being easily erodible) represent important coastal sand sources in their own right (Owens, 1974; McCann, 1979; Armon & McCann, 1977, 1979). In parts of the North American Great Lakes region, tills (Krumbein, 1933; Dreimanis, 1976; Scott, 1976) and lake sediments (Eyles *et al.*, 1983) exposed in coastal bluffs are deficient in both gravel and sand, yielding relatively little material for beach development (Davidson-Arnott, 1986; Philpott, 1986).

Another result of high sediment discharge in early paraglacial time is the extensive distribution on parts of the inner continental shelf of sands derived from wasting of major ice masses such as the Laurentide Ice Sheet (e.g. Boyd, Scott & Douma, 1988). Darby (1990) has argued that the elemental composition of Fe–Ti oxide minerals (primarily ilmenite) in sands of the US Atlantic Shelf points to the Hudson River as the dominant source of sands as far south as Cape Hatteras (North Carolina). This implies very large discharge and sediment supply from the Hudson River, which 'may have rivalled ... much larger rivers like the Mississippi or Amazon ... , at least during brief glacial melting events'. In a partly analogous context along the Québec North Shore of the Gulf of St. Lawrence, Syvitski & Praeg (1989) recognized an acoustic stratigraphic unit interpreted as early paraglacial, representing the melting of land-based ice caps and the rapid progradation of outwash deltas under conditions of abrupt glacial-isostatic rebound (falling relative sea level). In other areas, such as the former Champlain Sea basin, in the St. Lawrence Lowland of eastern Canada, large marine outwash deposits now lie well above present sea level (e.g. Rust & Romanelli, 1975; Sharpe, 1988).

Paraglacial settings and coastal outcomes

Paraglacial settings are arguably the most diversified set of coastal environments treated in this volume. It is a challenge to select sufficient and appropriate examples to represent the range of paraglacial coasts. This treatment is necessarily selective. Following a review of glacial erosion features (fjords and other overdeepened basins) and coastal sedimentation within them, we proceed to a consideration of coastal development related to various types of glaciogenic sediment source and their spatial and temporal distribution. Numerical models are applied in two cases in order to clarify the effects of major sediment supply and delivery processes.

Coasts dominated by glacial erosion: overdeepened basins and coastal barrens

Fjords are glacially scoured valleys partially submerged below sea level (Fig. 10.1). They may include one or more overdeepened basins, which are commonly confined at their down-valley limits by shallower sills. Glacially overdeepened basins also occur in other forms, including: fjord-like lake basins (cf. Eyles, Mullins & Hine, 1990), converted from fjords to lakes by isostatic uplift and emergence of the fjord sill; submarine valley complexes (Fig. 10.7), wherein subglacial meltwater is sometimes invoked as the erosional agent (e.g. Boyd *et al.*, 1988); and broader basins formed by differential

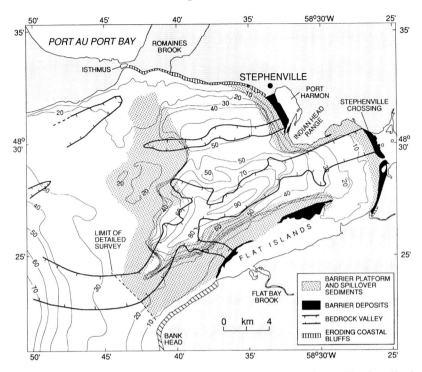

Figure 10.7. Overdeepened basins in St. George's Bay, southwest Newfoundland (redrawn after Shaw & Forbes, 1990b, 1992). Basin depths are greatest (>90 m off Flat Island and >60 m off Stephenville) over partially filled submarine valleys more than 180 m deep that extend beneath the barriers at Stephenville and Stephenville Crossing (cf. Fig. 5 of Shaw & Forbes, 1990b). The 12 km long Flat Island spit and the 3.5 km beach-ridge barrier complex at Stephenville are fed from eroding coastal bluffs along the north and south sides of the bay and rest on thick subaqueous barrier platforms developed along and across the basins (cf. Shaw & Forbes, 1992).

excavation of weaker lithologies and subsequently flooded to form arms of the sea (Miller, Mudie & Scott, 1982; Piper *et al.*, 1983). Two distinct forms of coastal sedimentation in glacial basins (fjord-head delta deposits and basin-margin barrier platforms) are treated in separate sections below.

Some coasts lie within areas of extensive glacial or glaciofluvial erosion, the result of which is a rock-bound coast of moderate relief, almost completely devoid of sediment (Fig. 10.4a). Coastal barrens of this type are known, for example, from parts of the Scandinavian and Greenland coasts and from a number of areas in eastern Canada. The latter include areas of scoured granites and metasediments on the Atlantic coast of Nova Scotia (Owens & Bowen, 1977) and Newfoundland (Forbes, 1984).

Sedimentation in fjord-head and proximal basin settings

Fjords by their very nature, being deep and elongated, create a huge storage capacity for sediments delivered to them. Syvitski *et al.* (1987) estimated that 24% of all the sediment transported from the land to the world ocean over the last 100 ka now resides within fjord basins. This estimate reflects both the storage capacity of silled basins and the paraglacial delivery of first- and second-cycle glaciogenic sediments. The enormous volumes of sediment that lie within a fjord basin are not obvious. Saguenay Fjord, a north-shore tributary to the St. Lawrence Estuary, contains 120 km^3 of fill. The rate of sediment delivery at the present time can account for only 7% of sediment deposition during the Holocene. This points to an earlier paraglacial episode of high sediment delivery (Syvitski & Praeg, 1989).

Fjord deposits reflect these high rates of sediment accumulation and provide a highly resolvable stratigraphic record of the processes affecting sediment delivery. Their interbasin variability reflects the wide range of coastal and hinterland parameters affecting both sediment supply and mechanisms of dispersal within the basin. These include glacial history (styles and rates of glacier movements and time since deglaciation), fluvial conditions (including seasonal distribution of runoff and partitioning of sediment between bedload and suspended load), geography, oceanography, climate and ecology (including basin shape, fetch, tidal characteristics, flushing dynamics, temperature, wind, sea ice, biomass), and tectonic and geotechnical conditions (including sea-level changes, earthquake frequency, and slope failures).

Many fjords are dominated by a large hinterland drainage basin delivering runoff to the head of the fjord. Fluvial water and sediment discharge produce a fjord-head delta (Fig. 10.8). The surface morphology of the delta is controlled by discharge characteristics, tides, and wave energy, among other factors. The land–sea interface may take the form of a braided channel complex, a single outlet channel, several distributaries, or tidal channels dissecting mud flats. Delta-front features can include sand and gravel barriers and river-mouth bars. These topset deposits seldom represent more than a small fraction of the sediment within the delta and seaward in the fjord itself, where sediment thickness can exceed 500 m.

The geomorphology of fjord-head sandurs, such as those described from Baffin Island (Church, 1972, 1978; Syvitski & Hein, 1991), reflects both the abrupt decrease in sediment delivery during the early to mid Holocene and the rapid emergence of the land resulting from isostatic rebound. As the ice sheet melted out of the local drainage basins, both sediment availability and the glacial meltwater component of fluvial discharge were reduced rapidly, leaving a runoff pattern reflecting the cold, snowmelt-dominated, semi-arid

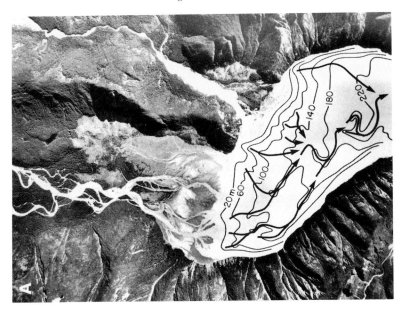

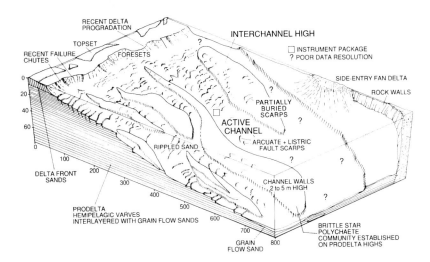

Figure 10.8. (*a*) Vertical air photograph of fjord-head Klinaklini River delta and side-entry Franklin delta in Knight Inlet, British Columbia, on the Pacific coast of Canada. Note distributary network with bars, alluvial flats, and intertidal flats on subaerial delta. Note also amalgamated fan morphology of Franklin delta and steep valley sidewall channels delivering sediment directly into the fjord. Arrows denote submarine channels that direct sand down the proximal prodelta slope. Reproduced from Syvitski & Farrow (1989). (*b*) Schematic drawing of bedload-dominated prodelta environments at head of Itirbilung Fiord, Baffin Island. Reproduced from Syvitski & Farrow (1989).

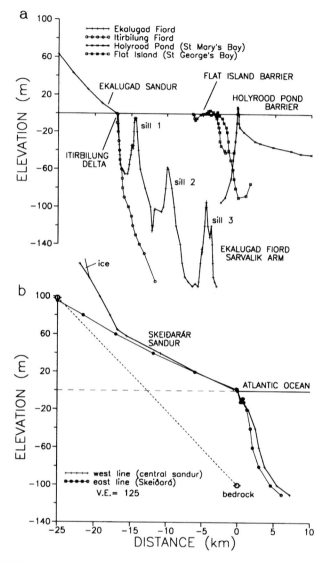

Figure 10.9. Representative topographic and bathymetric profiles for selected large-scale systems building into deep basins (ocean to the right in all cases; V.E. = vertical exaggeration): (*a*) Fjords and other over-deepened basins: Ekalugad Fiord, Baffin Island, eastern Canadian Arctic: sandur and fjord-head delta building into Tasiujaq Cove behind breached moraine (sill 1), with two, more distal, sill-confined basins to seaward in north arm of fjord (after Knight & Church, 1970; Church, 1972); Itirbilung Fiord, Baffin Island: fjord-head delta and prodelta slope (cf. Fig. 10.8*b*); not shown are the four sills at depths of 200, 350, 400, and 270 m located 7.5, 16, 24, and 46 km, respectively, down-fjord (see Fig. 8 of Syvitski & Hein, 1991); Holyrood Pond, St. Mary's Bay, southeast Newfoundland: large, high, sand–gravel barrier resting on fjord

conditions prevailing at the present time (Church, 1974). The extent of active delta progradation beyond raised early Holocene terraces reflects in large part the size of the hinterland draining to the fjord. Trapping of sediment in lake basins and the depth of the marine receiving basin both play a role, among other factors, in determining the size of the modern sandur surface, but these are of a secondary nature.

In some basin settings, wave-built ridges can be well preserved on emerging deltas. On an actively prograding fjord-head delta, however, channel migration and rapid progradation combine to reduce the effectiveness of wave processes in the fjord (Syvitski & Farrow, 1983). Furthermore, the depth of the receiving basin puts much of the fluvial sediment out of reach of wave action or tidal currents. Thus, once the sediment reaches the seafloor, it tends to remain in place unless removed into deeper water by submarine slides, slumps, or gravity currents (Prior *et al.*, 1982; Prior, Bornhold & Coleman, 1983; Syvitski & Farrow, 1989). The suspended load is dispersed along the surface of the fjord within a buoyant plume. The latter can extend over 100 km from the river mouth within the restricted confines of a fjord, with corresponding effects on the extent of hemipelagic deposits. The process can be likened to a perforated conveyor belt, with the highest rates of sedimentation near the river mouth, decreasing exponentially down the fjord (Syvitski *et al.*, 1985, 1987). The bedload is dumped along the length of the river valley with seaward decrease in river gradient (Fig. 10.9a; Church, 1972), after which any remaining bedload is deposited at the river mouth (Kostaschuk, 1985). Because of the laterally confined nature of the delta front between the fjord walls (Fig. 10.8a), the bedload accumulates as steep delta foresets within the first few hundred metres of the river mouth. Alongshore currents are absent, a factor that contributes (along with the high rates of bedload deposition) to both oversteepening and overloading of the sand and gravel sediment pile. Delta-front failures typically occur in a series of chutes, commonly 10 to 30 m wide, that wander down the delta front (Fig. 10.8b). The chutes can provide an indication of the failed volume. Where they are large, with significant relief

Caption for fig. 10.9 (*cont.*).
sill (Forbes, 1984; Forbes & Taylor, 1987; Shaw *et al.*, 1990); fjord basin (Holyrood Pond) to left, St. Mary's Bay to right; Flat Island spit, St. George's Bay, southwest Newfoundland (Shaw & Forbes, 1987, 1990b, 1992): sand and gravel beach-ridge spit complex and sandy barrier platform developed along the side of a deep coastal basin (see Fig. 10.7). (*b*) Two parallel profiles about 10 km apart across Skeiðarársandur (Fig. 10.11), southeast Iceland (after Boothroyd & Nummedal, 1978; Hine & Boothroyd, 1978; Nummedal *et al.*, 1987a). Ice of Skeiðarárjökull (Fig. 10.11), a piedmont outlet glacier of Vatnajökull, is shown at upper left.

(10 to 30 m), they attest to the existence of turbidity currents, which can travel many tens of kilometres. The proximal part of the prodelta (the first few kilometres) is often a zone of bypassing, where the gravity flows do not deposit their sediment load but often erode the seafloor, enlarging the channels that confine the flow (Fig. 10.8). Typically, the channel form disappears as the seafloor slope falls below 1°. The sediment load is then deposited as the flow decelerates and spreads across the seafloor.

Submerged fjord-head outwash deltas and shoreface platforms, such as those found along the south coast of Newfoundland (Flint, 1940; Forbes & Shaw, 1989), also attest to a rapid decrease in sediment supply in early postglacial time. These features reside in an area of mid to late Holocene submergence (Forbes, Shaw & Eddy, 1993). In the absence of significant modern sediment discharge from hinterland basins, they now form narrow estuaries with shallow, steep-fronted, relict-delta platforms. The lips of these platforms are found in present water depths down to about 30 m.

Numerical models can help to elucidate the relationships between sedimentation processes, environmental forcing and resulting depositional architecture. DELTA2 is a process–response model that simulates delta growth in a fjord-head setting (Syvitski et al., 1988). In this model, river sediment is discharged to the coast as bedload or suspended load and is spread into the sea by a combination of up to four processes: 1. delta-front progradation through bedload dumping at the river mouth; 2. hemipelagic sedimentation of suspended particles from the river plume; 3. proximal delta-front bypassing by turbidity currents and cohesionless debris flows; and 4. downslope diffusive processes (wave or tidal current reworking or small slides) that work to smear previously deposited sediment into deeper water (Syvitski et al., 1988; Syvitski, 1989). The model computes the seafloor surface at different time intervals and the accumulation rates produced by the various depositional mechanisms. Input data include the seasonal river discharge, suspended sediment and bedload, the shape of the river channel, and the dimensions of the basin to be filled with sediment. The model can be used to predict sediment fill under complex sea-level fluctuations and over irregularly shaped basins.

Fig. 10.10 presents a DELTA2 simulation of sedimentation in Itirbilung Fiord, Baffin Island (cf. Figs. 10.8b 10.9a), a bedload-dominated basin. Input parameters were taken from Syvitski & Hein (1991) and include two periods of sediment delivery: 1. the Hypsithermal period between 7.0 and 5.6 ka BP, a time of warm dry summers, when valley glaciers from the Foxe Ice Sheet ablated and receded rapidly (Church, 1978); and 2. the Neoglacial period from 5.6 to 0.1 ka BP, a moister time but with much cooler summers. The Hypsithermal condition was represented using an annual bedload transport

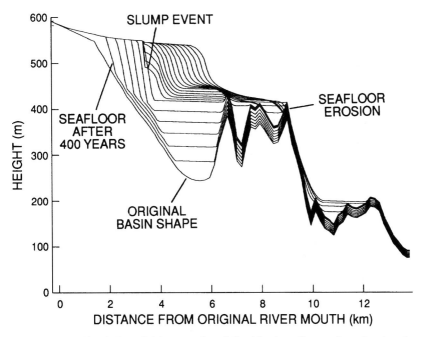

Figure 10.10. Simulation of delta growth and fjord-basin sedimentation, showing the Holocene record of successive seafloor profiles every 400 years in Itirbilung Fiord, Baffin Island (for modelling details see Syvitski & Daughney, 1992). Two distinct climatic intervals, involving different fluvial discharge and oceanographic conditions, are reflected in the simulated sediment record: 1. an early paraglacial episode of high sediment supply lasting 2400 years; 2. a 'postglacial' interval of lower sediment input continuing for 5600 years (see text for further details).

(sand) amounting to 0.5 Mt, primarily in the spring freshet, and annual suspended load (mud) of 0.17 Mt, introduced from the spring through to the autumn. The early paraglacial influx was associated with an exponentially decreasing drop of 50 m in relative sea level (primarily resulting from glacioisostatic rebound). The Neoglacial period had much lower rates of sediment input (bedload and suspended load estimated at $0.016 \, \mathrm{Mt \, a^{-1}}$ and $0.013 \, \mathrm{Mt \, a^{-1}}$, respectively), occurring largely in the late spring and early summer. During this time, relative sea level fell in a roughly linear fashion over a range of 10 m, and the delta front prograded slowly into a much more energetic basin (more open water, more wave activity, and stronger currents).

The resulting simulation (Fig. 10.10) shows the computed profile of the seafloor along an axial line normal to the delta front at 400-year intervals. The model output shows the rapid delta-front progradation during the Hypsithermal interval and the concomitant generation of sediment gravity flows into an otherwise quiet basin. The transition from Hypsithermal to Neoglacial is associated with a major adjustment of the delta front in the form of a large slump. This event temporarily slows the progradation of the delta and causes the formation of a terrace. Increased deepwater currents during the Neoglacial period cause subsequent erosion of hemipelagic deposits that mantle some of the bedrock highs in the outer reaches of the fjord. High-resolution seismic reflection profiles collected along the axis of Itirbilung Fiord show stratigraphy similar to that produced by the model (Syvitski & Hein, 1991).

Exposed outwash coasts

The difference between deeply embayed basins receiving sediment from active or relict glacial outwash sources (the fjord-head deltas described above) and partially embayed or open coasts fronting outwash plains is superficially striking, reflecting the influence of wave processes in unprotected settings. Under these circumstances, beach and barrier facies develop along the sandur front (Héquette & Ruz, 1990; Shaw, Taylor & Forbes, 1990). Distributary migration and longshore dispersal of sediment under waves may contribute jointly to the maintenance of a relatively simple shoreline form. Although minor sinuosity may initiate some longshore cell structure, this is inhibited by the large sediment supply in active proglacial environments.

The sandur coast of southeast Iceland (King, 1956; Hine & Boothroyd, 1978; Nummedal, Hine & Boothroyd, 1987a) is an eminent example, perhaps the largest active feature of its kind. It represents proglacial conditions of very high sediment supply from piedmont glaciers at the southern margin of Vatnajökull (Fig. 10.11). The sediment flux (typically about 0.4×10^6 tons per year) is enhanced by volcanogenic glacier-burst (jökulhlaup) floods that generate discharges as high as $100\,000\,\mathrm{m^3\,s^{-1}}$, total flow volumes ranging up to $15\,\mathrm{km^3}$, and sediment transport estimated at about 29×10^6 tons for a burst discharge of $3.5\,\mathrm{km^3}$ in 1954 (Rist, 1957; Nummedal *et al.*, 1987a). These conditions have produced extensive sand and gravel outwash (sandur) plains filling broad depressions in the underlying Tertiary and Quaternary basaltic rocks. The Skeiðarársandur (Fig. 10.9*b*) has built seaward to a broad front 20 to 30 km south of the present ice margin and covers an area in excess of $1000\,\mathrm{km^2}$. Braided channel networks cross the sandur (Boothroyd & Nummedal, 1978), attaining widths as great as 10 km (Skeiðará) at the shoreline, where they break through the barrier beach at a number of distributary outlets. The subaerial slope of the sandur, 0.010 (0.57°) or more near the ice margin, decreases to less than 0.003 distally. The much steeper shoreface slope, up to 0.026 (~5°) off the Skeiðará outlet (Fig. 10.9*b*),

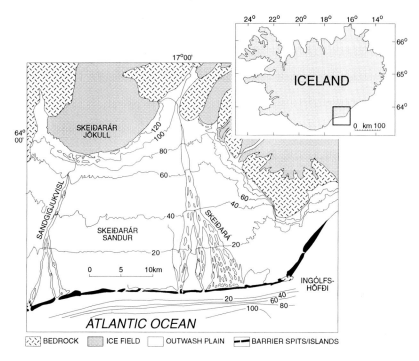

Figure 10.11. Outwash plain of Skeiðarársandur, southeast Iceland, and barriers developed along its seaward margin (redrawn after Nummedal *et al.*, 1987a).

diminishes westward to 0.020 (~1°) and decreases seaward to 0.006 at the 100 m isobath. Maximum clast size (*A*-axis) decreases down the Skeiðará channel system from about 320 mm near the ice front to approximately 10 mm near the coast (Fig. 6 of Boothroyd & Nummedal, 1978); the dominant sediment type over the distal sandur (below about 50 m elevation) is sand. Seismic refraction indicates that depth to bedrock at the shoreline is about 100 m at the mouth of Skeiðará, rising to sea level 10 km to the east at Ingólfshöfði (Figs. 10.9*b* and 10.11; Nummedal *et al.*, 1987a).

This coast is subjected to a highly energetic wave climate resulting from northeastward movement and intensification of winter storms, which generate southwesterly swell followed by higher waves (significant wave height, up to 16 m) from the southwest through southeast (Nummedal, 1975; Hine & Boothroyd, 1978; Viggósson *et al.*, 1994). The waves generate substantial longshore drift during the winter months, reworking the seaward margin of the outwash plain to produce a set of broad sandy barrier islands. These are delimited by distributary outlets and impound wide wind-tidal flats. The barriers increase in width from 200 m to about 750 m to the west (Hine &

Boothroyd, 1978). They are characterized by a broad storm ridge or berm rising to approximately 5 m, supporting widespread aeolian bedforms (transverse dunes), fronted by a lower berm during the summer months, and by nearshore bars off the major distributaries (Fig. 10.9*b*; Hine & Boothroyd, 1978; Nummedal *et al.*, 1987a). Despite the great quantity of sediment delivered to the coast by meltwater runoff and glacier bursts, short-lived progradation of about 200 m is the maximum observed in association with any one event off Skeiðará; over the past century, the mean position of the shoreline has been essentially stable in that area (Nummedal *et al.*, 1987a). This implies rapid reworking of littoral sediment alongshore and downslope, consistent with the widening barriers and broader submarine slope to the west (Fig. 10.9*b*). In terms of the large-scale architecture of the system, however, wave-formed structures are of minor importance. The sandur deposits constitute a very large, seaward-prograding, coarsening-upward clastic wedge, prograding seaward across a broad front. The large barrier-beach complex forms only a minor surface decoration.

The Skeiðarársandur has formed in a laterally unconfined, high-wave-energy, and recently stable sea-level setting. The outwash coast of the Gulf of Alaska is a roughly analogous case (Boothroyd & Nummedal, 1978). Where relative sea levels are rising and topography provides some lateral confinement, the resulting morphology can include transgressive barrier structures behind which tidal estuaries may develop (Fig. 3 of Shaw *et al.*, 1990). Where relative sea level has been falling throughout most of the Holocene, the coastal morphology may be dominated by superficial reworking of the outwash surface into a succession of quasi-parallel beach ridges (Fig. 10.4*d*; Blake, 1975), locally amalgamated or separated by shallow depressions, the morphology of which records short-lived details of longshore transport, coastal cell development, spit extension and recurve development, and other common features of sediment-limited gravel beaches (Fig. 10.4*b*; Héquette & Ruz, 1990).

Basin-margin coasts with large sediment supply

The coast of St. George's Bay, southwest Newfoundland, provides a case study of coastal evolution in the presence of a large paraglacial sediment supply close to a former ice margin and glacially overdeepened coastal basins. The shores of the bay are marked by extensive coastal bluffs, typically 20 to 30 m high, partially stabilized in some parts and actively eroding in others. These eroding sections reveal thick sequences of ice-contact diamicts, proximal outwash sands and gravels, and glaciomarine muds of Late

Wisconsinan age (MacClintock & Twenhofel, 1940; Brookes, 1974, 1987; Grant, 1987, 1991). They represent the most-extensive exposure of glaciogenic sediments to be found on the Newfoundland coast. Erosional bluffs extend for 10 km along the north shore of the bay and 40 km to the southwest along the south shore (Fig. 10.7). Along with a depositional complex at least 90 m thick (probably thicker over buried valleys), forming a broad sill across the middle of the bay, and even thicker sequences within the valleys, these deposits record a major late-glacial depocentre in the area (Shaw & Forbes, 1990b). Two basins landward of the sill, the deepest extending to more than 90 m water depth under present conditions, are underlain by partially filled bedrock depressions with the appearance of subglacial tunnel valleys. These contain thick stratified sequences overlying acoustically unstratified sediments. The top of the unstratified unit occurs as deep as 190 m below present sea level, beneath the present-day beaches at Stephenville and Stephenville Crossing (Fig. 10.7).

Relative sea level stood at 44 m above present at the time of initial ice recession about 13.7 ka BP (Brookes, Scott & McAndrews, 1985; Grant, 1991), but dropped to a minimum at about 25 m below present c. 9.5 ka BP (Forbes *et al.*, 1993). St. George's Bay is open to the prevailing wind from the southwest, with potential fetches in excess of 700 km across the Gulf of St. Lawrence. Wave-formed ripples with wavelengths up to 3 m developed in pebble gravels (modal grain sizes of 2 to 29 mm) over a large area of the sill (Fig. 15 of Shaw & Forbes, 1990b), attest to wave conditions at least equivalent to $H_S > 3.8$ m at peak periods of 11 s (cf. Forbes & Boyd, 1987). Significant wave heights of this magnitude have a return interval of less than 1 year (Ouellet & Llamas, 1979), although wave energy is significantly limited by sea ice from late December to early May (Forbes & Taylor, 1994; see also Chapter 9).

Along the southeast side of St. George's Bay, the 12 km long Flat Island barrier (Fig. 10.7) forms a large, drift-aligned, sand and gravel spit complex (Shaw & Forbes, 1987, 1992). It rests on a sandy subaqueous barrier platform, partially filling the adjacent basin and providing a foundation for subaerial barrier growth. The platform has a pronounced break in slope (Figs. 10.9*a* and 10.12) that rises to 5 m water depth at the distal end of the spit, converging distally on the beach from an initial width of 2 km and depth of 25 m at the proximal (southwest) end. This platform is analogous to spit platforms described elsewhere, in particular one from a raised paraglacial shoreline sequence in Denmark (Nielsen, Johannessen & Surlyk, 1988), but is larger and thicker than other reported examples. It consists of a clinoform prism of sand, up to 50 m thick, draped along the south flank of the southern basin in St.

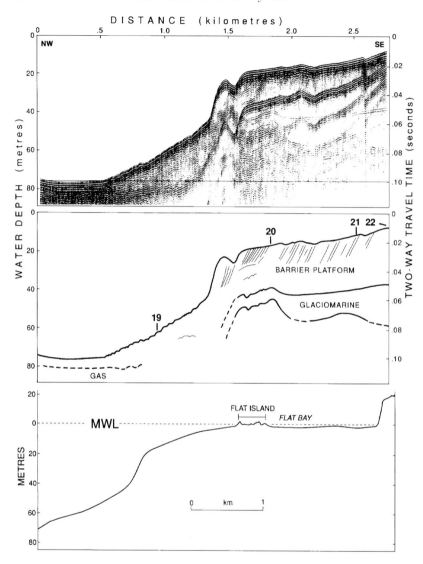

Figure 10.12. Seismic reflection (sparker) profile through outer part of Flat Island barrier platform (Figure 10.7) and underlying glaciomarine sediments in an area of subaqueous slope failure on the seaward face of the platform. Note seaward-dipping clinoform structures underlying platform surface at right and slump structures on the seaward slope. Numbers on profile refer to sediment samples. Grain sizes range from muddy sand with a mode of about 3 φ (mean 4 φ) at sample 19 to well-sorted medium sand (modal size about 1.6 φ) on the upper shoreface at sample 22. Bottom panel shows a cross-section through the Flat Island barrier and subaqueous platform at a much-reduced vertical exaggeration relative to the same profile in Figure 10.9*a*. Reproduced from Shaw & Forbes (1992).

George's Bay. The seaward-dipping beds within the platform have typical dip angles of about 6°. The seaward face in an area showing evidence of recent failure (Fig. 10.12) has a slope of about 14° at the top, running out to a gentler slope of 3° at the base and the sediment size ranges from medium sand on the upper shoreface to muddy fine sand on the lower slope (Shaw & Forbes, 1992). Gravel occurs on the beach and in the trough of the nearshore bar system. The subaerial barrier (Fig. 10.13*a*) consists of three distinct segments (Shaw & Forbes, 1987):

- a narrow proximal section that widens downdrift to include a number of partially truncated beach ridges aligned at an angle to the present shoreline, with thin salt marsh accumulating along the landward shore and in depressions between the gravel beach ridges;
- a central section that was breached by extensive washover during the early 1950s, developing a wide flood-tidal shoal on the landward side and interrupting the littoral drift;
- a broad distal section, more than 1 km wide, consisting of several discordant sequences of gravel beach ridges capped with aeolian sand.

Another platform structure is present along the north shore of St. George's Bay, extending across the northern valley at Stephenville, around a rocky promontory (Indian Head, Fig. 10.7), and feeding sediment to the barrier at Stephenville Crossing. This barrier rests on a submerged early Holocene delta sequence graded to a sea level about 25 m below present (Fig. 6 of Shaw & Forbes, 1990b). In the Stephenville embayment north of Indian Head, the barrier platform has provided the foundation for progradation of a gravel beach-ridge complex almost 4 km in length and 1 km wide (Grant, 1975; Shaw & Forbes, 1992). Yet another large clinoform sediment body has developed on the landward side of the sill, building landward into the inner basins. This spillover unit results from reworking of sediments primarily at lower sea levels (most of the sill was subaerial for a time in the early Holocene), but continues to grow under present conditions of active reworking on the sill. The spillover wedge has a steep basin face with a slope of 10° to 20°, is 20 to 50 m thick, consists of landward-dipping units of sand and gravel overlying glaciomarine muds in the basin, and extends for more than 15 km across the bay on the inner side of the sill (Shaw & Forbes, 1992).

Coastal sediment complexes can develop on shallow or emergent fjord sills, impounding lakes or estuaries (cf. Pickrill, Irwin & Shakespeare, 1981). The barrier at Holyrood Pond in southeast Newfoundland (Fig. 10.9*a*; Forbes, 1984, 1985; Shaw *et al.*, 1990) is an example of beach development on a shallow fjord sill. This sand and gravel barrier, more than 200 m wide and over

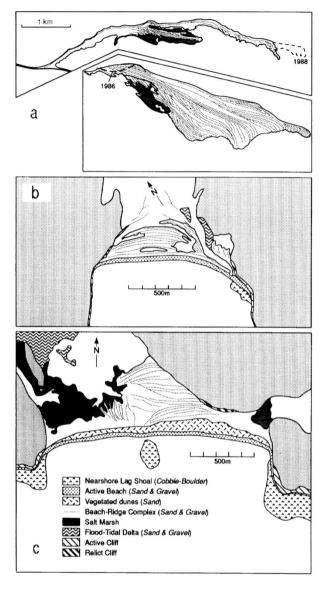

Figure 10.13. Beach-ridge morphology showing evidence of progradation (drawn from air photographs at various scales): (*a*) Proximal and distal sections of Flat Island barrier in St. George's Bay (after Shaw *et al.*, 1990; see Fig. 10.7). (*b*) Prograded beach-ridge complex being overrun by transgressive storm ridge, Shoal Bay, southeast Newfoundland (see Forbes, 1984). (*c*) Beach-ridge pattern at Lawrencetown Beach (Fig. 10.15) on the Eastern Shore of Nova Scotia, showing early progradation and evidence of trailing spits from former drumlin anchor off the middle of the present beach (see Boyd *et al.*, 1987; Forbes *et al.*, 1990).

7 m high, is derived from a complex body of ice-contact deposits exposed alongshore (Eyles & Slatt, 1977). It is part of a sweeping 5.5 km bayhead beach, of which about 2 km forms the barrier across Holyrood Pond. The 'pond' is a steep-sided fjord trough some 100 m deep and about 20 km long. The sill is rock-cored with a thin cover of till. It lies just below present sea level, providing a foundation for the barrier. The latter consists of amalgamated landward-dipping washover fans spilling into the fjord (broadly similar to the St. George's Bay spillover complex) and a thinner, seaward-dipping, high-energy beachface. The latter often has a steep, reflective, lower-berm foreshore slope (~0.15 or 8.5°; vertical in the presence of an ice-foot) and a lower-angle (0.07 or 4°) upper slope. Vertical accretion of more than 2 m has been observed on the upper beachface in association with development of large swash bars with wavelengths up to 100 m (Forbes, 1985) in coarse sand and very fine gravel. The vertical drop on the landward side of the barrier is considerably greater than relief on the ocean side (Fig. 10.9*a*), where the shoreface slope flattens out at about 35 to 40 m water depth. The shoreface is largely erosional, with a thin veneer of sand or wave-rippled gravel overlying stony mud interpreted as a glacial diamict (Forbes, 1984).

The capacity of systems such as the Holyrood Pond barrier or St. George's Bay sill to maintain themselves under rising sea level is limited. The continuing existance of the Holyrood Pond barrier has been aided by survival of the glaciogenic sediment supply at the site. Ongoing appropriation of sediment from this source enables the barrier to grow horizontally and vertically as relative sea level rises. On the other hand, beach gravel capping a 17 m deep sill at a site 20 km north of Holyrood Pond attests to early drowning of a barrier there (Forbes, 1984). Similarly, the sill and spillover system in St. George's Bay is now subaqueous (Fig. 10.7; Shaw & Forbes, 1992).

Despite the impressive size of the subaerial barriers at Stephenville and Flat Island, the sandy subaqeous platforms contain more than 90% of the total sediment accumulation in the barrier complexes. These features demonstrate the importance of antecedant morphology (in particular, of shoreface slope and basin relief) in determining the morphological and stratigraphic development of coastal deposits. They also demonstrate the large storage capacity of coastal basins and the relative importance of subaqueous sediment volumes in these settings, as noted above in the section on fjord-head deposition.

Drumlin coasts: source-switching and time-varying supply

Drumlins are elliptical hills of subglacial origin (Embleton & King, 1968; Davies & Stephens, 1978). They are usually less than 60 m high, have typical lengths of 0.5 to 2 km, and length/breadth ratios of 1 to 4 or more (Stea & Brown, 1989). Most drumlins are composed of till or other ice-contact deposits. They occur together in groups of a few hundred to a few thousand: 10000 drumlins occupy an area of about 12600 km^2 in northern New York state (Embleton & King, 1968), while another 6000 are present across Lake Ontario to the north (Chapman & Putnam, 1966).

Where drumlins emerge at sea level through isostatic uplift in partially protected marine settings, as observed for example along the coast of Dolphin and Union Strait in the Canadian Arctic (Potschin, 1989), poorly sorted gravel beach deposits form roughly concentric patterns defining successive stages of emergence. The extent of marine modification is limited and varies inversely with the elevation and time since initial emergence. On the other hand, where rising relative sea level results in progressive inundation and erosional transgression of a drumlin field, almost complete reworking of glaciogenic sediments can occur (e.g. Johnson & Reed, 1910; Guilcher, 1962; Piper *et al.*, 1986; Boyd *et al.*, 1987; Carter *et al.*, 1989, 1990b, 1992; Carter & Orford, 1991; Orford, Carter & Forbes, 1991a; Orford *et al.*, 1991b). Highly-organized coastal sedimentary systems can develop in such circumstances. However, these systems are vulnerable not only to changes in sea level and other environmental variables such as storm frequency and intensity, but also to reduction in sediment supply when feeder drumlins are consumed or the erosion is halted by barrier growth and cliff stabilization. In the present context, the themes we wish to emphasize are sediment supply control, the diversity of barrier form, the episodic nature of barrier transformation as the transgression proceeds, and the interaction between barrier development on the outer coast and the evolution of backbarrier estuarine systems.

Under conditions of marine transgression typical of drumlin coasts in parts of Nova Scotia and western Ireland (Taylor *et al.*, 1986; Carter *et al.*, 1989), individual drumlins or sets of drumlins successively acquire and relinquish control of coastal development on time scales of the order of a few hundred to a few thousand years (Forbes & Taylor, 1987; Carter *et al.*, 1989, 1992). Johnson (1925) was probably the first to describe the manner in which sediment supply from discrete drumlin sources of finite volume determines successive growth, failure, landward translation and re-establishment of barrier beaches (cf. Boyd *et al.*, 1987; Carter *et al.*, 1987; Forbes & Taylor, 1987). This evolved into the concept of an erosional front that propagates landward under transgression, successively attacking, consuming, and overtaking drumlin sediment sources (Carter & Orford, 1988).

A distinction can be drawn between cases where drumlins are distributed over a shallow basin (as in Boston Harbour, Mahone Bay and Clew Bay) and those where the Holocene transgression has flooded long narrow valleys embedded in drumlin fields (e.g. parts of the Eastern Shore of Nova Scotia). In the former case – drumlin islands scattered over a shallow flooded basin – erosion may extend over a wide front that penetrates deeply into the archipelago (Fig. 10.14). This depends to a large extent on the potential for island linking, which in turn is partly dependent on the distance between drumlins (which function both as sediment sources and anchor points) and on the depth of the intervening water (Fig. 4 of Carter *et al.*, 1987), among other factors. The proximity and placement of neighbouring drumlins also affect the wave-field perturbation (Carter & Orford, 1988). Where conditions are unfavourable for island linking, as in Mahone Bay (Piper *et al.*, 1986),

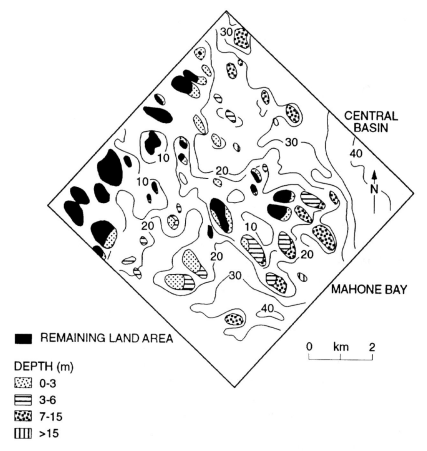

Figure 10.14. Drumlins and erosional lags in Mahone Bay, South Shore of Nova Scotia (after Piper *et al.*, 1986). Open ocean to south; erosional front moving from bottom to top of figure.

sediment released by wave erosion of drumlins is distributed over the nearby seafloor.

As cliffs and boulder lags develop at drumlin headlands (Fig. 10.4*b*; Carter *et al.*, 1990b), the latter exert a growing influence on the nearshore wave field (Carter, Jennings & Orford, 1990a). The drumlin headlands create an irregular embayed coastline. Sediment supply and littoral sediment budgets vary widely along the coast from compartment to compartment. With the development of barrier beaches between drumlin headlands, the coast-normal penetration of erosion is reduced and more nearly approximates a curvilinear front. Sediment dispersal then comes to be controlled by the nature and extent of barrier development and associated wave transformation. Although landward migration of barriers at the outer coast may respond in part to variations in the rate of sea-level rise, progressing in-phase with the transgression (Orford *et al.*, 1991a), other factors including storm frequency, sediment supply, and barrier morphology and dynamics (Forbes *et al.*, 1990; Orford *et al.*, 1991b) are also important in determining the migration rate. Rates of drumlin cliff recession vary with time since initial attack (sediment production per unit recession distance and development of a dissipative boulder lag platform both increase with time in the early stages), and in relation to beach development in the vicinity, weather conditions, and lithological variability (Stea & Fowler, 1979; Sonnichsen, 1984), among other factors.

Sediment derived from drumlin sources includes a wide range of grain size from clay to boulders (Sonnichsen, 1984; Carter *et al.*, 1990b). While boulders form lag shoals and boulder frames in the vicinity of the cliffs, finer gravel and sand are moved alongshore into adjacent beach systems. Beach and barrier forms range from prograded sandy or gravelly beach-ridge complexes (in cases of sediment abundance) to trailing spits and drift-aligned fringing beaches (often with complex cell structure related to local sediment deficits; Carter & Orford, 1991), to swash-aligned barriers of various types (Forbes *et al.*, 1990). Muds accumulate in deeper parts of basins such as Mahone Bay or in backbarrier estuaries. Most boulders remain as lag shoals within the original footprint of the source drumlin. The depths of these shoals record approximate mean sea level at the time of incision (Fig. 10.14).

Where sediment supply is adequate and other factors allow, beaches prograde, even under rising sea level. Lawrencetown Beach (Figs. 10.13*c* and 10.15), downdrift of extensive till cliffs, has accumulated a beach-ridge complex more than 600 m wide, recording several phases of growth, including the development and later closure of a tidal inlet at the west end of the beach (Boyd & Honig, 1992). At the other extreme, smaller, gravel-dominated barriers linked to restricted sources (e.g. Fisherman's Beach and Story Head

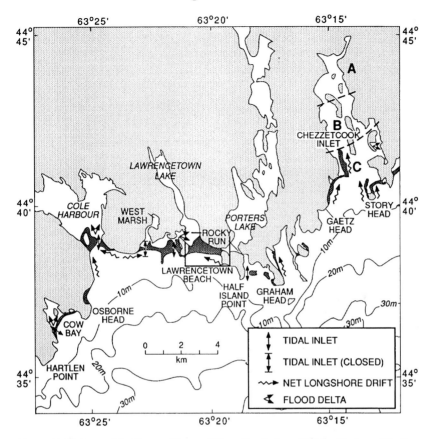

Figure 10.15. Part of the Eastern Shore of Nova Scotia from Halifax Harbour (Hartlen Point) to Chezzetcook Inlet, showing major barrier systems, tidal inlets and flood deltas, with directions of net longshore drift and subdivisions A, B, C of Chezzetcook Inlet (Carter *et al.*, 1992). Box shows location of Fig. 10.13c. Note also location of Story Head (Fig. 10.4b).

Beach, Fig. 10.4b) display symptoms of sediment starvation, ranging from longshore cell development on drift-aligned spits (Carter *et al.*, 1990b) to rapid landward migration, stretching, and incipient breaching on swash-aligned structures under appropriate circumstances of substrate, exposure, sea-level rise, and incident wave conditions (Forbes *et al.*, 1991b). The swash-aligned Story Head barrier has been moving landward at an average rate of about $8\,\mathrm{m\,a}^{-1}$ for nearly 40 years, incising a shallow, subtidal, backbarrier platform and almost outrunning the drift-aligned trailing ridge connecting the barrier to its headland source. As a result, a tidal breach developed in December 1986 at the end of the trailing ridge (arrow in Fig. 10.4b). Cut off from its principal sediment source, the swash-aligned barrier is susceptible to rapid failure or realignment over the coming few years or decades. Sediment residing in barriers such as this may be abandoned on the shoreface and inner shelf (Forbes & Boyd, 1989; Forbes *et al.*, 1991b) or it may be recycled landward to feed new barriers as the erosional front advances.

The beaches described above represent several distinctive barrier types developed within the common framework of marine transgression through a drumlin field. Forbes *et al.* (1990) recognized at least five barrier types in a preliminary review and classification based on the drumlin coasts of Nova Scotia. These were:

type 1 – prograded beach-ridge complexes (both sand and gravel);
type 2 – high, stable, steep-faced, gravel barriers, often with nearshore sand aprons;
type 3 – low, unstable, gravel barriers subject to frequent overwash;
type 4 – fringing barriers and drift-aligned spits;
type 5 – large sand barriers, typically with some aeolian dune development.

We return to this classification in a later discussion, where we also examine the potential for barrier transformation from one type to another.

The degree of enclosure of backbarrier basins varies from minimal to complete: from large open bays such as Boston Harbour (Johnson & Reed, 1910; Rosen & Leach, 1987), to impounding barriers with lagoons draining by seepage (Carter *et al.*, 1984). Within partially enclosed estuaries, the upper limit of the transgression (the reach of high water at extreme tides) may lead the erosional front by a distance of the order of a few kilometres. The advance of the transgression and landward movement of the erosional front are correlated and co-directed, but with a time-varying lag. In larger estuaries enclosed by mixed sand–gravel barriers, flood-delta complexes develop landward of tidal inlets, while flood-tidal sand sheets and basin muds accumulate in the central estuarine basin or basins. Lawrencetown Lake and West Marsh (Figs. 10.13c and 10.15) are examples of this type (Taylor *et al.*, 1986; Boyd & Honig, 1992; Shaw *et al.*, 1993). More open estuaries are subject to a mix of wave and tidal processes. Chezzetcook Inlet (Fig. 10.15) represents this type, in which sediment accumulation is initiated farther up-estuary, and can take the form of marsh accretion, channel levée growth, lateral flood-delta expansion, and incipient beach and barrier formation, among others (Carter *et al.*, 1989, 1992; Orford *et al.*, 1991b; Shaw *et al.*, 1993). In particular, Scott (1980) and Carter *et al.* (1989, 1992) have documented the rapid growth of flood-delta splays in Chezzetcook Inlet over the past few decades and related this development to instability and sediment release at the outer coast. Apart from barrier instability at Story Head on the east, significant changes have also occurred on the west side of the outer estuary, where the drumlin core of Cape Antrim has been essentially eliminated. The Cape Antrim barrier has moved landward at an average rate of about $6\,\mathrm{m\,a}^{-1}$, eventually encountering a formerly protected drumlin and initiating erosion of a new headland (Fig. 10.15; Fig. 6 of Forbes *et al.*, 1990).

As relative sea level has risen along the Eastern Shore of Nova Scotia throughout the Holocene, the infilling of estuarine embayments has proceeded in much the same manner as outlined above. Detailed seismic reflection profiling, sidescan sonar imaging, direct observations and coring on the inner shelf have revealed preserved estuarine sediments in former valley depressions out to present water depths of at least 45 m (e.g. Forbes & Boyd, 1989; Forbes *et al.*, 1991a). The data indicate that estuarine basin muds, flood-delta sands, and salt-marsh peats are preserved on the shelf, whereas barrier structures are absent. Sand is present only in thin patches and appears to be almost completely recycled landward with the transgression. Much of the seafloor consists of bedrock outcrop or coarse cobble-boulder lags over till, but large areas of finer gravel are also present, especially in the vicinity of former valley embayments where gravel barriers may have existed. We hypothesize that much of this finer gravel originated from drumlin sources, was recycled alongshore into beaches and barriers as observed at the present coast, and has subsequently been abandoned on the shelf as the shoreline moved landward. A mechanism for abandonment has been described by Forbes *et al.* (1991b). Active wave-formed ripples (Forbes & Boyd, 1987) demonstrate that gravel is frequently mobilized on the shelf to water depths of at least 40 m, suggesting that abandoned beach gravels could easily be dispersed over the seafloor, erasing any remnant barrier morphology and generating the extensive flat and rippled gravel surfaces we observe today.

Embayed coast with diminishing sediment supply

At numerous sites in southeastern Canada, gravel-dominated bayhead barriers take the form of high (type 2) storm ridges at the seaward margins of prograded (type 1) gravel beach-ridge sets (Figs. 10.4c and 10.13b; cf. Forbes, 1985; Shaw & Forbes, 1987; Forbes *et al.*, 1989, 1990; Shaw *et al.*, 1990; Stea, Forbes & Mott, 1992). In a few cases, packages of old beach ridges can be found behind sandy barriers (Fig. 10.13c). More commonly, sand is present only in the nearshore or as an aeolian cap over a gravel structure. Where a high gravel ridge is present, its crest elevation is typically much higher than that of the next highest ridge behind. Commonly, the active (most seaward) ridge has a radius of curvature much larger than that of the oldest ridges on the landward side (Fig. 10.13*b*).

This morphology – sets of low, curvilinear, gravel beach ridges lying behind high, narrow, relatively straight, active storm ridges – may result from various combinations of rising relative sea level, changing wave climate, and reduced sediment supply. In the simplest cases, we have proposed that diminishing sediment supply is the principal control, resulting in a switch from seaward progradation to landward migration (Shaw & Forbes, 1987). The transgressive storm ridge then incorporates reworked sediment from the seaward margin of the beach-ridge complex, particularly at the ends of the beach where the straightening arc intersects older ridges most acutely; accretes to a higher level as reduced sediment supply leads to washover and crestal buildup; and may continue to build up under a sequence of progressively rarer and larger storm-wave events. Slowly rising relative sea level may be important, not only in determining the seaward increase in crest elevation of the progradational ridges, but also in controlling the sediment supply on which seaward progradation depends.

Several small bayhead barrier and beach-ridge complexes thought to have developed in this way are found in embayments on the southern Avalon Peninsula in southeast Newfoundland. These include barriers at O'Donnells, Shoal Bay, Biscay Bay, and Portugal Cove South (Figs. 10.4*c* and 10.13*b*; Shaw & Forbes, 1987). The range of beach-ridge elevations and differences in elevation between the seaward ridge and the next highest ridge behind (Fig. 10.16*a*) reflect varying exposure (Fig. 10.16*b*) and other factors, but all show the typical pattern described above. Although a small deposit of outwash sand and gravel is present at Portugal Cove South and may have been an important source of beach material at that site, the shoreline is predominantly rocky in all four embayments and surficial sediments on the adjacent slopes consist almost exclusively of thin ground moraine (Henderson, 1972). Although tide-gauge data from St. John's suggest a slight rise in sea level during the past few decades (Shaw & Forbes, 1990a), radiocarbon dates on organic materials from the gravel beach-ridge plain at Placentia (Shaw & Forbes, 1987, 1988) suggest a very limited (probably decelerating) rise of relative sea level on the Avalon Peninsula during the past 2000 years or so, believed to be the approximate time since initial emplacement of the back-barrier ridge sets at Biscay Bay (Shaw *et al.*, 1990) and the other sites in the area.

Rising relative sea level, which controls the level of wave attack, can be a major determinant of access to erodible deposits in paraglacial settings. It is critical to an understanding of beach development in the embayments of the

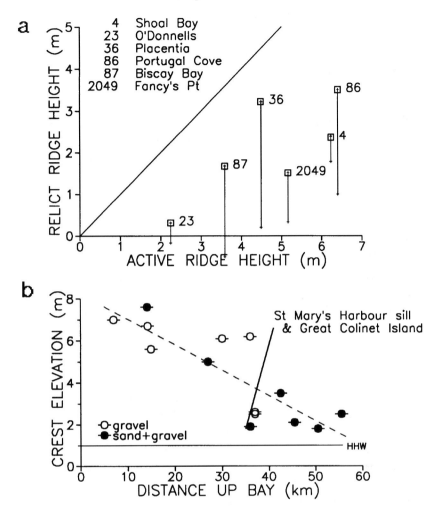

Figure 10.16. (*a*) Crest heights (maximum and minimum) of relict prograded beach ridges in relation to heights of active storm ridges for a selection of sites in southeast Newfoundland (Shoal Bay, Fig. 10.13*b*; O'Donnells; Placentia; Portugal Cove South, Fig. 10.4*c*; Biscay Bay – Forbes, 1985; Shaw & Forbes, 1987), and Nova Scotia (Fancy's Point Beach – Forbes *et al.*, 1990; Shaw *et al.*, 1993). Mean sea level datum; solid line represents height equivalence; regression of maximum relict ridge height (H_R) on storm-ridge height (H_A) suggests the relationship $H_R = 0.6 H_A - 0.5$. (*b*) Reduction in barrier-crest elevation northward into St. Mary's Bay, southeast Newfoundland (Forbes, 1985), showing two interpretations, both reflecting reduced wave energy in protected settings: 1. there are two populations, separated by Great Colinet Island and St Mary's Harbour sill, which partially enclose the inner bay (to right of solid line); 2. barrier crest elevations decrease in a quasi-linear fashion (broken line) with distance from the mouth of the bay.

southern Avalon Peninsula. Numerical experiments using idealised linear valley-side slopes of rock, mantled with a uniform thickness of till, can help to elucidate the effect of decelerating sea-level rise on sediment supply to the bayhead beach (Fig. 10.17). Assuming that all unconsolidated sediment resting on bedrock at or below sea level is removed by waves between time steps ($\Delta t = 100$ years), the sediment supply to the bayhead sink will vary linearly with the rate of sea-level rise, the valley slope, and the thickness of glaciogenic deposits. Holding slope and till thickness constant, the sediment supply diminishes as the rate of relative sea-level rise decreases. In the examples shown (Fig. 10.17), the rate of sea-level rise ($r > 0$) was constrained to decrease as

$$dr/dt = (1 - k)\,r$$

where $k = 0.05$ is an arbitrary constant, and the resulting relative sea-level curve depends on the initial rate of sea-level rise (r_0) and initial sea level (z_0) at

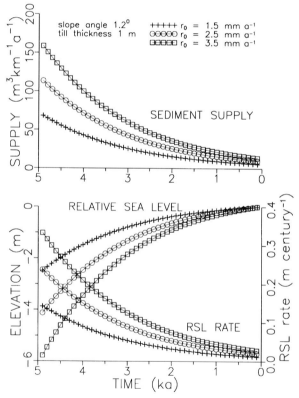

Figure 10.17. Numerical simulation of sediment supply from an evenly distributed veneer of till as a function of sea-level changes and linear valley slope in a rectangular embayment (see text for details). r_0 is rate of sea-level rise.

an arbitrary time (here taken to be 5 ka BP). Adopting initial rates of relative sea-level rise of 1.5, 2.5, and 3.5 mm a^{-1}, and corresponding initial sea levels of approximately 2.6, 4.3, and 6.1 m below present, we obtain rates of sediment supply (per kilometre of shoreline) decreasing respectively from 68, 113, and 158 m^3 a^{-1} at 4.9 ka BP to less than 13 m^3 a^{-1} at present, assuming a valley slope of 1.2° and uniform till thickness of 1 m (Fig. 10.17). Increasing the valley slope to 2° (with r_0 = 2.5 mm a^{-1} and all other parameters the same) decreases the initial supply to 68 m^3 a^{-1} at 4.9 ka BP and less than 6 m^3 a^{-1} at present (i.e. the steeper slope yields a lower volume of sediment per unit rise in relative sea level).

Because the depth of water along the bay axis increases seaward, the rate of sediment supply would have to increase in order to maintain a constant rate of progradation as the beach builds out. Under constant sea-level rise and sediment supply, the progradation would eventually cease. A diminishing rate of sea-level rise (Fig. 10.17) would exacerbate the effect, leading eventually to a cessation of seaward construction (Fig. 10.18). Although the precise threshold conditions are unknown, a switch from progradation to transgressive (landward) reworking is likely to ensue. This is believed to have occurred at a number of the sites in southeast Newfoundland.

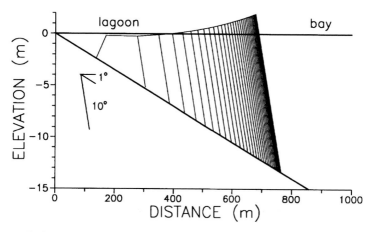

Figure 10.18. Numerically simulated beach-ridge stratigraphy along the axis of a 5 km deep embayment, assuming initial relative sea-level rise of 1.5 mm a^{-1} from −2.38 m at 3.4 ka BP (the 3.5 mm a^{-1} curve in Fig. 10.17), till thickness of 1 m, and valley side slope of 1.2°. Valley walls open seaward at an angle of 29° and valley axis slope is 1°. Beach is assumed to be linear and orthogonal to valley axis. Runup on a 10° beachface slope is assumed to build ridge crests to 2 m above ambient relative sea level. Time lines are at equal intervals of the order of 100 years. Increasing prism and decreasing sediment supply combine to produce a rapid deceleration and eventual cessation of seaward growth. Backbarrier beach ridges rise seaward and the oldest ridges are now submerged in the lagoon.

Boulder accumulations in sediment-starved settings

Boulders are prominent constituents of glacial deposits in many areas, occurring as isolated erratics or large clasts in till, sometimes many tens of kilometres from the nearest source outcrop (cf. Fig. 26 of Stea *et al.*, 1992). As discussed earlier, erosion of boulder-rich tills at the coast can result in the development of boulder lag platforms, composed of blocks too large to be removed by wave action and many smaller boulders subject to limited reworking (Carter *et al.*, 1990b). Where relative sea level is rising, these boulder accumulations are abandoned on the inner shelf at various levels, depending on the time of formation (Fig. 10.14; Piper *et al.*, 1986; Forbes & Boyd, 1989). Although much of the boulder fraction in glacial deposits eroded at the coast remains in place, boulders may be either dispersed or concentrated under various combinations of waves, currents, and sea-ice action. Fortuitously, many formerly glaciated mid- to high-latitude areas with large supplies of boulder-size debris are also exposed to cold-climate effects, including sea ice, resulting in the development of ice-rafted and ice-pushed accumulations that constitute a distinctive class of paraglacial deposits (Forbes & Taylor, 1994). This is not to imply that all ice-rafted boulder deposits are paraglacial: boulder-strewn tidal flats, for example, need not necessarily depend on a paraglacial source for their boulder supply (Dionne, 1981).

Coastal boulder accumulations drawn largely from glacial deposits or isolated erratics occur in a number of areas and take several forms. On the Swedish island of Gotland (Baltic Sea), ice-rafted crystalline erratics are found in irregular accumulations on beaches and rock platforms (Philip, 1990). Additions to these deposits observed during ice-rafting events are thought to come from sources on the adjacent seafloor. The shores of the St. Lawrence Estuary in Canada are richly adorned with boulders on tidal flats, postglacial terraces, and rock platforms (e.g. Brochu, 1954; Dionne, 1962, 1972a, 1972b, 1979; Guilcher, 1981). Dionne (1972b) demonstrated that ice-rafting processes have been more effective than glacial transport in moving crystalline boulders across the St. Lawrence from the north shore to the south. Along the south shore of the estuary, he observed that crystalline boulders of north-shore provenance amounted to 2.7% of the total boulder count in glacial deposits, 33% on wave-washed till surfaces, 61% at the present shoreline, and 75% on raised marine terraces. A large proportion of the ice-rafted boulders are nevertheless presumed to have been eroded initially from glacial deposits, although some may have originated directly from bedrock. This is also the case in other areas where boulder barricades, boulder pavements, and bouldery tidal flats (Figs. 10.4*e,f*) have been observed, including Newfoundland and Labrador (Tanner, 1939; Rosen, 1979; Forbes, 1984), Ungava Bay (Lauriol &

Gray, 1980), Hudson Bay and vicinity (Dionne, 1978), southeast Baffin Island (McCann, Dale & Hale, 1981), and the sub-Antarctic islands (Hansom, 1983). In western Newfoundland, coastal boulder accumulations are found on rock platforms and at the base of gravel beaches below partially stabilized till bluffs, in a context similar to that described by Philip (1990).

Boulders can also be dispersed by waves and currents under suitable circumstances. At Story Head, on the Nova Scotia coast, where an extensive boulder lag platform has developed in front of the eroding till cliff (Fig. 10.4*b*; Carter *et al.*, 1990b), clast mobility is limited by interlocking and the large displacement moments required to move individual boulders over adjacent clasts of similar size. Nevertheless, some escape and move downdrift, forming a boulder frame at the base of the trailing ridge and spit. Boulders reaching the swash-aligned barrier are driven up the beachface by swash surges, moving on a substrate of finer pebble and cobble gravel, and become incorporated in the barrier crest and washover deposits (Forbes *et al.*, 1991b). In this way, a small proportion of boulders keep pace with the transgression; the great majority, however, are overtaken and abandoned on the shelf.

Discussion

Much of the emphasis in the foregoing sections has been on the nature and disposition of glaciogenic deposits as sediment sources for the coast. It is also clear that postglacial changes in relative sea level play an important and ubiquitous role in the evolution of any coast and the stratigraphy of associated deposits. This role is enhanced in some paraglacial situations, where relative sea-level changes also determine the availability of sediments (Fig. 10.17) or the timing and location of sediment reworking (Figs. 10.10 and 10.18).

Even the delivery of fluvial sediments can be affected by sea-level changes. In areas subject to continuing isostatic uplift, progressive downcutting of river channels leads to the reworking and liberation of sediment stored in raised fjord-head and valley terrace deposits (Church, 1972, 1978). In this way, relative sea-level changes may substitute for basin scale in prolonging the postglacial release of sediment from alluvial basins (cf. Church & Slaymaker, 1989), just as they extend the process in other shoreline settings (Forbes & Taylor, 1987). Nevertheless, computations of sediment release from raised Holocene delta terraces at the head of Cambridge Fiord, northeast Baffin Island (Table 10.1), show that the contributions from scavenging of terrace deposits declined dramatically (approximately 75%) over the first 1000 years following deglaciation.

Our numerical simulation of barrier progradation (Fig. 10.18) is based on a

Table 10.1. *Estimates of delta sediment reworking as a function of emergence rate and associated channel incision, from terrace volumes removed from raised outwash delta deposits at head of Cambridge Fiord, northern Baffin Island*

Terrace	Elevation (m)	Volume removed ($10^6 \, m^3$)	Time interval (ka BP)	Δt (ka)	Rate ($10^7 \, kg \, a^{-1}$)
1	50	4.5	6.10–5.90	0.20	3.6
2	45	2.8	5.90–5.60	0.30	1.5
3	38	3.0	5.60–5.25	0.35	1.4
4	32	2.6	5.25–4.80	0.45	0.9
5	25	4.8	4.80–0.00	4.80	0.2

After Syvitski *et al.*, (1984).

simple volumetric accounting procedure and ignores important aspects of nearshore wave and sediment dynamics, beach morphodynamics, shoreface sediment exchange, and the possibility of variations in sea-level rise or wave climate, among other factors. At Fancy's Point Beach on the Atlantic coast of Nova Scotia (Forbes *et al.*, 1990), a barrier and beach-ridge complex similar to the Avalon Peninsula examples has developed in an area of drumlin terrain. From a suite of low beach ridges, partially submerged at the back of the barrier, the profile rises abruptly seaward to a set of high-level ridges showing evidence of recent progradation (Stea *et al.*, 1992; Shaw *et al.*, 1993). This morphology may reflect a more-complex history of variation in sediment supply from local drumlins and relict barriers to seaward, variations in relative sea-level rise, and possibly more than one phase of progradation interrupted by a transgressive phase. One hypothesis is that increased sediment input and renewed progradation against a high-level ridge may have been a response to higher wave energy during the Little Ice Age. Evidence of non-stationarity in sea-level trends on a variety of time scales ranging from 10 to 1000 years (e.g. Shaw & Forbes, 1990a; van de Plassche, 1991; Scholand, Varekamp & Thomas, 1992) and of changes in storm characteristics affecting wave climate on similar scales (e.g. Lamb & Johnson, 1961; Neu, 1984) suggests the probability of complex interactions involving variable sediment supply and beach morphodynamics. Indeed, observations from Story Head Beach (Fig. 10.4*b*) indicate that variable rates of relative sea-level rise and intermittent storminess over the past 40 years were responsible for large variations in washover transport on a decadal scale (Orford *et al.*, 1991a), but may also have caused an abrupt destabilization of the barrier at some time in the early 1950s (Forbes *et al.*, 1991b).

In this context, there is scope for development and refinement of the barrier classification presented earlier in the section on drumlin coasts. Some further subdivision may be desirable, including recognition of the distinction between spits and fringing barriers, both drift- and swash-aligned, and differentiation of sand, mixed sand–gravel, and gravel beaches on the basis of modal morphodynamic state, tidal inlet relations, dune conditions, or other factors. In addition, the classification should recognize typical transitions or evolutionary paths from type to type. Forbes *et al.* (1990, 1991b) suggested, for example, that high, reflective, type-2 barriers may be catastrophically destabilized and transformed into low, unstable, washover-dominated, type-3 barriers under particular circumstances. Orford *et al.* (1991b) proposed additional paths in a 'process-oriented typology' that combined aspects of the previous classification with concepts of headland source and anchor control and a primary differentiation between drift- and swash-aligned forms (see Fig. 1.4). We suggest that a further development needs to incorporate multiple sediment source types (drumlin headland, pre-existing coastal deposits to seaward, and underlying or adjacent tidal and estuarine sequences), the possibility of varying evolutionary paths (some with higher transition probabilities than others), the relative proportions and mobility of sand and gravel, and the importance of tidal inlet processes, among other factors.

Conclusions

Sediment supply limitation and a highly localized distribution of sources along the coast are major determinants of coastal evolution in paraglacial settings. Sediments are derived primarily from glaciogenic deposits, although direct contributions from fluvial sources or bedrock erosion are important locally. Paraglacial sediment recycling (through incision of deltaic and other deposits as relative sea level falls or by landward reworking in a transgressive context) is an important component in some situations. Proglacial systems receiving sediment from glacial meltwater remain active in some high-latitude settings. The outwash coasts of the Gulf of Alaska and southern Iceland are among the largest glacially influenced depositional coasts in the world. Relict complexes of this type have been described from the Cape Cod and Long Island region of the eastern USA (Johnson, 1925) and from the northern Gulf of St. Lawrence (Syvitski & Praeg, 1989).

Physiographic setting is another important arbiter of coastal form, determining the extent of sheltering, the opportunity for integration of sediment transport and deposition along the coast, and the storage potential of coastal basins. In areas of rapid sedimentation into deep water on an open coast (as in the case of the Skeiðarársandur), the system may expand accommodation space over time. In other cases, large glacial basins may be completely filled (Pickrill *et al.*, 1981). In general, however, glacially overdeepened basins provide enormous unused storage capacity, despite holding a large proportion of postglacial sediment transfers from the land to the sea (Syvitski *et al.*, 1987). Although complex barrier systems and other coastal facies have developed in areas of large active (proglacial) or relict (ice-marginal) sediment supply, such as the southeast Iceland and southwest Newfoundland examples in this review, these impressive coastal deposits represent a very small proportion of the total sediment body where deep water is present close to shore. Knowledge of the entire depositional system and architecture may be critical to an understanding of Holocene coastal evolution in such settings (Shaw & Forbes, 1992).

The limited sediment supply in many paraglacial settings imposes distinctive patterns on beach, barrier, and backbarrier development. While some systems show symptoms of slow supply depletion (Figs. 10.13*b,c*, 10.18), others may experience cycles of supply fluctuation over time scales of the order of 1000 years (Fig. 10.4*b*). In both cases, abrupt changes in coastal stability can occur over a few years or during storms, as local sediment sources (such as drumlin cliffs) or conduits (littoral transport guides such as the trailing ridge at Story Head) are turned on or off and as individual components – barrier complexes, tidal inlets, or estuarine systems – adjust to changing circumstances. Drumlin coasts provide a wealth of evidence for behaviour of this kind and numerous opportunities for quasi-experimental investigations of rapid coastal evolution.

Paraglacial coasts encompass an enormous variety of environments and products. We have attempted to describe the major types and to highlight key issues. These include major erosional features of glacial origin, fjord-head and proximal basin sedimentation, open-coast outwash progradation, coastal development in basin-margin settings, drumlin–coast interactions, and embayed coast deposition under falling sediment supply. Key issues affecting paraglacial coastal development include the wide range of grain sizes in sediment supplied to the coast, the proximity of large sediment supplies to deep basins (particularly in early postglacial time), the complex controls on paraglacial sediment delivery to the coast, and the vulnerability of systems with limited sediment supply to small changes in energy or mass inputs.

Acknowledgements

Permission for reproduction of copyright materials is gratefully acknowledged as follows: Springer-Verlag New York (Fig. 10.2), Macmillan Magazines [*Nature*] (Fig. 10.3), the Royal Society of Canada (Fig. 10.6), the Geological Society of London (Fig. 10.8), and Elsevier Science Publishers (Fig. 10.12).

The ideas developed in this review have their roots in the classical works of early pioneers such as Douglas Johnson. They have evolved in our own work through years of shared experience and stimulating discussions with many friends and colleagues, to whom we offer our sincere thanks and appreciation. In particular, we wish to acknowledge the contributions of the late Bill Carter, and of Julian Orford, John Shaw, and Bob Taylor, whom we also thank for constructive reviews of the manuscript. This is contribution no. 16993 of the Geological Survey of Canada.

References

Allen, J.R.L. (1990). The Severn Estuary in southwest Britain: its retreat under marine transgression, and fine-sediment regime. *Sedimentary Geology*, **66**, 13–28.

Amos, C.L. & Long, B.F.N. (1980). The sedimentary character of the Minas Basin, Bay of Fundy. In *The coastline of Canada* ed. S.B. McCann, pp. 123–52. Geological Survey of Canada, Paper 80-10.

Amos, C.L. & Zaitlin, B.A. (1985). Sub-littoral deposits from a macrotidal environment, Chignecto Bay, Bay of Fundy. *Geo-Marine Letters*, **4**, 161–9.

Armon, J.W. & McCann, S.B. (1977). Longshore sediment transport and a sediment budget for the Malpeque barrier system, southern Gulf of St. Lawrence. *Canadian Journal of Earth Sciences*, **14**, 2429–39.

Armon, J.W. & McCann, S.B. (1979). Morphology and landward sediment transfer in a transgressive barrier island system, southern Gulf of St. Lawrence, Canada. *Marine Geology*, **31**, 333–44.

Blake, W., Jr. (1975). Radiocarbon age determinations and postglacial emergence at Cape Storm, southern Ellesmere Island, Arctic Canada. *Geografiska Annaler*, **57A**, 1–71.

Boothroyd, J.C. & Nummedal, D. (1978). Proglacial braided outwash: a model for humid alluvial-fan deposits. In *Fluvial sedimentology*, ed. A.D. Miall, pp. 641–68. Canadian Society of Petroleum Geologists, Memoir 5.

Boulton, G.S. (1976). The development of geotechnical properties in glacial tills. In *Glacial till: an interdisciplinary study*, ed. R.F. Legget, pp. 292–303. Royal Society of Canada, Special Publication 12.

Boyd, R., Bowen, A.J. & Hall, R.K. (1987). An evolutionary model for transgressive sedimentation on the Eastern Shore of Nova Scotia. In *Glaciated coasts*, ed. D.M. FitzGerald & P.S. Rosen, pp. 87–114. San Diego: Academic Press.

Boyd, R. & Honig, C. (1992). Estuarine sedimentation on the Eastern Shore of Nova Scotia. *Journal of Sedimentary Petrology*, **62**, 569–83.

Boyd, R., Scott, D.B. & Douma, M. (1988). Glacial tunnel valleys and Quaternary history of the outer Scotian Shelf. *Nature*, **333**, 61–4.

Brochu, M. (1954). Un problème des rives du Saint-Laurent: blocaux erratiques observés à la surface des terrasses marines. *Revue de Géomorphologie dynamique*, **5**, 76–82.

Brookes, I.A. (1974). *Late-Wisconsin glaciation of southwestern Newfoundland (with special reference to the Stephenville map-area)*. Geological Survey of Canada, Paper 73-40. 31 pp. & maps.

Brookes, I.A. (1987). Late Quaternary glaciation and sea-level change in southwest Newfoundland, Canada. In *Centennial field guide*, vol. 5, ed. D.C. Roy, pp. 445–50. Geological Society of America, Northeastern Section.

Brookes, I.A., Scott, D.B. & McAndrews, J.H. (1985). Postglacial relative sea-level change, Port au Port area, west Newfoundland. *Canadian Journal of Earth Sciences*, **22**, 1039–47.

Bruun, P. (1962). Sea–level rise as a cause of shore erosion. *Journal of Waterways and Harbors Division*, American Society of Civil Engineers, **88**, 117–30.

Cant, D.J. (1991). Geometric modelling of facies migration: theoretical development of facies successions and local unconformities. *Basin Research*, **3**, 51–62.

Carr, A.P. (1969). Size grading on a pebble beach: Chesil Beach, England. *Journal of Sedimentary Petrology*, **39**, 297–311.

Carter, R.W.G., Forbes, D.L., Jennings, S.C., Orford, J.D., Shaw, J. & Taylor, R.B. (1989). Barrier and lagoon coast evolution under differing relative sea-level regimes: examples from Ireland and Nova Scotia. *Marine Geology*, **88**, 221–42.

Carter, R.W.G., Jennings, S.C. & Orford, J.D. (1990a). Headland erosion by waves. *Journal of Coastal Research*, **6**, 517–29.

Carter, R.W.G., Johnston, T.W. & Orford, J.D. (1984). Stream outlets through mixed sand and gravel coastal barriers: examples from southeast Ireland. *Zeitschrift für Geomorphologie, N.F.*, **28**, 427–42.

Carter, R.W.G. & Orford, J.D. (1988). Conceptual model of coarse clastic barrier formation from multiple sediment sources. *Geographical Review*, **78**, 221–39.

Carter, R.W.G. & Orford, J.D. (1991). The sedimentary organisation and behaviour of drift–aligned gravel barriers. In *Proceedings, Coastal Sediments 91 (Seattle)*, pp. 934–48. New York: American Society of Civil Engineers.

Carter, R.W.G. & Orford, J.D. (1993). The morphodynamics of coarse clastic beaches and barriers: a short- and long-term perspective. In *Coastal morphodynamics*, ed. A. Short, *Journal of Coastal Research*, Special Issue **15**, 158–79.

Carter, R.W.G., Orford, J.D., Forbes, D.L. & Taylor, R.B. (1987). Gravel barriers, headlands and lagoons: an evolutionary model. In *Coastal sediments '87*, vol. 2, pp. 1776–92. New Orleans: American Society of Civil Engineers.

Carter, R.W.G., Orford, J.D., Forbes, D.L. & Taylor, R.B. (1990b). Morphosedimentary development of drumlin-flank barriers in a zone of rapidly rising sea level, Story Head, Nova Scotia. *Sedimentary Geology*, **69**, 117–38.

Carter, R.W.G., Orford, J.D., Jennings, S.C., Shaw, J. & Smith, J.P. (1992). Recent evolution of a paraglacial estuary under conditions of rapid sea-level rise: Chezzetcook Inlet, Nova Scotia. *Proceedings of the Geologists' Association*, **103**, 167–85.

Chapman, L.J. & Putnam, D.F. (1966). *The physiography of southern Ontario*, 2nd edn. Toronto: University of Toronto Press. 386 pp.

Church, M. (1972). *Baffin Island sandurs: a study of Arctic fluvial processes*. Geological Survey of Canada, Bulletin 216. 208 pp.

Church, M. (1974). Hydrology and permafrost with reference to northern North America. In *Permafrost hydrology*, pp. 7–20. Ottawa: Canadian National Committee, International Hydrological Decade, Environment Canada.

Church, M. (1978). Palaeohydrological reconstructions from a Holocene valley fill. In *Fluvial sedimentology*, ed. A.D. Miall, pp. 743–72. Canadian Society of Petroleum Geologists, Memoir 5.

Church, M. & Gilbert, R. (1975). Proglacial fluvial and lacustrine environments. In *Glaciofluvial and glaciolacustrine sedimentation*, ed. A.V. Jopling & B.C. McDonald, pp. 22–100. Society of Economic Paleontologists and Mineralogists, Special Publication 23.

Church, M. & Ryder, J.M. (1972). Paraglacial sedimentation: a consideration of fluvial processes conditioned by glaciation. *Geological Society of America Bulletin*, **83**, 3059–72.

Church, M. & Slaymaker, O. (1989). Disequilibrium of Holocene sediment yield in glaciated British Columbia. *Nature*, **337**, 452–4.

Dalrymple, R.W., Knight, R.J., Zaitlin, B.A. & Middleton, G.V. (1990). Dynamics and facies model of a macrotidal sand-bar complex, Cobequid Bay–Salmon River estuary (Bay of Fundy). *Sedimentology*, **37**, 577–612.

Darby, D.A. (1990). Evidence for the Hudson River as the dominant source of sand on the US Atlantic Shelf. *Nature*, **346**, 828–31.

Davidson-Arnott, R.G.D. (1986). Erosion of the nearshore profile in till: rates, controls and implications for shore protection. In *Proceedings, Symposium on Cohesive Shores* (Burlington), pp. 137–49. Ottawa: National Research Council.

Davies, G.L.H. & Stephens, N. (1978). *The geomorphology of the British Isles: Ireland*. London: Methuen. 250 pp.

Davis, W.M. (1890). Structure and origin of glacial sand plains. *Geological Society of America Bulletin*, **1**, 195–202.

Davis, W.M. (1896). The outline of Cape Cod. *Proceedings, American Society of Arts and Science*, **31**, 303–32.

Dionne, J-C. (1962). Note sur les blocs d'estran du littoral sud du Saint-Laurent. *Canadian Geographer*, **7**, 69–77.

Dionne, J-C. (1972a). Caractéristiques des schorres des régions froides, en particulier de l'estuaire du Saint-Laurent. *Zeitschrift für Geomorphologie, Suppl. Bd.*, **13**, 131–62.

Dionne, J-C. (1972b). Caractéristiques des blocs erratiques des rives de l'estuaire du Saint-Laurent. *Revue de Géographie de Montréal*, **26**, 125–52.

Dionne, J-C. (1978). Le glaciel en Jamésie et en Hudsonie, Québec subarctique. *Géographie physique et Quaternaire*, **32**, 3–70.

Dionne, J-C. (1979). Les blocs d'estran à Saint-Fabien-sur-Mer, estuaire maritime du Saint-Laurent, Québec. *Maritime Sediments*, **15**, 5–13.

Dionne, J-C. (1981). A boulder-strewn tidal flat, North Shore of the Gulf of St. Lawrence, Québec. *Géographie physique et Quaternaire*, **35**, 261–7.

Dreimanis, A. (1976). Tills: their origin and properties. In *Glacial till: an interdisciplinary study*, ed. R.F. Legget, pp. 11–49. Royal Society of Canada, Special Publication 12.

Dyke, A.S., Morris, T.F. & Green, D.E.C. (1991). *Postglacial tectonic and sea level history of the central Canadian Arctic*. Geological Survey of Canada, Bulletin 397. 56 pp.

Embleton, C. & King, C.A.M. (1968). *Glacial and periglacial geomorphology*. London: Edward Arnold. 608 pp.

Emery, K.O. (1955). Grain size of marine beach gravels. *Journal of Geology*, **63**, 39–49.

Esmark, J. (1827). Remarks tending to explain the geological history of the Earth. *Edinburgh New Philosophical Journal for 1826*, 107–21.

Eyles, C.H., Eyles, N. & Lagoe, M.B. (1991). The Yakataga Formation: a late Miocene to Pleistocene record of temperate glacial marine sedimentation in the Gulf of Alaska. In *Glacial marine sedimentation: paleoclimatic significance*, ed. J.B. Anderson & G.M. Ashley, pp. 159–80. Geological Society of America, Special Paper 261.

Eyles, N., Eyles, C.H. & Day, T.E. (1983). Sedimentologic and palaeomagnetic characteristics of glaciolacustrine diamict assemblages at Scarborough Bluffs, Ontario, Canada. In *Tills and related deposits*, ed. E.B. Evenson, Ch. Shlüchter & J. Rabassa, pp. 23–45. *Proceedings, INQUA Symposia on the Genesis and Lithology of Quaternary Deposits* (USA 1981/ Argentina 1982). Rotterdam: Balkema.

Eyles, N., Mullins, H.T. & Hine, A.C. (1990). Thick and fast: sedimentation in a Pleistocene fiord lake of British Columbia, Canada. *Geology*, **18**, 1153–57.

Eyles, N. & Slatt, R.M. (1977). Ice-marginal sedimentary, glacitectonic, and morphologic features of Pleistocene drift: an example from Newfoundland. *Quaternary Research*, **8**, 267–81.

FitzGerald, D.M. & Rosen, P.S., eds. (1987). *Glaciated coasts*. San Diego: Academic Press. 364 pp.

Flint, R.F. (1940). Late Quaternary changes of level in western and southern Newfoundland. *Geological Society of America Bulletin*, **51**, 1757–80.

Flint, R.F. (1971). *Glacial and Quaternary geology*. New York: Wiley. 892 pp.

Forbes, D.L. (1984). Coastal geomorphology and sediments of Newfoundland. Geological Survey of Canada, Paper 84-1B, 11–24.

Forbes, D.L. (1985). Placentia Road and St. Mary's Bay: field–trip guide to coastal sites in the southern Avalon Peninsula, Newfoundland. In *Proceedings, Canadian Coastal Conference 1985* (St. John's). pp. 587–605. Ottawa: National Research Council.

Forbes, D.L. & Boyd, R. (1987). Gravel ripples on the inner Scotian Shelf. *Journal of Sedimentary Petrology*, **57**, 46–54.

Forbes, D.L. & Boyd, R. (1989). Submersible observations of surficial sediments and seafloor morphology on the inner Scotian Shelf. In *Submersible observations off the east coast of Canada*, ed. D.J.W. Piper, pp. 71–81. Geological Survey of Canada, Paper 88–20.

Forbes, D.L., Boyd, R. & Shaw, J. (1991a). Late Quaternary sedimentation and sea level changes on the inner Scotian Shelf. *Continental Shelf Research*, **11**, 1155–79.

Forbes, D.L. & Shaw, J. (1989). *Cruise report 88018(E)*, Navicula *operations in southwest Newfoundland coastal waters: Port au Port Bay, St. George's Bay, La Poile Bay to Barasway Bay and adjacent inner shelf*. Geological Survey of Canada, Open File 2041. 57 pp.

Forbes, D.L., Shaw, J. & Eddy, B.G. (1993). Late Quaternary sedimentation and the postglacial sea-level minimum in Port au Port Bay and vicinity, west Newfoundland. *Atlantic Geology*, **29**, 1–26.

Forbes, D.L. & Taylor, R.B. (1987). Coarse-grained beach sedimentation under paraglacial conditions, Canadian Atlantic coast. In *Glaciated coasts*, ed. D.M. FitzGerald & P.S. Rosen, pp. 51–86. San Diego: Academic Press.

Forbes, D.L. & Taylor, R.B. (1994). Ice in the shore zone and the geomorphology of cold coasts. *Progress in Physical Geography*, **18**, 59–89.

Forbes, D.L., Taylor, R.B., Orford, J.D., Carter, R.W.G. & Shaw, J. (1991b). Gravel barrier migration and overstepping. *Marine Geology*, **97**, 305–13.

Forbes, D.L., Taylor, R.B. & Shaw, J. (1989). Shorelines and rising sea levels in eastern Canada. *Episodes*, **12**, 23–8.

Forbes, D.L., Taylor, R.B., Shaw, J., Carter, R.W.G. & Orford, J.D. (1990). Development and stability of barrier beaches on the Atlantic coast of Nova Scotia. In *Proceedings, Canadian Coastal Conference 1990* (Kingston). pp. 83–98. Ottawa: National Research Council.

Fulton, R.J. (1989). Foreword. In *Quaternary geology of Canada and Greenland*, ed. R.J. Fulton, pp. 1–11. Geological Survey of Canada, *Geology of Canada*, no. 1 (also Geological Society of America, *Geology of North America*, K–1).

Grant, D.R. (1975). Recent coastal submergence of the Maritime Provinces. *Proceedings, Nova Scotian Institute of Science*, **27**, 83–102.

Grant, D.R. (1987). *Quaternary geology of Nova Scotia and Newfoundland*. 12th INQUA Congress, Field Excursion A3/C3. Ottawa: National Research Council. 62 pp.

Grant, D.R. (1989). Quaternary geology of the Atlantic Appalachian region of Canada. Chapter 5 in *Quaternary geology of Canada and Greenland*, ed. R.J. Fulton, pp. 393–440. Geological Survey of Canada, *Geology of Canada*, no. 1 (also Geological Society of America, *Geology of North America*, K–1).

Grant, D.R. (1991). *Surficial geology, Stephenville–Port aux Basques, Newfoundland*. Geological Survey of Canada, Map 1737A, scale 1:250 000.

Guilcher, A. (1962). Morphologie de la baie de Clew (comté de Mayo, Irlande). *Bulletin de l'Association de géographes français*, no. **303–4**, 53–65.

Guilcher, A. (1981). Cryoplanation littorale et cordons glaciels de basse mer dans la région de Rimouski, côte sud de l'estuaire du Saint-Laurent, Québec. *Géographie physique et Quaternaire*, **35**, 155–69.

Hansom, J.D. (1983). Ice-formed intertidal boulder pavements in the sub-antarctic. *Journal of Sedimentary Petrology*, **53**, 135–45.

Harland, W.B. & Herod, K.N. (1975). Glaciations through time. *Geological Journal*, Special Issue **6**, 189–216.

Henderson, E.P. (1972). *Surficial geology of Avalon Peninsula, Newfoundland*. Geological Survey of Canada, Memoir 368, 121 pp, Map 1320A, scale 1:250 000.

Héquette, A. & Ruz, M–H. (1990). Sédimentation littorale en bordure de plaines d'épandage fluvioglaciaire au Spitsberg nord-occidental. *Géographie physique et Quaternaire*, **44**, 77–88.

Hine, A.C. & Boothroyd, J.C. (1978). Morphology, processes, and recent sedimentary history of a glacial–outwash plain shoreline, southern Iceland. *Journal of Sedimentary Petrology*, **48**, 901–920.

Jansen, E. & Sjøholm, J. (1991). Reconstruction of glaciation over the past 6 Myr from ice-borne deposits in the Norwegian Sea. *Nature*, **349**, 600–3.

Jennings, S. & Smyth, C. (1987). Coastal sedimentation in east Sussex during the Holocene. *Progress in Oceanography*, **18**, 205–41.

Johnson, D.W. (1915). The nature and origin of fjords. *Science*, **41**, 537–43.

Johnson, D.W. (1919). *Shore processes and shoreline development*. New York: John Wiley and Sons. 584 pp.

Johnson, D.[W.] (1925). *The New England–Acadian shoreline*. New York: Wiley. 608 pp. (Facsimile edition 1967, New York and London: Hafner.)

Johnson, D.W. & Reed, W.G., Jr. (1910). The form of Nantasket Beach. *Journal of Geology*, **18**, 162–89.

Jopling, A.V. & McDonald, B.C., eds. (1975). *Glaciofluvial and glaciolacustrine sedimentation*. Society of Economic Paleontologists and Mineralogists, Special Publication 23. 320 pp.

Kelley, J.T. (1987). An inventory of coastal environments and classification of Maine's glaciated shoreline. In *Glaciated coasts*, ed. D.M. FitzGerald & P.S. Rosen, pp. 151–76. San Diego: Academic Press.

King, C.A.M. (1956). The coast of southeast Iceland near Ingólfshöfdi. *Geographical Journal*, **122**, 241–6.

Kirk, R.M. (1980). Mixed sand and gravel beaches: morphology, processes and sediments. *Progress in Physical Geography*, **4**, 189–210.

Knight, R.J. & Church, M. (1970). Tasiujaq Cove, Ekalugad Fiord, Baffin Island. In *1970 Data Record Series*, vol. 1, pp. 31–62. Ottawa: Canadian Oceanographic Data Centre, Department of Energy, Mines and Resources.

Kostaschuk, R.A. (1985). River mouth processes in a fjord–delta, British Columbia. *Marine Geology*, **69**, 1–23.

Krumbein, W.C. (1933). Textural and lithological variations in glacial till. *Journal of Geology*, **41**, 382–408.

Lamb, H.H. & Johnson, A.I. (1961). Climatic variation and observed changes in the general circulation. *Geografiska Annaler*, **43**, 363–400.

Lauriol, B. & Gray, J.T. (1980). Processes responsible for the concentration of boulders in the intertidal zone in Leaf Basin, Ungava. In *The coastline of Canada*, ed. S.B. McCann, pp. 281–92. Geological Survey of Canada, Paper 80-10.

MacClintock, P. & Twenhofel, W.H. (1940). Wisconsin glaciation of Newfoundland. *Geological Society of America Bulletin*, **51**, 1729–56.

Massari, F. & Parea, G.C. (1988). Progradational gravel beach sequences in a moderate- to high-energy, microtidal marine environment. *Sedimentology*, **35**, 881–913.

Matthews, J.B. (1981). The seasonal circulation of the Glacier Bay, Alaska, fjord system. *Estuarine, Coastal and Shelf Science*, **12**, 679–700.

McCann, S.B. (1979). Barrier islands in the southern Gulf of St. Lawrence, Canada. In *Barrier islands from the Gulf of St. Lawrence to the Gulf of Mexico*, ed. S.P. Leatherman, pp. 29–63. New York: Academic Press.

McCann, S.B., Dale, J.E. & Hale, P.B. (1981). Subarctic tidal flats in areas of large tidal range, southern Baffin Island, eastern Canada. *Géographie physique et Quaternaire*, **35**, 183–204.

McLaren, P. (1982). *The coastal geomorphology, sedimentology and processes of eastern Melville and western Byam Martin Islands, Canadian Arctic Archipelago*. Geological Survey of Canada, Bulletin 333. 39 pp.

Miall, A.D. (1983). Glaciomarine sedimentation in the Gowganda Formation (Huronian), northern Ontario. *Journal of Sedimentary Petrology*, **53**, 477–91.

Miller, A.A.L., Mudie, P.J. & Scott, D.B. (1982). Holocene history of Bedford Basin, Nova Scotia: foraminifera, dinoflagellate and pollen records. *Canadian Journal of Earth Sciences*, **19**, 2342–67.

Moseley, M.E., Wagner, D. & Richardson, J.B. III. (1992). Space shuttle imagery of recent catastrophic change along the arid Andean coast. In *Paleoshorelines and prehistory: an investigation of method*, ed. L.L. Johnson, pp. 215–35. Boca Raton, FL: CRC Press.

Neu, H.J.A. (1984). Interannual variations and longer-term changes in the sea state of the North Atlantic from 1970 to 1982. *Journal of Geophysical Research*, **89**, 6397–402.

Nielsen, L.N., Johannessen, P.N. & Surlyk, F. (1988). A late Pleistocene coarse-grained spit-platform sequence in northern Jylland, Denmark. *Sedimentology*, **35**, 915–37.

Nummedal, D. (1975). Wave climate and littoral sediment transportation on the southeast coast of Iceland. *Proceedings, International Congress of Sedimentologists* (Nice), Theme 6, 127–36.

Nummedal, D., Hine, A.C. & Boothroyd, J.C. (1987a). Holocene evolution of the south-central coast of Iceland. In *Glaciated coasts*, ed. D.M. FitzGerald & P.S. Rosen, pp. 115–50. San Diego: Academic Press.

Nummedal, D., Pilkey, O.H. & Howard, J.D., ed. (1987b). *Sea-level fluctuation and coastal evolution*. Society of Economic Paleontologists and Mineralogists, Special Publication 41. 267 pp.

Ouellet, Y. & Llamas, J. (1979). Complément et analyse des hauteurs des vagues dans le golfe du St-Laurent. *Le naturaliste canadien*, **106**, 123–39.

Orford, J.D., Carter, R.W.G. & Forbes, D.L. (1991a). Gravel barrier migration and sea-level rise: some observations from Story Head, Nova Scotia. *Journal of Coastal Research*, **7**, 477–88.

Orford, J.D., Carter, R.W.G. & Jennings, S.C. (1991b). Coarse clastic barrier environments: evolution and implications for Quaternary sea–level interpretation. *Quaternary International*, **9**, 87–104.

Owens, E.H. (1974). A framework for the definition of coastal environments in the southern Gulf of St. Lawrence. In *Offshore geology of eastern Canada*, vol. 1, ed. B.R. Pelletier, pp. 47–76. Geological Survey of Canada, Paper 74-30.

Owens, E.H. & Bowen, A.J. (1977). Coastal environments of the Maritime Provinces. *Maritime Sediments*, **13**, 1–31.

Philip, A.L. (1990). Ice-pushed boulders on the shores of Gotland, Sweden. *Journal of Coastal Research*, **6**, 661–76.

Philpott, K.L. (1986). Coastal engineering aspects of the Port Burwell shore erosion damage litigation. In *Proceedings, Symposium on Cohesive Shores* (Burlington), pp. 309–338. Ottawa: National Research Council.

Pickrill, R.A., Irwin, J. & Shakespeare, B.S. (1981). Circulation and sedimentation in a tidal-influenced fjord lake: Lake McKerrow, New Zealand. *Estuarine, Coastal and Shelf Science*, **12**, 23–37.

Piper, D.J.W., Letson, J.R.J., DeIure, A.M. & Barrie, C.Q. (1983). Sediment accumulation in low-sedimentation, wave-dominated, glaciated inlets. *Sedimentary Geology*, **36**, 195–215.

Piper, D.J.W., Mudie, P.J., Letson, J.R.J., Barnes, N.E. & Iuliucci, R.J. (1986). *The marine geology of the inner Scotian Shelf off the South Shore, Nova Scotia*. Geological Survey of Canada, Paper 85-19. 65 pp.

Posamentier, H.W., Jervey, M.T. & Vail, P.R. (1988). Eustatic controls on clastic deposition. I Conceptual framework. In *Sea-level research: an integrated approach*, ed. C.K. Wilgus, B.S. Hastings, C.G.St.C. Kendall, H.W. Posamentier, C.A. Ross, & J.C. van Wagoner, pp. 109–24. Society of Economic Paleontologists and Mineralogists, Special Publication 42.

Potschin, M.B. (1989). Drumlin fields of the Bernard Harbour area, Northwest Territories. Geological Survey of Canada, Paper 89-1D, 113–117.

Prior, D.B., Bornhold, B.D., Coleman, J.M. & Bryant, W.R. (1982). Morphology of a submarine slide, Kitimat Arm, British Columbia. *Geology*, **10**, 588–92.

Prior, D.B., Bornhold, B.D. & Coleman, J.M. (1983). *Geomorphology of a submarine landslide, Kitimat Arm, British Columbia*. Geological Survey of Canada, Open File 961, 5 sheets (scale 1:10,000).

Reimnitz, E., Hayden, E., McCormick, M. & Barnes, P.W. (1991). Preliminary observations on coastal sediment loss through ice rafting in Lake Michigan. *Journal of Coastal Research*, **7**, 653–64.

Rist, S. (1957). Skeiðarárhlaup, 1954. *Jökull*, 7, 30–6.

Rosen, P.S. (1979). Boulder barricades in central Labrador. *Journal of Sedimentary Petrology*, **49**, 1113–24.

Rosen, P.S. & Leach, K. (1987). Sediment accumulation forms, Thompson Island, Boston Harbor, Massachusetts. In *Glaciated coasts*, ed. D.M. FitzGerald & P.S. Rosen, pp. 233–50. San Diego: Academic Press.

Rust, B.R. & Romanelli, R. (1975). Late Quaternary subaqueous outwash deposits near Ottawa, Canada. In *Glaciofluvial and glaciolacustrine sedimentation*, ed. A.V. Jopling & B.C. McDonald, pp. 177–92. Society of Economic Paleontologists and Mineralogists, Special Publication 23.

Ruz, M-H., Héquette, A. & Hill, P.R. (1992). A model of coastal evolution in a transgressed thermokarst topography, Canadian Beaufort Sea. *Marine Geology*, **106**, 251–78.

Ryder, J.M. (1971). The stratigraphy and morphology of para-glacial alluvial fans in south-central British Columbia. *Canadian Journal of Earth Sciences*, **8**, 279–98.

Scholand, S.J., Varekamp, J.C. & Thomas, E. (1992). Paleo-environments in salt marsh sequences: geochemical and foraminiferal studies. *Eos, Transactions American Geophysical Union*, **73**, Spring Meeting Supplement, 154–5.

Scott, D.B. (1980). Morphological changes in an estuary: a historical and stratigraphic comparison. In *The coastline of Canada*, ed. S.B. McCann, pp. 199–205. Geological Survey of Canada, Paper 80-10.

Scott, D.B., Boyd, R., Douma, M., Medioli, F.S., Yuill, S., Leavitt, E. & Lewis, C.F.M. (1989). Holocene relative sea-level changes and Quaternary glacial events on a continental shelf edge: Sable Island Bank. In *Late Quaternary sea-level correlation and applications*, ed. D.B. Scott, P.A. Pirazzoli & C.A. Honig, pp. 105–19. Dordrecht: Kluwer.

Scott, D.B. & Greenberg, D.A. (1983). Relative sea-level rise and tidal development in the Fundy tidal system. *Canadian Journal of Earth Sciences*, **20**, 1554–64.

Scott, J.S. (1976). Geology of Canadian tills. In *Glacial till: an interdisciplinary study*, ed. R.F. Legget, pp. 50–66. Royal Society of Canada, Special Publication 12.

Sharpe, D.R. (1988). Glaciomarine fan deposition in the Champlain Sea. In *The Late Quaternary development of the Champlain Sea basin*, ed. N.R. Gadd, pp. 63–82. Geological Association of Canada, Special Paper 35.

Shaw, J. & Forbes, D.L. (1987). Coastal barrier and beach-ridge sedimentation in Newfoundland. In *Proceedings, Canadian Coastal Conference 1987* (Québec). pp. 437–54. Ottawa: National Research Council.

Shaw, J. & Forbes, D.L. (1988). Crustal warping and sediment supply as controls on recent coastal development in Newfoundland. *Maritime Sediments and Atlantic Geology*, **24**, 211.

Shaw, J. & Forbes, D.L. (1990a). Short- and long-term relative sea-level trends in Atlantic Canada. In *Proceedings, Canadian Coastal Conference 1990* (Kingston). pp. 291–305. Ottawa: National Research Council.

Shaw, J. & Forbes, D.L. (1990b). Late Quaternary sedimentation in St. George's Bay, southwest Newfoundland: acoustic stratigraphy and seabed deposits. *Canadian Journal of Earth Sciences*, **27**, 964–83.

Shaw, J. & Forbes, D.L. (1992). Barriers, barrier platforms, and spillover deposits in St. George's Bay, Newfoundland: paraglacial sedimentation on the flanks of a deep coastal basin. *Marine Geology*, **104**, 119–40.

Shaw, J., Taylor, R.B. & Forbes, D.L. (1990). Coarse clastic barriers in eastern Canada: patterns of glaciogenic sediment dispersal with rising sea levels. *Proceedings, Skagen Symposium. Journal of Coastal Research*, Special Issue **9**, 160–200.

Shaw, J., Taylor, R.B. & Forbes, D.L. (1993). Impact of the Holocene transgression on the coastline of Nova Scotia, Canada. *Géographie physique et Quaternaire*, **47**, 221–38.

Shepard, F.J. (1963). *Submarine geology*, 2nd edn. New York: Harper and Row. 557 pp.

Shipp, R.C., Staples, S.A. & Ward, L.G. (1987). Controls and zonation of geomorphology along a glaciated coast, Gouldsboro Bay, Maine. In *Glaciated coasts*, ed. D.M. FitzGerald & P.S. Rosen, pp. 209–31. San Diego: Academic Press.

Sonnichsen, G. (1984). *The relationship of coastal drumlins to barrier beach formation along the Eastern Shore of Nova Scotia*. Dalhousie University, Centre for Marine Geology, Technical Report 6. 66 pp.

Stanley, D.J. (1968). Reworking of glacial sediments in the North West Arm, a fjord-like inlet on the southeast coast of Nova Scotia. *Journal of Sedimentary Petrology*, **38**, 1224–41.

Stea, R.R. & Brown, Y. (1989). Variation in drumlin orientation, form and stratigraphy relating to successive ice flows in southern and central Nova Scotia. *Sedimentary Geology*, **62**, 223–40.

Stea, R.R., Forbes, D.L. & Mott, R.J. (1992). *Quaternary geology and coastal evolution of Nova Scotia*. GAC–MAC Joint Annual Meeting (Wolfville), Field Excursion A-6 Guidebook. Halifax: Geological Association of Canada, Department of Earth Sciences, Dalhousie University. 125 pp.

Stea, R.R. & Fowler, J.H. (1979). *Minor and trace element variations in Wisconsinan tills, Eastern Shore region, Nova Scotia*. Nova Scotia Department of Mines and Energy, Paper 79-4, 30 pp.

Stone, G.H. (1899). *The glacial gravels of Maine and their associated deposits*. U.S. Geological Survey, Monograph 34. 499 pp.

Stravers, J.A., Syvitski, J.P.M. & Praeg, D.B. (1991). Application of size sequence data to glacial–paraglacial sediment transport and sediment partitioning. In *Principles, methods, and application of particle size analysis*, ed. J.P.M. Syvitski, pp. 293–310. Cambridge University Press.

Swift, D.J.P., Stanley, D.J. & Curray, J.R. (1971). Relict sediments on continental shelves: a reconsideration. *Journal of Geology*, **79**, 322–46.

Syvitski, J.P.M. (1989). Modelling the fill of sedimentary basins. In *Statistical applications in the earth sciences*, ed. F.P. Agterberg & G. Bonham-Carter, pp. 505–15. Geological Survey of Canada, Paper 89-9.

Syvitski, J.P.M., Asprey, K.W., Clattenburg, D.A. & Hodge, G.D. (1985). The prodelta environment of a fjord: suspended particle dynamics. *Sedimentology*, **32**, 83–107.

Syvitski, J.P.M., Burrell, D.C. & Skei, J.M. (1987). *Fjords: processes and products*. New York: Springer-Verlag. 379 pp.

Syvitski, J.P.M. & Daughney, S. (1992). DELTA2: Delta progradation and basin filling. *Computers and Geosciences*, 18, 839–97.

Syvitski, J.P.M. & Farrow, G.E. (1983). Structures and processes in bayhead deltas: Knight and Bute Inlet, British Columbia. *Sedimentary Geology*, **36**, 217–44.

Syvitski, J.P.M. & Farrow, G.E. (1989). Fjord sedimentation as an analogue for small hydrocarbon-bearing fan deltas. In *Deltas: sites and traps for fossil fuels*, ed. M.K.G. Whateley & K.T. Pickering, pp. 21–43. Geological Society (London), Special Publication 41.

Syvitski, J.P.M., Farrow, G.E., Taylor, R., Gilbert, R. & Emory-Moore, M. (1984). SAFE: 1983 delta survey report. In *Sedimentology of Arctic fjords experiment: HU83-028 data report, 2. Canadian Data Report of Hydrography and Ocean Sciences*, **28**, 18.1–18.91.

Syvitski, J.P.M. & Hein, F.J. (1991). *Sedimentology of an arctic basin: Itirbilung Fiord, Baffin Island, Northwest Territories*. Geological Survey of Canada, Paper 91-11. 66 pp.

Syvitski, J.P.M. & Praeg, D.B. (1989). Quaternary sedimentation in the St. Lawrence Estuary and adjoining areas, eastern Canada: an overview based on high resolution seismo-stratigraphy. *Géographie physique et Quaternaire*, **43**, 291–310.

Syvitski, J.P.M., Smith, J.N., Boudreau, B. & Calabrese, E.A. (1988). Basin sedimentation and the growth of prograding deltas. *Journal of Geophysical Research*, **93**, 6895–908.

Tanner, V. (1939). Om de blockrika strandgördlarna (boulder barricades) vid subarktiska oceankuster. *Terra*, **51**, 157–65.

Taylor, R.B. (1978). The occurrence of grounded ice ridges and shore ice piling along the northern coast of Somerset Island, N.W.T. *Arctic*, **31**, 133–49.

Taylor, R.B., Carter, R.W.G., Forbes, D.L. & Orford, J.D. (1986). Beach sedimentation in Ireland: contrasts and similarities with Atlantic Canada. Geological Survey of Canada, Paper 86-1A, 55–64.

Taylor, R.B. & Forbes, D.L. (1987). Ice-dominated shores of Lougheed Island: type examples for the northwest Queen Elizabeth Islands, arctic Canada. In *Proceedings, Canadian Coastal Conference 1987* (Québec). pp. 33–48. Ottawa: National Research Council.

van de Plassche, O. (1991). Late Holocene sea-level fluctuations on the shore of Connecticut inferred from transgressive and regressive overlap boundaries in salt-marsh deposits. *Journal of Coastal Research*, Special Issue **11**, 159–79.

Viggósson, G., Sigurðarson, S., Halldórsson, A. & Brunn, P. (1994). Stabilisation of the tidal entrance at Hornafjörður, Iceland. In *Proceedings, Hornafjörður International Coastal Symposium* (Höfn). pp. 271–300. Kópavogur: Icelandic Harbour Authority.

Walcott, R.I. (1972). Past sea levels, eustasy and deformation of the earth. *Quaternary Research*, **2**, 1–14.

Wang, Y. & Piper, D.J.W. (1982). Dynamic geomorphology of the drumlin coast of southeast Cape Breton Island. *Maritime Sediments and Atlantic Geology*, **18**, 1–27.

Wightman, D.M. (1976). The sedimentology and paleotidal significance of a late Pleistocene raised beach, Advocate Harbour, Nova Scotia. Unpublished M.Sc. dissertation, Dalhousie University, Halifax. 157 pp.

11

Coastal cliffs and platforms

G.B. GRIGGS AND A.S. TRENHAILE

Introduction

Our individual perceptions of the coastline are clearly related to our local coastal geomorphology. On global and regional scales coastal morphology often correlates closely with tectonic setting (Inman & Nordstrom, 1971). Frequently, major geomorphic contrasts exist between the subdued coastlines along passive continental margins and the rugged coastlines along convergent plate boundaries. Along most coastlines, modern (active) coastal landforms are similar to their Pleistocene counterparts, which suggests that current tectonic and coastal processes have been fairly uniform over extended periods of recent geological time (Lajoie, 1986; see also Chapter 12).

Many exposed coastlines along passive continental margins, such as the southeast and Gulf coasts of the United States or parts of the northeast coast of Australia, for example the Gold Coast, consist of a low-relief coastal plain, bordered offshore by a wide continental shelf (see Chapters 2 and 4). These relatively stable coasts are also characterized primarily by meso- and macrotidal conditions, energy dissipation and depositional landforms, such as wide sandy beaches and dunes (see Chapter 4), as well as offshore barrier islands or bars (or a barrier reef in the case of northeastern Australia; see Chapter 8).

In marked contrast to the Atlantic coast of the United States and Canada, the Pacific Coasts of North and South America are active margins, and are either present or former collision coasts, where two plates have collided. Even a casual glance at almost any individual segment of this coastline will immediately reveal striking differences in coastal landforms and geological history. In addition to seismicity and vulcanism, collision coasts or active margins are characterized by deep-sea trenches offshore, narrow continental shelves, coastal mountains and often uplifted marine terraces. Erosional landforms such as steep sea cliffs, rocky headlands, sea stacks and islands produce a very different coastline than many trailing-edge coasts with their depositional landforms. Thus, coastal cliffs and platforms, the focus of this chapter, commonly have primary origins which are related directly to their large-scale tectonic setting.

Having made these generalizations, and recognizing the importance of tectonic setting (which was first recognized by Suess (1888) over a hundred years ago), it is important to point out that rugged coasts with rocky cliffs are not restricted to collision coasts. There are high cliffs, for example, around the British Isles, and in northern France, southeastern Australia and eastern Canada. All of these coastlines would be appropriately termed passive margins or trailing edges. The presence of rocky cliffs, sea stacks and arches is very common along these shorelines in part because the structural grain of the land is often at a high angle to the coast, which leads to exposure of rocks of variable resistance alongshore. Additionally, however, factors such as limited alluvial sediment contributions or loss of material into estuaries or embayments may lead to the development of rocky cliff coasts. Although the structural grain of collision or Pacific-type coasts, in contrast, is parallel to the shore, the tectonic activity and uplift which characterizes these coastlines typically produces the sea cliffs, terraces, headlands, and other characteristic features enumerated above.

Coastal cliffs, as a result of their tectonic setting, structural and lithologic framework and recent geologic history, vary widely in their height and morphology. Although it is common to think of cliffs as being solely a product of marine erosion, terrestrial or subaerial processes can be equally important in affecting the landforms exposed along the shoreline at any particular location. In most such locations it is the combination of marine and terrestrial processes which creates the distinctive coastal slopes.

Rock coasts

A large proportion of the coasts of the world is rocky, and even many sandy and cobble beaches are backed by rock cliffs or underlain by shore platforms. The prominence of such features is not reflected, however, in the modern process-orientated coastal literature, where most emphasis has been placed on beaches and other systems that respond rapidly to changing environmental conditions. While chemical and physical analyses, geochronometric dating, physical and mathematical modelling, and careful measurement of processes and erosion rates are providing valuable insights, we can still often only speculate on the mode of development of rock coasts.

Processes

Rock coasts are the legacy of marine and subaerial processes that have been operating for thousands of years (Trenhaile, 1987). The type, intensity, and focus of these processes have varied with shifts in relative sea level, and with temporal and spatial changes in climate, exposure, and rock type. Although some attempt has been made to identify and measure the results of the processes operating on rock coasts, scientists are still largely ignorant of their precise nature and relative importance. This is unfortunate, as coastal cliffs are being populated at an increasing rate as more people in the world's developed countries move to the coast. The study of processes on rock coasts, especially the acquisition of quantitative data, has been hindered by the very slow rates of change, the role of high-intensity – low-frequency events, exposed and often dangerous environments for wave measurement and submarine exploration, the lack of access to precipitous or heavily vegetated cliffs, poor research funding, and the small body of active researchers in this area. In any case, even if we completely understood the nature of contemporary erosive processes, it would still be difficult to explain the morphology of coasts that often retain the vestiges of environmental conditions that were quite different from today.

Wave action

Mechanical wave action is the primary erosional agent in most storm-wave and more vigorous swell-wave environments. In polar, tropical and other relatively low-energy environments, where waves are generally much weaker, they still have an important role in removing the products of physical and chemical weathering, simply because they are the dominant source of energy for transporting material in the coastal zone.

Mechanical wave erosion is accomplished by a number of processes, but we usually have to gauge their relative importance from ambiguous morphological evidence. The presence of large, angular debris and fresh rock scars has convinced many workers that wave quarrying is usually the dominant erosive mechanism in vigorous wave environments. Wave quarrying usually requires the alternate presence of air and water. Water hammer, high shock pressures generated against structures by breaking waves (up to 690 kPa has been recorded), and probably most importantly, the compression of pockets of air in rock crevices, are most effective in a narrow zone extending from just below the still water level up to the wave crest (Sanders, 1968). Although abrasive processes are not as closely associated with the water level, their efficacy rapidly decreases with depth below the water surface.

Abrasion or corrasion takes place as sand or other coarse granular material is washed, dragged or rolled across a rock surface or when this material is thrown against a coastal cliff or bluff. The importance of this process at any particular site is related to the presence of tools (sand, gravel or pebbles) and the wave energy available to move these tools around. Abrasion tends to produce much smoother surfaces than the quarrying processes described above, although where lithologic inhomogeneities or structural weaknesses occur, abrasion can produce local scouring or grooving of the bedrock. An extreme example of this focused abrasion is in the formation of potholes where large pebbles or cobbles abrade or carve circular depressions on intertidal rock platforms. The abrasional process is limited in its shoreward extent by the elevation to which waves can throw or wash material up against the flanking seacliff. Abrasion is most effective where the sediment cover in the littoral zone is sufficiently thin (up to perhaps 10 cm) such that the available wave energy is able to agitate or dislodge particles down to the underlying bedrock (Robinson, 1977).

Most mathematical models suggest that the pressures exerted on vertical structures by standing, breaking, and broken waves are highest at, or slightly above, the still water level (Trenhaile, 1987). The zones of maximum wave pressure and most effective erosional processes are therefore essentially coincident, occurring at or near to the water surface. Consequently, the long-term level of greatest erosion must be related to the neap high and low tidal levels, the elevations most often occupied by the still water level. Wave energy is probably greater during the high tidal periods because of greater water depths and reduced friction. The highest and most effective waves occur during storms, however, when the water surface is raised above its still water or tidal level. It is at these times that wave impact and runup reach farthest inland and the uppermost reaches of the shoreline or seacliff are attacked most vigorously. The zone of greatest wave erosion is therefore probably above the neap high tidal level, particularly in microtidal environments and where hard rocks resist all but the most vigorous storm waves that operate at elevated, supratidal levels.

Chemical and salt weathering

Cliffs and intertidal shore platforms subjected to alternate wetting and drying by ocean spray, splash, and tides are suitable environments for many chemical and physical weathering processes. They are probably only major erosive mechanisms in sheltered areas and on particularly susceptible rocks in cooler regions, but they assume an important and sometimes dominant role in warmer climates.

Chemical and salt weathering tend to operate together and it is usually very difficult to distinguish the results in the field. They contribute to the general weathering of cliffs and shore platforms, the development of tafoni and honeycombs, and to the suite of processes collectively referred to as water layer levelling (So, 1987; Matsukura & Matsuoka, 1991). There is no evidence, however, of permanent saturation above the low tidal level, or of its corollary, an abrupt transition in the intertidal zone from less-resistant, weathered rock above the saturation level, to lower, more-resistant, unweathered rock below; concepts that have dominated the Australasian literature on shore platforms for almost a century (Bartrum, 1916).

The nature and efficacy of the destructional processes operating on limestone coasts are contentious issues. Although surface sea water is normally saturated or supersaturated with calcium carbonate, marine karren in the intertidal and splash-spray zones of limestone coasts is similar to the variety of sharp pinnacles, ridges, grooves and circular basins etched on land by meteoric water. Carbonate solution may occur in pools at night, however, when faunal respiration, which is not taken up by algae during the hours of darkness, reduces the pH of the water and causes calcium carbonate to be transformed into more soluble bicarbonate. Nevertheless, other biochemical processes may inhibit or prevent solution through the coating of rock surfaces with dissolved organic substances, and the building, by organic substances, of complexes with calcium ions. It has been estimated that solution is responsible for about 10% of the total erosion of the coast of Aldabra Atoll in the Indian Ocean (Trudgill, 1976), but alternative bioerosional mechanisms are more important in western Ireland (Trudgill, 1987) and on Grand Cayman Island (Spencer, 1988), and chemical solution does not even occur in the rock pools of the northern Adriatic (Schneider, 1976).

Bioerosion

Many workers have argued that coastal 'solutional' features are primarily the result of bioerosional activities by a variety of marine organisms. The removal of the substrate by direct organic activity is probably most important in tropical regions, where there are an enormously varied marine biota and calcareous substrates that are particularly susceptible to biochemical and biomechanical processes (Spencer, 1988; Fischer, 1990).

Algae, fungi, and lichen are pioneer colonizers in the inter- and supratidal zones. In addition to their role as rock borers and their effect on water chemistry, microflora allow subsequent occupation by gastropods, chitons, echinoids and other grazing organisms. Grazers abrade rock surfaces by mechanical rasping as they feed on epilithic and endolithic microflora. In the mid-tidal zone of Aldabra Atoll, for example, grazing organisms are responsible for about one-third of the surface erosion where sand is available for abrasion, and as much as two-thirds where sand is absent (Trudgill, 1976). Other erosion is accomplished by barnacles, sipunculoid and polychaete worms, gastropods, echinoids, bivalve molluscs, Clionid sponges and numerous other rock borers, particularly in the lower portions of the intertidal zone. Borers directly remove rock material, but also, by enlarging crevasses and creating weaknesses in the rock structure, they render the residual rock more susceptible to wave action and, by increasing the surface area, to weathering.

There is a fairly large body of information on bioerosional rates, but its reliability and applicability vary enormously (Trenhaile, 1987; Spencer, 1988). Nevertheless, the overall erosion rate on both vertical and horizontal limestone surfaces in a variety of environments, has often been found to be between about 0.5 and 1 mm a^{-1}, which may reflect the maximum boring rate of endolithic microflora (Schneider & Torunski, 1983).

Expansion–contraction mechanisms

The expansion and contraction of certain clay minerals such as the montmorillonite/smectite group is probably an important erosive mechanism in shales and other argillaceous rocks subjected to tidally induced cycles of wetting and drying. Clay-rich rocks can also be weakened by temperature-dependent wetting and drying resulting from the attraction of the positively charged ends of water molecules to the negatively charged surfaces of clay particles contained within small rock capillaries.

However, we neither know the extent to which the traditional field evidence for freeze–thaw action can be explained by the temperature-dependent adsorption of water, nor do we understand the precise nature of such 'frost' action. The two mechanisms may act together, the suction generated by unfreezable adsorbed water, trapped in rock pores by expanding ice, causing free or bulk water to migrate to the freezing front. The pressures generated by this mechanism may be a more-important cause of frost shattering than the approximately 9% expansion of water upon freezing (Matsuoka, 1988).

Although we are still unsure of the responsible processes, we do have a general sense of the conditions most suitable for their operation. Cool coastal regions appear to be almost optimum environments. Rocks in the intertidal and spray and splash zones can attain high levels of saturation (Trenhaile & Mercan, 1984), and they experience many more frost cycles than those further inland. More than 200 freeze–thaw alternations can occur in a year in the intertidal zone in coastal Maine, for example, compared with 30–40 inland (Kennedy & Mather, 1953). Tidally induced frost cycles occur when rocks are alternately frozen when exposed to freezing air temperatures, and thawed when submerged in water above freezing. Rocks thaw very rapidly when inundated by rising tides, but they may need several hours to freeze once they have been exposed by falling tides (Robinson & Jerwood, 1987). Whether critical saturation levels can be maintained in the rocks over this period remains to be determined.

Mass movement

Fresh rock faces and accumulations of debris attest to the importance of rock falls on many coasts (Fig. 11.1). They are more common than deep-seated slides, but most are fairly small. Falls occur in well-fractured rocks, especially over notches cut into the cliff foot by waves, chemical solution, or bioerosional agencies. Large coastal landslides require suitable geological conditions (Grainger & Kalaugher, 1987). Argillaceous and other easily sheared rocks with low bearing strength are particularly susceptible to translational sliding, as are alternations of permeable and impermeable strata, and massive rocks overlying incompetent materials (Pitts, 1986). Rotational slumps usually occur

Figure 11.1. Rockfall from coastal bluffs in central California along shore-parallel and shore-normal joint sets in sedimentary rocks (note person at lower right for scale).

in thick, fairly homogeneous deposits of clay, shale or marl (Flageollet & Helluin, 1987). Most deep-seated mass movements may be attributed to undercutting at the base of the slope and the build-up of groundwater. They therefore tend to take place during or shortly after periods of prolonged and/or intense precipitation, or as a result of the accumulation of water from human activities (including septic systems, runoff disruption, and landscape irrigation). Coastal cliff failures may therefore be rather instantaneous events in the case of rock falls due to wave undercutting, or seismic shaking, or may take place slowly in the case of deep-seated mass movements initiated by excess groundwater.

Shoreline erosion and retreat

Factors affecting the form and rate of shoreline erosion

The rate of coastal cliff erosion and the resulting landforms are related to both the cliff-forming materials and the physical processes to which they are exposed, as well as the time over which these processes have operated. The tectonic and climatic histories of any particular coastline are additional variables which have major effects on the relief, the rock types, stratigraphy and structure present, as well as weathering processes and vegetation development.

Rock type

The type of rocks exposed in coastal cliffs or platforms is an extremely important factor in determining their rate of erosion. In general, crystalline rocks (igneous and metamorphic) are more resistant to marine and subaerial erosion than sedimentary rocks, and consolidated or cemented sedimentary rocks are more resistant than those which are unconsolidated. Where rocks of variable resistance are interbedded (sandstones and shales, for example) and exposed on the shoreline, we can expect the erosion rates and resulting landforms to express these differences. Although it is difficult to separate the importance of lithologic variables from other factors controlling shoreline erosion, such studies are needed in order to quantify the importance of these individual variables. Within a distance of 60 km along the shoreline of central California, a variety of materials including unconsolidated Pleistocene dune sands, Pliocene sandstones, Miocene mudstones, and Mesozoic granites are exposed in the coastal bluffs. Erosion rates in these materials range from $2.5 \, \text{m a}^{-1}$ in the dune sands to imperceptible rates in the granites.

Structural weaknesses

Internal structural weaknesses, typically joints, fractures and faults, exert a primary control on both the morphology and rate of cliff erosion in many areas. These weaknesses provide zones of accelerated wave erosion which often account for the presence of arches and caves in these cliffs (Fig. 11.2). Although it appears that this process is often underappreciated, it is abundantly clear that the overall planform of many coastal cliffs, as well as the smaller scale irregularities, are often intimately tied to joint orientation and spacing. While a cliff may appear to consist of a massive or thick-bedded mudstone or siltstone, closer inspection reveals that it is the presence of joints, which may owe their origin and orientation to the recent tectonic history of the area, which ultimately determines the size and geometry of the blocks which fail or are dislodged from the cliff. Even the weakest, but mechanically intact and unweathered, rock could theoretically withstand pressures generated by cliffs over a thousand metres in height. These heights are never attained, however, because the critical height is limited by the presence of joints, faults, and other mechanical defects, rather than the integral strength of the rock itself (Terzaghi, 1962).

Figure 11.2. Caves eroded at base of cliffs by wave action along structural weaknesses in sedimentary rocks along the central California coast.

Along the shoreline of northern Monterey Bay, California, the young sedimentary rocks (dominantly mudstones and siltstones) are jointed in both shore-parallel and shore-normal directions. Failure typically occurs as large (up to 3 m thick) joint-bounded blocks fall to the beach below following undercutting of the base of the bluff by wave action (Fig. 11.1). In this instance, the strength or the degree of cementation or consolidation of the bedrock itself (as determined by typical soil or rock engineering methods) is of considerably less significance in affecting the rates of shoreline erosion and the resulting cliff configuration, than is the distribution and orientation of the joint sets and the degree to which they are developed.

Exposure to wave action

A combination of the regional wave climate or severity and frequency of wave attack, and the degree of exposure of the sea cliff to waves, is of fundamental importance in determining the rate of cliff retreat. Because of the irregular nature of storm-wave attack, we typically witness cliff retreat as an episodic and largely unpredictable process. Little erosion or retreat may occur for years and then, under the right combination of cliff weakening or weathering (perhaps due to excess groundwater or pore pressures) and wave attack and tidal elevation, major failure and significant erosion may take place virtually overnight.

Local variations in nearshore and offshore topography often serve to focus or disperse wave energy such that adjacent sections of coast may experience widely varying degrees of wave attack and therefore erosion rates. Within central Monterey Bay on the California coast, for example, the head of Monterey Submarine Canyon extends nearly to the shoreline. Wave energy is refracted away from the canyon head, due to its greater depths, such that wave heights are typically several times greater immediately north or south of the canyon than in the area immediately landward. Similarly, the distribution and relief of nearshore bedrock platforms leads to a complex pattern of wave attack at the shoreline.

The presence or absence of a protective beach, whether permanent or seasonal, may provide the most important buffer to a cliff from direct wave attack. There are those areas where no beach ever forms, due to a combination of offshore topography, relief of the coastline, resistant nature of the rocks, and lack of sufficient sand. In these instances, wave attack of the cliffs is a year-round process. In other locations, beaches exist seasonally, and cliff exposure to wave attack is only intermittent. There are also those sites where, either due to recent uplift or isostatic rebound or the presence of a very wide equilibrium beach, cliff attack is relatively infrequent, despite the wave conditions.

Relative sea-level rise

The relationship between sea-level rise and the uplift or subsidence of the coastline and the resulting landforms will be covered more thoroughly in the next section, on marine platforms and terraces. It is important to note here that the relative rise or fall in sea level can exert a major impact on the nature of the coastal cliffs. For example, where tectonic uplift or isostatic rebound is occurring at a more rapid rate than eustatic sea-level rise, then the coastal cliffs can be isolated from wave attack and preserved as relict features. At this stage subaerial processes may act to degrade the slope. Conversely, where the relative sea-level rise is positive, the cliffs should be reached more frequently by waves and therefore subject to greater erosion rates.

Although many factors can influence sea level, leading to a noisy record, various studies utilizing tide-gauge data find an average global rate of sea-level rise of $1–2 \, \mathrm{mm \, a^{-1}}$, over the last 100 years (Gornitz, 1991). Along the coast of southeast Alaska in the Juneau and Skagway areas, however, due to postglacial rebound, the sea level is dropping relative to the coastline at rates of $13 \, \mathrm{mm \, a^{-1}}$ to over $17 \, \mathrm{mm \, a^{-1}}$. Coastal cliffs are thus being uplifted and preserved at a relatively rapid rate. The northern British Isles were covered with ice during the glacial episodes of the Pleistocene, and were correspondingly depressed, while southern England was ice free. Postglacial rebound has produced a regional tilting, with Scotland rising in the north while the south coast of England is sinking, leading to slow inundation. Relative sea-level change thus varies significantly from north to south and has produced differing effects on the coastal cliffs.

Tidal variation

Global differences in daily tidal ranges vary from nearly imperceptible to over 12 metres. The range in elevation between high and low tides determines the extent of the coastal cliffs or bluffs which are alternately exposed and covered with seawater and impacted by waves. Through the processes which create intertidal platforms and terraces, this factor is potentially very significant, as discussed in the next section. Water table weathering to the zone of permanent saturation, which includes the softening and weakening of the rocks through weathering and solution, allows for the erosional processes (including hydraulic impact) to be effective over an increasingly broader zone as the vertical range between high and low tide increases.

Human activity

As coastal populations worldwide have migrated to the coastline, either seasonally or permanently, the impacts of those humans and their development have begun to alter both coastal landforms and cliff stability. Buildings, utilities and coastal protection structures have been built on cliff tops, on the faces of the cliffs themselves, as well as on the fronting beach. There are many densely populated coastal areas where little of the natural cliffs can any longer be seen, as they have been completely armoured with protective materials (Fig. 11.3).

Heavy construction on bluff tops, as well as landscaping and the required irrigation or watering required to support that vegetation, has added several times the normal average annual precipitation along the coastline of southern California (Griggs & Savoy, 1985). The net result is to increase the pore pressures in the cliff materials, thereby decreasing their strength and accelerating the cliff failure process. In addition, the runoff from the impervious surfaces accompanying urbanization of coastal bluff top areas has typically been directed into culverts or drains which have focused runoff on the bluff face, thereby increasing local cliff erosion or stability (Fig. 11.4).

Coastal engineering structures have had several effects on the development and stability of seacliffs. Large structures, such as breakwaters and jetties, typically induce upcoast impoundment of littoral drift, thereby widening the beach and protecting the cliffs from direct wave attack. As a consequence,

Figure 11.3. Intensively developed and protected coastal bluffs along the coastline of southern California, where little natural bluff is exposed.

Figure 11.4. These 22 metre high cliffs continue to fail due to a combination of rock weakening by surface runoff and groundwater seepage, undercutting by wave action, and mass movement along joint sets. The debris at the base of the cliff was due to failure caused by the 1989 Loma Prieta earthquake. The apartment units were subsequently demolished due to cracking of the bluff.

however, the downcoast beaches are initially starved and this beach loss leads to an increase in exposure to wave attack and accelerated cliff erosion can be expected (Griggs & Johnson, 1976). Local protection structures such as seawalls and revetments are constructed to control wave-induced cliff retreat, and as such can temporarily halt cliff erosion and stabilize coastal cliffs or bluffs. The type of structure utilized and its height, depth, lateral extent and durability are important factors in determining the effectiveness of these shoreline erosion control devices (Fulton-Bennett & Griggs, 1986).

Documenting rates of cliff erosion

It has been recognized by coastal reseachers working with sea cliff evolution and erosion that cliff or bluff retreat is usually an episodic process. Most of the major episodes of cliff erosion occur during the simultaneous occurrence of high tides and large storm waves. At these times waves can reach high enough up on the shoreline to attack those areas which are less frequently inundated and the material which has been weakened progressively through weathering can be dislodged and removed. The sequence of processes may include beach scour followed by direct cliff or bluff attack, undercutting of the base of the cliff followed by collapse of the overlying unsupported material, or simply hydraulic quarrying of blocks or rock which were stable during conditions of lower wave energy. Terrestrial processes, such as landsliding, slumping, or rock falls triggered independently of wave attack, may also be extremely important, and even dominant in areas where the base of the bluff is protected by a coastal engineering structure or seawall (Fig. 11.5).

Although we often use or publicize average annual cliff retreat or erosion rates, in reality we are simply comparing the position of the cliff edge at different points in time (whether from historic maps or aerial photographs, or actual field measurements), and dividing by the total number of years between these data points, to derive an 'average' erosion rate. Cliffs or bluffs may remain superficially unchanged for years, and then due to the right combination of bluff saturation, or tidal level and wave attack, several metres may fail instantaneously. Averaging this loss over the time interval between major storms produces an average rate which may vary from millimetres per year along resistant granitic coasts to metres per year in unconsolidated sediments.

Most studies of sea cliff erosion have relied on measurements from historic aerial photographs or maps, as these typically provide the longest data base. Vertical stereoscopic aerial photographs are common tools used by coastal geologists and geomorphologists. In populated coastal areas of the United States, for example, the aerial photographic record may extend back 50 or 60 years. A span this long typically includes a representative interval of storm and calmer weather conditions. Thus, the range in actual erosion rates derived from sequential measurements of the location of the cliff edge, relative to some baseline or benchmark (a road or structure, for example) over the time span of the photos, can produce a reasonably accurate picture of the pattern of cliff retreat.

The extent of the data base utilized (aerial photographs, maps, or ground measurements), the resolution of this data base (e.g. aerial photo scale and clarity), the skill or experience of the investigator, as well as the historical

Figure 11.5. These 25 metre high coastal bluffs continued to fail through mass movements during heavy rainfall despite their isolation from wave action.

representativeness of the time interval spanned by the data base, are all important factors which affect the reliability of calculated cliff erosion rates. Wide variations can result depending upon the length of the historical record utilized, or the particular segment of coast analyzed, for example, due to long-term climatic or storm-frequency variations, or due to the alongshore differences in the geological materials and their resistance or susceptibility to erosion. The long-term hazards of constructing on geologically active coastal cliffs where long-term erosion rates have not been carefully evaluated, or where average erosion rates from some nearby areas have been extrapolated to the site, can be very costly (Figs. 11.4 and 11.5).

One of the few attempts to quantify both the erosional processes and the rates of cliff retreat in order to develop a quantitative model is the work of

Sunamura (1975, 1977, 1982). Sunamura used a wave tank which included a model cliff made of sand and cement to obtain a relationship between process and response in cliff form which was then compared with field data. Because Sunamura was primarily interested in the direct erosion of the cliff, he concentrated his attention on the formation of the notch which formed at the base of the cliff. The distribution of wave pressure applied to the cliff face by the waves was calculated theoretically and measured empirically and the results were compared with the depth of actual notch erosion. Sunamura was able to show that for waves breaking directly on the cliff face there is a clear relationship between wave pressure and erosion, but that for broken waves, erosion occurs at a much slower rate.

Variations in cliff/platform morphology

Cliff profiles

Steep or undercut cliffs are typical of wave-dominated environments, whereas convex slopes develop where the climate and wave regime are more conducive to subaerial weathering and erosion. Geological factors, including structural weaknesses, stratigraphic variations, and the attitudes or orientations of the bedding, also exert a strong influence on the shape and gradient of cliffs (Emery & Kuhn, 1980; Trenhaile, 1987). Although there is a complex relationship between the shape of cliff profiles and the dip of bedding and/or joint planes in the rock (Terzaghi, 1962), cliffs tend to be steepest in rocks either horizontally or vertically bedded, especially if, as in the chalk of southern England and northern France, they are also lithologically very homogeneous (Fig. 11.6). Slopes are generally more moderate in rocks which dip seawards or landwards. Seaward dips often produce dip slopes that are typically quite smooth (Fig. 11.7), but when dips are landward, many different beds can be exposed and differential erosion may produce a very irregular cliff face. The resistance and thickness of the sedimentary strata and the position of the weakest members are also of great importance. A weak stratigraphic unit exposed to wave action at the base of a cliff, for example, often results in formation of a notch and eventually collapse of the overlying, unsupported material.

    Composite cliffs have more than one major slope element. They include bevelled (hog's back, slope-over-wall) cliffs with convex or straight seaward-facing slopes above steep, wave-cut faces, and multi-storied cliffs with two or more steep surfaces separated by more gentle slopes. Some composite cliffs reflect geological influences or the combined effects of

Figure 11.6. Steep cliffs cut into the horizontally bedded chalks along the Normandy coast of northern France.

Figure 11.7. Failure along dip slopes in shale along the northern California coast has produced a smooth cliff profile (arrow shows dip direction).

subaerial and marine processes. Others, however, are the result of Pleistocene changes in climate and sea level. It has been proposed that wave-cut cliffs, abandoned during the last glacial stage, were gradually replaced by the upward growth of convex slopes developing beneath the accumulating talus. With removal of the debris since the sea rose to its present position, marine erosion has subsequently either trimmed the base of the convex slopes to form composite cliffs, or, where erosion has been more rapid, completely removed it to form steep, wave-cut cliffs (Trenhaile, 1987).

A case study

The development of coastal cliffs has been modelled numerically over two glacial–interglacial cycles, using a range of values to represent cliff height (25 to 100 m), rates of marine erosion (0.01 to $0.001 \, \mathrm{m\,a^{-1}}$) and subaerial weathering (0.01 to $0.0001 \, \mathrm{m\,a^{-1}}$) in cool coastal regions. The mathematical form of the model is broadly similar to Fisher's (1866) model, which has been adopted and extended by a number of other workers. It was found that (Fig. 11.8):

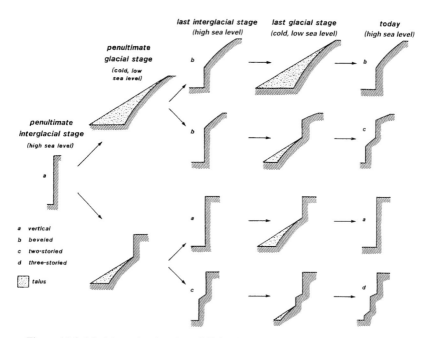

Figure 11.8. Model results showing cliff development through two glacial–interglacial cycles, rates of marine and subaerial erosion differ.

1. Vertical cliffs formed at the end of the first and second glacial–interglacial cycles, when low subaerial weathering rates were combined with fast wave erosion.
2. Bevelled profiles developed at the end of the first cycle in almost 80% of the 72 model runs. This form generally reappeared at the end of the second cycle, although it was replaced by two-storied cliffs in four runs. Bevelled profiles were least likely to develop when weathering rates were low.
3. Two-storied profiles developed at the end of the first cycle in runs with slow subaerial weathering and wave erosion. These profiles always added another story at the end of the second cycle, and further cycles would continue to add others, until an equilibrium form had been attained.
4. The probability of formation of bevelled and multi-storied cliffs increases with the height of the cliff.

The prolonged period of subaerial degradation required for composite cliff development, according to traditional concepts, could only have occurred in areas that lay beyond the ice margins and above the marine limit in the last glacial stage. Much of the coast of northern North America and Europe, however, was under ice or below sea level at that time. To study the effect of these factors on cliff development, a survey was made of the coast of Atlantic Canada, based upon maps, aerial photographs and field work. Steep coastal slopes were classified as dominantly wave-cut, composite (bevelled and multi-storied), or essentially unmodified glacial surfaces.

Palaeozoic sedimentary and volcanic rocks underlie most of the area, but there are also extensive outcrops of Archaean and especially Palaeozoic intrusions. Almost all of Atlantic Canada experiences fairly vigorous wave action, although sea ice protects most of the coast in winter. The coast also lies within the optimum climatic zone for atmospherically and tidally induced frost cycles (Trenhaile, 1987).

Modelling suggests that wave erosion rates increase with the rate that sea level rises and decrease with the rate that it falls (Trenhaile, 1989). The rate and direction of postglacial changes in relative sea level varied in Atlantic Canada according to local rates of isostatic recovery. Relative sea level has therefore risen in southerly areas that were near the ice margins, and fallen in more northerly areas that were under fairly thick ice.

Most cliffs in Atlantic Canada are between 3 and 30 m in height, although they occasionally range up to 100 m or more in the resistant metamorphic and igneous rocks of upland areas. There are wave-cut cliffs in areas of low relief. They occur in sheltered environments in fairly weak Triassic and Palaeozoic sedimentary rocks, but only in exposed areas in older, more resistant rocks.

Steep glacial slopes, little modified by wave action, are generally in resistant
Cambrian and Precambrian rocks, especially where glacial ice flowed parallel
to the coast.

Most of the region was buried beneath ice in the late Wisconsin, and many
of the small unglaciated areas, most notably in western Newfoundland, were
depressed below sea level. Therefore, despite the widespread occurrence of
high cliffs, rapid weathering, and other suitable conditions for composite cliffs,
they are largely restricted to northern Cape Breton Island and a few other areas
in the southern portion of the region that lay beyond the ice margins and above
the marine limit (Fig. 11.9). Nevertheless, the fact that they are found in a few
areas that were under ice or sea water in the late Wisconsin suggests that they
can be formed in other ways, although one can only speculate at present on
some possible origins:

1. Postglacial development could have occurred in ice marginal areas above
   the marine limit where, as in the upper Bay of Fundy, there was early
   deglaciation of frost susceptible rocks (Fig. 11.9).

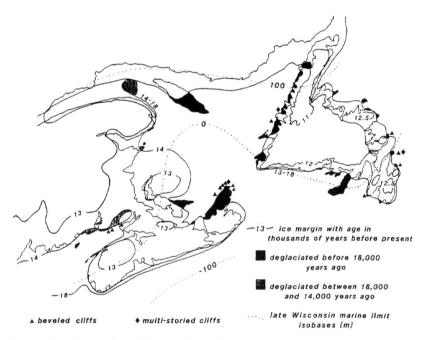

Figure 11.9. Composite cliffs in Atlantic Canada and late-Wisconsin glacial and
marine limits.

2. In unglaciated areas below the marine limit where emergence was gradually succeeded by submergence, debris could have accumulated for a few thousand years when relative sea level was low, and then been removed as it rose to its present level. As it is unlikely that talus could have reached the top of high, resistant cliffs in the short time available, however, this explanation could only account for the formation of multi-storied profiles.

Although wave erosion at the base of steep glaciated slopes could produce composite profiles, the general lack of these profiles in Atlantic Canada suggests that the rocks have been too resistant for significant marine erosion, especially where relative sea level has been falling.

Shore platforms

The erosion of the coastline and the subsequent removal of the rock and sediment debris by nearshore currents cause the progressive retreat or recession of the shoreline. This leaves behind the remnant of the old slope or shoreline marked by the lowest level to which the erosion reached, the shore platform. These platforms have been recognized globally for decades and several differing arguments have been put forward regarding their shape and evolution, and the relationship of these to the causal factors or physical processes.

Shore platforms are usually nearly flat or slope gently seaward, and they range up to about a kilometre in width. Cross-shore profiles vary from linear to concave to more complex or composite profiles with steeper inner edges and flatter outer edges (Bradley & Griggs, 1976). These platforms may extend from approximately the high tide level, at the base of the receding cliff, to an elevation below and beyond the low tide level in the nearshore zone. In recent years the genetically neutral term 'shore platform' has for the most part replaced the term wave-cut platform, thereby avoiding the implication that these flat features owe their origins exclusively to wave abrasion. The most widely held view is that a variety of different processes operate on these platform surfaces and that the resulting morphology reflects these differences. Nevertheless, it has been shown that there is a strong relationship between the gradient and other aspects of the morphology of shore platforms, and tidal range, in a wide variety of environments and rock types (Trenhaile, 1980, 1987).

A high-water or structural storm ledge frequently occurs at the junction between the inner edge of the shore platform and the base of the sea cliff along the tectonically active west coast of California. Similar benches, which include the '2 metre' bench of the Pacific, have been described in many other parts of the world (Trenhaile, 1971, 1987). This storm ledge or platform is a geographically restricted feature, at most only 25 to 50 metres in width, and best developed on a cliffed coast consisting of well-bedded sedimentary rocks which are nearly flat lying or dip gently seaward (Fig. 11.10). Typically, but not always, the platform forms along a single slightly more resistant bedding plane. This feature appears to owe its origin to a combination of storm-wave attack at high tide and perhaps water table weathering.

Although arguments have been put forward in the past that this two metre bench provided evidence for a slightly higher eustatic sea level in the very recent past (Fairbridge, 1961), careful analysis and process observations along the shoreline where these features exist indicate that they are modern features forming under present storm wave and sea-level conditions. Debris from the flanking sea cliff commonly accumulates on the back edge of the terrace, but storm-wave attack under high tide conditions will sweep this material away and clear the platform surface. Undercutting of the sea cliff may also take place during these events, and when combined with terrestrial processes, lead to

Figure 11.10. A storm ledge or platform along the central coast of California which has been eroded into thin-bedded mudstones which dip gently seaward.

ongoing sea cliff retreat. The outer edge of the bench or platform is constantly under wave attack, and in order for an equilibrium profile to be preserved, the sea cliff and the platform must be, over the long term, eroded at the same rate. If either horizontal or vertical platform erosion were to occur at a greater rate than sea cliff retreat, then at a constant sea level the platform should soon disappear completely. On the other hand, if platform erosion were slower than cliff retreat, the cliff would soon be isolated from wave attack and would become a relict feature.

Uplifted marine terraces

Elevated marine terraces are common along collision coasts where uplift is taking place. The coastlines of New Guinea, New Zealand, as well as the west coasts of Central and North America, are examples of locations where uplifted marine terraces are particularly well-developed. These terraces typically resemble a flight of stairs, commonly less than a kilometre in width, which ascend to elevations of several hundred metres above present sea level. Each terrace consists of a nearly horizontal or gently seaward dipping erosional or depositional platform backed by a steep or degraded, relict sea cliff along its landward margin (Fig. 11.11). Based on modern nearshore process observations, the shoreline angle, or the intersection of the relict platform and

Figure 11.11. A sequence of four uplifted marine terraces along the California coast, with a storm platform along the base of the sea cliff.

the relict sea cliff, provides a good approximation of the location and elevation of the abandoned or former shoreline, and hence a relative-sea-level highstand (Lajoie, 1986).

Over the past 25 years, a general consensus has developed that a sequence of uplifted Pleistocene marine terraces is the geologic and geomorphic record of repeated glacio-eustatic sea-level highstands superimposed on a rising coastline. Thus, a rising coastline is a continuous strip chart on which relatively brief sea-level highstands were successively recorded as erosional or depositional landforms (Lajoie, 1986). While earlier studies of these uplifted marine terraces focused on surface morphology and the sedimentary deposits overlying these abrasional platforms, more recent work has concentrated on the significance of these terrace sequences as tools to help unravel the recent tectonic history of the associated coastlines. Using a combination of radiometric dates on the lowermost terraces (where fossils may still be preserved), the sea-level rise curve developed from oxygen isotope stratigraphy from deep-sea cores, and the present elevation of each terrace, it has been possible to determine the uplift rates for particular segments of coastline. Dividing the absolute amount of uplift experienced by each successive terrace since it was formed at some palaeo sea level, by the age of that terrace, allows us to calculate an average uplift rate (Veeh & Chappell, 1970; Bloom *et al.* 1974; Chappell, 1983; Lajoie, 1986).

Conclusions

Rock coasts are dynamic landscape elements that are adjusting to the contemporary morphogenic environment. As rates of change are generally very slow, however, many coasts retain vestiges of former environmental conditions that have been inherited, with little contemporary modification, from interglacial stages when the sea level was similar to that of today. The degree to which coasts are contemporary or inherited features depends upon rates of erosion, which are determined by the intensity of the erosional processes and the resistance of the rock, and the amount of time that these processes have operated. We are still unable, however, to identify and quantify the erosive processes operating on rock coasts, and the geological characteristics that determine the resistance of rocks to these processes. Mathematical modelling can be used to simulate the long-term development of rock coasts, but reliable models must also be based upon reliable field data. Despite a dramatic increase in the number of coastal workers in the last few decades, however, there has probably been an absolute, as well as a relative, decline in the number concerned with rock coasts. We are therefore still dependent, in many areas, on theories developed in the early part of this century, and unfortunately there appears to be little prospect of eliminating many of the deficiencies in our knowledge in the near future.

References

Bartrum, J.A. (1916). High water rock platforms: a phase of shoreline erosion. *Transactions of the New Zealand Institute,* **48**, 132–4.

Bloom, A.L., Broecker, W.S., Chappell, J.M., Matthews, R.K., & Mesolella, K.J. (1974). Quaternary sea-level fluctuations on a tectonic coast: new 230Th/234U dates from the Huon Peninsula, New Guinea. *Quaternary Research,* **4**,185–205.

Bradley, W.C. & Griggs, G.B. (1976). Form, genesis, and deformation of some central California wave-cut platforms. *Bulletin of the Geological Society of America,* **87**, 433–49.

Chappell, J.M. (1983). A revised sea-level record for the last 300,000 years from Papua, New Guinea. *Journal of Geophysical Research,* **14**, 99–101.

Emery, K. O. & Kuhn, G. G. (1980). Erosion of rock coasts at La Jolla, California. *Marine Geology,* **37**, 197–208.

Fairbridge, R. (1961). Eustatic changes in sea-level. *Physics and Chemistry of the Earth,* **4**, 99–185.

Fischer, R. (1990). Biogenetic and nonbiogenetically determined morphologies of the Costa Rican Pacific coast. *Zeitschrift für Geomorphologie,* **34**, 313–21.

Fisher, O. (1866). On the disintegration of a chalk cliff. *Geological Magazine,* **3**, 354–6.

Flageollet, J. C. & Helluin, E. (1987). Morphological investigations of the sliding areas along the coast of Pays d'Auge, near Villerville, Normandy, France. In *International geomorphology,* vol. 1, ed. V. Gardiner, pp. 477–86. Chichester, John Wiley.

Fulton-Bennett, K.W. and Griggs, G.B. (1986). *Coastal protection structures   and their effectiveness.* Joint publication of California State Dept. Boating and Waterways and University of California, Santa Cruz, Institute of Marine Sciences. 48 pp.

Gornitz, V. (1991). Global coastal hazards from future sea-level rise. *Palaeogeography, Palaeoclimatology, Palaeoecology,* **89**, 379–98.

Grainger, P. & Kalaugher, P. G. (1987). Intermittent surging movements of a coastal landslide. *Earth Surface Processes and Landforms,* **12**, 597–603.

Griggs, G.B. & Johnson, R.E. (1976). The effects of the Santa Cruz Small Craft Harbor on coastal processes in northern Monterey Bay, California. *Environmental Geology,* **1**, 229–312.

Griggs, G.B., & Savoy, L.E. (1985). *Living with the California coast.* Durham, NC: Duke University Press. 393 pp.

Inman, D.L. & Nordstrom, C.E. (1971). On the tectonic and morphologic classification of coasts. *Journal of Geology,* **79**, 1–21.

Kennedy, T.B. & Mather, K. (1953). Correlation between laboratory freezing and thawing and weathering at Treat Island, Maine. *Proceedings of the American Concrete Institute,* **50**, 141–72.

Lajoie, K.R. (1986). Coastal tectonics. In *Active tectonics,* pp. 95–124. Washington, DC: National Academy Press.

Matsukura, Y. & Matsuoka, N. (1991). Rates of tafoni weathering on uplifted shore platforms in Nojima-Zaki, Boso Peninsula, Japan. *Earth Surface Processes and Landforms,* **16**, 51–6.

Matsuoka, N. (1988). Laboratory experiments on frost shattering of rocks. *Science Reports of the Institute of Geosciences, University of Tsukuba* (Japan), **9A**, 1–36.

Pitts, J. (1986). The form and stability of a double undercliff: an example from south-west England. *Engineering Geology,* **22**, 209–16.

Robinson, L.A. (1977). Erosive processes on the shore platform of northeast Yorkshire, England. *Marine Geology,* **23**, 339–61.

Robinson, L. A. & Jerwood, L. C. (1987). Sub-aerial weathering of chalk shore platforms during harsh winters in southeast England. *Marine Geology,* **77**, 1–14.

Sanders, N.K. (1968). The development of Tasmanian shore platforms. PhD thesis. University of Tasmania.

Schneider, J. (1976). Biological and inorganic factors in the destruction of limestone coasts. *Contributions to Sedimentology,* **6**, 1–112.

Schneider, J. & Torunski, H. (1983). Biokarst on limestone coasts, morphogenesis and sediment production. *Marine Ecology,* **4**, 45–63.

So, C. L. (1987). Coastal forms in granite, Hong Kong. In *International geomorphology,* ed. V. Gardiner, pp. 1213–29. Chichester: John Wiley.

Spencer, T. (1988). Limestone coastal morphology. *Progress in Physical Geography,* **12**, 66–101.

Suess, E. (1888). *The Face of the Earth II* (English Translation by H.B.C. Sollas, 1906, Oxford: Oxford University Press).

Sunamura, T. (1975). A laboratory model for wave-cut platform formation. *Journal of Geology,* **83**, 389–97.

Sunamura, T. (1977). A relationship between wave-induced cliff erosion and erosive force of wave. *Journal of Geology,* **85**, 613–18.

Sunamura, T. (1982). A predictive model for wave-induced cliff erosion, with application to Pacific coasts of Japan. *Journal of Geology,* **90**, 167–78.

Terzaghi, K. (1962). Stability of steep slopes on hard unweathered bedrock. *Geotechnique,* **12**, 251–70.

Trenhaile, A.S. (1971). Lithological control of high-water rock ledges in the Vale of Glamorgan, Wales. *Geografiska Annaler,* **53A**, 59–69.

Trenhaile, A.S. (1980). Shore platforms: a neglected coastal feature. *Progress in Physical Geography,* **4**, 1–23.

Trenhaile, A. S. (1987): *The geomorphology of rock coasts* . Oxford:Clarendon/ Oxford University Press. 384 pp.

Trenhaile, A. S. (1989). Sea-level oscillations and the development of rock coasts. In *Applications in coastal modeling,* eds. V. C. Lakhan & A. S. Trenhaile, Amsterdam: Elsevier. pp. 271–95.

Trenhaile, A. S. & Mercan, D. W. (1984). Frost weathering and the saturation of coastal rocks. *Earth Surface Processes and Landforms,* **9**, 321–31.

Trudgill, S. T. (1976). The marine erosion of limestone on Aldabra Atoll, Indian Ocean. *Zeitschrift für Geomorphologie, Suppl. Bd.,* **26**, 164–200.

Trudgill, S. T. (1987). Bioerosion of intertidal limestone, Co. Clare, Eire. 3. Zonation, process and form. *Marine Geology,* **74**, 111–21.

Veeh, H.H. & Chappell, J. (1970). Astronomical theory of climatic change: support from New Guinea. *Science,* **167**, 862–5.

12

Tectonic shorelines

P. A. PIRAZZOLI

Sea level is the common and unifying element of coastal tectonics.
K.R. Lajoie, 1986

Introduction

Various classifications of coasts have been attempted in the literature, in some cases inspired by the plate tectonics model (e.g. Inmam & Nordstrom, 1971). According to Rice (1941) and to the American Geological Institute (1960), the term 'tectonic' is defined in a very wide sense, 'designating the rock structure and external forms resulting from the deformation of the earth's crust'. This definition implies that certain processes, which are not always considered as 'tectonic', must also be taken into account. It is now widely accepted that phenomena like glacioisostasy and hydroisostasy have produced a vertical deformation of the earth's crust in virtually all coastal areas over the last 20 ka (Clark, Farrell & Peltier, 1978). Moreover, sediment deposition near coasts and especially in delta areas contributes, together with erosion processes and volcanic eruptions, to the modification of loads exerted on the earth's crust, causing additional vertical deformation (see Chapter 3). Consequently, all the coasts of the world can be considered as more or less tectonic and none are vertically stable. Instead of proposing a new classification, the present chapter aims therefore to identify various kinds of vertical movement which may affect any of the types of coast considered elsewhere in this volume, with, however, more emphasis on seismo-tectonic processes. Certain effects of vertical movements on coastal evolution are also considered.

Main causes of vertical displacements in coastal areas

As sea level is an equipotential surface of the gravity field, it can be influenced by any factor affecting gravity: from space (astronomical effects), from the atmosphere (meteorological changes), in the oceans (hydroisostasy, steric changes) at the earth's surface (glacioisostasy, volcano-isostasy, sedimento-isostasy) and in the earth's interior (thermo-isostasy, density changes).

Geoid changes

Over the long term (several million years), important vertical displacements at the surface of the earth can be produced by density changes in the interior. Satellite ranging has revealed that sea-surface topography does not correspond exactly to that of a rotational ellipsoid, and that there are many bumps and depressions with a relief of up to 200 m (Gaposchkin, 1973; Marsh & Martin, 1982). This is probably the consequence of an uneven mass distribution inside the earth. As the earth's interior is not solid, relative movements between masses with different densities have certainly occurred in the past and will probably continue to do so as long as the earth has not completely cooled. Each density change modifies the earth's gravity field. The water surface, being liquid, will adjust to the new gravity field almost instantaneously. The lithosphere on the other hand, being denser than sea water and liable to be deformed and warped, although being solid, will also adjust, but with a less absolute deformation occurring over a much longer time period. The difference in vertical displacement and the time lag between liquid and solid adjustment may cause transgression or regression phenomena on a regional or even continental scale (Mörner, 1976). Changes in the earth's rate of rotation and in the tilt of the rotation axis may also produce transgressions and regressions across wide areas (Fairbridge, 1961; Barnett, 1983; Mörner, 1988; Sabadini, Doglioni & Yuen, 1990; Peltier, 1990). Very little information is available, however, on case studies of such changes in the past, though vertical displacements deduced from fossil shorelines have in some cases been ascribed to geoidal changes (e.g. Martin et al., 1985; Nunn, 1986).

Thermo-isostatic and volcano-isostatic deformation

It is now generally accepted that as the ocean crust spreads away from its point of origin along submarine ridges, it cools and thickens, thereby increasing in density. As a result the seafloor subsides isostatically, gradually submerging oceanic islands as they are carried laterally. Over several million years thermo-isostatic submergence will be of the order of kilometres (Fig. 12.1a). In tropical waters subsidence is increased by the load of coral reefs, which have to grow vertically to maintain their sea-level position. The normal evolution of an oceanic island, in the absence of eustatic changes in sea level, would therefore be a gradual, continuous submergence and, at the same time, the rapid erosion at the surface of all emerged parts caused by subaerial weathering and wave action. In tropical waters such processes would support the classical sequence ranging from a high island with fringing coral reefs to an island with a barrier reef, and finally to an atoll (Darwin, 1842; see also

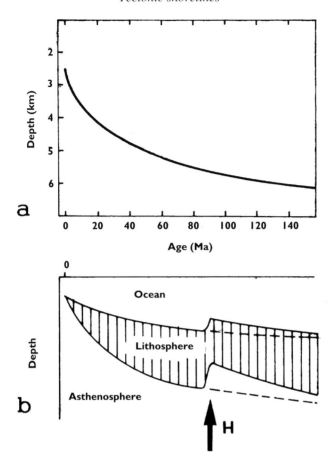

Figure 12.1. (*a*) The relation between depth and age of normal oceanic crust (adapted from Menard, 1986). (*b*) Schematic section of the lithosphere along a hot spot trace, the heat flow *H* causes temporary thinning and isostatic uplift of the lithosphere; after the hot spot zone has been left behind, lithospheric subsidence, caused by cooling and thickening, will occur again. (Adapted from Detrick & Crough, 1978; no scale.)

Chapter 7). Average subsidence rates have been estimated at 0.2 mm a^{-1} since 60 Ma in the Marshall Islands (Menard & Ladd, 1963) and 0.12 mm a^{-1} since the Pliocene in Mururoa Atoll (Labeyrie, Lalou & Delibrias, 1969) (for a review of results of deep drilling in atolls, see Guilcher, 1988, pp. 68–75). The normal evolutionary trend of a oceanic island implies, therefore, gradual subsidence and relatively rapid erosion of exposed rocks. That is why non-carbonate rocks older than a few million years are so rare in mid-plate oceanic areas. If an oceanic island is not subsiding, it is anomalous and requires explanation.

Where the oceanic crust approaches a hot spot, which according to Menard (1973) corresponds to a fixed asthenospheric bump, the normal cooling process is reversed, so that the crust is heated, becomes less dense and thinner, and thus rises. During transport away from the hot spot there is renewed cooling and subsidence (Menard, 1973; Detrick & Crough, 1978) (Fig. 12.1*b*). Near hot spots, extrusion of lava often occurs. The resulting load produces on the lithosphere an isostatic depression under the volcanic pile, over a distance generally less than 150 km from the load barycentre, and a peripheral raised rim at a distance of between some 150 and 300 to 330 km (McNutt & Menard, 1978) (Fig. 12.2), i.e. volcano-isostatic effects similar to the glacioisostatic effects of an ice sheet with equivalent mass (see below). When the translation movement of an oceanic plate over a hot spot has produced a line of islands, and/or two or more hot spots form an alignment in the direction of plate movement, interaction between isostatically depressed areas and uplifted rims of nearby islands is possible and may produce complicated sequences of vertical deformation, with repeated phases of upheaval and sinking (Coudray & Montaggioni, 1982; Scott & Rotondo, 1983a,b; Montaggioni, 1989).

A good example of these processes is given by the interaction between the Society Islands chain, its active hot spot and certain nearby Tuamotu atolls, in the south Pacific. Anaa is a closed atoll located on the western margin of the Tuamotu Archipelago, presently about 200 km east of the hot spot area of the Society Islands (Fig. 12.3). In Anaa, exposed ancient reefs dating from the Last Interglacial period (Veeh, 1966; Pirazzoli et al., 1988), which are usually submerged in other Tuamotu atolls, are well preserved in the northeastern part of the reef rim, where they reach an elevation of about 4 m above present sea level (Fig. 12.4). There is no evidence of older exposed reefs. During the Holocene, the sea reached here a level about 1.35 ± 0.1 m higher than at present, i.e. 0.3–0.5 m higher than in other Tuamotu atolls. These data indicate that Anaa started uplifting recently (probably less than 300 ka ago) and that its relative uplift rate, in relation to other Tuamotu atolls remote from active hot spot areas, can be estimated to have been at most 0.1 mm a^{-1} since the Last

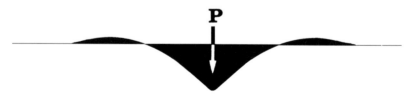

Figure 12.2. A point load *P* causes a flexure on a floating elastic plate. In the case of the oceanic lithosphere loaded by a volcano, the flexure will consist of a subsidence zone, which reaches its maximum under the load barycentre, and an arching at some distance.

Interglacial period, and about 0.1 mm a^{-1} during the late Holocene. This slight uplifting trend can easily be explained by thermal rejuvenation phenomena (thermo-isostasy) while Anaa is approaching the uplifted side of the asthenospheric bulge of the Society Islands hot spot (Pirazzoli *et al.*, 1988). On the other hand, as the distance of Anaa from the barycentre of the most recent important volcanic load on the lithosphere (which corresponds to the Tahiti and Moorea volcanoes, see Fig. 12.3*a*) is over 350 km, volcano-isostasy has probably not yet contributed to the recent uplift trend.

The Pacific plate translation carries Anaa in a northwest direction, at a speed of about 110 mm a^{-1}. Following this trajectory, Anaa is approaching the

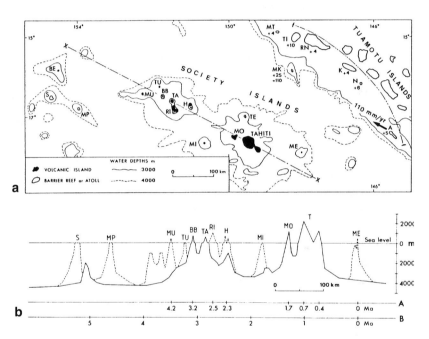

Figure 12.3. (*a*) Location map of the Society and northwest Tuamotu Islands. The hot spot of the Society Islands is now active near Mehetia (ME). Elevated reefs in the Tuamotus are found only between the f–f curve and Tahiti; their present-day emergence (in metres) is indicated near the names of the atolls: A = Anaa; MK = Makatea; N = Niau; K = Kaukura; RN = Rangiroa; TI = Tikehau; MT = Mataiva. The maximum elevation is observed in Makatea: 20–25 m for mid-Pleistocene reefs, and about 110 m for the summit of the island. MO = Moorea; TE = Tetiaroa; MI = Maiao;H = Huahine; RI = Raiatea; TA = Tahaa; BB = Bora Bora; TU = Tupai; MU = Maupiti; MP = Mopelia; S = Scilly; BE = Bellinghausen. (*b*) Topographic profile across the Society Islands; continuous line: profile along X–X (see *a*); broken line: projections of nearby volcanic complexes on X–X; A: average K/Ar ages of the volcanic islands; B: linear age variation corresponding to a uniform velocity of the Pacific plate of 110 mm a^{-1} (adapted from Pirazzoli & Montaggioni, 1985).

Figure 12.4. Reefs dating from the Last Interglacial reach about +4 m at Anaa Atoll (Tuamotu Islands, Pacific Ocean) (from Pirazzoli & Montaggioni, 1985).

hot spot area at a rate of 45 mm a^{-1} and will continue to do so for the next 1.3 Ma, when it will reach the minimum distance of about 140 km from that area. If the 0.1 mm a^{-1} uplift rate does not change and vertical erosion can be ignored, Anaa will then rise by thermo-isostasy to about 135 m in altitude. Meanwhile, however, new important volcanic loads may be constructed above the hot spot in a position nearer to Anaa than the Tahiti–Moorea load, thus creating a new volcano-isostatic arch, which might increase the uplift of Anaa.

Makatea Island, which was located about 3 Ma ago in the present position of Anaa in relation to the Society Islands hot spot area, displays the pattern that Anaa is likely to show in the future. Makatea is a raised atoll (Fig. 12.5) reaching an altitude of 113 m (Montaggioni, 1985). If a possible first thermo-isostatic uplift interrupting the normal subsidence trend of the island is disregarded (about 14 Ma ago, moving near the Pitcairn hot spot, as suggested by Montaggioni, 1989), Makatea entered the uplifting side of the Society Islands hot spot swell, just as Anaa is doing now, slightly earlier than 3 Ma ago. During the Pleistocene, when the huge volcanic masses of Moorea and Tahiti erupted (mainly between 2.0 and 0.3 Ma ago), Makatea was located near the axis of their isostatic arch. This caused an additional uplift of the island. Moving away from the hot spot area, Makatea remained on the isostatic arch

Figure 12.5. Raised limestone cliffs in Makatea Island (Tuamotu Islands, Pacific Ocean). Arrows indicate two former shorelines: at +25 m (dated more than 200 ka), and at +5/+8 m (dated 100 to 140 ka) (Montaggioni, 1985) (photo L. F. Montaggioni, adapted).

of Tahiti and Moorea (McNutt & Menard, 1978; Lambeck, 1981), which is still viscoelastically active: subsidence due to incipient lithospheric cooling is therefore still compensated in Makatea by volcano-isostatic uplift, resulting in upheaval rates of the order of 0.05 mm a^{-1}, both for the Late Pleistocene and for the late Holocene (Pirazzoli & Montaggioni, 1985). Viscoelastic effects are expected to decrease with time, however, while lithospheric cooling will continue. The uplift phase of Makatea, which has predominated during the last 3 Ma, is therefore coming to an end, whereas a new uplift trend is beginning in Anaa Atoll.

Similar thermo- and volcano-isostatic phenomena have been studied in the chain of the Cook–Austral Islands (Lambeck, 1981; Calmant & Cazenave, 1986; Pirazzoli & Veeh, 1987; Woodroffe *et al.*, 1990; Pirazzoli & Salvat, 1992), Henderson and Pitcairn (Spencer, 1989; Spencer & Paulay, 1989), the Hawaii Islands (Hamilton, 1957; Walcott, 1970; Scott & Rotondo, 1983a,b; Campbell, 1986) and can be applied to most raised oceanic islands remote from plate boundaries, and even to seamounts (Cazenave *et al.*, 1980).

Glacioisostatic and hydroisostatic deformation

One of the best known causes of lithospheric deformation is glacial isostasy. Global in its many side effects, according to geophysical models (Clark, Farrell & Peltier, 1978; Nakada & Lambeck, 1987; Peltier, 1990, 1991), it may exceed the growth of a mountain range or the filling of a sedimentary basin in terms of the mass involved (Pirazzoli & Grant, 1987). As shown by Daly (1934) (Fig. 12.6), the load of an ice sheet deforms the earth's crust. The resulting subsidence beneath the ice makes deeper material flow away, and raises an uplifted rim at a certain distance. When the ice sheet melts, an unloading occurs, resulting in uplift beneath the melted ice; the marginal rim will consequently tend to subside and move towards the centre of the vanishing load. In the oceans the load of melted water will make the ocean floor subside (hydroisostasy) (Daly, 1934; Bloom, 1971). This will cause a flow of deep material from beneath the oceans to beneath the continents. The latter will tend therefore to rise, with reactivation of seaward flexuring at the continental edge (Bourcart, 1949; Clark et al., 1978). From a tectonic point of view, coasts are therefore very unstable zones, which may follow the isostatic trend prevailing

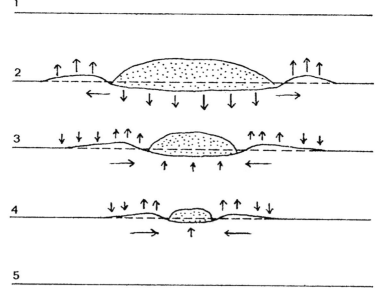

Figure 12.6. Crustal deformations caused by the melting of an ice cap. Stage 1 shows undeformed land surface; stage 2 shows ice sheet at maximum extent and crustal response; stages 3 and 4 represent melting of ice sheet; and stage 5 indicates return to initial state. (Adapted from Daly, 1934.)

in the nearby ocean, or that in the nearby continent, depending on the local topography and coastal configuration, the width of the continental shelf, the water depth, etc.

Part of the above isostatic movements are elastic, i.e. contemporaneous to loading and unloading. However, because of the viscosity of the earth's material, part of the movements may continue for several thousand years after loading or unloading has stopped. Although most of the continental ice sheets had disappeared by 8000 years BP in Canada and Fennoscandia, uplift and subsiding movements have continued to be active there ever since and will probably continue to be active for several more millennia, although at decreasing rates (Fig. 12.7).

Using reconstructions of the Late Pleistocene–early Holocene melting history of polar ice sheets and consequent glacio- and hydroisostatic effects around the globe, geophysical models may predict, with assumptions, trends of vertical movements caused by the last deglaciation in any coastal sector of the world (e.g. Clark *et al.*, 1978; Nakada & Lambeck, 1987; Peltier, 1991; Nakada, Yonekura & Lambeck, 1991). These models are therefore very useful for the reconstruction of possible postglacial coastal evolution in areas with insufficient field data.

Vertical deformation near plate boundaries

Approaching tectonic plate boundaries, which are generally areas of high seismicity, vertical movements become more complex, with various local or regional geodynamic factors superimposed, often with predominant effects, on glacio- and hydroisostatic factors.

Subduction zones

Underthrusting side: Near a subduction zone, the behaviour of the under-thrusting plate will differ notably from that of the overthrusting one. On the underthrusting side, the oceanic plate is most likely subjected to arching phenomena, in order to make possible the change from a horizontal translation movement to a subduction beneath the overriding plate. This arching implies a wave-like flexuring of the lithosphere, with first a slow gradual uplift, as in the case of an oceanic island approaching a hot spot, then a gradual subsidence at accelerating rates. When approaching subduction trenches, however, very few islands are at the right distance from an oceanic trench to verify the existence of a wave-like flexuring pattern. In the Loyalty Islands, east of New Caledonia, the occurrence of marine terraces of the same age at different altitudes in various islands, depending on the distance of these islands from the New

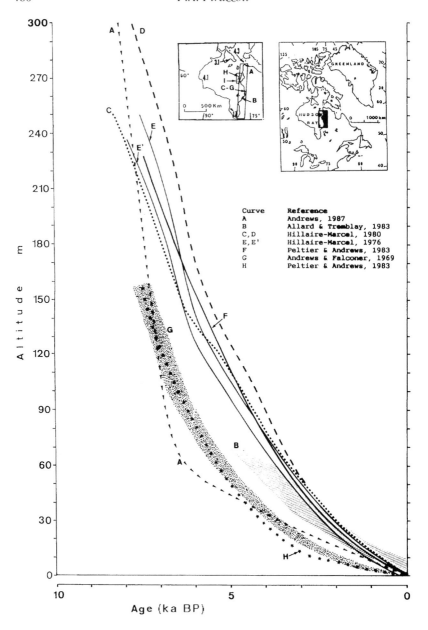

Figure 12.7. Relative sea-level curves obtained from the eastern shores of Hudson Bay (Canada) indicate very rapid uplift (of glacioisostatic origin) in the early Holocene, at rates decreasing gradually during the late Holocene. For comments, see Pirazzoli (1991).

Hebrides Trench, was interpreted by Dubois, Launay & Recy (1974) as being due to pre-subduction arching, probably the first report of this phenomenon in the literature. More recently, elevated geomorphological features in Christmas Island, Indian Ocean, 200 km southwest of the axis of the Sunda Trench (Nunn, 1988; Woodroffe, 1988) and in the Daito (Borodino) Islands, 150 km east of the Ryukyu Trench (Ota & Omura, 1992), have been ascribed to similar arching phenomena.

Overthrusting side: On the overthrusting side, rapid rates of vertical displacement are frequent. Uplift in particular may be caused by (i) piling up above the oceanic plate, on the inner side of the trench, of sediments too light to be subducted, which will raise the overthrusting edge isostatically; (ii) elastic rebound phenomena linked to the subduction; (iii) tilting of lithosphere blocks or other tectonic processes; and (iv) volcanic activities.

Tectonic trends often appear to be incremental and gradual over the long term, although in reality they frequently consist of sudden coseismic vertical movements, occurring at the time of earthquakes of great magnitude, separated by more or less long periods of quiescence or even of slow interseismic movement contrary to the coseismic movement. Fast aseismic displacements have also been reported. Active folding is frequent and it is detected generally on the basis of geomorphological and geodetic methods. In coastal areas, short wavelength folding is clearly recognizable in folded marine terraces. Among many possible examples, the southern tip of Taiwan displays remarkable undulations of the middle Holocene coral-reef terrace along a distance of about 24 km, with elevations ranging from 5 to 36 m above present sea level (Liew & Lin, 1987). Repetitive up-and-down movements are also possible, as well as accelerations in the rates of vertical displacement.

Examples of coastal coseismic uplift in subduction areas deduced from ancient shorelines are reported frequently in the literature. Along the western coast of the Muroto Peninsula, Shikoku, Japan, there are three main superimposed coastal terraces, all of which decrease in height to the north. The middle (Fig. 12.8) and lower terraces were formed in the Last Interglacial age and in the middle Holocene, respectively. These terraces are mainly abrasional in origin, but in some places they are depositional, showing that they were built under transgressional sea levels (Yoshikawa, Kaizuka & Ota, 1964). Tectonic movements in this area are characterized by acute uplift tilting landward, associated with great earthquakes and chronic subsidence tilting seaward during interseismic periods. In Fig. 12.9, changes in height obtained by repeated levellings are shown for the periods 1895 to 1929 and 1929 to 1947.

Figure 12.8. Remnants of Last Interglacial marine terraces at about +150 m near Hane River, approx. 17 km northwest of Muroto Cape, Shikoku, Japan (photo P.A. Pirazzoli 1974).

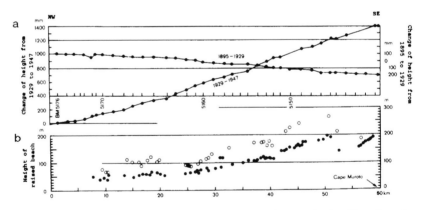

Figure 12.9. (*a*) Vertical change of bench marks surveyed by precise levelling along the Muroto Peninsula (Shikoku, Japan) (upper graph) and (*b*) vertical distribution of height of former shorelines on terraces ascribed to the Last Interglacial (solid circles) and to the penultimate interglacial (open circles) (lower graph). The amounts of vertical change of bench marks during the preseismic period from 1895 to 1929 are shown in relation to the present sea level (scale on right) and the vertical changes of bench marks during the period 1929 to 1947 (which include the effects of the 1946 Nankai Earthquake) are shown relative to bench mark 5176 on the left side of the graph (scale on left) (adapted from Ota, 1975).

The former period falls between two large earthquakes; the amount of subsidence during an interseismic period – on average 120 years here – was estimated at 0.95 m from geodetic and historical records. The latter levelling period includes the 1946 Nankai Earthquake, when the tip of the peninsula was uplifted 1.20 m. The mean uplift rate is estimated to be $2\,\mathrm{mm\,a^{-1}}$ at the southern tip of the peninsula. At this rate the Last Interglacial shoreline would be expected to be at 240 m in altitude and the Holocene terrace at 12 m, reasonably comparable to the observed heights of 190 m and 13 m, respectively (Yoshikawa, Kaizuka & Ota, 1981).

Various examples of coastal coseismic subsidence have been reported from the Pacific coast of North America (Gulf of Alaska, Washington and Oregon). In the Gulf of Alaska, where the Pacific plate is being subducted beneath the American plate, the great earthquake of 27 March 1964 (magnitude $\geqslant 8.4$), which had its epicentre in the Prince William Sound area, was characterized by vertical crustal movements over a region of at least $200\,000\,\mathrm{km^2}$ (some 700 to 800 km long and 150 to 300 km wide) with trends roughly parallel to the Gulf of Alaska coast and the Aleutian volcanic arc and trench. Average uplift within the seaward zone is 2 to 3 m, except along a narrow zone where combined crustal warping and faulting have resulted in an uplift of 11.5 m on land and possibly more than 15 m on the sea floor. Subsidence in the adjacent zone to the north averages about 1 m and reaches a maximum of 2.2 m. There is no abrupt change of level between the two zones, but rather a northward and northwestward tilting around the zero isobase, with a maximum slope of about $0.11\,\mathrm{m\,km^{-1}}$ (Plafker & Rubin, 1967).

In the area affected by the 1964 Alaska earthquake, postseismic uplift occurs where coseismic subsidence was observed, and postseismic subsidence occurs where coseismic uplift was observed (Savage & Plafker, 1991). At Anchorage in particular, where the 1964 earthquake caused subsidence of 0.71 m, repeated levellings and sea-level measurements indicate as much as 0.55 m of land uplift in the decade following the earthquake. This deformation is explained by Brown *et al.* (1977) as due to creep along the down dip extension of the fault which ruptured in 1964. As noted by Savage & Plafker (1991), however, the immediate postseimic response is damped out within the first decade, and the subsequent vertical displacement rates appear to be fairly steady.

Geological evidence along the coast of the Gulf of Alaska reveals a complex history of Holocene tectonic displacements, in which areas of postglacial net emergence (as much as 55 m since 7650 years BP) or submergence (at least 90 m) largely coincide with areas of significant earthquake-related uplift and subsidence. In the uplifted zone, for example,

Middleton Island displays a series of marine terraces indicating that 40 m of relative emergence has occurred in at least five major upward pulses during the last 4470 ± 250 years BP. A sixth terrace was formed on the island as a result of 3.3 m of uplift during the 1964 earthquake (Plafker & Rubin, 1967).

Along the outer coast of Washington State, near the boundary between the Juan de Fuca plate and the North American plate in the Cascadia subduction zone, coastal subsidence commonly accompanies a great subduction earthquake, the coseismic subsidence occurring mainly in an onshore belt, flanked by a zone of coseimic uplift mostly offshore (Atwater, 1987). In northern Oregon, Darienzo & Peterson (1990) found buried marsh deposits indicating a series of episodic, abrupt subsidences of the marsh surfaces to low intertidal levels. The stratigraphy shows evidence of sudden subsidence displacements of 1.0 to 1.5 m alternating with gradual uplift displacements of the order of 0.5 to 1.0 m. This was interpreted as reflecting coseismic strain release (abrupt subsidence) following interseismic strain accumulation (gradual uplift). Recurrence intervals between subsidence events in this area range from possibly less than 300 years to at least 1000 years, with the last dated event likely to have taken place in 300 to 400 years BP. Overlaps of radiocarbon ages for at least four subsidence events suggest that events may have been synchronous over at least 200 km of the central part of the Cascadia subduction zone.

Similar events are known to have happened in many seismically active coastal areas, especially in the Circum-Pacific region (Ota, 1991), in central Chile (Kaizuka *et al.*, 1973), Japan (Yonekura, 1972, 1975; Nakata *et al.*, 1979; Ota, 1985), Vanuatu (Taylor *et al.*, 1980; Jouannic, Taylor & Bloom, 1982), and New Zealand (Wellman, 1967; Berryman, Ota & Hull, 1989; Ota, Miyauchi & Hull, 1990; Ota, Hull & Berryman, 1991a) (Fig. 12.10).

Although uplift or subsidence rates are commonly reported as linear in the long term, case studies of Late Quaternary accelerations in uplift movements also exist. In Kikai Island (Ryukyus, Japan), which is entirely capped by a sequence of marine terraces, the uppermost terrace, at over 200 m in altitude, has been dated by *in situ* corals at about 125 ka. This and other dates from younger, lower terraces have enabled Konishi, Omura & Nakamichi (1974) to infer an uplift rate of 1.5–2.0 mm a^{-1} since the Last Interglacial. However, reef-proper facies of the same upper terrace formation were also dated to 200 ka or more by ^{230}Th/^{234}U, and 400 to 600 ka by electron spin resonance (ESR) (Omura *et al.*, 1985; Koba *et al.*, 1985). These new results suggest that the tectonic trend has changed, between 200 and 125 ka, to a faster uplift movement.

Other examples from the Mediterranean are even more drastic, with a long

Figure 12.10. View west over the rapidly emerging coastline at Turakirae Head along the southern coast of the North Island of New Zealand (41.4° S, 174.9° E). In the foreground: four of the five large storm beach ridges preserved along the coast (paths follow the upper surface of three of them) can be seen dipping westward (towards Wellington City). The second lowest ridge represents the storm beach that formed prior to more than 3 metres of uplift that accompanied the $M \geqslant 8.0$ earthquake on January 23, 1855. The lowest ridge has formed during the last 137 years. The age of higher and older storm beach ridges is poorly known, but the highest ridge probably formed about 6500 years ago, when rising Holocene sea level reached its highest and present position. Each of the four uplifted ridges is inferred to represent a large earthquake similar in magnitude to that experienced in 1855 (Wellman, 1967; Berryman *et al.*, 1992). In the background: the age of higher level terraces farther west, tilted in a northwesterly direction, is also unknown; marine terraces and gravels have been identified more than 300 metres above present sea level, and all may have been deposited during the last c. 130 000 years (Ota, Williams & Berryman, 1981). (Photo by Lloyd Homer, Institute of Geological and Nuclear Sciences, Lower Hutt, New Zealand.)

subsidence period producing a series of always higher transgressions towards the end of the Pliocene, which caused the development of extensive marine platforms in Calabria (Carobene & Dai Pra, 1990), followed by subsequent rapid Pleistocene uplift, bringing marine sediments as high as 1360 m since the Calabrian (Dumas *et al.*, 1980) or 157 m since the Last Interglacial (Dumas *et al.*, 1988). The uplift movements would be produced by periodic earthquakes of the type which occurred near Messina in 1908, with an average return time of

1000 to 1500 years (Valensise & Pantosti, 1992). In the Aegean, a similar change of trend happened slightly later, with transgression (subsidence) predominant until the lower Pleistocene and regression (uplift) since that time (Keraudren, 1975); the amplitude of this fluctuation was about 500 m in the island of Karpathos (Keraudren & Sorel, 1984).

Collision zones

Where a continental lithosphere exists on both sides of a plate boundary, subduction will generally be prevented by the low density of lithospheric material and collision or transform processes will occur with vertical movements which can became highly irregular. The island of Rhodes, Greece, where various distorting movements took place after the late Pliocene, may be a case in point. Though these predominant movements were 'a general tilting movement with the raising up of the northeast part and the sinking of the southwest', 'the island did not react to this stress as a sole rigid body, but was disrupted in several blocks separated by normal faults' (Mutti, Orombelli & Pozzi, 1970, p. 71). A recent study of emerged Holocene shorelines (Pirazzoli *et al.*, 1989) has shown that the east coast of the island consists of at least eight small crustal blocks (up to twelve kilometres in length) each exhibiting a different tectonic behaviour. The number of emerged Holocene shorelines in each block varies from one to seven and a slightly submerged shoreline can also be found in the north part of the island (as well as various indications of submergence along the west coast). Relative sea-level changes in the northernmost crustal block of the island are shown in Fig. 12.11. The uppermost shoreline (A_1), now at about +3.4 m (Fig. 12.12), was reached by the sea around 6000 years BP. Shortly before 4000 years BP, a 1.3 m coseismic subsidence movement brought the sea surface to A_3, where typical sea-level erosion notches were carved into the limestone cliffs. A few centuries later, however, a reversed coseismic movement brought the shoreline back to the previous position at +3.4 m, where it remained almost stable for about one thousand years. Soon after 2280 ± 110 years BP (probably at the time of the 222 BC earthquake which destroyed the statue of Colossus in the harbour of Rhodes), a coseismic uplift suddenly displaced the relative sea level from about +3.4 m to about -0.4 m (A_8). On the limestone layers which emerged at that time, several coastal quarries were worked in Roman times, which are now slightly submerged. Subsidence movements took place during the next fifteen centuries, leaving marks of three shorelines (A_6, A_5 and A_4), which are now emerged, followed by renewed uplift during the last 400 years. Other examples of coseismic vertical movements deduced from fossil shorelines in the eastern

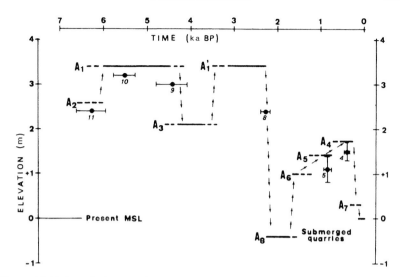

Figure 12.11. Late Holocene relative sea-level changes in the northernmost part of the east coast of Rhodes Island (Greece), from the city of Rhodes to Kalitea. Numbered filled circles with horizontal/vertical bars correspond to radiocarbon-dated samples with uncertainty ranges. See text for details. (From Pirazzoli *et al.*, 1989.)

Figure 12.12. The notch with a hammer corresponds to shoreline A₁, at about +3.4 m at Kalitea (northern part of Rhodes Island). It marks the sea-level position at the time the Colossus was erected in Rhodes Harbour. In 222 BC a great earthquake is known to have ruined the Colossus. The same earthquake, according to geomorphological and radiometric evidence, is likely to have caused a coseismic uplift, displacing the relative sea level from +3.4 m to about −0.4 m in this part of the island (photo P.A. Pirazzoli).

Mediterranean have been reported from Crete (Thommeret *et al.*, 1981), Turkey (Kelletat & Kayan, 1983; Pirazzoli *et al.*, 1991) and Euboea Island (Stiros *et al.*, 1992).

Transform zones

The best known example of a transform fault zone crossing a coastal area is probably that of the San Andreas Fault, which extends south to north over a distance of about 1300 km between the Gulf of California and the Mendocino Fracture Zone. In the Santa Cruz area, six marine terraces indent a 60 km stretch of coastal topography. Their raised shell beds were first described by Darwin (1891). According to Valensise & Ward (1991), these terraces have been raised by repeated, dominantly horizontal slip earthquakes on the San Andreas Fault, which could recur every three to six centuries, uplifting the terraces at average rates between $0.13 \, \text{mm a}^{-1}$ and $0.35 \, \text{mm a}^{-1}$. In several areas along the San Andreas Fault evidence of uplift is missing, however, and in southern California, a recent aseismic uplift has been reported (Castle, Church & Elliott, 1976).

Volcanic zones

In volcanic areas, very rapid rates of vertical displacement can be reached. Kaizuka, Miyauchi & Nagaoka (1983) recognized over 20 steps of marine terraces, up to 120 m above sea level; in Iwo (Sulphur) Island, a volcano situated 1200 km south of Tokyo, the average uplift rate was more than $100 \, \text{mm a}^{-1}$ over several centuries, reaching $200 \, \text{mm a}^{-1}$ in some parts of the island (Kaizuka, 1992). However, the best-known case study of irregular and fast vertical movements is that of the Phlegraean Fields caldera near Naples, Italy. Here, burrows of *Lithophaga* in the columns of the so-called Temple of Serapis (a Roman market probably built in the second century BC near the Pozzuoli Harbour), are today found at a level of several metres above present sea level (Fig. 12.13). Geomorphological, archaeological and historical data suggest a complex relative sea-level history, produced by alternating subsidence and uplift. The subsidence rate seems to average $10 \, \text{mm a}^{-1}$ in the long term; after the Temple was constructed, subsidence reached a maximum of 12 m in the tenth century AD, followed by an uplift of 7 m during the Middle Ages, culminating (at a rate of 350 to $500 \, \text{mm a}^{-1}$) in the Monte Nuovo eruption in 1538 AD, and further subsidence (Suess, 1900; Günther, 1903; Flemming, 1969; Yokoyama, 1971; Grindley, 1974). Tide gauge data from Pozzuoli Harbour show two brief periods (1970 to 1973 and 1982 to 1984) of rapid (up to $800 \, \text{mm a}^{-1}$) uplift, reaching a total of 3.2 m between 1968 and 1984 (Berrino *et al.*, 1984; Lajoie, 1986; Luongo & Scandone, 1991), followed by renewed subsidence, which was still ongoing in October 1991.

Figure 12.13. The temple of Serapis at Pozzuoli (Naples) in October 1991. Note the dark band on the columns produced by molluscan borings (the upper limit of which varies, according to Suess (1900), between 5.68 m and 5.98 m above the paving of the temple) and the fact that the floor of the temple is almost completely dry (photo P. A. Pirazzoli). The same floor still appears submerged in photographs taken in 1982, 1983 and 1984 (Vita-Finzi, 1986, pp. 10–11).

Effects of tectonics on coastal evolution

As shown above, rates and duration of tectonic movements can be very variable and it is obvious that the morphodynamic evolution of a coastal area which would be uplifted several metres in a sudden jerk, with a recurrence period of several thousand years, will be different from that of the same area uplifting gradually and continuously at the rate of $1\,\mathrm{mm\,a}^{-1}$. In particular, dated tectonic events (e.g. coseismic uplift movements) can be used as a basis for assessing development rates of constructional and destructional forms and shore-forming processes (Kelletat, 1991).

The evolution of a shoreline depends mainly on the sea-level position, the type of rock exposed, the shore profile and the local climatic, biological, sedimentologic and hydrologic processes which can produce erosion or accretion (Carter, 1988). Any relative sea-level change will not only displace the shoreline kinematically at a new level, but also modify the zones exposed to wave action and consequently longshore sediment transport and deposition.

Tectonics (in the broad sense) is one of the two main components of relative sea-level changes, the other one being eustasy. During the Quaternary period, eustatic changes consisted of sea-level oscillations at rates of the order of $10\,\mathrm{mm\,a^{-1}}$ during periods of about $10\,\mathrm{ka}$, interspaced with stillstands during periods of a few thousand years. This means that on a tectonically stable coast (assuming that such a coast exists), each main sea-level oscillation would always affect the same vertical coastal zone. Each new shoreline would consequently destroy the preceding one at the same level, or bury it inland if accretion predominates so that such coasts would consist of broad coastal plains bordered offshore by wide continental shelves and gentle continental slopes. These coastlines would be characterized, according to Lajoie (1986), by low-scale depositional landforms such as broad sandy beaches, offshore barrier bars and low to moderate topographic relief.

The effect of a tectonic trend of uplift or subsidence, on the other hand, will be to displace vertically the coastal range and continually create new landforms. There is, however, a considerable difference between transitory isostatic rebound (such as that induced by glacioisostasy or hydroisostasy) and sustained tectonic deformation, since the former produces impacts very similar to those of eustatic changes, whereas the displacements caused by the latter are generally more durable. Nevertheless, near ice sheet areas, glacio-isostatic changes can be much greater than eustatic changes (Fig. 12.7) and the morainic material supplied by each new glaciation is so abundant that an almost completely new situation will generally be found by the sea after each oscillation, as in the case of sustained tectonic deformation.

More generally, uplift will tend to produce rugged coastlines and increased erosion rates on the new relief. Deepening of valleys and stream beds by river courses will bring more sediments to the coast and feed beaches where regressional sedimentary sequences will produce progradational shorelines with the development of shore-parallel sets of beach ridges (Carter, 1988). The superimposition of long-term uplift trends on eustatic fluctuations will favour the formation of stepped marine and river terraces and the development of underground streams.

Subsidence on the other hand will tend to bury valleys and accumulate sediments in coastal plains, where the reduced river slopes will favour the development of peaty and marshy areas. Inland migration of shorelines will result in land loss by submergence. In tropical areas, coral reefs will tend to move away from the shore and construct barrier reefs. Outside tropical areas, oceanic islands will gradually disappear, becoming submerged seamounts. The evolution of shorelines is therefore closely dependent on the evolution of local tectonic processes.

References

Allard, M. & Tremblay, G. (1983). La dynamique littorale des îles Manitounuk durant l'Holocène. *Zeitschrift für Geomorphologie, Suppl. Bd.*, **47**, 61–95.

American Geological Institute (1960). *Glossary of geology and related sciences*, 2nd edn. Washington, DC.

Andrews, J.T. (1987). Glaciation and sea level: a case study. In *Sea surface studies: a global view*, ed. R.J.N. Devoy, pp. 95–126. London: Croom Helm.

Andrews, J.T. & Falconer, G. (1969). Late glacial and post-glacial history and emergence of the Ottawa Islands, Hudson Bay, Northwest Territories: evidence on the deglaciation of Hudson Bay. *Canadian Journal of Earth Science*, **6**, 1263–76.

Atwater, B.F. (1987). Evidence for great Holocene earthquakes along the outer coast of Washington State. *Science*, **236**, 942–4.

Barnett, T.P. (1983). Recent changes in sea level and their possible causes. *Climatic Change*, **5**, 15–38.

Berrino, G., Corrado, G., Luongo, G. & Toro, B. (1984). Ground deformation and gravity changes accompanying the 1982 Pozzuoli uplift. *Bulletin Volcanologique*, **47-2**, 187–200.

Berryman, K.R., Hull, A.G., Beanland, S., Pillans, B. & Deely, J. (1992). Excursion to Cape Turakirae. In *IGCP 274 Programme and Abstracts, Annual Meeting, Wellington, New Zealand 1992*, pp. 64–9. Wellington: Geological Society of New Zealand miscellaneous publication 65A.

Berryman, K.R., Ota, Y. & Hull, A.G. (1989). Holocene paleoseismicity in the fold and thrust belt of the Hikurangi subduction zone, eastern North Island, New Zealand. *Tectonophysics*, **163**, 185–95.

Bloom, A.L. (1971). Glacial-eustatic and isostatic controls of sea level since last glaciation. In *The Late Cenozoic glacial ages*, ed. K.K. Turekian, pp. 355–79. New Haven, CT: Yale University Press.

Bourcart, J. (1949). *Géographie du fond des mers*. Paris: Payoy.

Brown, L.D., Reilinger, R.E., Holdahl, S.R. & Balazs, E.I. (1977). Postseismic crustal uplift near Anchorage, Alaska. *Journal of Geophysical Research*, **82**, 3369–78.

Calmant, S. & Cazenave, A. (1986). The effective elastic lithosphere under the Cook-Austral and Society islands. *Earth and Planetary Science Letters*, **77**, 187–202.

Campbell, J.F. (1986). Subsidence rates for the southeastern Hawaiian Islands determined from submerged terraces. *Geo-Marine Letters*, **6**, 139–46.

Carobene, L. & Dai Pra, G. (1990). Genesis, chronology and tectonics of the Quaternary marine terraces of the Tyrrhenian coast of northern Calabria (Italy): their correlation with climatic variations. *Il Quaternario*, **3**, 75–94.

Carter, R.W.G. (1988). *Coastal environments*. London: Academic Press.

Castle, R.O., Church, J.P. & Elliott, M. (1976). Aseismic uplift in southern California. *Science*, **192**, 251–3.

Cazenave, A., Lago, B., Dominh, K. & Lambeck, K. (1980). On the response of the ocean lithosphere to sea-mount loads from *Geos 3* satellite radar altimeter observations. *Geophysical Journal of the Royal Astronomical Society*, **63**, 233–52.

Clark, J.A., Farrell, W.E. & Peltier, W.R. (1978). Global changes in postglacial sea level: a numerical calculation. *Quaternary Research*, **9**, 265–78.

Coudray, J. & Montaggioni, L. (1982). Coraux et récifs coralliens de la province indo-pacifique: répartition géographique et altitudinale en relation avec la tectonique globale. *Bulletin de la Société Géologique de France*, **24**, 981–93.

Daly, R.A. (1934). *The changing world of the ice age.* New Haven: Yale University Press.

Darienzo, M.E. & Peterson, C.D. (1990). Episodic tectonic subsidence of late Holocene salt marshes, northern Oregon central Cascadia margin. *Tectonics,* **9**, 1–22.

Darwin, C. (1842). *The structure and distribution of coral reefs.* London: Smith, Elder & Co.

Darwin, C. (1891). *Geological observations.* London: Smith, Elder and Co.

Detrick, R.S. & Crough, S.T. (1978). Island subsidence, hot spots and lithospheric thinning. *Journal of Geophysical Research,* **83**, 1236–44.

Dubois, J., Launay, J. & Recy, J. (1974). Uplift movements in New Caledonia–Loyalty Islands area and their plate tectonics interpretation. *Tectonophysics,* **24**, 133–50.

Dumas, B., Guérémy, P., Hearty, P.J., Lhénaff, R. & Raffy, J. (1988). Morphometric analysis and amino acid geochronology of uplifted shorelines in a tectonic region near Reggio Calabria, south Italy. *Palaeogeography, Palaeoclimatology, Palaeoecology,* **68**, 273–89.

Dumas, B., Guérémy, P., Lhénaff, R. & Raffy J. (1980). Terrasses quaternaires soulevées sur la façade calabraise du Détroit de Messine (Italie). *Comptes Rendus de l'Académie des Sciences,* Paris, D, **290**, 739–42.

Fairbridge, R.W. (1961). Eustatic changes in sea level. *Physics and Chemistry of the Earth,* **4**, 99–185.

Flemming, N.C. (1969). Archaeological evidence for eustatic change of sea level and earth movements in the western Mediterranean during the last 2000 years. *Geological Society of America, Special Paper,* **109**, 1–125.

Gaposchkin, E.M. (1973). *Smithsonian standard Earth. SAO Special Report, 353.* Washington, DC: Smithsonian Institution, Astrophysical Observatory.

Grindley, G.W. (1974) Relation of volcanism to earth movements, Bay of Naples, Italy. In *Proceedings of the Symposium on 'Andean and Antarctic Volcanology Problems',* Santiago, September 1974, pp. 598–612. Naples: Giannini.

Guilcher, A. (1988). *Coral reef geomorphology.* Chichester: Wiley.

Günther, R.T. (1903). Earth-movements in the Bay of Naples. *Geographical Journal,* **22**, 121–49 + 269–89.

Hamilton, E.L. (1957). Marine geology of the southern Hawaiian ridge. *Bulletin of the Geological Society of America,* **68**, 1011–26.

Hillaire-Marcel, C. (1976). La déglaciation et le relèvement isostatique sur la côte Est de la baie d'Hudson. *Cahiers de Géographie du Québec,* **20**, 185–220.

Hillaire-Marcel, C. (1980). Multiple component postglacial emergence, Eastern Hudson Bay, Canada. In *Earth rheology, isostasy and eustasy,* ed. N.A. Mörner, pp. 215–30. New York: Wiley.

Inman D.L. & Nordstrom, C.E. (1971). On the tectonic and morphologic classification of coasts. *Journal of Geology,* **79**, 1–21.

Jouannic, C., Taylor, F.W. & Bloom, A.L. (1982). Sur la surrection et la déformation d'un arc jeune: l'arc des Nouvelles-Hébrides. In *Contribution à l'Etude Géodynamique du Sud-Ouest Pacifique,* Travaux et Documents ORSTOM, 147, pp. 223–46. Paris: ORSTOM.

Kaizuka, S. (1992). Coastal evolution at a rapidly uplifting volcanic island: Iwa-Jima, western Pacific Ocean. *Quaternary International,* **15/16**, 7–16.

Kaizuka, S., Matsuda, T., Nogami, M. & Yonekura, N. (1973). Quaternary tectonic and recent seismic crustal movements in the Arauco Peninsula and its environs, central Chile. *Geographical Reports of the Tokyo Metropolitan University,* **8**, 1–49.

Kaizuka, S., Miyauchi, T. & Nagaoka, S. (1983). Marine terraces, active faults and tectonic history of Iwo-jima (in Japanese, with English abstract). *Ogasawara Research*, **9**, 13–45.

Kelletat, D. (1991). The 1550 BP tectonic event in the eastern Mediterranean as a basis for assessing the intensity of shore processes. *Zeitschrift für Geomorphologie, Supplement*, **81**, 181–94.

Kelletat, D. & Kayan, I. (1983). First C^{14} datings and Late Holocene tectonic events on the Mediterranean coastline, west of Alanya, southern Turkey (in Turkish, with English abstract). *Türkiye Jeoloji Kurumu Bülteni*, **C26**, 83–7.

Keraudren, B. (1975). Essai de stratigraphie et de paléogéographie du Plio-Pléistocène égéen. *Bulletin de la Société Géologique de France*, **17**, 1110–20.

Keraudren, B. & Sorel, D. (1984). Relations entre sédimentation, tectonique et morphologie dans le Plio-Pléistocène de Karpathos (Grèce): mouvements verticaux et datations radiométriques. L'Antropologie (Paris), **88**, 49–61.

Koba, M., Ikeya, T., Miki T. & Nakata, T. (1985). ESR ages of the Pleistocene coral reef limestones in the Ryukyu Islands, Japan. In *First International Symposium on ESR Dating, Ube City, Japan*, Tokyo: Ionics Publishing Co. 12 pp.

Konishi, K., Omura, A. & Nakamichi, O. (1974). Radiometric coral ages and sea-level records from the late Quaternary reef complexes of the Ryukyu Islands. In *Proceedings Second International Coral Reef Symposium, Brisbane*, pp. 595–613.

Labeyrie, J., Lalou, C. & Delibrias, G. (1969). Etude des transgressions marines sur l'atoll de Mururoa par la datation des différents niveaux de corail. *Cahiers du Pacifique*, **13**, 59–68.

Lajoie, K.R. (1986). Coastal tectonics. In *Active Tectonics*, Studies in Geophysics, pp. 95–124. Washington DC: National Academy Press.

Lambeck, K. (1981). Lithospheric response to volcanic loading in the Southern Cook Islands. *Earth and Planetary Science Letters*, **55**, 482–96.

Liew, P.M. & Lin, C.F. (1987). Holocene tectonic activity of the Hengchun Peninsula as evidenced by the deformation of marine terraces. *Memoirs of the Geological Society of China*, **9**, 241–59.

Luongo, G. & Scandone, R., eds. (1991). Campi Flegrei. *Journal of Volcanology and Geothermal Research*, **48**, 1–227.

Marsh, J.G. & Martin, T.V. (1982). The Seasat altimeter mean sea surface model. *Journal of Geophysical Research*, **87**, 3269–80.

Martin, L., Flexor, J.M., Blitzkow, D. & Suguio, K. (1985). Geoid change indications along the Brazilian coast during the last 7000 years. In *Proceedings Fifth International Coral Reef Congress*, Tahiti, vol. 3, pp. 85–90.

McNutt, M. & Menard, H.W. (1978). Lithospheric flexure and uplifted atolls. *Journal of Geophysical Research*, **83**, 1206–12.

Menard, H.W. (1973). Depth anomalies and the bobbing motion of drifting islands. *Journal of Geophysical Research*, **78**, 5128–37.

Menard, H.W. (1986). *Islands*. Scientific American Library. New York: Freeman & Co.

Menard, H.W. & Ladd, H.P. (1963). Oceanic islands, sea-mounts, guyots and atolls. In *The sea*, ed. M.N. Hill, pp. 365–87. New York: Interscience Publishers.

Montaggioni, L.F. (1985). Makatea Island, Tuamotu Archipelago. In *Proceedings of the Fifth International Coral Reef Congress, Tahiti*, vol. 1, pp. 103–58.

Montaggioni, L.F. (1989). Le soulèvement polyphasé d'origine volcano-isostasique: clef de l'évolution post-oligocène des atolls du Nord-Ouest des Tuamotus (Pacifique central). *Comptes Rendus de l'Académie des Sciences, Paris*, II, **309**, 1591–8.

Mörner, N.A. (1976). Eustasy and geoid changes. *Journal of Geology*, **84**, 123–51.

Mörner, N.A. (1988). Terrestrial variations within given energy, mass and momentum budgets: paleoclimate, sea level, paleomagnetism, differential rotation and geodynamics. In *Secular solar and geomagnetic variations in the last 10,000 years*, ed. F.R. Stephenson & A.W. Wolfendale, pp. 455–78. Dordrecht: Kluwer Academic Publishers.

Mutti, E., Orombelli, G. & Pozzi, R. (1970). Geological studies on the Dodecanes Islands (Aegean Sea). IX. Geological map of the island of Rhodes (Greece), explanatory notes. *Annales Géologiques des Pays Helléniques (Athens)*, **22**, 77–226.

Nakada, M. & Lambeck, K. (1987). Glacial rebound and relative sea-level variations: a new appraisal. *Geophysical Journal of the Royal Astronomical Society*, **90**, 171–224.

Nakada, M., Yonekura, N. & Lambeck, K. (1991). Late Pleistocene and Holocene sea-level changes in Japan: implications for tectonic histories and mantle rheology. *Palaeogeography, Palaeoclimatology, Palaeoecology*, **95**, 107–22.

Nakata, T., Koba, M., Jo, W., Imaizumi, T., Matsumoto, H. & Suganuma, T. (1979). Holocene marine terraces and seismic crustal movement. *Science Reports of the Tohoku University., 7th Series (Geography)*, **29**, 195–204.

Nunn, P.D. (1986). Implications of migrating geoid anomalies for the interpretation of high-level fossil coral reefs. *Geological Society of America Bulletin*, **97**, 946–52.

Nunn, P.D. (1988). Plate boundary tectonics and oceanic island geomorphology. *Zeitschrift für Geomorphologie, Suppl. Bd.*, **69**, 39–53.

Omura, A., Tsuji, Y., Ohmura, K. & Sakuramoto, Y. (1985). New data on Uranium-series ages of Hermatypic corals from the Pleistocene limestone on Kikai, Ryukyu Islands. *Transactions and Proceedings of the Palaeontological Society of Japan*, NS, **139**, 196–205.

Ota, Y. (1975). Late Quaternary vertical movement in Japan estimated from deformed shorelines. *Royal Society of New Zealand, Bulletin*, **13**, 231–9.

Ota, Y. (1985). Marine terraces and active faults in Japan with special reference to coseismic events. In *Tectonic geomorphology*, ed. M. Morisawa & J.T. Hack, pp. 345–66. Boston: Allen & Unwin.

Ota, Y. (1991). Coseismic uplift in coastal zones of the western Pacific rim and its implications for coastal evolution. *Zeitschrift für Geomorphologie, Suppl. Bd.*, **81**, 163–79.

Ota, Y., Hull, A.G. & Berryman, R. (1991a). Coseismic uplift of Holocene marine terraces in the Pakarae River area, eastern North Island, New Zealand. *Quaternary Research*, **35**, 331–46.

Ota, Y., Miyauchi, T. & Hull, A.G. (1990). Holocene marine terraces at Aramoana and Pourerere, eastern North Island, New Zealand. *New Zealand Journal of Geology and Geophysics*, **33**, 541–6.

Ota, Y. & Omura A. (1992). Contrasting styles and rates of tectonic uplift of coral reef terraces in the Ryukyu and Daito Islands, southwestern Japan. *Quaternary International*, **15/16**, 17–29.

Ota, Y., Williams, D.N. & Berryman, K.R. (1981). *Late Quaternary tectonic map of New Zealand 1: 50 000. Parts Sheets Q27, R27 and R28, Wellington*. Wellington: New Zealand Geological Survey.

Peltier, W.R. (1990). Glacial isostatic adjustment and relative sea-level change. In *Sea-level change*, Geophysics Study Committee, National Research Council, pp. 73–87. Washington, DC: National Academy Press.

Peltier, W.R. (1991). The ICE-3G model of late Pleistocene deglaciation: construction, verification, and applications. In *Glacial isostasy, sea-level and mantle rheology*, NATO Advanced Science Institutes Series, C, 334, ed. R. Sabadini, K. Lambeck & E. Boschi, pp. 95–119. Dordrecht: Kluwer Academic Publishers.

Peltier, W.R. & Andrews, J.T. (1983). Glacial geology and glacial isostasy of the Hudson Bay region. In *Shorelines and isostasy*, ed. D.E. Smith & A.G. Dawson, pp. 285–319. London: Academic Press.

Pirazzoli, P.A. (1991). *World atlas of Holocene sea-level changes*. Elsevier Oceanography Series, 58. Amsterdam: Elsevier.

Pirazzoli, P.A. & Grant, D.R. (1987). Lithospheric deformation deduced from ancient shorelines. In *Recent plate movement and deformation*, ed. K. Kasahara, pp. 67–72. American Geophysical Union & Geological Society of America, Geodynamics Series, 20.

Pirazzoli, P.A., Koba, M., Montaggioni, L.F. & Person, A. (1988). Anaa (Tuamotu Islands, central Pacific): an incipient rising atoll? *Marine Geology*, **82**, 261–9.

Pirazzoli, P.A., Laborel, J., Saliège, J.F., Erol, O., Kaian, I. & Person, A. (1991). Holocene raised shorelines on the Hatay coasts (Turkey): paleoecological and tectonic implications. *Marine Geology*, **96**, 295–311.

Pirazzoli, P.A. & Montaggioni, L.F. (1985). Lithospheric deformation in French Polynesia (Pacific Ocean) as deduced from Quaternary shorelines. In *Proceedings Fifth International Coral Reef Congress, Tahiti*, vol. 3, pp. 195–200.

Pirazzoli, P.A., Montaggioni, L.F., Saliège, J.F., Segonzac, G., Thommeret, Y. & Vergnaud-Grazzini C. (1989). Crustal block movement from Holocene shorelines: Rhodes Island (Greece). *Tectonophysics*, **170**, 89–114.

Pirazzoli, P.A. & Salvat, B. (1992). Ancient shorelines and Quaternary vertical movements on Rurutu and Tubuai (Austral Isles, French Polynesia). *Zeitschrift für Geomorphologie*, **36**, 431–51.

Pirazzoli, P.A. & Veeh, H.H. (1987). Age ^{230}Th/^{234}U d'une encoche émergée et vitesses de soulèvement quaternaire à Rurutu, îles Australes. *Comptes Rendus de l'Académie des Sciences, Paris*, II, **305**, 919–23.

Plafker, G. & Rubin, M. (1967). Vertical tectonic displacements in south-central Alaska during and prior to the great 1964 earthquake. *Journal of Geosciences*, Osaka City University, **10**, 53–66.

Rice, C.M. (1941). *Dictionary of geological terms*. Ann Arbor: Edwards.

Sabadini, R., Doglioni, C. & Yuen, D.A. (1990). Eustatic sea level fluctuations induced by polar wander. *Nature*, **345**, 708–10.

Savage, J.C. & Plafker, G. (1991). Tide gage measurements of uplift along the south coast of Alaska. *Journal of Geophysical Research*, **96**, 4325–35.

Scott, G.A.J. & Rotondo, G.M. (1983a). A model for the development of types of atolls and volcanic islands on the Pacific lithospheric plate. *Atoll Research Bulletin*, **260**, 1–33.

Scott, G.A.J. & Rotondo, G.M. (1983b). A model to explain the differences between Pacific plate island atoll types. *Coral Reefs*, **1**, 139–50.

Spencer, T. (1989). Tectonic and environmental histories in the Pitcairn Group, Paleogene to Present: reconstructions and speculations. *Atoll Research Bulletin*, **322**, 1–22.

Spencer, T. & Paulay, G. (1989). Geology and geomorphology of Henderson Island. *Atoll Research Bulletin*, **323**, 1–18.

Stiros, S.C., Arnold, M., Pirazzoli, P.A., Laborel, J., Laborel, F. & Papageorgiou S. (1992). Historical coseismic uplift in Euboea Island (Greece). *Earth and Planetary Science Letters*, **108**, 109–17.

Suess, E. (1900). *La Face de la Terre* (French translation of: Das Antlitz der Erde), t. II, pp. 598–638. Paris: Colin.

Taylor, F.W., Isacks, B.L., Jouannic, C., Bloom, A.L. & Dubois, J. (1980). Coseismic and Quaternary vertical tectonic movements, Santo and Malekula Islands, New Hebrides Island arc. *Journal of Geophysical Research*, **85**, 5367–81.

Thommeret, Y., Thommeret, J., Laborel, J., Montaggioni, L.F. & Pirazzoli, P.A. (1981). Late Holocene shoreline changes and seismo-tectonic displacements in western Crete (Greece). *Zeitschrift für Geomorphologie, Suppl. Bd.*, **40**, 127–49.

Valensise, G. & Pantosti, D. (1992). A 125 kyr-long geological record of seismic source repeatability: the Messina Straits (southern Italy) and the 1908 Earthquake (Ms $7\frac{1}{2}$). *Terra Nova*, **4**, 472–83.

Valensise, G. & Ward, S.N. (1991). Long-term uplift of the Santa Cruz coastline in response to repeated earthquakes along the San Andreas Fault. *Bulletin of the Seismological Society of America*, **81**, 1–11.

Veeh, H.H. (1966). Th^{230}/U^{238} and U^{234}/U^{238} ages of Pleistocene high sea level stand. *Journal of Geophysical Research*, **71**, 3379–86.

Vita-Finzi, C. (1986). *Recent earth movements*. London: Academic Press.

Walcott, R.I. (1970). Flexure of the lithosphere at Hawaii. *Tectonophysics*, **9**, 435–46.

Wellman, H.W. (1967). Tilted marine beach ridges at Cape Turakirae, New Zealand. *Journal of Geosciences*, Osaka City University, **10**, 123–9.

Woodroffe, C.D. (1988). Vertical movement of isolated oceanic islands at plate margins: evidence from emergent reefs in Tonga (Pacific Ocean), Cayman Islands (Caribbean Sea) and Christmas Island (Indian Ocean). *Zeitschrift für Geomorphologie, Suppl. Bd.*, **69**, 17–37.

Woodroffe, C.D., Stoddart, D.R., Spencer, T., Scoffin, T.P. & Tudhope, A.W. (1990). Holocene emergence in the Cook Islands, South Pacific. *Coral Reefs*, **9**, 31–9.

Yokoyama, I (1971). Pozzuoli event in 1970. Nature, **229**, 532–3.

Yonekura, N. (1972). A review on seismic crustal deformations in and near Japan. *Bulletin of the Department of Geography, University of Tokyo*, **4**, 17–50.

Yonekura, N. (1975). Quaternary tectonic movements in the outer arc of southwest Japan with special reference to seismic crustal deformations. *Bulletin of the Department of Geography, University of Tokyo*, **7**, 19–71.

Yoshikawa, T., Kaizuka, S. & Ota, Y. (1964). Mode of crustal movement in the late Quaternary on the southeast coast of Shikoku, southwestern Japan (in Japanese with English abstract). *Geographical Review of Japan*, **37**, 27–46.

Yoshikawa, T., Kaizuka, S. & Ota, Y. (1981). *The landforms of Japan*. Tokyo: University of Tokyo Press.

13

Developed coasts

K.F. NORDSTROM

Introduction

Modification of the coast by humans has occurred over countless centuries (Walker, 1984; 1988), but direct alteration of exposed ocean coasts on a massive scale is a relatively recent phenomenon. Large-scale alterations began in the nineteenth century, when there was a need to accommodate vessels of large burden and deep water drafts and there was steam power to enable large modifications (Marsh, 1885; De Moor & Bloome, 1988; Terwindt, Kohsiek & Visser, 1988). Development of the coastlines of the world has accelerated in the last few decades (Wong, 1985, 1988; Koike, 1988; Moutzouris & Marouikian 1988; Cencini & Varani, 1988). Marco Island, on the west coast of Florida, progressed from a wilderness to a fully developed shoreline in fewer than 20 years (Reynolds, 1987). Many coastal communities in the USA experienced dramatic growth in this period, with construction of high-rise condominiums (Carter, 1982; Leatherman, 1987; Schmahl & Conklin, 1991). By the mid-1970s 1687 km (or 37%) of the ocean frontage of the coastal barriers on the US Atlantic and Gulf coasts were occupied by buildings, roads and related features (Lins, 1980). Strip development near the ocean shoreline has historically dominated the land conversion process in coastal communities (Mitchell, 1987), placing the location of much of the development where it is readily affected by wave and wind processes. Shore protection structures are emplaced seaward of these developments where they can have the greatest impact on shoreline processes and the most dynamic coastal landforms.

The proportion of coastline that is now protected against flooding and erosion or that is significantly affected by human development is estimated at 12.9% in Italy (Cencini & Varani, 1988), 20.9% in Korea (Park, 1988), 38.5% in England (Carr, 1988), 40.0% in Japan (Koike, 1988), and nearly the entire coast of Belgium is protected by seawalls (De Moor & Bloome, 1988). Many communities that are only partially stabilized are *en route* to total stabilization

477

(Pilkey & Wright, 1988), and undeveloped shorelines that are adjacent to developed and protected shorelines are often profoundly affected by human-induced sediment starvation (Leatherman, 1984; Nordstrom, 1987a). Communities that have been severely damaged by coastal storms are often rebuilt to larger proportions (Fischer, 1989) (Fig. 13.1). Some areas have been redesigned and rebuilt as human artifacts and bear litttle resemblance to the coast that formerly existed (Nagao, 1991; Nagao & Fujii, 1991).

Many of the interactions between human activities and coastal processes have been studied intensively. The engineering literature on the relationship between structures, coastal processes and beach change is vast, although much of this research is concerned with design of structures rather than the effects of structures after they are emplaced. There is a large body of literature on the effects of jetties on adjacent beaches (Kieslich, 1981; Dean, 1988; Hansen & Knowles, 1988) and a growing body of literature of field-orientated studies examining the effects of shore protection strategies at small temporal and spatial scales, including the effects of groynes (Sherman *et al.*, 1990; Bauer *et al.*, 1991), seawalls (Kraus & Pilkey, 1988; Plant & Griggs, 1992) and beach nourishment (Schwartz & Bird, 1990). Many studies have been conducted into recreational impacts on coastal dunes (Vogt, 1979). Study of the effects of buildings or activities of shorefront residents on the morphology of the coast is

Figure 13.1. Gulf Shores, Alabama, 1981, showing reconstruction after Hurricane Frederick. The slogan during reconstruction 'Gulf Shores: Bigger and Better' reflects the common practice of increasing the level of development after storm damage.

also represented, with most geomorphic studies on effects on coastal dunes (e.g. Gares, 1983, 1990; Nordstrom & McCluskey 1985; Nordstrom, McCluskey & Rosen, 1986; Nordstrom, 1988a). Insight into coastal changes at larger spatial scales is provided in case studies of effects of a specific type of protection strategy over several kilometres of shoreline or in several different locations, including assessments of groyne fields (Everts, 1979), seawalls and bulkheads (Fulton-Bennett & Griggs, no date) and beach nourishment (Leonard, Clayton & Pilkey, 1990; Houston, 1991a). There are also case studies of regional effects of several different kinds of protection structure (Morton, 1979; Carter, Benson & Guy, 1981; Carter, Monroe & Guy,1986; Zabawa, Kerhin & Bayley, 1981; Nordstrom, 1988b; Hall & Pilkey, 1991), and the effect of storms on human structures (Griggs & Johnson, 1983; Finkl & Pilkey, 1991).

The understanding of the relationship between human activities and coastal evolution is far from adequate despite these previous studies. Most investigations have a purpose other than to determine geomorphological evolution; studies on beach nourishment, for example, usually focus on rates of sediment loss or economic considerations, rather than on landform assemblages associated with the new sediment (Houston, 1991b; Schmahl & Conklin, 1991). The lack of scientific interest in the evolution of coasts parallels the lack of interest in the evolution of many other developed physical systems (Gregory, 1985). Many geomorphologists may not want to study human developed systems because: they are accustomed to examining landscape evolution at greater time scales than evolve under developed conditions; they are deterred by the magnitude of the problem of isolating cause and effect; or they think that geomorphological principles are inappropriate where the dominant agent of landform change is earth-moving machinery. In such cases, human alterations are viewed as an aberration, rather than an integral component of landscape evolution.

Fundamental studies of the evolution of natural coasts are of considerable scientific interest, but managerial interests require baseline information to assess the numerous physical, chemical and environmental problems along our shorelines, and it would be a mistake to neglect areas that have already been developed. Coasts that are subject to rapid and largely uncontrolled occupation are often the most at risk from environmental change in the near future, yet paradoxically, it is these coasts that we often know least about. Although many people visit developed shorelines, the scientist often rebuffs them, preferring to seek out sites that are, superficially, more 'natural'. This chapter attempts, in a small way, to redress the balance of the rest of this volume, by exploring the geomorphological evolution of developed coasts.

The role of humans in the evolution of a developed coast is examined here by comparing landform characteristics on an open-ocean barrier island coast prior to extensive development with landform characteristics and evolutionary trends under developed conditions. Specific examples are provided of the evolution of the New Jersey coast (Fig. 13.2). A barrier island setting was selected because of the strong dependence of geomorphic features on coastal processes. New Jersey was selected because this location has the longest history of development and stabilization of any barrier island coast in the USA (Mitchell, 1987; Pilkey & Wright, 1988; Hall & Pilkey, 1991). Information on the development process is provided for the entire exposed ocean coast, but evolutionary trends are examined in detail on the barrier spits and islands (termed barriers). Human adjustment dominates the landscape on some of the barriers (Figs. 13.3 and 13.4) or has profoundly altered processes, sediment budgets or landforms on other barriers. The barriers in New Jersey share common characteristics with other coastal barriers in the USA, including

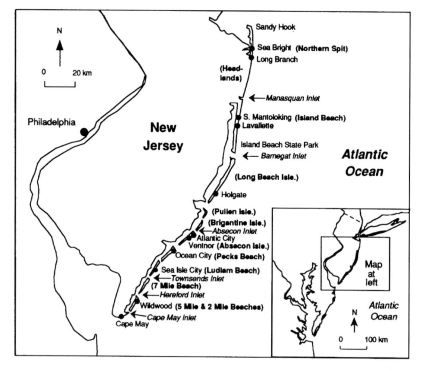

Figure 13.2. New Jersey shoreline study area. Dots represent locations of towns; arrows identify locations of inlets; names in bold print in parenthesis identify barriers for which data are presented in Table 13.1.

Figure 13.3. Atlantic City, NJ, on Absecon Island, showing pronounced human influence on the coastal landcape of an intensively developed barrier island.

Figure 13.4. The barrier south of Sea Bright, NJ, showing the seawall that prevents an inlet from forming at a location where inlets have formed and closed frequently in the past.

Table 13.1. *Characteristics of the New Jersey shoreline in 1885–86 and on the most recent maps and photos*

Location and year of map	Dune characteristics		Width of upland (m)		Minimum width of contiguous upland/marsh (m)
	Highest elevation	Maximum inland extent	Maximum	Minimum	
North spit					
1886	7.6	375	1120	30	30
1981	ND	ND	390 (1193)	60	60
Headlands					
	NA	NA	NA	NA	NA
Island Beach					
1886	4.6	610	920	130	140
1989	7.6 (10.7)	120 (580)	1180 (570)	240 (97)	240 (140)
Long Beach Island					
1886	9.1	600	720	140	140
1972	10.7 (4.0)	270 (280))	1310 (730)	180 (270)	180 (360)
Pullen Island					
1886	3.0	260	740	280	280
1989	6.1	260	760	0	234
Brigantine Island					
1885	4.6	470	530	100	260
1989	6.1 (3.0)	370 (130)	2070 (320)	340 (105)	340 (315)
Absecon Island					
1885	8.5	560	1270	130	230
1989	None	None	2980	190	190
Pecks Beach					
1886	6.1	370	920	150	550
1972, 1989	4.6 (4.0)	190 (480)	1350 (490)	290 (280)	710 (280)

Ludlam Beach					
1886	5.5	600	720	60	340
1972	8.2	200	940	150	290
7 Mile Beach					
1886	13.1	680	960	270	500
1972	14.6	280	980 (210*)	390 (20*)	450 (100*)
5 Mile Beach					
1886	4.6	690	770	320 (50)	870
1972	None	None	2120	790	790
2 Mile Beach					
1886	4.6	190	480	160	420
1972	4.6	390	1120	390	1340
Cape May					
1886	None	None	280	60	160
1972	None	None	1034	720	720

All numbers from 1885–6 are taken from undeveloped locations. Numbers from most recent maps and photos are for developed portions of the barriers; numbers in parenthesis are taken in undeveloped portions of developed barriers. All distances have been rounded to nearest 10 m. Elevations have been converted from feet to metres.

Most recent widths of upland (all land above marsh elevation) and upland/marsh were taken from 1986 air photos at sites where the most recent map is 1972.

*The undeveloped part of the barrier here at Stone Harbor Point has been eliminated since 1986 due to erosion (see text).

Rockaway, New York, Ocean City, Maryland, Miami Beach, Florida and Galveston, Texas, that have been developed for amenity housing, beach-orientated recreation and related services. Most contain little agriculture or woodland, and most lack industrial plants or commercial ports for use other than fishing (Mitchell, 1987).

Methodology

The effects of development are assessed by comparing the dimensions, configurations, topography, surface cover and mobility of the New Jersey barriers prior to major human occupance with conditions represented on the most recent maps and air photos. The first accurate maps of topography along the New Jersey coast were drawn in 1885–86 using data from surveys conducted the previous decade. The maps are at a scale of 1:63 360; marsh areas are delineated and the contour interval of 5 feet (1.52 m) permits evaluation of dune characteristics. Data on characteristics of the coastline between 1885–86 and 1934 were also taken from 1:63 360 scale maps. Data for conditions after 1934 were determined using 1:9600 and 1:24 000 scale vertical air photos and 1:24 000 scale topographic maps. Information on dates of construction of shore protection structures was taken from federal documents (US Army Corps of Engineers (USACOE), 1954, 1957, 1990; US Congress, 1976) and historical studies (Gares, 1983). The effect of a major storm on the developed coastal barriers was determined by identifying changes caused by the 6–7 March 1962 northeaster, which was the most damaging storm on record, and comparing the effects of the storm with subsequent human modifications to reconstruct the coastal landscape. Information on the likelihood of implementation of future protection strategies was derived from recent planning documents (New Jersey Department of Environmental Protection (NJDEP), 1981; USACOE, 1989a, b, c) supplemented by interviews with key personnel in the State and the US Army Corps of Engineers.

Characteristics of the New Jersey coast

The New Jersey ocean coast (Fig. 13.2) is approximately 205 km long and consists of a northern barrier spit; a headland 4.5 to 7.5 m high, composed of unconsolidated sediments; a southern barrier spit and barrier island complex; and a short southern headland near Cape May that is composed of Pleistocene sand and gravel. Mean tidal range is about 1.3 m along the open coast (National Oceanic and Atmospheric Administration (NOAA), 1991). The rates of sea-level rise at Sandy Hook and Atlantic City, based on trends between 1940 and

1970, are 4.6 mm a^{-1} and 2.8 mm a^{-1} (NJDEP, 1981). Dominant winds blow from the northwest, although storm waves generated by northeasterly winds account for the dominance of southerly drift south of Barnegat Inlet (Fig. 13.2). The northern spit and headlands are sheltered from the full effects of these waves by Long Island, northeast of the study area (Fig. 13.2, inset), and net drift is to the north. Littoral drift rates increase to the north and south of a nodal zone located between Manasquan and Barnegat Inlets (Ashley, Halsey & Buteux, 1986). The drift rate at Sandy Hook is estimated at 382 000 m^3 a^{-1} to the north (Caldwell, 1967); the drift rate at Cape May is estimated at 190 000 m^3 a^{-1} to the south (US Congress, 1976). Annual average significant wave height is 0.8 m with an average wave period of 8.3 s (Thompson, 1977). Grain sizes of native beach sediments decrease from coarse sand in the northern headlands to fine sand in the south (McMaster, 1954).

Human modifications on the New Jersey coast

Conditions prior to extensive human development

Native Americans used the shoreline to fish and gather shells during the warmer months, but they do not appear to have occupied the barriers (Koedel,1983). Only a few hunting cabins, homesteads and boarding houses existed on the islands before construction of railroads in the mid and late nineteenth century; the barrier islands were used for livestock grazing, timber, whaling, farming and shell-fishing (Sea Isle City, 1982; Koedel, 1983). Atlantic City was the first major barrier island resort, being built up rapidly after a rail line was completed in July 1854.

Most of the barriers were low and narrow prior to development and they were backed by marsh deposits colonized principally by *Spartina alterniflora* and *Spartina patens*. Some of the marshes appear to have formed on locations of overwash of sediment from the ocean to the bay or former flood deltas. The upland portions of the barriers were at locations of dunes, former spit recurves and zones of overwash that covered former marsh surfaces. The widths of the upland portions of the barriers (Table 13.1, columns 4 and 5) varied from a maximum of over 1000 m at former spit recurves to minimums of <100 m where wave processes had recently closed inlets through longshore sediment transport. Multiple dune ridges were common on portions of several of the barriers. The foredune appeared to take the form of a broad-based ridge along much of Absecon Island and on Seven Mile Beach (Fig. 13.2) where dunes were highest (Table 13.1, column 2), but large portions of most barriers were characterized by isolated hummocky dunes and 28.3% of the barriers appeared to have no dunes at all. Dunes were least well-developed near active and recently active inlets that had closed naturally. Vegetation on the upland portion of the barriers was characterized by lush growth of cedar, holly and other trees and a variety of grasses (Sea Isle City, 1982). The coastal barriers appeared to be highly mobile prior to human development. State Geologist G.H. Cook, cited in USACOE (1957), noted in 1882 that stumps of trees felled with a metal axe and hoof prints of shod horses in peat that were found exposed on the beach at Long Beach Island after a storm indicated that the island migrated across its width in no less than 200 years. Recent dating of peat buried offshore of Ocean City indicates a 2000 m inland shift of that barrier in the past 5730 years (Psuty, 1986).

Construction of buildings and support infrastructure

Growth of coastal resorts was rapid following construction of the railroads. Atlantic City had a permanent population of nearly 5500 in 1880 (Funnel, 1975), and there were 650 hotels and boarding houses by 1891 (Reynolds, 1981). Railroad lines extended along 57.8% of the New Jersey shoreline in 1885–86, although roads extended along only 19.1% of the barriers. Four of the barriers lacked access by either road or rail at that time. The barrier island resort industry developed rapidly in the early twentieth century, due primarily to increased use of private automobiles (Koedel, 1983).

The first buildings were constructed on the upland portions of the barriers at inlets and at locations where railway lines from the mainland first made contact with the barriers. New isolated communities appeared, and growth then extended outward from these locations, both alongshore on the upland portion of the barriers and bayward onto the marsh surface. Dunes were graded to a flatter form to facilitate construction of buildings and roads in the very first stages of development. Most of the natural vegetation was destroyed in the process of this conversion.

Inspection of maps reveals that local filling of the marsh behind the upland occurred on several barriers between 1886 and 1902. The marsh at Ocean City was filled all the way to the edge of the backbay by 1907. The marsh behind the barrier north of Cape May was completely covered to a landward distance of over 700 m by 1913. Dredging of channels into the backbarrier marsh to accommodate boats was first accomplished in the larger settlements south of Ocean City and occurred between 1905 and 1913. Filling of the bay landward of the barriers to accommodate new buildings was never widespread, but fill was used to facilitate road access to bridges to the mainland.

The sequence of development observed at the locations that developed early was followed at later time periods on other barriers. Conspicuous filling of the marsh did not take place near Mantoloking (Fig. 13.2) until the period between 1914 and 1930, when there is the first evidence of lagoon development. Large-scale conversion of the marsh to lagoon developments did not occur on many of the barriers until after World War II, when changes became rapid and widespread (Fig. 13.5). Most of these projects used the existing marsh as substrate and placed materials dredged from the new waterways on the marsh surface. The shapes of the lagoon developments usually mimic the shapes of the marsh in plan view (Fig. 13.5), but the function of the backbarrier changed from a marsh to a residential area. Construction of lagoon housing is now severely restricted by regulations governing use of wetlands, but marshland is now found in only isolated enclaves on the back sides of the developed portions of barriers in New Jersey.

Native plant species are prevented from recolonizing on most of the barriers because of the human preference for using lawn grass and exotic shrubs and trees for landscaping. Many landowners have replaced all vegetation with gravel, using weed killer to maintain an unvegetated surface, and these locations are now barren (Fig. 13.6). Natural vegetation other than foredune communities only remains in a few natural zones that are maintained as preserves.

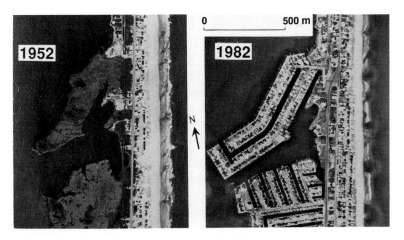

Figure 13.5. South Mantoloking, NJ, showing the construction of lagoon housing on marsh that was common practice prior to environmental regulations in the 1970s.

Figure 13.6. Developed barrier near Barnegat Inlet, NJ, showing the level of development characteristic of most of the barrier islands in New Jersey, the omnipresent groynes and the small dune used as protection. Photo. was taken just after the northeasterly storm March 29, 1984.

Activities at inlets

All but eleven of the inlets that existed in 1885–86 and all inlets that formed since that time were closed artificially by the US Army Corps of Engineers or kept from re-opening after natural closure to eliminate undesirable shoreline fluctuations, facilitate land transportation along the coast, or increase the hydraulic efficiency of nearby controlled inlets. Five of the natural inlets are now stabilized by jetties. Jetty construction confines flow between structures, resulting in abandonment of natural ebb and marginal flood channels, causing wave dominance of adjacent relict ebb tidal delta areas, and resulting in landward migration of bars and elimination of typical ebb tidal deltas (Hubbard, 1975; Hansen & Knowles, 1988). Jetties reduce the rate of shoreline change by preventing inlets from migrating. The average absolute rate of both erosion and accretion decreased dramatically at the inlets after jetties were constructed. Shore-parallel change in the throats of the inlets decreased from an average of $5.6\,\mathrm{m\,a}^{-1}$ to $0.0\,\mathrm{m\,a}^{-1}$ at the three inlets where jetties are closely spaced and from $10.7\,\mathrm{m\,a}^{-1}$ to $4.0\,\mathrm{m\,a}^{-1}$ at Barnegat and Absecon Inlets (Fig. 13.2), where the jetties are wide enough apart for beaches to form between them (Nordstrom, 1987b). The mobility of beaches on the ocean shoreline updrift and downdrift of jetties decreased from an average of $3.8\,\mathrm{m\,a^{-1}}$ to $2.4\,\mathrm{m\,a^{-1}}$ at the five inlets. The net change is unidirectional (accretion updrift, erosion downdrift) rather than bidirectional (erosion–accretion cycles) as occurred before construction of the jetties (Nordstrom, 1987b).

The ebb tidal delta of a jettied inlet can develop to a new equilibrium size (Dean & Walton, 1975; Marino & Mehta, 1988), allowing sediment to pass to the downdrift shoreline. Bypassing is not achieved on most New Jersey inlets with jetties because of the prominence of the structures (which often act as terminal groynes) and frequent channel dredging. Disposal of material dredged from inlets at sea represents a substantial loss of sediment from the longshore transport system (Dean 1988; Marino & Mehta, 1988). Much of the sediment dredged from Manasquan, Absecon and Cape May Inlets has been dumped at sea, although some of this sediment has been placed on adjacent beaches in the past. The volumes dredged from these three inlets following jetty construction have exceeded $5.8 \times 10^6\,\mathrm{m^3}$ (Nordstrom, 1987b).

Shorelines are developed adjacent to four of the inlets that are not stabilized by jetties, and two of these unjettied inlets are maintained by dredging. Dredging at these inlets has changed the amount of sediment transferred across inlets and has influenced the location of accretion and erosion on adjacent shorelines by either changing the location of tidal channels or maintaining them in place, depending on human preference. Maintenance dredging of an existing channel configuration is the more common. This practice keeps the channel from fluctuating as widely as it would under natural conditions, and it reduces the periodicity of, or virtually eliminates, erosion/deposition cycles associated with breaching of the ebb tidal delta. The mobility of developed inlet shorelines in New Jersey can be less than occurred under natural conditions, even in the absence of jetties or dredging, because bulkheads adjacent to inlets truncate shoreline displacement during erosional phases of cycles of accretion and erosion or prevent breaches in the barriers updrift of inlets (Nordstrom, 1988b).

Shore protection projects

Efforts were made to control erosion as early as 1847 at Cape May and 1857 at Atlantic City (USACOE, 1957). The first documented case of groyne construction on the New Jersey shoreline was at Sandy Hook in the 1860s (Bearss, 1976). Structures resembling seawalls (but termed breakwaters) existed at Sandy Hook in 1863 and bulkheads were employed there as early as 1878 (Gares, 1983). Bulkheads existed at Sea Bright in 1880 (Psuty, 1988).

Fig. 13.7 reveals trends in the implementation of shore protection projects since 1900. The data are based on inventories by the Corps of Engineers (USACOE, 1954, 1957, 1990) and include local and state projects as well as federal projects. The totals represent a conservative estimate of the structures emplaced, because other narratives make reference to structures that are not reported in the Corps inventories, and many structures identified in the inventories that were repaired are not counted in the total. The trends reveal a

Groynes

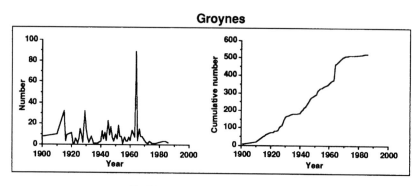

Bulkheads, seawalls, revetments

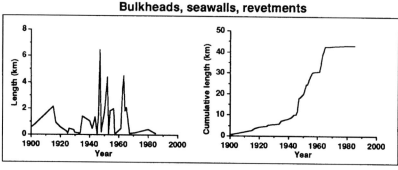

Beach fill

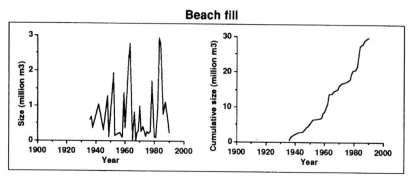

Figure 13.7. Implementation of shore protection projects since 1900.

temporal shift in preference for each of the three strategies, from extensive use of groynes, through a period of extensive construction of shore-parallel structures (bulkheads, seawalls, revetments), to the present emphasis on beach nourishment. A considerable number of projects using all three methods were employed following the storm of 6–7 March 1962 (Fig. 13.7), but groynes and seawalls have been employed sparingly since the 1960s. Beach nourishment is now the preferred alternative, but the groynes and shore-parallel structures remain in place and still have a pronounced effect on shoreline change. Dune building programmes have been implemented in many communities, and there have been demonstration projects for small-scale, non-traditional techniques of erosion control. The effects of the demonstration projects on coastal evolution have been negligible, but dune building programmes have had a significant impact on coastal change and are discussed below.

Shore-parallel structures

Over 43 km of New Jersey shoreline are protected by shore-parallel structures (USACOE, 1990). The majority of them are timber bulkheads, usually 3.0 to 4.0 m above mean low water (MLW), but stone and concrete seawalls and revetments are common. The seawall at Sea Bright (Fig. 13.4) is the largest structure, with a top elevation of 5.8 m above MLW.

Bulkheads are also common on the bay shoreline (Fig. 13.5) because the unconsolidated sand used as fill material to provide substrate for house construction on marshes is mobilized readily by bay waves and currents. These structures are not represented in Fig. 13.7. They are poorly documented because design criteria and details of emplacement are lacking and public funds are rarely required for their construction; the cost of construction is within the financial capability of homeowners, and permission to construct bulkheads was easy to obtain.

Groynes

The groynes are constructed of timber, quarrystone or a combination of timber and stone. They vary greatly in length and height. Some are extremely effective at trapping sediment, such as the 240 m terminal groyne downdrift of the Cape May groyne field and the 245 m terminal groynes just north of Hereford Inlet. Many groynes are over 3 m above MLW at the outer end, but more recent groynes, especially in the southern part of the state, have been built or rebuilt to allow some sand to bypass them and are as low as 0.6 m at the outer end.

Groynes reduce the seasonal fluctuations in beach profile (Everts, 1979) and lower the rate of beach retreat within the groyne field, but they also cause accelerated erosion downdrift. The degree of sand starvation downdrift of groynes, like that downdrift of jetties, is dependent on the efficiency of the groyne system in trapping sand entering from updrift and the amount of sediment deflected seaward as well as the rate of net transport to gross transport and the length and height of the terminal structure (Everts, 1979).

Downdrift effects of structures

Significant regional erosion effects have occurred downdrift of jetties at Manasquan Inlet and Cape May Inlet where the average rate of shoreline retreat has been about $3\,\mathrm{m\,a}^{-1}$ (Everts, 1979). Erosion downdrift of Barnegat and Absecon Inlets has been offset by beach nourishment operations, and bulkheads and groynes have been constructed downdrift of all jettied inlets to mitigate adverse erosion. Accelerated erosion has occurred downcoast of the terminal groynes at Sandy Hook, the developed southern end of Long Beach Island, Stone Harbor Point (just north of Hereford Inlet), and Cape May City. The starvation downdrift of the Cape May groyne field is about 2500 m long with a retreat rate of about $6\,\mathrm{m\,a}^{-1}$ (Everts, 1979). The erosion rate downdrift of groynes and bulkheads constructed at Stone Harbor Point was $36\,\mathrm{m\,a}^{-1}$ over the 12 year period following their construction (Nordstrom, 1988b) and the 1.8 km long spit that existed south of the terminal groyne in 1972 is now gone.

Beach nourishment

Data on past beach nourishment operations funded by the State and local communities are difficult to find and information about privately funded projects is often not available at all in the public domain (Pilkey & Clayton, 1989), yet existing reports indicate that at least 101 nourishment operations have been conducted in New Jersey, involving over $29.8 \times 10^6\,\mathrm{m}^3$ of sediment (USACOE, 1990). Much of the fill emplaced before the 1970s was obtained from the backbays. More-recent operations have used sediment from offshore and from inlets. Some of the fill materials have been placed offshore of the beach foreshore in an attempt to induce natural feeding but have had little impact on the sediment budget of the beach (US Congress, 1976).

Nourished areas are often readily distinguishable from adjacent areas because of their great width, particularly during the first few years of emplacement, and the presence of coarse sediments on the surface of the backbeach. Accelerated deflation occurs, but dunes are rarely allowed to form in intensively developed areas; drift accumulations on the beach are removed to retain wide, flat recreation platforms; and drift accumulations behind the beach are removed because sand inundates boardwalks and buildings. The fill is quickly reworked to a steeper upper beach, but eventually, the erosion rate slows, and the beach returns to a gentler profile similar to pre-fill conditions. Nourishment provides a temporary alteration of the process of deposition, the rate of change, and the location, size, shape, and internal structure of the landform. The periodic implementation of beach fill projects introduces a cycle of beach evolution that has a time scale determined by human activities.

Dunes

Dunes are now well established in more than half of the shorefront communities. Most of the communities that have dunes have a budget for dune building, and many communities have implemented regulations to prevent construction of houses in these locations. The State of New Jersey requires communities to conduct dune-building programmes as a condition for aid for hazard mitigation (Mauriello & Halsey, 1987), and dunes may be conspicuous in locations where they would not occur under natural processes nor under human influence, except for legal requirements (Fig. 13.3). These management factors increase the likelihood that a dune will form and persist (Nordstrom & Gares, 1990), but cycles of erosion and accretion as well as the form, composition and position of the dune may reflect the specific management practice employed locally. Wave erosion of the seaward face of the dune may be more rapid than it was when the dune was farther back from the beach under undeveloped conditions, but restoration of the dune is rapid, because it is aided by human efforts.

In some locations, the building-up and levelling of the dune follows a seasonal cycle but solely under the influence of human processes; dunes are created in the autumn to provide a barrier against storm waves and are eliminated in the summer to accommodate bathers. In some cases, dunes are a by-product of beach cleaning operations; the beach surface is scraped and piled at the landward margin of the beach. Paradoxically, the actions of beach cleaning (removal of vegetation litter, with its shoots and seeds) eliminates one of the conditions needed to initiate natural development of foredunes. Dunes created by earth-moving equipment from beach sediments include size fractions that could not be moved by eolian processes. The coarse sediments may form a surface lag, preventing further deflation by eolian processes, which gives the landform an unnatural appearance.

Dunes that are built with the aid of sand fences and vegetation plantings are larger than those that would form behind narrow beaches under natural conditions. Conversely, dunes are absent in several resort communities, where beaches are hundreds of metres wide, because dunes are graded. Relative to natural dunes, dunes in human-modified environments may be smaller or larger for a given source width or grain size because of active or passive human modifications. Once dunes in developed areas are shaped according to human needs, attempts are made to protect them in place to retain their utility. Thus, initial mobility is high, but the dunes become less mobile than their undeveloped counterparts (Nordstrom, 1990). The location of the dune on the beach profile may be different from the location under natural conditions because dune position is dictated by human preference rather than the interplay between vegetation growth, sediment supply and wave erosion.

Small-scale effects

Small-scale human-altered landforms may not affect long-term coastal evolution but they are of interest to geomorphologists, engineers, planners and managers. Buildings alter wind flow and locally change the location of accretion and scour on beaches and dunes (Nordstrom & McCluskey, 1985; Nordstrom *et al.*, 1986). The large buildings on some portions of the New Jersey shoreline cause a reversal of the regional wind flow and result in onshore transport during offshore winds; the buildings can create a deflation surface and accelerate scour of the backbeach at considerable distances from the buildings (Nordstrom, 1987a; Gundlach & Siah, 1987). Wind-blown sand accumulates against groynes, bulkheads, boardwalks and buildings, creating a diverse dune landscape, characterized by shapes and locations distinct from those that would occur under natural conditions.

Human alterations may be as subtle as the introduction of new heavy mineral assemblages in beach fill. This change may be trivial in terms of dynamic coastal evolution, but alteration of the character of the previous natural fill may eventually prevent interpretation of provenance and ancient clastics (Galvin, 1991) and obscure interpretation of natural coastal evolution.

The introduction of exotic species has had a pronounced impact on the evolution of the coastal landscape in some portions of the USA. The most widespread alteration of dune characteristics attributed to the introduction of exotic species has occurred along most of the Pacific coast. Here, European beach grass (*Ammophila arenaria*), artificially planted to stabilize inlet shorelines, spread rapidly, creating a higher, more linear and better vegetated foredune than existed previously. The new dune is a more complete trap to sand blown landward of the beach, and it has profoundly altered the morphology and biota of the littoral zone (Cooper, 1958; Wiedemann, 1984). The impact of exotic species is less pervasive in New Jersey. Developed areas are frequently barren of vegetation, both native and exotic. The introduced reedgrass (*Phragmites australis*) has colonized portions of the backbarrier above mean high water level. Japanese sedge (*Carex cobomugi*) is found in several areas in the dunes. These species appear conspicuously different from the natural vegetation, but data are insufficient to determine whether there is a morphologic response associated with them that would affect the evolution of landforms.

Effects of storms and post-storm construction

The effects of the northeasterly storm of 6–7 March 1962 and subsequent human activities demonstrate the relative roles of natural processes and human agency in determining the evolution of the New Jersey barriers. This storm was classified as an unusually severe extratropical cyclone. Wind speeds were relatively low, with maximum gusts of $25.5 \, \mathrm{m \, s^{-1}}$, but the fetch distance for wave generation was about 1600 km. Visual estimates of the height of breaking waves range from 6.1 to 9.1 m (USACOE, 1962). The increase in ocean water level due to storm surge was 1.2 m. The major reason for widespread erosion of beaches and dunes and damage to buildings was the duration of the storm over five high tides. Tidal flooding was common on all the barriers and accounted for most of the economic losses. Waves and overwash had very different effects on different portions of the barriers. The greatest geomorphological changes and damages to buildings were on Long Beach Island (Fig. 13.2), where beaches were narrow and dunes were low prior to the storm. A veneer of fresh sand existed everywhere on the surface of this barrier following the storm, but the major overwash fans appeared where street ends were located. Overwash penetrated into the bay, even where there were bulkheads on the bayside (Fig. 13.8). The barrier was breached in five places. Overwash also reached the bay on Ludlam Beach north of Sea Isle City, but did not reach the bay on the other barriers. Many oceanside bulkheads on other barriers failed during the storm, but there was little wave effect landward of locations where bulkheads remained intact (USACOE, 1962).

The storm caused considerable property damage, but little lasting geomorphological effect. All breaches that occurred in developed areas as a result of the storm were closed artificially, starting 9 March, after the storm abated (USACOE, 1962). There was little new substrate for salt marsh and little effect on island widening by natural processes. Dredging of lagoons to facilitate boating resulted in loss of potential for new salt marsh substrate to form. Overwash fans on the subaerial portions of the island increased ground elevation, but these locations were not colonized by natural vegetation. There was greater inland penetration of overwash across the surfaces of shore-perpendicular streets, but this sand was removed to facilitate transportation. The loss of sand from the beach and dunes on the oceanside was compensated for by the creation of a new dune using sediments dredged from borrow areas

Figure 13.8. Effects of the March 1962 storm on Long Beach Island. The overwash
channels and deposits in the backbay are similar to those that would occur under natural
conditions. Photo. courtesy of the US Army Corps of Engineers.

in the bay. The new dune was linear and shore-parallel, and bore little
resemblance to the hummocky dune that characterized many areas prior to
development and that would have characterized natural dune growth adjacent
to the overwash channels that were created during the storm (Fig. 13.8).
Construction of new buildings and facilities occurred at a rapid pace after the
storm; many of the new structures were more elaborate than those constructed
prior to 1962 (US Congress, 1976), and the new seaward-most construction
line was at the same location it was prior to the storm. The net effect of the
storm in altering the location of the shoreline, re-creating natural landforms or
re-initiating new cycles of natural landform evolution was negligible. The
large number of shore protection projects implemented after the storm
(Fig. 13.7) reduced the likelihood of pronounced geomorphological effects of
future storms.

The normal sequence of events following storm alteration of natural
systems is creation of a post-storm landscape with subsequent evolution of
landforms according to natural processes. In human-altered systems, however,
the post-storm landscape is quickly restored to characteristics suitable to
human perception of the value of the resource. Severe damage to beachfront
homes by storms in 1878 and 1884 at Atlantic City did not prevent post-storm

reconstruction (USACOE, 1990). Even in the nineteenth century, the scale of alterations resulting from storms was small enough that human action could restore buildings and infrastructure before natural processes could re-establish natural landforms that could dominate the landscape or control coastal evolution. As a result of past precedent, and taking into account the development pressure and current level of investment in many barrier islands, it is likely that future structures damaged by storms will be replaced (USACOE, 1989a) or there will be an increase in the level of development and protection.

Human alterations and the coastal landscape

Evolution through past modifications

Offshore depth contours of 1.8, 3.7 and 5.5 m below MLW have generally moved landward on the New Jersey coast since the mid nineteenth century (USACOE, 1957), but the upper foreshores have undergone less erosion since development, and several barriers have an accretional trend (Everts & Czerniak 1977; Dolan, Hayden & Heywood, 1978; Galvin, 1983). As a result, there has been little onshore migration of the intertidal shoreline over the past century.

Beach berm widths now vary from <15 m in many locations to a maximum of about 400 m near Hereford Inlet (USACOE, 1990). The maximum elevations of isolated dunes in developed areas and in presently undeveloped areas are generally higher than they were in 1886, but they do not extend as far inland (Table 13.1). Dunes do not exist in many areas, but these locations are now protected by bulkheads and seawalls. There are numerous locations where dunes exist but are no higher than 1.5 m above the elevation of the backbeach and no wider than 15 m (USACOE, 1990). These locations would be eliminated or washed over in a major storm like the one that occurred in March 1962. Despite this limitation to their human utility value, the dunes are a more continuous barrier than they were prior to development in 1886, and they are more likely to restrict barrier island mobility.

The greatest change in the dimensions of the barriers due to development is the dramatic increase in the width of upland (both maximum and minimum) through filling of the marsh. The minimum widths in Table 13.1 were measured at the narrowest portion of each barrier in each of the two time periods, although these locations are usually not at the same place. The minimum width of contiguous upland and marsh is a measure of the potential for breaching of the barrier. This potential has decreased at all barriers except

Absecon Island and Ludlam Beach, although the narrowest portion of Absecon Island is now protected by bulkheads on both sides of the island. The lower width at Five Mile Beach under developed conditions (Table 13.1) does not represent a narrowing of the island through time. The narrowest location is now at the location of the former inlet that separated Five Mile Beach from Two Mile Beach and thus reflects a dramatic increase in barrier dimensions rather than a loss. The minimum widths of the barriers in presently undeveloped enclaves (in parenthesis in Table 13.1) may be greater or lesser than in 1886, depending on the nature of events at the adjacent inlets. The dramatic decrease in minimum width on the undeveloped portion of Seven Mile Beach is due to erosion resulting from human-induced sediment starvation at Stone Harbor Point.

Phases in the evolution of barrier islands through time are presented in idealized form in Figs. 13.9 and 13.10. Absolute dates are not specified because different barriers reach a given phase at different times because of

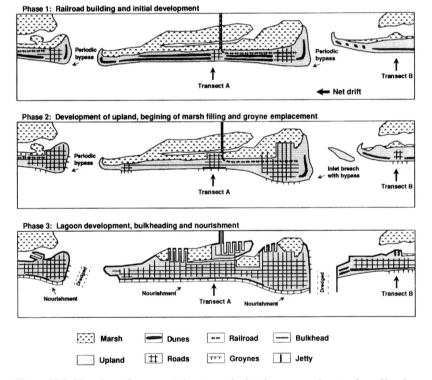

Figure 13.9. Plan view of representative stages in development and protection of barrier islands through time. Transects A and B are locations represented in profiles in Fig. 13.10.

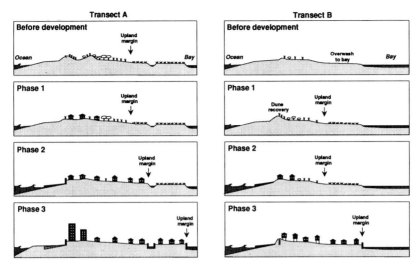

Figure 13.10. Profile view of representative stages in development and protection of barrier islands through time. Transects A and B are locations represented in plan view in Fig. 13.9.

differences in proximity to major population centres and ease of access. The activities include: elimination of dunes to accommodate initial construction; fill of marsh to increase developable land and cutting of lagoons to accommodate boats; protection using groynes and bulkheads; construction of jetties and implementation of dredging projects to maintain a static shoreline and a predictable navigation channel at inlets; and beach nourishment to provide protection and maintain a recreation platform. These activities have resulted in an increase in the width of the islands, a reduction in the rate of erosion, and a reduction in the role that natural processes play in altering the configuration of the islands or the likelihood that they will migrate.

Differences through time represent differences in intensity of development, which result in selection of different protection alternatives. Several communities in New Jersey are now protected by a combination of beach nourishment, groynes, bulkheads and dunes. Most communities employ at least two of these methods of protection. The level of development at Transect A in Figs. 13.9 and 13.10 has been sufficiently intensive to justify use of large-scale nourishment operations to protect buildings and infrastructure. Human actions along Transect B indicate a reliance on less costly public expenditures and include dune building and flood proofing (accomplished by raising buildings) with a bulkhead providing back-up protection.

Planned future actions

Groynes are not likely to be employed extensively in the future, although protection projects for potential implementation in New Jersey identify both construction of new groynes and maintenance of existing groynes (USACOE, 1989a, b, c). Large structures that are planned for future implementation include a weir-type breakwater and deposition basin at Absecon Inlet (Fig. 13.2) and a new north jetty and weir at Townsends Inlet along with extension of the south jetty. Sand bypass at these inlets would reduce the rate of net change on both sides of the inlet. The progression of human alterations at these inlets would take them from a condition of cyclic change, with high rates of mobility, through a phase of unidirectional change with slightly lower rates of shoreline mobility, to a phase of limited or no cyclic change with low rates of mobility.

Most of the planned expenditures for shore protection projects are for beach nourishment operations. A new project began recently at Ocean City, and projects are planned for the vicinity of Barnegat Inlet, Sea Isle City and Sea Bright. The Sea Bright project would restore a beach to a location that has not had one in decades. The creation of new dunes is not always specified in design plans. Dunes are included in plans for Brigantine Island. These dunes would have a top design elevation of 4.6 m (15 feet) and the lowest elevation of the entire protected proportion of the island would be as high as the highest elevation in 1885 (Table 13.1).

Scenarios of future change

The Seabright and Cape May seawalls are considered the final stage of the progression of human alterations to maintain a fixed position on a migrating shoreline (Pilkey, 1981). In this stage, beaches and dunes are replaced by bulkheads and seawalls; shoreline position is static; sediment availability is minimal; and the waves are incapable of creating an equilibrium feature. Nordstrom (1988b) presents a scenario of change for unjettied inlets that has similarities to Pilkey's, in that it is based on a reliance on structural protection and is driven by limitations in sediment availability. The two scenarios are similar to the no-action scenario used by the Corps of Engineers to determine the benefit of future protection projects. The assumptions include: 1. the present bulkheads or groynes will be maintained or improved; 2. sediment budgets will be diminished in locations where the structures interfere with sediment transport; 3. shoreline mobility will be reduced in areas protected by structures; 4. inlets that are now dredged will continue to be dredged, and existing channels will be maintained in place. Scenarios like these describe a closed system of coastal evolution. They are grounded, at a basic level, in the physical processes operative on developed coastal barriers, but they are simplistic in evaluating future human responses. Rejuvenation of features characteristic of a natural landscape can occur with massive inputs of beach nourishment or changes in regulations that prohibit or limit construction or reconstruction of protection structures. Activities in New Jersey indicate that rejuvenation is occurring at a massive scale through implementation of large-scale beach nourishment projects. These operations re-establish the potential for an open system (cyclic) model of beach change. The re-establishment of the potential for cyclic change is the result of conscious human decisions (and the availability of massive funding). Whether natural geomorphic features will form on or adjacent to new beaches is dependent on human values for the resource. Although restored beaches are likely to evolve under natural cycles of wave action in the short term (at least between nourishment operations), transfers between the beach and locations farther inland may be prevented by human actions. A lack of linkage between the beach and the remainder of the barrier can occur under both the beach nourishment scenario and the seawall scenario.

Titus (1990) combines physical processes and human action, including beach nourishment, in four potential scenarios of change for developed barriers in response to an increased rate of sea-level rise. These may be compared to the two responses postulated by Psuty (1986) for the New Jersey barriers (without regard to human action) that include drowning in place and migration through overwash, wind transport and inlet sedimentation. The scenarios for developed barriers (Titus, 1990) include: 1. no protection, leading to eventual abandonment (although development compatible with a dynamic barrier is still possible); 2. engineered retreat that mimics natural retreat by artificially filling the bay sides of barriers while the oceanside erodes; 3. raising the barrier in place by placing sand on the beach and concomitantly raising buildings and support infrastructure; and 4. constructing seawalls and flood protection structures around the barrier, creating a ring-levée enclave. These options were examined in regard to Long Beach Island, NJ (Fig. 13.2) because its level of development (similar to that portrayed in Fig. 13.6) placed it near the middle of the spectrum in regard to development density. The first two scenarios would mimic the migration model of Psuty (1986); the second two follow the model for drowning in place, although drowning would be prevented by human efforts.

Titus uses a greater number of assumptions and more explicit assumptions about human actions than are usually employed in models of evolution for developed coasts, considering factors such as preference and precedent, rights of property owners, legal implications, economic motivation, environmental interests and constraints of government programmes. The most telling finding of the economic evaluation is that the rental value for Long Beach Island exceeds the protection cost for the most expensive option (raising the barrier), leaving the relevant question 'how', not 'whether' to protect the island. Given the economics of protecting Long Beach Island, many developed barriers could justify protection for almost any conceivable rate of sea-level rise or unit cost of sand for nourishment (Titus, 1990). The response of the barriers to accelerated sea-level rise may mimic the natural response by migrating or remaining in place as water level rises, but the role of human agency in the process could be greater than at present.

The likelihood that future shore protection projects will be implemented increases with increasing levels of development. The Corps of Engineers found that protection of Brigantine Island was not economically justified in 1957, but protection was justified in 1989 (USACOE, 1989a). Federal funding increases the likelihood of human alterations and their scale. The sequence of human alterations on the New Jersey coast indicates that the occupation of the coastal fringe is widespread, inevitable, and incontrovertible under present management practice (Nordstrom, 1990).

Discussion

Human action may be considered as an external factor that perturbs or changes the natural system or as part of the system (Phillips, 1991). Human action can be considered an exogenous factor if the temporal or spatial scale under consideration is an order of magnitude greater than the temporal or spatial scale of human influence, or an endogenous factor if the scales under study and the scales of human agency are similar in magnitude (Phillips, 1991). Human alterations on the New Jersey shoreline are endogenous in that they are are persistent, recurrent, and are more frequent than changes made by natural processes or equal or exceed the spatial scale of natural changes.

There are three basic approaches used to investigate the effect of human activities on coastal systems. These may be described as the overlay method, the no-action method and the active human-input method. The overlay method compares and contrasts a developed area with an undeveloped area that is assumed to have the same process controls. The undeveloped area may be the same location prior to development (the approach used here), or it may be a different location. Use of the same location prior to development may result in databases that differ in methods of collection, level of detail, or quality of measurement. Use of an undeveloped shoreline segment that is distant from the developed shoreline is problematic because of differences in setting, such as wave climate or tidal regime. An undeveloped shoreline segment that is adjacent to a developed segment may have a similar wave climate and tidal regime but may be affected by the developed segment through sand starvation (as at Stone Harbor Point). Care must be taken to ensure that a segment that lacks buildings or infrastructure actually represents a natural condition. Use of sand fences, vegetation plantings and introduction of exotic species by humans has converted thousands of kilometres of dune landscape in the USA (Cooper, 1958; Wiedemann, 1984; Godfrey & Godfrey, 1973). Many hundreds of kilometres of these coastlines have no human structures to provide clues to human influence. The degree of naturalness of the undeveloped coast should be specified in all comparative studies.

The no-action method assumes that the kind of shoreline change that occurred in the recent past will continue unabated by local actions. This method is useful to evaluate potential changes in the absence of future adjustments so that these potential changes can be used as the basis for selecting the optimum solution for protection (Nordstrom, 1988b), but it is unrealistic for long-term prediction because of the likelihood of future human action.

Pre-requisites for successful experimental designs in geomorphological investigations include understanding of appropriate processes and responses and selection of appropriate means of measuring and analyzing the pertinent parameters (Sherman, 1989). The active human input approach is the most realistic to use for investigation of shorelines like New Jersey, where future changes depend on human decisions to implement numerous or large-scale structural or non-structural solutions. Predictions must be based on probabilities of human action, calibrated with knowledge of physical processes in a multiple scenario format, requiring that coastal scientists know planning and management policies (Nordstrom, 1987c) and the likelihood that they will be implemented or altered during the time span represented in models of coastal evolution.

Human influence is sufficiently important to change both the space and time frameworks within which geomorphological processes should be examined (Trofimov, 1987). Human processes alter the previous rate of shoreline change and the periodicities of landform evolution, and they change the location, size, shape, and internal structure of the landforms and the tendency to undergo cyclic development in both the short and long term. Prediction according to existing geomorphological principles of natural processes and coastal evolution in New Jersey is restricted to locations seaward of the berm crest on nourished shorelines and seaward of the seaward-most line of pioneer vegetation on most of the non-nourished shorelines. The history of landform changes in developed systems is determined by human action, and the linkage between natural process and form is obscured (Nordstrom, 1990). Although human alterations obscure the effects of natural processes, they still provide the best clues to the future evolution of coastal systems where human influence is an endogenous factor.

Conclusions

The incompatibility of many human alterations with the way coastal barriers change through natural processes is obvious. What is not obvious is the relevance of natural models of coastal evolution to coastal barriers that have passed the critical threshold beyond which return to a naturally functioning system is not an acceptable management option. Traditional models of coastal evolution based on natural processes may be appropriate to anticipate future effects on undeveloped barriers and barriers where the scale of human activities is too small to overcome natural changes, but the precedent established on the barriers in New Jersey and many other developed barriers in the USA indicate that there may be fewer of these locations in the future.

References

Ashley, G.M., Halsey, S.D. & Buteux, C.B. (1986). New Jersey's longshore current pattern. *Journal of Coastal Research*, **2**, 453–63.

Bauer, B.O., Allen, J.R., Nordstrom, K.F. & Sherman, D.J. (1991). Sediment redistribution in a groin embayment under shore-normal wave approach. *Zeitschrift für Geomorphologie, Suppl. Bd.*, **81**, 135–48.

Bearss, E. (1976). Historic resources study: Sandy Hook defenses 1857–1948. Draft report prepared for the Denver Service Center, National Park Service. Denver, CO: National Park Service.

Caldwell, J.M. (1967). Coastal processes and beach erosion. *CERC Reprint 1/67*. Ft. Belvoir, VA: US Army Corps of Engineers Coastal Engineering Research Center.

Carr, A.P. (1988). UK – England. In *Artificial structures and shorelines*, ed. H.J. Walker, pp. 137–44. Dordrecht: Kluwer Academic Publishers.

Carter, C.H., Benson, D.J. & Guy, D.E., Jr. (1981). Shore protection structures: effects on recession rates and beaches from the 1870's to the 1970's along the Ohio shore of Lake Erie. *Environmental Geology*, **3**, 353–62.

Carter, C.H., Monroe, C.B. & Guy, D.E., Jr. (1986). Lake Erie shore erosion: the effect of beach width and shore protection structures. *Journal of Coastal Research*, **2**, 17–23.

Carter, R.W.G. (1982). Condominiums in Florida. *Geography*, **62**, 41–3.

Cencini, C. & Varani, L. (1988). Italy. In *Artificial structures and shorelines*, ed. H.J. Walker, pp. 193–206. Dordrecht: Kluwer Academic Publishers.

Cooper, W.S. (1958). The coastal sand dunes of Oregon and Washington. *Geological Society of America Memoir*, 72.

Dean, R.G. (1988). Sediment interaction at modified coastal inlets: processes and policies. In *Hydrodynamics and sediment dynamics of tidal inlets,* ed. D.G. Aubrey & L. Weishar, pp. 412–39. New York: Springer-Verlag.

Dean, R.G. & Walton, T.L,. Jr. (1975). Sediment transport processes in the vicinity of inlets with special reference to sand trapping. In *Estuarine research*, vol. 2, ed. L.E. Cronin, pp. 129–50. New York: Academic Press.

De Moor, G. & Bloome, E. (1988). Belgium. In *Artificial structures and shorelines,* ed. H.J. Walker, pp. 115–26. Dordrecht: Kluwer Academic Publishers.

Dolan, R., Hayden, B. & Heywood, J. (1978). Analysis of coastal erosion and storm surge hazards. *Coastal Engineering*, **2**, 41–54.

Everts, C.H. (1979). Beach behavior in the vicinity of groins: two New Jersey field examples. In *Coastal Structures 79,* pp. 853–67. New York: American Society of Civil Engineers.

Everts, C.H. & Czerniak, M.T. (1977). Spatial and temporal changes in New Jersey beaches. In *Coastal Sediments 77, pp. 444–59.* New York: American Society of Civil Engineers.

Finkl, C. & Pilkey, O.H., eds. (1991). *The impacts of Hurricane Hugo: September 10–22, 1989. Journal of Coastal Research*, SI 8.

Fischer, D.L. (1989). Response to coastal storm hazard: short-term recovery versus long-term planning. *Ocean and Shoreline Management*, **12**, 295–308.

Fulton-Bennett, K. & Griggs, G.B. (no date). *Coastal protection structures and their effectiveness.* Sacramento, CA: California Department of Boating and Waterways.

Funnell, C.E. (1975). *By the beautiful sea.* New York: Alfred Knopf.

Galvin, C. (1983). Sea level rise and shoreline recession. In *Coastal Zone 83,* pp. 2684–705. New York: American Society of Civil Engineers.

Galvin, C. (1991). Native sand beaches: a disappearing research resource. *SEPM News*, **3** (3), 3.

Gares, P.A. (1983). Historical analysis of shoreline changes at Sandy Hook spit. In *Applied coastal geomorphology at Sandy Hook, New Jersey,* ed. K.F. Nordstrom, J.R. Allen, D.J. Sherman, N.P. Psuty, L.D. Nakashima & P.A. Gares, pp. 15–50. New Brunswick, NJ: Rutgers University Center for Coastal and Environmental Studies Technical Report CX 1600-6-0017.

Gares, P.A. (1990). Eolian processes and dune changes at developed and undeveloped sites, Island Beach, New Jersey. In *Coastal dunes: form and process*, ed. K.F. Nordstrom, N.P. Psuty & R.W.G. Carter, pp. 361–80. Chichester: John Wiley.

Godfrey, P.J. & Godfrey, M.M. (1973). Comparison of ecological and geomorphic interactions between altered and unaltered barrier island systems in North Carolina. In *Coastal geomorphology*, ed. D.R. Coates, pp. 239–58. Binghamton, NY: State University of New York.

Gregory, K.J. (1985). *The nature of physical geography.* London: Edward Arnold.

Griggs, G.B. & Johnson, R.E. (1983). Impact of 1983 storms on the Coastline. *California Geology*, **36**, 163–74.

Gundlach, E.R. & Siah, S.J. (1987). Cause and elimination of the deflation zones along the Atlantic City (New Jersey) shoreline. In *Coastal Zone 87,* pp. 1367–69. New York: American Society of Civil Engineers.

Hall, M.J. & Pilkey, O.H. (1991). Effects of hard stabilization on dry beach width for New Jersey. *Journal of Coastal Research*, **7**, 771–85.

Hansen, M. & Knowles, S.C. (1988). Ebb-tidal delta response to jetty construction at three South Carolina inlets, In *Hydrodynamics and sediment dynamics of tidal inlets,* ed. D.G. Aubrey& L. Weishar, pp. 364–81. New York: Springer-Verlag.

Houston, J.R. (1991a). Beachfill performance. *Shore and Beach*, **59**, 15–24.

Houston, J.R. (1991b). Rejoinder to: discussion of Pilkey and Leonard (1990) [*Journal of Coastal Research*, 6(4) 1023 *et seq.*] and Houston (1990) [*Journal of Coastal Research*, 6(4) 1047 *et seq.*]. *Journal of Coastal Research*, **7**, 565–77.

Hubbard, D.K. (1975). Morphology and hydrodynamics of the Merrimack River ebb-tidal delta. In *Estuarine research*, vol. 2, ed. L.E. Cronin, pp. 253–66. New York: Academic Press.

Kieslich, J.M. (1981). *Tidal inlet response to jetty construction*. GITI Report 19. Vicksburg, MI: US Army Engineer Waterway Experiment Station.

Koedel, C.R. (1983). History of New Jersey's barrier islands. In *New Jersey's barrier islands: an ever-changing public resource*, pp. 17–28. New Brunswick, NJ: Rutgers University Center for Coastal and Environmental Studies.

Koike, K. (1988). Japan. In *Artificial structures and shorelines*, ed. H.J. Walker, pp. 317–30. Dordrecht: Kluwer Academic Publishers.

Kraus, N.C. & O.H. Pilkey, eds. (1988). Effects of seawalls on the beach. *Journal of Coastal Research*, SI 4.

Leatherman, S.P. (1984). Shoreline evolution of North Assateague Island, Maryland. *Shore and Beach*, **52**, 3–10.

Leatherman, S.P. (1987). Approaches to coastal hazard analysis: Ocean City, Maryland. In *Cities on the beach*, ed. R.H. Platt, S.G. Pelczarski & B.K.R. Burbank, pp. 143–54. Chicago: University of Chicago Department of Geography Research Paper 224.

Leonard, L., Clayton, T. & Pilkey, O.H. (1990). An analysis of replenished beach design parameters on U.S. east coast barrier islands. *Journal of Coastal Research*, **6**, 15–36.

Lins, H.F. (1980). Patterns and trends of land use and land cover in Atlantic and Gulf coast barrier islands. Reston, VA: *US Geological Survey Professional Paper* 1156.

Marino, J.N. & Mehta, A.J. (1988). Sediment trapping at Florida's east coast inlets. In *Hydrodynamics and sediment dynamics of tidal inlets*, ed. D.G. Aubrey & L. Weishar, pp. 284–96. New York: Springer-Verlag.

Marsh, G.P. (1885). *Earth as modified by human action*. New York: Charles Scribner's Sons.

Mauriello, M.N. & Halsey, S.D. (1987). Dune building on a developed coast. In *Coastal Zone 87*, pp. 762–79. New York: American Society of Civil Engineers.

McMaster, R.L. (1954). *Petrography and genesis of the New Jersey beach sands*. Trenton, NJ: New Jersey Geological Survey Bulletin 63.

Mitchell, J.K. (1987). A management-oriented, regional classification of developed coastal barriers. In *Cities on the beach*, ed. R.H. Platt, S.G. Pelczarski & B.K.R. Burbank, pp. 31–41. Chicago: University of Chicago Department of Geography Research Paper 224.

Morton, R.A. (1979). Temporal and spatial variations in shoreline changes and their implications, examples from the Texas gulf coast. *Journal of Sedimentary Petrology*, **49**, 1101–12

Moutzouris C.I. & Maroukian, H. (1988). Greece. In *Artificial structures and shorelines*, ed. H.J. Walker, pp. 207–15. Dordrecht: Kluwer Academic Publishers.

Nagao, Y., ed. (1991). *Coastlines of Japan*. New York: American Society of Civil Engineers.

Nagao, Y. & Fujii, T. (1991). Construction of man-made island and preservation of coastal zone. In *Coastlines of Japan*, ed. Y. Nagao, pp. 212–26. New York: American Society of Civil Engineers.

National Oceanic and Atmospheric Administration (NOAA) (1991). *Tide Tables 1991: East Coast of North and South America.* Washington, DC: US Department of Commerce, NOAA.

New Jersey Department of Environmental Protection (NJDEP) (1981). *New Jersey Shore Protection Master Plan.* Trenton, NJ: New Jersey Department of Environmental Protection.

Nordstrom, K.F. (l987a). Shoreline changes on developed coastal barriers. In *Cities on the beach*, ed. R.H. Platt, S.G. Pelczarski & B.K.R. Burbank, pp. 65–79. Chicago: University of Chicago Department of Geography Research Paper 224.

Nordstrom, K.F. (1987b). Management of tidal inlets on barrier island shorelines. *Journal of Shoreline Management*, **3**, 169–90.

Nordstrom, K.F. (1987c). Predicting shoreline changes at tidal inlets on a developed coast. *Professional Geographer*, **39**, 457–65.

Nordstrom, K.F. (1988a). Dune grading along the Oregon coast, USA: a changing environmental policy. *Applied Geography*, **8**, 101–16.

Nordstrom, K.F. (1988b). Effects of shore protection and dredging projects on beach configuration near tidal inlets in New Jersey. In *Hydrodynamics and sediment dynamics of tidal inlets,* ed. D.G. Aubrey& L. Weishar, pp. 440–54. New York: Springer-Verlag.

Nordstrom, K.F. (1990). The intrinsic value of depositional coastal landforms. *Geograpical Review*, **80**, 68–81.

Nordstrom, K.F. & Gares, P.A. (1990). Changes in the volume of coastal dunes in New Jersey, USA. *Ocean and Shoreline Management,* **13**, 1–10.

Nordstrom, K.F. & McCluskey, J.M. (1985). The effects of houses and sand fences on the eolian sediment budget at Fire Island, New York. *Journal of Coastal Research*, **1**, 39–46.

Nordstrom, K.F., McCluskey, J.M. & Rosen, P.S. (l986). Aeolian processes and dune characteristics of a developed shoreline. In *Aeolian geomorphology*, ed. W.G. Nickling, pp. 131–47. Boston: Allen and Unwin.

Park, D.W. (1988). Korea – South. In *Artificial structures and shorelines*, ed. H.J. Walker, pp. 311–16. Dordrecht: Kluwer Academic Publishers.

Phillips, J.D. (1991). The human role in earth surface systems: some theoretical considerations. *Geographical Analysis*, **23**, 316–31.

Pilkey, O.H. (1981). Geologists, engineers, and a rising sea level. *Northeastern Geology*, **3/4**, 150–8.

Pilkey, O.H. & Clayton, T.D. (1989). Summary of beach replenishment experience on U.S. east coast barrier islands. *Journal of Coastal Research*, **5**, 147–59.

Pilkey, O.H. & Wright, H.L., III (1988). Seawalls versus beaches. *Journal of Coastal Research*, SI 4, 41–64.

Plant, N.G. & Griggs, G.B. (1992). Interactions between nearshore processes and beach morphology near a seawall. *Journal of Coastal Research*, **8**, 183–200.

Psuty, N.P. (1986). Impacts of impending sea-level rise scenarios: the New Jersey barrier island responses. *Bulletin of the New Jersey Academy of Sciences*, **31**, 29–36.

Psuty, N.P. (1988). USA – New Jersey and New York. In *Artificial structures and shorelines*, ed. H.J. Walker, pp. 573–80. Dordrecht: Kluwer Academic Publishers.

Reynolds, W.J. (1981). The social and geomorphic forces that shaped New Jersey's shore. Unpublished seminar paper. New Brunswick, NJ: Department of Geography, Rutgers University.

Reynolds, W.J. (1987). Coastal structures and long term shore migration. In *Coastal Zone 87*, pp. 414–26. New York: American Society of Civil Engineers.

Schmahl, G.P. & Conklin, E.J. (1991). Beach erosion in Florida: a challenge for planning and management. In *Coastal Zone 91*, pp. 261–71. New York: American Society of Civil Engineers,

Schwartz, M.L. & Bird, E.C.F. (1990). Artificial beaches. *Journal of Coastal Research*. SI 6.

Sea Isle City (1982). *Sea Isle City Centennial 1882–1982*. Sea Isle City, NJ: City Hall.

Sherman, D.J. (1989). Geomorphology: praxis and theory. In *Applied geography: issues, questions, and concerns*, ed. M.S. Kenzer, pp. 115–31. Dordrecht: Kluwer Academic Publishers.

Sherman, D.J., Bauer, B.O., Nordstrom, K.F. & Allen, J.R. (1990). A tracer study of sediment transport in the vicinity of a groin. *Journal of Coastal Research*, **6**, 427–38.

Terwindt, J.H.J., Kohsiek, L.H.M. & Visser, J. (1988). The Netherlands. In *Artificial structures and shorelines*, ed. H.J. Walker, pp. 103–14. Dordrecht: Kluwer Academic Publishers.

Thompson, E.F. (1977). *Wave climate at selected locations along U.S. coasts*. Technical Report 77–1. Ft. Belvoir, VA: US Army Corps of Engineers Coastal Engineering Research Center.

Titus, J.G. (1990). Greenhouse effect, sea level rise, and barrier islands: case study of Long Beach Island, New Jersey. *Coastal Management*, **18**, 65–90.

Trofimov, A.M. (1987). On the problem of geomorphological prediction. *Catena Supplement*, **10**, 193–97.

US Army Corps of Engineers (1954). *Beach erosion control report on cooperative study (survey) Atlantic coast of New Jersey, Sandy Hook to Barnegat Inlet*. Philadelphia: US Army Corps of Engineers, Philadelphia District.

US Army Corps of Engineers (1957). *Beach erosion control report on cooperative study (survey) of the New Jersey coast Barnegat Inlet to the Delaware Bay entrance to the Cape May canal*. Philadelphia: US Army Corps of Engineers, Philadelphia District.

US Army Corps of Engineers (1962). *Coastal storm of 6–7 March 1962: post flood report*. Philadelphia: US Army Corps of Engineers, Philadelphia District.

US Army Corps of Engineers (1989a). *Benefits reevaluation study: Brigantine Island, New Jersey*. Philadelphia: US Army Corps of Engineers, Philadelphia District.

US Army Corps of Engineers (1989b). *Benefits reevaluation study: Absecon Island, New Jersey*. Philadelphia: US Army Corps of Engineers, Philadelphia District.

US Army Corps of Engineers (1989c). *Benefits reevaluation study: Townsends Inlet/Seven Mile Island, New Jersey*. Philadelphia: US Army Corps of Engineers, Philadelphia District.

US Army Corps of Engineers (1990). *New Jersey shore protection study: report of limited reconnaissance study*. Philadelphia: US Army Corps of Engineers, Philadelphia District.

US Congress (1976). *New Jersey coastal inlets and beaches Hereford inlet to Delaware Bay entrance to Cape May canal*. House Document 94–641. Washington, DC: US Government Printing Office.

Vogt, G. (1979). Adverse effects of recreation on sand dunes: a problem for coastal zone management. *Coastal Zone Management Journal*, **6**, 37–68.

Walker, H.J. (1984). Man's impact on shorelines and nearshore environments: a geomorphological perspective. *Geoforum*, **15**, 395–417.

Walker, H.J. (1988). Artificial structures and shorelines: an introduction. In *Artificial structures and shorelines*, ed. H.J. Walker, pp. 1–8. Dordrecht: Kluwer Academic Publishers.

Wiedemann, A.M. (1984). *The ecology of Pacific northwest coastal sand dunes: a community profile*. Washington, DC: US Department of the Interior Fish and Wildlife Service.

Wong, P.P. (1985). Artificial coastlines: the example of Singapore. *Zeitschrift für Geomorphologie, Suppl. Bd.,* **57**, 175–92.

Wong, P.P. (1988). Singapore. In *Artificial structures and shorelines*, ed. H.J. Walker, pp. 383–92. Dordrecht: Kluwer Academic Publishers.

Zabawa, C.F., Kerhin, R.T. & Bayley, S. (1981). Effects of erosion control structures along a portion of the northern Chesapeake Bay shoreline. *Environmental Geology*, **3**, 201–11.

Index

accommodation space, 42, 52, 98–9, 107–8, 121, 126, 138–40, 158, 173
Adelaide River, 193–200, 202–3, 212
aeolianite, 234, 235, 244, 252, 287
Alaska, 344, 348, 352–7, 373, 394, 435, 463–4
Aldabra Atoll, 289, 429–30
allocyclicity, 97–9
alluvial fan, 353, 374
Amanzimnajama, Lake, 227
Anaa, Tuamotus, 289, 454–8
anabranch, 202
antecedent karst, 273, 287–8, 313
antecedent topography, 14, 34, 42, 122, 127, 133, 187, 192–4, 204, 208, 227, 246, 249, 344, 357, 361, 380, 399
anthropogenic influence, 21–4, 320, 477–509
  on cliffs, 435–7
  on deltas, 112–4
  on reefs, 320, 328–30
Antrim, Cape, 405
Arctic coastal plain, 341–72
armouring, 11, 383
Atchafalaya Delta, 98, 110–11, 113
Atlantic City, 481, 484–6, 489, 496–7
atoll, 5–6, 267–302, 304–5, 452–7
  Aldabra, 289, 429–30
  Cocos (Keeling) Islands, 267–294
  Funafuti, 267–94
  Midway, 272
  Mururoa, 272, 453
  structure, 268–73, 304–5, 452–7
  Tuamotu, 272, 282, 289–90, 454–7
autocyclicity, 97–9
avalanching, 227
Avalon Peninsula, 406
Avulsion, 97–8, 113, 203

back barrier, 225, 230, 235, 400
  channels, 232
  flats, 237 = = 8
  marsh, 486
backwater swamp, 188, 191, 213
barrier, 16–18, 20–1, 48–52, 54–60, 70–3, 123–86, 189, 191, 221, 244–69, 397, 399–409
  bayhead, 378, 405–6
  beach, 362, 392, 394
  breach, 47, 55, 225, 232, 244, 495, 497
  cell-confined, 232

clastic, 221
cuspate, 221, 237
erosion, 244–9
fringing, 404, 412
gravel, 10–11, 14, 16–18, 159, 378, 397–405
looped, 237
mainland-attached, 250
mainland beach, 159–60
overstepping, 48, 54, 133, 244, 250–2
overwash, 158, 231, 243–4, 251, 361, 485, 495
Pleistocene, 230, 247
pocket, 237
prograding, 58, 131, 140–1, 152–6, 254
receded, 159
regressive, 140–5, 234, 253–6
retreat, 234, 497–9
stationary, 156–60
transgressive, 61, 127–43, 146–7, 149–52, 165, 168–78, 250–3
translation, 48, 132–4, 245
barrier island, 124, 127, 140, 155, 158, 250, 353–5, 361–2, 393, 480, 484, 497–9
  drowned, 127–8, 365
Bay of Funday, 105, 210, 383, 444
beach morphodynamics, 9–10, 47, 68, 124
beach nourishment, 478–9, 491–3, 499–501
beach ridge, 97–8, 104, 253, 394, 397–8, 405, 407, 470
beachridge plain, 11, 99, 152, 192, 204–6, 225, 398, 402
beachrock, 252, 286
Beaufort Sea, 341, 346–8, 352–64
bedload, 196, 201
Bermuda, 234, 244
big swamp phase, 18–20, 46–7, 208–11
bioerosion, 429–30
bluff, 341, 350, 353–4, 359, 380, 383, 394–5, 431, 437, 439
  retreat, 344, 357, 362, 365, 380
boulder barricade, 379, 410
boulder pavement, 410
boundary conditions, 37–40, 188–94
  process, 38–40
  spatial, 37–8
breakwater, 500
Breton Island, Cape, 443–5
Bruun rule, 4, 73, 133–4, 250
bulkhead, 479, 489, 491, 495, 498–500
Byron, Cape, 166

Calabria, 464–6
calcification, 306–10
California, 431, 433–5, 441, 445–8
Cambridge Fiord, 411
Cape Antrim, 405
Cape Breton Island, 443–5
Cape Byron, 166
Cape Tribulation, 312, 326–7, 329
carbonate production, 40, 246, 321, 324–5
carbonate solution, 267, 272–5, 287–8, 313, 429, 454
Caribbean, 330–2
Carpentaria, Gulf of, 209–11
catastrophe theory, 14, 310
cay, 5–6, 285–93, 325
channelisation, 235, 240
chaos, 13, 33–4
chenier ridge, 20, 48, 121, 192, 204–6, 210
Chezzetcook Inlet, 404
Clarence estuary, 189–92, 195–6
cliff, 14, 127, 129, 317, 425–50
    abrasion, 428
    anthropogenic influence, 435–7
    bevelled, 443
    chemical weathering, 428–9
    composite, 440–2
    drumlin, 402, 413
    equilibrium profile, 447
    failure, 14, 344, 431–2, 434
    mass movement, 431–2, 439
    multi-storied, 443
    profiles, 440–2
    processes, 427–32
    rate of retreat, 432–40
    relict, 447–8
    salt weathering, 428–9
    tectonic setting, 425–6
    two-storied, 443
    vertical, 443
    wave action, 427–8, 434
coastal lake, saline, 167–8, 192, 220, 235–6
    see also estuary
Cocoas (Keeling) Islands, 267–94
collision coasts, 425–6, 466–8
continentality, 304
continuity equation, 304
Cook Islands, 289–90
coral reef, *see* reef
coseismicity, 461–4
crevassing, 99
cyclone *see* hurricane

Daly River, 192–212
dams, 23, 112–13
delta, 87–120, 222, 353, 376, 387, 391, 412
    anthropogenic influence, 112–14
    Atchafalaya, 98, 110–11, 113
    bayhead, 108, 110

birdfoot, 93, 103, 191
classification, 103
complex, 97–8, 105, 107
cycle, 97–8
deformation, 97, 108
DELTA2, 390–1
density contrasts, 90–3
ebb tidal, 11, 58, 124, 167–8, 174, 221–2, 229, 488–9
facies model, 103–5
flood tidal, 11, 58, 167, 221–2, 229, 243–4, 404, 485
fjord-head, 377–8, 386–91
front, 88–9, 99, 101–2
Ganges–Brahmaputra, 98, 105
Gilbert-type, 90
Huanghe, 92, 112
lacustrine, 108
lagoon, 229, 242
lobe, 97–8, 105, 107
Mackenzie, 346, 358
Mahakam, 102, 105
Mississippi, 93, 97–101, 103, 105, 109–14
Niger, 95
Nile, 23, 87, 112–13
outwash, 380, 390–1, 412
plain, 88–9, 99–101, 222
processes, 390–1
Rhone, 98, 104
river-dominated, 88, 103–5
São Francisco, 94, 96, 104, 125
shelf-phase, 105–12
shoal-water, 105–12
succession, 99
switching, 47, 97–8
tide-dominated, 95–6, 103–5
wave-dominated, 94, 96, 103–5
developed coasts, 477–509
dissipative beach, 47, 402
distributary-mouth bars, 92–3, 101–2
dredging, 489, 495, 499–500
drowned river valley, 125, 187, 192, 256
drumlin, 16–18, 373–4, 377, 399–405
    archipelago, 373
    island, 401
    headland, 402, 413
dune, 13, 124, 132–58, 221, 227, 229, 244, 478–9, 482–3, 485–6
    cliff-top, 127, 149, 164
    hummocky, 485
    multiple ridges, 485
    Pleistocene barrier, 229–30
    rebuilding, 491–4
    transgressive, 149–52
    vegetation, 149, 152, 225, 486, 494–5
dynamic equilibrium, 5–6, 45, 62, 131, 291, 365–6

earthquake, 14, 463–4, 466–8
El Niñom 39, 328
equilibrium, 42, 44–5
  chaotic, 45
  dynamic, 5–6, 45, 62, 131, 291, 365–6
  metastable, 45
  potential, 62
  steady-state, 5–6, 45, 291–3
equilibrium profile, 4–6, 44–5, 134–5, 350, 357, 447
erratics, 410
esker, 380
estuarine funnel, 188, 198, 200, 204–7
estuarine plain, 18–20, 187–218, 203
estuary, 18–20, 49–51, 138, 140, 145, 167–8, 187–218, 235–6
  back-barrier, 138–40, 400
  barrier, 49–51, 167–8, 189–92, 235–6
  Bay of Funday, 167–8, 189–92, 235–6
  Bay of Funday, 105, 210, 383, 444
  blind, 220
  Clarence, 189–92, 195–6
  drowned-valley, 167, 174
  Gironde, 210
  macrotidal, 18–20, 105, 187–218
  St Lawrence, 386, 410
  saline coastal lake, 167–8, 192, 220, 235–6
  Severn, 210
  tide dominated, 18–20, 105, 187–218
etchplanation, 303
evaporite, 243

feedback, 20, 33, 42–51, 139, 247
  negative, 33, 42–6
  positive, 33, 46–51
fish kill, 229
Fitzroy River, 229
fjord, 373–4, 377–8, 384–6, 388
Flat Island, 385, 388, 395–6, 398
flocculation, 242
floodplain, 188, 203, 213
fluid dynamics, 33, 43, 53
Fly River, 194
foredune, 132, 149, 152, 155–9
Funafuti Atoll, 267–94
funnel, *see* estuarine funnel
funnelling coefficient, 198–200, 202

Galveston Island, 253
Ganges–Brahmaputra delta, 98, 105
geoid changes, 452
geomorphological convergence, 57
Gilbert River, 210
Gironde estuary, 210
glacial diamict, 374, 382, 394, 399
glaciogenic sediment, 382–4, 386, 394, 397, 411–14

glacioisostasy, *see* isostasy
glaciomarine sediments, 237, 374, 382, 394
Grand Cayman Island, 429
Great Barrier Reef, 290–1, 303–39
Great Lakes, 383
ground-penetrating radar, 154–5
groyne, 478–9, 489, 491–2, 499–500
Gulf of Carpentaria, 209–11
gypsum, 230
gyttja, 227, 234, 237

Halimeda, 310, 312–13, 316, 321–4, 328, 332
Holyrood Pond, 397, 399
homeostasis, 20
honeycombs, 429
hot spot, 454
Huanghe Delta, 92, 112
human influence, *see* anthropogenic influence
hurricane, 13, 15–16, 244, 291–3
hysteresis, 42–3

ice, 341–67
  anchor, 357, 366
  block, 350
  bottomfast, 346, 354–5
  first year, 343, 346–7
  floe, 348, 350
  frazil, 348–9, 357, 366
  glacier, 346
  gouging, 349–50
  ground, 344, 376
  ice-keel turbate, 357, 365
  ice-rafted deposits, 410
  island, 346
  landfast, 350
  pack, 346, 350
  pile-up, 350–2
  pore, 344
  push, 355, 365, 410
  ride-up, 350–2, 364
  scouring, 349–50, 357, 364–6
  sea ice, 341, 346–52, 410
  sea-ice break-up, 342
  sea-ice freeze up, 342, 346–7, 355, 366
  slush, 348
  wallow, 348, 355, 366
  wedge, 344
Iceland, 392
incised valley, 227, 230, 238
inheritance, 21, 34, 38, 42, 54–7, 122, 124, 128, 168–70, 178, 198–200, 202, 212, 227, 245, 303, 319, 344, 448
inlet closure, 229, 237, 240, 244–5
  artificial, 488–9, 495–6

isostasy, 451–9
  glacioisostasy, 21, 25, 39, 187, 277, 451, 458–9, 470
  hydroisostasy, 39, 277, 279, 451, 458–9, 470
  rebound, 470
  thermo-isostasy, 452–9
  volcano-isostasy, 452–9
Itirbilung Fiord, 390–1

jetty, 478, 488–9, 499–500
jokulhlaup, 392

kame, 380
karren, 429
karst, 272–5, 287–8, 313, 429, 454
  thermokarst, 344, 353, 359, 361, 363
King Point, 359, 360, 362
Kiribati, 281–5, 289–91
Kosi lagoon, 226–9, 238
Kugmallit Bay, 358

Lagoa dos Patos, brazil, 242
lag gravel, 162, 382–3, 405
lagoon, 138–40, 145, 219–65, 270–7, 283–5, 404–9
  atoll, 270–7, 283–5
  back-barrier, 124, 127–8, 138–40, 222, 404
  barrier reef, 319
  chemical precipitation, 243
  closed, 220
  deficit, 256–8
  definition, 220
  Delaware, 254
  equilibrium, 256–7
  estuarine, 167–8, 192, 220, 235–6
  evolution, 219–65
  facies, 230, 235, 237
  isolation, 244
  Kosi, 226–9, 238
  Late Pleistocene, 229
  Maine, 237–238
  Mdoti, 232
  Mhlanga, 232
  open, 220
  partly-closed, 220
  processes, 241–245
  segmentation, 227, 230, 241–2, 248
  Siyai, 225, 240, 243, 255
  south-east African, 222–233
  surplus, 256–258
  uMgababa, 231–2
landslides, 431–2
large Scale Coastal Behaviour, 13–14, 34–6, 64–75, 123, 170, 174, 178
Lawrencetown beach, 398
Lawrencetown Lake, 404

levée, 92, 99, 113, 204
lithospheric flexure, 456–7
littoral drift, *see* longshore drift
Long beach island, 486, 492, 495–6, 501
Long Island, 485
longshore drift, 104, 123, 134, 137–8, 158, 160, 165, 178, 192, 221–2, 225, 229, 235, 244, 254, 362, 393, 485
low wooded island, 312
Loyalty Islands, 459, 461
Lucke model, 222–3, 239, 256
lutocline, 196

McArthur River, 195–6
Mackenzie Delta, 358, 359, 361
Mackenzie River, 346, 358
macrotidal estuary, 187–218
Mahakam Delta, 102, 105
Makatea Island, 456–7
Makawulani, Lake, 229
Maldives, 289–90
mangrove, 208–11, 232, 243, 249
Marco Island, 477
Markovian inheritance, 34, 38, 42, 54–7
marsh, 222, 234, 237, 243, 248–9, 397, 485–7, 497, 499
Marshall Islands, 272, 289–90, 453
Mary River, 192, 195–7, 203–6, 212–13
Mdoti lagoon, 232
meander, 188, 196–207, 212
  cuspate, 192
  cutoff, 201
  estuarine, 192, 199–202, 212
  inherited, 202, 212
  sinuous, 194, 198–200, 212
  tract, 208
Mediterranean, 3, 464–6
megaripples, 162
Mhlanga lagoon, 232
microatoll, 283, 284, 326
Middleton Island, 463–4
Midway Atoll, 272
Mississippi Delta, 93, 97–101, 103, 105, 109–14
model, 16–21, 33–76
  computer, 16, 122–3, 133
  conceptual, 16, 68
  deterministic, 66
  facies, 103–5, 194, 230, 250
  fold catastrophe, 14
  forward, 35, 73
  inverse, 35, 73–4
  Lucke, 222–3, 239, 256
  mathematical, 68, 448
  Nichols, 256
  numerical, 69
  parametric morphological behaviour, 70–3

process–response, 103–5, 390
simulation, 70–6, 123, 133–78
Monterey Bay, 434
moraine, 373–4, 380, 406
morphodynamics, 3, 8–10, 33–86, 124, 168–70, 196, 207
Moruya Beach, 63, 153, 156
motu, 5–6, 285–93
Mpungwini, Lake, 229
Mtwalume lagoon, 232
Muroto peninsula, 461–3
Mururoa Atoll, 272, 453
Myall Lakes, 150–152
Myrmidon Reef, 307–9

Natal, 222–34, 238, 248, 254, 257
New Jersey coast, 480–504
New South Wales, 11, 50, 53, 58–64, 121–86, 189–92, 235–6, 239, 252, 254
Newfoundland, 163, 385–90, 394–414, 443–5
Niger Delta, 95
Nile Delta, 23, 87, 112–13
non-linear behaviour, 13–14, 18, 33–4, 43, 52–4
nonstationarity, 43, 48, 57
Nova Scotia, 16–18, 48, 137, 385, 399–414
nutrients, 306–10, 327–8

Ocean City, 480, 486
Ord River, 18, 209
outwash, 374, 377–80, 382, 406
plain, 377, 380, 392–4, 413–14
overthrusting plate, 461–6
overwash, *see* washover
overstepping, 17–18, 48, 133, 244, 250–2

palaeochannel, 188, 201, 209
paradigm shift, 1–2
paraglacial coasts, 373–424
definition, 373–6
parasequence, 107
patterned ground, 353
permafrost, 341–72, 376
pingo, 353
Pingok Island, 354–5
plate boundaries, 459–68
plate tectonics, 451–70
point bar, 199, 201, 209
polar coasts, 341–72
Pompey Complex, 312–15
prodelta, 88–9, 99, 102, 387

Raine Island, 322–3
ravinement surface, 111, 133, 137, 162, 362, 365
reaction time, 9
recurrence interval, 64, 293

reef, 267–302, 303–39
algal-dominated, 310
anthropogenic influence, 320, 328–30
atoll, *see* atoll
barrier reef, 268–9, 303–39, 452
catch-up, 277–85
continental shelf, 303–39
fringing reef, 268–9, 304, 313, 314–15, 319, 322, 326, 330–2, 452
give-up, 277–9, 330–2
Great barrier Reef, 290–1, 303–39
keep-up, 277–9, 324, 330–2
Last interglacial, 273–5
linear, 312
mid-shelf, 314–15
mining, 328
Myrmidon, 307–9
platform, 312
remnant, 313
ribbon, 312, 317, 324
shelf, 304
reef flat, 312, 319–20, 325
reef islands, 5–6, 283, 285–94, 325
reflective beach, 47, 355, 359
regolith stripping, 303, 324
relaxation time, 58–64, 293, 374–7
response time, 58–65
return interval, *see* recurrence interval
revetment, 491
Rhodes Island, 466–7
Rhone Delta, 98, 104
ria, 187
Richards Islands, 347, 362
rock fall, 431–2
rubble rampart, 5–6, 292–3
Ryukyus Islands, 464

sabkha, 220
St George's Bay, 385, 394, 397, 399
St Lawrence Estuary, 386, 410
St Lawrence, Gulf of, 383–4, 395
St Lucia Lake, 228–30, 238, 243
Saguenay Fjord, 386
salinisation, 204–5, 212–13
salinity stratification, 229
salt wedge, 93, 101, 196
San Andreas Fault, 468
sandur, 386–7, 392
São Francisco Delta, 94, 96, 104, 125
scroll bar, 199, 201
scroll plain, 192, 201
sea level, 1, 8, 16–21. 39, 47–9, 59–63, 70–4, 105–12, 124–60, 167, 176, 187–8, 208–9, 211–14, 240, 249, 250–7, 268–9, 273–85, 287–91, 294, 304–6, 310, 316–9, 332–4, 377, 380–1, 395, 399, 406–9, 435, 442–8, 451–2, 458–9, 460, 469–70, 484, 501–2

sea stacks, 426
seawall, 477–9, 491, 500
sediment budget, 11–14, 41, 123, 134, 137–8, 169–70, 357, 362, 492
sediment supply, 22–4, 33, 127, 128, 145, 147, 194, 238, 240, 246
seismic profiling, 71, 107, 122, 141–3, 154–5, 227, 229, 274–5, 318–19, 357, 361–2, 392–3, 396, 405
self organisation, 49–51
self regulation, *see* feedback
Sepik River, 193, 195
sequence stratigraphy, 67, 107–12, 211
Serapis, Temple of, 468–9
Severn Estuary, 210
shore platform, 425, 445–50
    relict, 447–8
shore protection project, 489–94, 496, 500–1
Sibaya, Lake, 245
Siberia, 341
sidescan sonar, 122, 349, 364
Siyai lagoon, 225, 240, 243, 255
Skeidararsandur, 392–3
Society Islands, 267
South Alligator River, 18–20, 46–7, 187–218
spit, 355, 359, 361, 394, 412
    barrier, 237, 480
    barrier island, 158
    cuspate, 227, 241
    drift-aligned, 378, 395, 402–4, 412–13
    growth, 221, 243
    headland, 128, 158–9
    paired, 241–2
    platform, 395
    recurved, 140, 155
    swash-aligned, 378, 395, 402, 412–13
    trailing, 402, 411
stamukhi shoal, 355
stationarity, 57, 65
step-response function, 62
Stephenville, 385, 395, 399
storm, 5–6, 9–10, 121, 131–2, 138, 158, 160, 162–4, 170, 239, 244, 286, 290–3, 352–3, 357, 366, 412, 479, 484, 495–7
storm block, 286
storm surge, 15–16, 131–2, 352–3, 358, 364, 366
Story head, 403, 405, 411–12
strandplain, 104–5
subduction zone, 459–66
subsidence, 268–9, 271–3, 452–3, 470
    atoll, 268–9, 271–3, 452–3
    beneath ice sheet, 458
    coseismic, 463–4
    interseismic, 463
    oceanic islands, 268–9, 271–3, 453

permafrost, 344–6
postseismic, 463
    thermal, 178
tafoni, 429
Taiwan, 461
talik, 344
tectonic shorelines, 39, 173, 451–75
thaw flow-slides, 344–5, 362
thermokarst, 344, 353, 356, 359, 361–5
threshold, 14, 18, 33, 42, 46–51, 304, 310, 329
    intransitive, 9
tidal asymmetry, 196, 198
tidal channels, 208, 234, 248, 312, 489
tidal currents, 234, 240–2
tidal flats, 229, 234, 237, 242–3, 248
tidal prism, 198, 204, 206, 213, 234, 240
till, 382–3, 400, 405, 410
time scale, 5, 35–6, 58–64, 124
    reaction time, 9
    relaxation time, 58–64, 293, 374–7
    response time, 58–64, 65
    lag, 61
tombolo, 237
Top End, Northern Territory, 18–20, 188–209
Torres Strait, 306
transform zones, 468
transgressive barriers, 127–43, 146–7, 165, 168–78, 250–3
tsunami, 15–16
Tuamotu Atolls, 272, 282, 289–90, 454–7
Tugela River, 224–5, 230, 254, 257
Tuktoyaktuk Peninsula, 349, 350, 358, 360, 362, 364
Tuncurry, 141–2, 152–5, 161, 174–6
tundra, 342, 354, 361
turbidite, 227, 382
turbidity, 306–7
turbidity cuurrents, 92, 176, 389–90

uMgababa lagoon, 231–2
underthrusting plate, 459–61
uplift, 454–6
    coseismic, 463–4
    postseismic, 463
uplifted marine terraces, 447–8, 463–4, 468

van Diemen Gulf, 18–20, 187–218

washover, 124, 131–40, 158–9, 231, 240, 243–4, 252, 361–2, 397, 399, 411–12, 485, 495
Warnbro Sound, 235, 244
water hammer, 427
water layer levelling, 429, 435

wave-dominated coasts, 121–86
wave quality, 316–30
wave quarrying, 427
West Marsh, 404
Wolfe Spit, 360–3

Yonge Reef, 322–3
Yukon, 345, 348

zonality, 303–4, 342–55, 357, 362, 365–7
Zululand, 222–34, 238, 248, 252